PROTEIN ENGINEERING IN INDUSTRIAL BIOTECHNOLOGY

PROTEIN ENGINEERING IN INDUSTRIAL BIOTECHNOLOGY

Edited by

Lilia Alberghina

Dipartimento di Biotecnologie e Bioscienze
Università degli Studi di Milano-Bicocca
Milan, Italy

CRC Press
Taylor & Francis Group
Boca Raton London New York

CRC Press is an imprint of the
Taylor & Francis Group, an **informa** business

CRC Press
Taylor & Francis Group
6000 Broken Sound Parkway NW, Suite 300
Boca Raton, FL 33487-2742

First issued in paperback 2019

ISBN-13: 978-90-5702-412-2 (hbk)
ISBN-13: 978-0-367-39897-2 (pbk)

British Library Cataloguing in Publication Data

A catalogue record for this book is available from the British Library.

**Visit the Taylor & Francis Web site at
http://www.taylorandfrancis.com**

**and the CRC Press Web site at
http://www.crcpress.com**

CONTENTS

PREFACE

Protein engineering by its iterative rational design is a powerful tool to both test general theories of protein structure and activity, as well as to develop more useful catalysts to be used in biotechnological processes or products.

Protein Engineering in Industrial Biotechnology aims to present a series of examples in which the application of protein engineering has successfully solved problems as diverse as the purification of recombinant proteins or the development of target molecules for drug discovery.

This book is organized in three sections, respectively: protein engineering for bioseparation, for biocatalysis and for health care. The more relevant industrial enzymes are covered: lipases, proteases, carboxypeptidases, glucanases and glucosidases, pectinolytic enzymes and enzymes for the bio-remediation of recalcitrant compounds. The interplay of solvent engineering to modulate the structure–to–activity relations is also discussed. A chapter is devoted to the application of protein engineering to biosensors.

The large potential applications of protein engineering to health care are also covered, from the development of new safe vaccines to therapeutic proteins. Specific attention has been devoted to new protein engineering in the development of target molecules for drug discovery.

The chapters have been written by an international team of experts from Europe, USA and Japan who have made major contributions in the field.

The present book aims to attract the interest of students in industrial biotechnology at the undergraduate and graduate level as well as that of everybody interested in basic research in protein structure, molecular genetics, bio-organic chemistry, biochemistry, agrobiotechnology, pharmaceutical sciences and medicine.

Lilia Alberghina

CONTRIBUTORS

Lilia Alberghina
Dipartimento di Biotecnologie e
    Bioscienze
Universitá degli Studi Milano-Bicocca
Via Emanueli 12
20126 Milan
Italy

Masuo Aizawa
Faculty of Bioscience and Biotechnology
Tokyo Institute of Technology
Nagatsuta, Midori-ku
Yokohama 226
Japan

Anne Belaïch
Laboratoire de Bioénergétique
    et Ingénierie des Protéines
UPR 9036, de l'IFR–1 du CNRS
31 Chemin Joseph-Aiguier
13402 Marseille Cedex 20
France

Riccardo Bertini
Centro Ricerche Dompé
67100 L'Aquila
Italy

Michael Boisclair
Mitotix, Inc.
One Kendall Square
Building 600
Cambridge, MA 02135
USA

Uwe T. Bornscheuer
Institut für Technische Biochemie
Universität Stuttgart
Allmandring 31
70569 Stuttgart
Germany

Paola Bossù
Centro Ricerche Dompé
67100 L'Aquila
Italy

Klaus Breddam
Department of Chemistry
Carlsberg Laboratory
Gamle Carlsberg Vej 10
2500 Copenhagen Valby
Denmark

Christian Cambillau
Architecture et Fonction des
    Macromolécules Biologiques
UPR 9039, de l'IFR–1 du CNRS
31 Chemin Joseph-Aiguier
13402 Marseille Cedex 20
France

Stéphane Canaan
Laboratoire de Lipolyse Enzymatique
UPR 9025, de l'IFR–1 du CNRS
31 Chemin Joseph-Aiguier
13402 Marseille Cedex 20
France

Giovanna Carpani
Genetic Engineering and Microbiology
    Laboratories
ENIRICERCHE S.p.A.
Via F. Maritano 26
San Donato Milanese
20097 Milan
Italy

Enrico Cernia
Department of Chemistry
University of Rome "La Sapienza"
P.le Aldo Moro 5
00185 Rome
Italy

Caroline E. Connor
Department of Pharmacology
   and Cancer Biology
Duke University Medical Center
Durham, NC 27710
USA

Jiří Damborsky
Laboratory of Structure and Dynamics
   of Biomolecules
Faculty of Science
Masaryk University
Czech Republic

Ann M. Davis
Affymax Research Institute
4001 Miranda Avenue
Palo Alto, CA 94304
USA

Giulio Draetta
Department of Experimental Oncology
European Institute of Oncology
Via Ripamonti 435
20141 Milan
Italy

Liliane Dupuis
Laboratoire de Lipolyse Enzymatique
UPR 9025, de l'IFR–1 du CNRS
31 Chemin Joseph-Aiguier
13402 Marseille Cedex 20
France

Lesley Farrington
Affymax Research Institute
4001 Miranda Avenue
Palo Alto, CA 94304
USA

Franco Felici
Dipartimento di Biologia
Università di Roma Tor Vergata
Via della Ricerca Scientifica
00173 Rome
Italy

Henri-Pierre Fierobe
BIP-CNRS
31 Chemin Joseph-Aiguier
13402 Marseille Cedex 20
France

Torben P. Frandsen
Carlsberg Laboratory
Department of Chemistry
Gamle Carlsberg Vej 10
2500 Copenhagen Valby
Denmark

Francesco Frigerio
Genetic Engineering and Microbiology
   Laboratories
ENIRICERCHE S.p.A.
Via F. Maritano 26
San Donato Milanese
20097 Milan
Italy

Giuliano Galli
Chiron S.p.A.
Via Fiorentina 1
53100 Siena
Italy

Guido Grandi
Chiron S.p.A.
Via Fiorentina 1
53100 Siena
Italy

Renata Grifantini
Chiron S.p.A.
Via Fiorentina 1
53100 Siena
Italy

Tetsuya Haruyama
Faculty of Bioscience and Biotechnology
Tokyo Institute of Technology
Nagatsuta, Midori-ku
Yokohama 226
Japan

Sophia Hober
Department of Biochemistry and
    Biotechnology
Royal Institute of Technology (KTH)
10044 Stockholm
Sweden

Dick B. Janssen
Department of Biochemistry
Groningen Biotechnology and
    Biomolecular Sciences Institute
University of Groningen
9747 AG Groningen
The Netherlands

Christian Jelsch
Laboratoire de Cristallographie et
    Modelisation des Matériaux
    Mineraux et Biologiques
LCM3B-CNRS
Boulevard des Aiguillettes
BP 239
54506 Vandoeuvre les Nancy Cedex
France

John Jenkins
Department of Food Macromolecular
    Science
Institute of Food Research
Earley Gate
Whiteknights Road
Reading RG6 6BZ
UK

Eiry Kobatake
Faculty of Bioscience and Biotechnology
Tokyo Institute of Technology
Nagatsuta, Midori-ku
Yokohama 226
Japan

Kerry J. Koller
Affymax Research Institute
4001 Miranda Avenue
Palo Alto, CA 94304
USA

Geja H. Krooshof
Department of Biochemistry
Groningen Biotechnology and
    Biomolecular Sciences Institute
University of Groningen
9747 AG Groningen
The Netherlands

Sonia Longhi
Architecture et Fonction des
    Macromolécules Biologiques
UPR 9039, de l'IFR–1 du CNRS
31 Chemin Joseph-Aiguier
13402 Marseille Cedex 20
France

Marina Lotti
Dipartimento di Biotecnologie e
    Bioscienze
Universitá degli Studi Milano-Bicocca
Via Emanueli 12
20126 Milan
Italy

Imma Margarit
Genetic Engineering and Microbiology
    Laboratories
ENIRICERCHE S.p.A.
Via F. Maritano 26
San Donato Milanese
20097 Milan
Italy

Olga Mayans
Department of Food Macromolecular
    Science
Institute of Food Research
Earley Gate
Whiteknights Road
Reading RG6 6BZ
UK

Donald P. McDonnell
Department of Pharmacology
 and Cancer Biology
Duke University Medical Center
Durham, NC 27710
USA

Anne Nicolas
Department of Crystal and Structural
    Chemistry
Bijvoet Center for Biomolecular
    Research
Utrecht University
Padualaan 8
3584 CH Utrecht
The Netherlands

Renzo Nogarotto
Genetic Engineering and Microbiology
    Laboratories
ENIRICERCHE S.p.A.
Via F. Maritano 26
San Donato Milanese
20097 Milan
Italy

Kjeld Olesen
Department of Chemistry
Carlsberg Laboratory
Gamle Carlsberg Vej 10
2500 Copenhagen Valby
Denmark

Cleofe Palocci
Department of Chemistry
University of Rome "La Sapienza"
P.le Aldo Moro 5
00185 Rome
Italy

Richard Pickersgill
Department of Food Macromolecular
    Science
Institute of Food Research
Earley Gate
Whiteknights Road
Reading RG6 6BZ
UK

Maria Grazia Pizza
Chiron S.p.A.
Via Fiorentina 1
53100 Siena
Italy

Jürgen Pleiss
Institut für Technische Biochemie
Universität Stuttgart
Allmandring 31
70569 Stuttgart
Germany

Rino Rappuoli
Chiron S.p.A.
Via Fiorentina 1
53100 Siena
Italy

Rick Rink
Department of Biochemistry
Groningen Biotechnology and
    Biomolecular Sciences Institute
University of Groningen
9747 AG Groningen
The Netherlands

Mireille Rivière
Laboratoire de Lipolyse Enzymatique
UPR 9025, de l'IFR–1 du CNRS
31 Chemin Joseph-Aiguier
13402 Marseille Cedex 20
France

Jean Louis Romette
AFMB/DISP
UPR 9039 du CNRS
ESIL Case 925
163 Avenue de Luminy
13288 Marseille Cedex 9
France

Enzo Scarlato
Chiron S.p.A.
Via Fiorentina 1
53100 Siena
Italy

Rolf D. Schmid
Institut für Technische Biochemie
Universität Stuttgart
Allmandring 31
70569 Stuttgart
Germany

Thomas G.M. Schmidt
Institut für Technische Biochemie
Universität Stuttgart
Allmandring 31
70569 Stuttgart
Germany

Claudia Schmidt-Dannert
Institut für Technische Biochemie
Universität Stuttgart
Allmandring 31
70569 Stuttgart
Germany

Arne Skerra
Institut für Bioanalytik GmbH
Rudolf Wissel Strasse 28
37079 Göttingen
Germany

Drummond Smith
Department of Food Macromolecular
  Science
Institute of Food Research
Earley Gate
Whiteknights Road
Reading RG6 6BZ
UK

Simonetta Soro
Department of Chemistry
University of Rome "La Sapienza"
P.le Aldo Moro 5
00185 Rome
Italy

Birte Svensson
Carlsberg Laboratory
Department of Chemistry
Gamle Carlsberg Vej 10
2500 Copenhagen Valby
Denmark

Emily Tate
Affymax Research Institute
4001 Miranda Avenue
Palo Alto, CA 94304
USA

Anna Tramontano
IRBM P. Angeletti
Via Pontina Km 30.600
00045 Pomezia
Rome
Italy

Mathias Uhlén
Department of Biochemistry and
  Biotechnology
Royal Institute of Technology (KTH)
10044 Stockholm
Sweden

Marco Vanoni
Dipartimento di Biotecnologie e
  Bioscienze
Universitá degli Studi Milano-Bicocca
Via Emanueli 12
20126 Milan
Italy

Robert Verger
Laboratoire de Lipolyse Enzymatique
UPR 9025, de l'IFR–1 du CNRS
31 Chemin Joseph-Aiguier
13402 Marseille Cedex 20
France

R.A.J. Warren
Department of Microbiology and
  Immunology and
Protein Engineering Network of
  Centres of Excellence
University of British Columbia
Vancouver, BC
Canada V6T 1Z3

Erik A. Whitehorn
Affymax Research Institute
4001 Miranda Avenue
Palo Alto, CA 94304
USA

Catherine Wicker-Planquart
Laboratoire de Lipolyse Enzymatique
UPR 9025, de l'IFR–1 du CNRS
31 Chemin Joseph-Aiguier
13402 Marseille Cedex 20
France

Kathryn Worboys
Department of Food Macromolecular
    Science
Institute of Food Research
Earley Gate
Whiteknights Road
Reading RG6 6BZ
UK

Yasuko Yanagida
Faculty of Bioscience and Biotechnology
Tokyo Institute of Technology
Nagatsuta, Midori-ku
Yokohama 226
Japan

1. PROTEIN ENGINEERING IN BASIC AND APPLIED BIOTECHNOLOGY: A REVIEW

LILIA ALBERGHINA and MARINA LOTTI

Dipartimento di Biotechnologie e Bioscienze, Università Degli Studi Milano-Bicocca via Emanueli 12, 20126 Milano, Italy

The major goal of protein engineering is the generation of novel molecules, intended as both proteins endowed with new functions by mutagenesis and completely novel molecules. This definition, which may sound broad and perhaps ambitious, in fact pinpoints one of the most promising developments in our ability to understand and control a protein's function. After the revolution introduced in protein science by the advent of genetic engineering, protein engineering can be considered as a second wave of innovation which is providing important breakthroughs in basic research and application, useful for studies on structure function relations and for exploitation in industry. Genetic engineering makes available unlimited amounts of purified proteins, whereas protein engineering produces tailor-made proteins redesigned such as to make them more suited to industrial requirements. On this basis it becomes evident that industrial biotechnology will enormously benefit of this possibility.

Natural targets of protein engineering are enzymes and several examples of modified catalysts have already been achieved and applied to industrial processes. However, protein engineering is not restricted to this field, since several non-enzyme proteins play important roles, for example, as drugs. Another fundamental area is that concerning antibodies, which may be planned both as specific carriers able to target drugs in the human body, as well as in the production of catalytic antibodies (abzymes) to be applied for reactions with non natural substrates.

Protein engineering is a complex and multidisciplinary field, where several different techniques and knowledge are applied in combination. The protein of interest needs to be first purified and characterized with regard to its functional properties, then to be cloned and overexpressed in a suitable host organism and subsequently to be modified so as to improve its performances. A variety of analytical as well as structural techniques will be then employed for its characterization, whereas fermentation technologies will support its large-scale production. Different enabling technologies contribute therefore to an interdisciplinary approach that is usually represented as the **protein engineering cycle** (Figure 1).

This chapter aims to outline the most innovative techniques as well as industrial applications of protein engineering, to provide the reader with a general, although necessarily non comprehensive, view on the field. More in general it attempts to "give the taste" of the opportunity provided by protein engineering to biotechnology, stressing the important and interactive linkage between basic science and application.

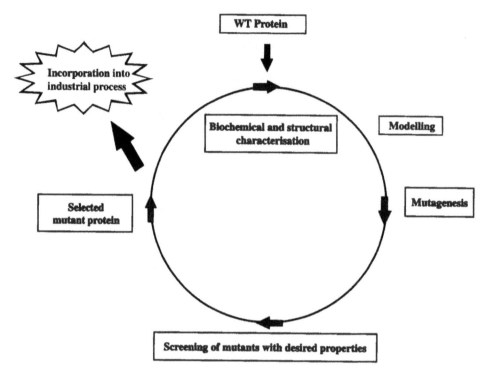

Figure 1 Cycle of protein engineering

ENABLING TECHNOLOGIES: HOW TO ENGINEER NEW PROTEIN FUNCTIONS

Diverse and complementary techniques form the knowledge and technological basis to engineer proteins and all of them must be mastered or at least be familiar to protein engineers. These basic techniques have been covered in recent and comprehensive books (Wrede and Schneider, 1994; Cleland and Craik, 1996;) and will therefore not be recalled hereafter. However, new techniques of mutagenesis/ screening were recently introduced, that expanded the possibility of modifying a protein's properties and deserve a brief comment (Table 1). Other essential enabling technologies are also outlined in the following.

Techniques of Mutagenesis

The starting point for every step of mutagenesis is a DNA sequence cloned from the original source or synthesized based on the protein sequence of interest. One or several amino acids may be substituted/inserted/deleted using a wide variety of methods. Two main conceptual approaches can be followed: i) site-directed mutagenesis, i.e. substitutions of nucleotides in the correspondence of pre-selected positions in the gene (polypeptide) ii) random mutagenesis.

Table 1 Experimental approaches for protein engineering

Rational design
Allows for the introduction of mutations targeted to specific protein sites. Requires a detailed knowledge of the protein structure and of structure-function relationships.

Molecular evolution
Does not require any knowledge on the protein structure and mechanism of action. It is based on the random generation of a vast number of mutants followed by screening for the desired functions

Generation of random libraries
Production of large collections of proteins, peptides of region thereof. Is often coupled with surface display to ease screening of the mutants

***De novo* protein design**
Generates novel structural scaffolds able to accommodate active sites or other protein functions

---

Rational design

Site directed mutagenesis (SDM) is a classical approach for protein engineers, exemplified by a broad scientific literature and by several contributions of this book. It consists in introducing a change in one or more amino acids and evaluating the effect of these pre-selected substitutions in the mutated product. By definition, this strategy, also called "rational mutagenesis", requires a prior knowledge of the role played by specific residues or regions of the protein. This means availability of the protein 3D structure, if possible also in complex with substrates, ligands, regulation elements or at least availability of sequence of proteins with related but not identical activity for comparison. In fact, sequence alignments may also provide support in the selection of the positions to mutate, in particular when the protein of interest belongs to a large and well characterized family of proteins. On the other hand computer-assisted technology may help in predicting the functional effect of the planned substitutions.

Rational design is being very widely employed to engineer in proteins new functions or influence their regulation. Important goals have been achieved by rational design, as it will be briefly summarized in the following paragraphs.

Non-rational (random) design

Random mutagenesis on whole protein sequences or parts thereof is the method of choice in all those cases where knowledge about the structure and function of the protein of interest is not sufficient to support a rational design approach. The most innovative techniques involve the generation of repertoires (libraries) of mutated sequences and procedures of mutation/selection that mimic the processes followed by nature during evolution. In both cases, a very high number of mutant variants is produced, so that the development and availability of sensitive and fast procedures of screening is vital to manage the experimental work.

Molecular evolution. Very recently, protein engineering has been revolutionized by new methods that mimic evolution, of course in much shorter time, in that they allow for the accumulation and selection of mutations beneficial to the property of

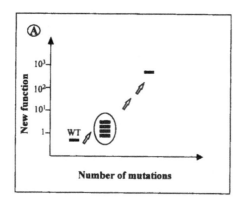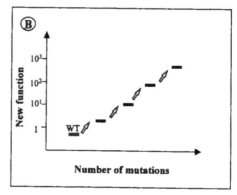

Figure 2 Comparison between two strategies for protein "evolution": DNA shuffling (A) and directed evolution (B) (modified from Arnold, 1996).

interest. These new approaches commonly referred to as *molecular evolution techniques,* allow to overcome the serious drawback represented by the fact that for several industrially important proteins the level of available knowledge is insufficient to rational design. Moreover, growing evidence suggests that many protein functions cannot be ascribed to single or few aminoacids but they rather depend on regions located far away from the active and regulatory sites, and therefore hard to be predicted *a priori*. Molecular evolution methods are based on the accumulation of beneficial mutations over several rounds of mutagenesis. Since the effect of every single mutation can be small, the availability of methods of screening sensitive and in addition applicable to large number of mutants, is an indispensable prerequisite to this goal. Evolutionary approaches can be grouped in two categories: DNA shuffling and directed evolution (Figure 2). Both methods rely on the well-established technology of polymerase-chain reaction (PCR), in its development known as "error-prone PCR", that allows for the introduction of a small preselected number of mutations in the DNA sequence.

DNA shuffling was developed by W. Stemmer in 1994 (Stemmer 1994 a and b). This technique is also known as "sexual PCR". It involves enzymatic fragmentation of DNA derived from different organisms previously randomly mutated as to introduce changes in the coding sequence. Fragmented DNA are then reassembled by error-prone PCR in the absence of primers, allowing their spontaneous reassemble and — at the same time — the introduction of new mutations. A final step of amplification with primers generates full-length products. This strategy has been applied to several cases. For example, the activity of β-lactamase towards the antibiotic cefotaxime was increased over 16,000 times through three cycles of sexual PCR, whereas the fluorescence of the green fluorescent protein was increased 45 fold (Crameri *et al*, 1996). The ambitious goal of modifying multigene determinants of complex metabolic functions was also achieved enhancing the activity of an arsenate locus from *Staphylococcus aureus* composed by three genes. This results introduced DNA shuffling in the field of environmental bioremediation (Crameri *et al*, 1997). In the same field, Kumamaru and colleagues recently evolved novel specificities in biphenyl dioxygenases (BPDox) by shuffling homologous genes from *Pseudomonas alcaligenes* and *P.cepacia*, mating yielded enzymes with enhanced 2,3-dioxygenation activity towards several biphenyl compounds and new oxygenation activity for aromatic hydrocarbons (Kumamaru *et al.*, 1998).

In vitro directed evolution is another powerful approach, developed by F. Arnold at Caltech, which, with repetitive cycles of mutagenesis/screening forces the evolution of the selected sequence towards the desired functions (Arnold, 1996). Differently from DNA shuffling, in this approach after every cycle of random mutagenesis, only a single variant showing functional improvements is selected and used as the parent for the next generation of mutants. A big advantage of this strategy is that it allows limiting the number of mutants to screen; on the other hand, by selecting a parent molecule at every step the researcher introduces a bias on the possible changes. This strategy has been successfully employed to engineer new functions and to make enzymes able to work in unusual environments. An esterase from *Bacillus subtilis* was modified as to become able to hydrolyze p-nitrobenzyl ester bonds in the presence of organic solvents, a reaction of interest for the pharmaceutical industry to remove protecting groups introduced during the synthesis of antibiotics of the class of cephalosporins. Specific activity was increased of 30-folds through four generations of random mutagenesis followed by DNA recombination (Moore and Arnold, 1996). Even the specific activity of subtilisin E in polar solvent was enhanced through the same approach (Chen and Arnold, 1993). Another property of paramount importance in biotechnology — enantioselectivity — was introduced by directed evolution in a lipase from *Pseudomonas aeruginosa* (Reetz *et al.*, 1997).

Combinatorial libraries and surface display

Another fundamental approach towards the generation of random mutants is the construction of "random libraries" representing peptides, cassette mutagenized proteins or antibodies, followed by the screening of variants with the desired functions. Random libraries might be obtained by i) random cassette mutagenesis or ii) by chemical synthesis. Whatever the origin, this generates very large repertoires of molecular diversity; which is evident if one considers that random mutagenesis on only three positions within a peptide will originate 8.000 (20³) different combinations. The technology of random libraries has proved to be potentially very powerful depending on the ability to screen simultaneously a vast number of variants. Screening is usually based on the affinity of the new variants for the target molecules (antigens, enzyme substrates etc), a procedure which is generally referred as "panning". A support of paramount importance to the development and application of combinatorial libraries was provided by the advent of techniques allowing the display of variant peptides or epitopes at the surface of phages or various cell types. If the sequence to be panned is fused with that of genes coding for proteins of the virus coat, mutagenized peptides will be exposed at the virus surface, without impairing its replication ability. Billions of mutated peptides and proteins can be displayed at the surface of phage particles and be selected by affinity chromatography. The library is thus screened for biological function, i.e. for its binding affinity or enzymatic activity. One of the main application of combinatorial libraries concerns the improvement of antibodies binding abilities. In this approach, the variable regions of antibodies are amplified by polymerase chain reaction and then mutated as to develop binding affinity for new antigens. Application to human antibodies is intensively explored for diagnosis and therapy. However phages are not the only available tool for protein display (see for a recent review Sathl and Uhlen, 1997). Thus for example, heterologous proteins have been targeted and anchored to the outer surface of yeast, mammalian and bacterial cells (Schreuder *et al.*, 1996). Among bacteria, both Gram negative (*Salmonella, Escherichia coli*) and Gram positive (*Staphylococci, Streptococci*) have been used, where gene fragments can be inserted into the genes coding for outer membrane proteins, lipoproteins, fimbria or flagellar proteins or whatever make fusion products accessible on the outer surface. Enzymes such as pullulanase and cellulase have been displayed on *E. coli* and staphylococci (Francisco *et al.*, 1992; Strauss and Goetz, 1996; Kornacker and Pugsley, 1990). Peculiar applications of surface technology can be found in the production of whole cells bioadsorbents for environmental purposes.

Application of the combinatorial approach to the identification of molecules with a particular biological activity is reported in the chapter by **Tramontanto and Felici** using a series of case studies ranging form the development of molecules that mimic carbohydrate antigenes and therefore useful for vaccines, to epitope mapping to the so-called "minibody", a small protein designed as to display antigen binding loops that can be randomised and used to select ligands.

The Impact of Structure Analysis and Prediction

Structures and modelling

Besides developments in molecular biology and biochemistry, biocomputing and methods for the determination of 3D structures, are indispensable tools in this field. Since the three-dimensional structure of a protein (enzyme, hormone, receptor) determines its biological function, it becomes important to have available a high resolution 3D structure, especially when the protein engineer aims to redesigning the protein function. Therefore, the importance of structural information for protein engineering cannot be overestimated, since the immediate challenge is to understand the rules relating the primary sequence to its 3D structure and to determine how structure is related to function. To date over 8300 entries are stored in the Brookhaven Protein Databank (PDB) (http://www.pdb.bnl.gov) enclosing proteins, nucleic acids and other biological macromolecules, representing ca. 800 unique structures. Recent advances in protein structure determination techniques originated the number of new structure solved every year to exponentially increase. A big number and a big deal of information; however only a small amount with respect to the variety of forms and shapes available in nature.

The most classical approach to protein structure is still X-ray diffraction, through which resolutions at less than 1Å have been obtained. Important advances were made in the field by using more intense X-ray sources, as cyclotron facilities, two dimensional area detectors coupled to computers making faster the obtainement and analysis of structural data. Rapid acquisition times provide the possibility to study transient phenomena. Difficulties intrinsic to this technique, in particular the necessity to obtain suitable crystals of the proteins to be studied, has been recently relieved by the use of robots to perform crystallization. It is now in use to perform protein engineering to facilitate protein crystallization, for example by introducing cysteines to obtain heavy atom derivative crystals isomorphous to the native protein (Price and Nagai, 1995).

A very comprehensive example about the role of crystallography in the elucidation of structure-function relationships in enzymes is provided in the chapter by **Longhi and colleagues** dealing with cutinase, a small hydrolytic enzyme presently under study for applications in biocatalysis and detergents.

More recently, advances in multidimensional NMR (Nuclear Magnetic Resonance) methodology made this technique competitive, especially because of the possibility to study proteins in solution, eventually under conditions near to the physiological ones. It is presently possible to determine by NMR structures of proteins in the range of 15–35 kDa Mr, some of which in complexes with ligands, what is of importance for the study of protein-drug and protein-protein complexes (for a recent review see Wagner, 1997; Clore and Gronenborn, 1998).

The role of biocomputing

Nature provided us with an astonishing variety of shapes and functions, most of which have still not be unraveled by researchers. Thus, even if a few hundreds of protein structures are solved, some other thousands still await to be studied. Fortunately, evolution seems to have selected a limited number of protein folds, where different sequences can be accommodated. This fact allows in many instances to extrapolate from the known structure information about those we still do not know. Therefore, knowledge-based structural prediction is becoming a common approach. Protein sequences are aligned with the sequences of homologous peptides of known structure and a structural framework is "modeled" by assembling pieces of structures derived from the most homologous proteins. This approach relies on the availability of structural databases and of software for sequence analysis and energy minimization as well as of computer graphic tools. One of the most innovative approaches in biocomputing is that known as **fold recognition**, a computational attempting to fish out the fold of a protein even when it has very little sequence homology to other sequences of the same structure (for a review see Westhead and Thornton, 1998).

Moreover development of computational methods in combination with computer graphics, presently accessible to most laboratories, introduced a significant element of rationality in protein engineering allowing to predict with an acceptable degree of confidence the effects of aminoacid substitutions. However, structure and sequence analysis still remains one of the bottlenecks in protein science (Table 2).

The impact of biocomputing on biotechnology is getting increasingly dramatic due to the enormous amount of information continuously provided by the genome sequencing projects. Currently some dozens of projects of sequencing are underway,

Table 2 Bottlenecks in the field of protein engineering

- Scarce information on the second genetic code, i.e. rules relating sequence to structure in proteins
- Limited number of known 3D structures
- Limited or not obvious information of the structure-function relations in proteins
- Difficulties in the management of information

Table 3 Whole genomes sequencing projects (archae, bacteria, eukaryote, not viruses)

20 COMPLETE PROKARYOTE GENOMES
1 COMPLETE EUKARYOTE GENOME
45 PROKARYOTE GENOMES IN PROGRESS
17 EUKARYOTE GENOMES IN PROGRESS

N.B. 15 complete genomes are available to public; *Aquifex aeolicus, Pyrococcus horikoshii, Bacillus subtilis, Treponema pallidum, Borrelia burgdorferi, Helicobacter pylori, Archaeoglobus fulgidus, Methanobacterium thermo, Escherichia coli, Mycoplasma pneumoniae, Synechocystis sp. PCC6803, Methanococcus jannaschii, Saccharomyces cerevisiae, Mycoplasma genitalium, Haemophilus influenzae.*

Another 6 are complete but still not availbale because of pending publication: *Rickettsia prowazekii, Pyrococcus abyssii, Bacillus sp. C-125, Pseudomonas aeruginosa, Ureaplasma urealyticum, Pyrobaculum aerophilum.*

Almost complete the following: *Pyrococcus furiosus, Mycobacterium tuberculosis H37Ru, Mycobacterium tuberculosis CSU93, Neisseria gonorrhea, Neisseria meningiditis, Streptococcus pyogenes.*

Several other sequencing projects are run by privat organisations

More information can be obtained at the web site: http//www.mcs.anl.gov/home/gaasterl/genomes.htlm

covering prokaryotic organisms, unicellular eukaryotes and complex eukaryotic genomes (Table 3). Enormous support is provided to researchers by the availability to the public of databanks and tools for analysis (Table 4).

De novo design

Differently from every approach mentioned before, where existing polypeptide templates are modified such as to introduce new properties, the aim of *de novo* protein design is to be able to design new (not pre-existing) sequences endowed with predefined structure and function (Hellinga, 1998). In this case the bottleneck is not a technical one, since synthesis of peptides can be now realised in several laboratories. Indeed the fundamental problem in *de novo* design is that the rules

Table 4 Some Web sites of interest to protein engineers

Structures and structure analysis	
Protein Data Bank, Brookhaven National Lab.	http://www.pdb.bnl.gov
SCOP-Structural classification of proteins	http://scop.mrc-lmb.cam.ac.uk/scop
RasMol	http://www.umass.edu/microbio/rasmol
Sequences and sequence analysis	
European Bioinformatic Institute, UK	http:// www2.ebi.ac.uk
European Molecular Biology Lab., G	http://www.embl-heidelberg.de
Expasy Molecular Biology Server	http://www.expasy.ch
NCBI Genebank	http://www.ncbi.nlm.nih.gov
Pedro's Biomolecular Research Tools	http://www.public.iastate.edu/

governing protein folding — what is often referred to as the second genetic code — are still not completely understood (Table 2). This makes it difficult both to predict protein structures from sequences and to design sequences that fold into a desired structure. The major difficulty arises from the enormous number of possible sequences to consider in the computational approach. As a result, designed structures have often a molten globule-like structure since they are not able to perform the tightly packed organization typical of native proteins. Recent developments obtained in automated protein design programs greatly extended the researcher's ability but still the complete design of novel biocatalysts exceeds the present know-how. However, small proteins able to adopt defined secondary structures have been designed (Quinn *et al.*, 1994; Dahyat and Mayo, 1997) and the number of sequence candidates to consider can be reduced by reducing the length of the sequence to be designed or introducing a bias in the search of experimental sequences, for example by phage display (Riddle *et al.*, 1997). One interesting achievement in this field was the construction of a hyperthermophilic variant of the B1 domain of Streptococcal protein G (Malakauskas and Mayo, 1998). Among the first practical applications of *de novo* design one might recall the design of a four helix bundle protein used as a source of rare amino acids when expressed in the rumen bacteria of dairy cows (Beauregard *et al.*, 1995) and the design of scaffolds for the introduction of active sites. Thus for example into de novo protein albebetin an interferon fragment was attached and the recombinant protein proved to effectively bind to the murine thymocyte receptor activating the blast transformation (Dolgikh *et al.*, 1996 a and b).

How to affect protein conformation: medium engineering and bioimprinting

It is worth mentioning that other approaches besides protein engineering, have recently improved our ability to control protein conformation and enzyme activity, i.e. medium enginering and molecular bioimprinting. Although these two approaches do not directly influence a protein's primary sequence, they produce conformational effects affecting its function, and are therefore properly considered in this book. The field of medium engineering developed based on the observation that the performances and even the specificity of enzymatic proteins can be modulated through the solvent where the reaction is made to occur. Enzymology in near-anhydrous environment, i.e. in organic solvents, is an important area in biotrasformation. Three major benefits are related to the use of organic solvents: i) the higher solubility of several organic substrates; ii) the higher stability of the enzymes, for example to the exposure to high temperatures required by some processes; iii) an increased selectivity observed in several cases and concering in particular selectivity towards enantiomeres compounds. The two latter properties have been suggested to depend on a certain degree of rigidity of the protein caused by the organic solvent itself (Wescott and Klibanov, 1994). In some cases, the advantages obtained largely compensated for the decrease in activity produced by the anydrous solvent (Klibanov, 1997). Often enzymes exposed to particular pH or ligands before liophylization and subsequent dissolution in organic solvents, prove to be able to retain specific properties, i.e affinity for the ligand, a property referred to as "memory". Based on this observation, a strategy for the activation of proteins has been developed which was called "molecular (bio)imprinting" and can be adapted to particular conditions where the reaction of interest has to take place (Rich and Dordick, 1997; Mingarro et al., 1995; Ramstrom et al., 1996). Supercritical fluids have also be exploited as non conventional reaction media (Ballesteros et al., 1995; Cernia et al., 1998). The contribution by **Cernia** and colleagues in this volume effectively illustrates how the specificity and activity of microbial lipases can be modulated through the use of organic solvents and supercritical fluids.

WHAT TO ENGINEER: TARGET MOLECULES AND PROPERTIES OF INTEREST

Engineering Enzymes

.The study of structure/function relations in naturally occurring enzymes can provide important clues of interest for protein engineers. However, the variety of strategies employed by nature in developing enzymes adapted for specific functions and environments is still far to be completely exploited. On the contrary, it is reasonable to suppose that we just scraped the surface of the diversity of solutions evolved. Biocatalysts purified, cloned and employed in industrial processes are possibly only those which were possible to screen from easily available sources that, in addition should have produced the desired enzyme in sufficient quantity.

Novel screening approaches are revealing the existence of a variety of unsuspected enzymatic activities. Even if the natural proteins might be not directly useful to industrial processes, they can provide precious informations useful to redesign conventional enzymes to perform special (novel) activities.

In this frame, we should recall innovative screening procedures set up to allow isolation of enzymes which escaped previous investigations. One of this methods is the so-called "expression cloning strategy", that is based on the expression of cDNA libraries from the selected source organism in host. In this approach, transformants are screened for several different activities, thus allowing to isolate new enzymes without relying on previous knowledge on the enzyme itself, because screening is completely activity-based. In this way Novo Nordisk was able to isolate over 130 fungal genes coding for potential industrial enzymes (Dalboege and Lange, 1998). In the same way, accumulation of more and more information about sequences of related enzyme, allows to develop nearly universal primers to be used in molecular screening programs. Recently, the ability to screen genomic DNA without the necessity to prior isolate the source organism has been reported, what obviously further extend our ability to obtain novel enzymes (Jacobsen, 1995).

The search of enzymes from extremophiles is also passing through a real boom. Environments extremes as for high or low temperatures, pressure, salt concentrations are sistematically exploited through highly concerted programs. The first commercial thermophilic enzyme has in fact been isolated over 30 year ago in thermal pools of the Yellowstone National Park. It is the DNA polymerase from *Thermus aquaticus* (Taq) which is now so widely used in PCR experiments. Extreme enzymes are seeked for use in processing of foods and chemicals or in some cases as leads for new antibiotics. The most important screening programs have been DEEPSTAR, that allowed to collect 1000 different organisms from deep-sea mud and 2500 strains from the Marianas Trench and a European Project in the frame of the EC-Biotech program. Interestingly, strains were isolated able to grow on toxic organic solvents, with interesting perspectives for use in environment bioremediation.

Much is expected from the analysis of sequence and structure of extreme enzymes, since it may provide clues in understanding factors influencing in particular the stability of proteins in particular solvents.

Properties of importance to engineer are stability (to temperature, pH, specific reaction conditions), specificity, regulation and eventually properties useful for the downstream processing of the recombinant proteins.

Engineering stability

Stability is a feature of critical importance in biotechnology and it is therefore pursued with intense efforts. Stability may concern: stability to temperature (food and chemicals processing), to pH (detergents), resistence to proteolytic cleavage (drugs). Attempts to understand factors affecting stability started quite early in protein science and have been based mainly on mutagenesis approaches or on the comparison between homologous sequences from, for example, mesophiles and extremophiles organisms (Fersht and Serrano, 1993).

Presently, several protein properties are recognized as having a stabilizing effect but we are still far from the possibility to control protein stability. Often, attempts to stabilize proteins by protein engineering have failed, since the introduction of stabilizing interactions destroyed other important interactions or caused a loss in function.

Several interactions of importance have been determined, each of these which give contributions which are quantitatively similar; the most significant are briefly outlined in the following.

- Disulfide bridges stabilize proteins by restricting the possible conformations of the unfolded proteins. However, engineering disulfide bridges has met alternate success, possibly because of the difficulty in predicting how the increased rigidity of the protein will be detrimental for the protein function. A significant success has been described in the stabilization of the T4 lysozyme by the Matthews group (Matthews B.W., 1995).
- Ionic interactions or salt bridges. Several thermophilic proteins have additional ion pairs as compared to their mesophilic homologous; however the effect of this interactions can be hardly forseen.
- α-helices stabilization is another common approach. Multiple mutations in the triosephosphate isomerase dimer have been introduced as to increase the helix propensity of some protein regions or to stabilize the 2D structure itself by interaction with the helix macrodipole. Stability was increased of about 3.0 Kcal/mol thus approaching the stability of the same protein from the thermophilic bacterium *Bacillus stearothermophilus* (Mainfroid *et al.*, 1996).
- Hydrophobic effect. Is the well known effect by which hydrophobic regions tend to be sequestered in the protein interior. Even the tight and highly organized packing of the protein core is important and in fact the native state of proteins show packing densities similar to those of organic crystals. Substitutions of large hydrophobic residues in the protein cores with smaller amino acids (large to small substitutions, LTS) have destabilizing effect, probably because cavities are

created within the protein structure. The opposite substitution (small to large, STL) and even double and compatible LTS and STL mutations have been considered as a possible way to improve stability. Recently, by this approach Shih *et al.* (1995) mutated three buried residues of chicken lysozyme obtaining to increase thermostability of about 0.6–1.3 Kcal/mol. Relying on the same effect but using random mutagenesis, Lee and colleagues (1996) were able to stabilize α1-antitrypsin of 1–3.6 Kcal/mol.

– Binding of ion metals or cofactors can increase structural stability when they bind with higher affinity to the folded than to the unfolded state. Engineering metal centers, in particular binding sites for metal ions, frequently Ca^{2+}, Zn^{2+}, Cu^{2+} aims to enhance protein stability, but also to control activity or ease purification (see later). The same concept of differently stabilized conformational states allows for control of protein activity; following this approach, histidine was engineered in the active site of trypsin to create a metal dependent switch of enzyme activity. Metal binding sites are also extremely useful to construct proteins that can be easily purified from complex media by interaction with immobilized chelating groups. Artificial metalloproteins are also explored as biosensors to detect the presence of free metals in solution. Moreover, modification of the activity of existing metalloenzymes is under study for use in bioremediation; in this case, the introduction of new binding pockets allows transformation of non-natural substrates, as it was recently obtain for monooxygenases modified such as to oxidize aryl sulfide and styrene. The most recent progresses in this field have been recently reviewed (Benson *et al.*, 1998).

Not only stability to temperature is important in industrial processes. The lipase from *Candida antarctica* B has been mutated, to increase the oxidation stability of the enzyme (Patkar *et al.*, 1998), whereas always a lipase for detergents was modified as to increase its stability to alkali and to proteases as required to improve its washing performances (Okkels *et al.*, 1996).

Stability in organic solvents or other unusual reaction media that might be necessary during industrial processes. The ability to use proteins in unusual or non-natural environments greatly expands their potential applications in biotechnology (Arnold, 1993).

Approaches to improve the stability of industrial enzymes when exposed to non-physiological conditions are reported in the chapter by **Grandi** *et al.*, describing how Bacterial neutral proteases, isoamylases and carbamylases have been subjected to either site-directed or random mutagenesis to improve their performances in industrial processes.

Engineering specificity

Application of protein engineering to the specificity of enzymes and other proteins concerns such a vast field of action, making it difficult to be summarized. It might be directed towards the fine tuning of enzyme specificity, as well as the introduction of novel properties on preexisting protein scaffolds and the introduction of regulatory sites or switches. Rational design, random mutagenesis as well as molecular evolution techniques, have all been applied in the effort of controlling protein specificity, an issue that we will try to focus on through a few significant examples (see for recent review Shao and Arnold, 1996; Vita, 1997).

A common approach towards redesigning enzyme specificity is that of using existing protein scaffolds to introduce new properties. Recently, a single amino acid substitution was sufficient to convert the protease papain to peptide nitrile hydratase (Dufour *et al.*, 1995).

A possibile alternative is to construct hybrid proteins where different catalytic and/or regulatory modules are incorporated (Nixon *et al.*, 1998).

Binding sites for metal ions often prove useful to regulate enzyme activity. Thus the introduction of a binding site for Cu^{2+} was sufficient to allow regulation of trypsin activity (Halfon and Craik, 1996).

Cofactor requirement can also be altered by mutagenesis. As an example we can quote the paper by Dean and coworkers that changed the dependence of *E.coli* isocitrate dehydrogenase from NADP to NAD by replacing amino acids in the adenosine-binding pocket (Chen *et al.*, 1995).

Several studies successfully obtained changes in the enzyme substrate specificity. Alignment of sequences of homologous enzymes with different specificities as well as structural information have proven useful as a guide to identify residues possibly involved in this property. Thus for example serine proteases have been subjected to extensive mutagenesis studies aimed to covert one protease in another endowed with different specificity (i.e. chymotrypsin and trypsin). Serine proteases have conserved folds and sequences but differ in the specificity towards the point were they cut the peptide, being large hydrophobic residues for chymotrypsin and a positively charged residue for trypsin. The story of engineering serine proteases has been very informative. In fact, mutation of all four different residues in the S1 site (binding for the residue proximal to the cutting site) was not sufficient to convert an enzyme in the other. Otherwise trypsin acquired chimotrypsin specificity after replacement of two surface loops which are not obviously involved neither in substrate binding nor in catalysis (Hedstrom *et al.*, 1992) Following these studies, a good degree of control on the proteases specificity has been gained. Subtilisin for example was much studied. More easily than in the previous studies, the specificity of the *B. amyloliquefaciens* subtilisin was changed so as to be similar to that of *B. licheniformis* subtilisin by changing only three residues (Wells *et al.*, 1987) and the ability to selectively cleave two consecutive basic residues was engineered by Ballinger and colleagues (1996) in the same enzyme.

Olesen and Breddam in their contribution to this book report about the production of mutants of yeast carboxypeptidase with altered P1 preference, with application for C-terminal modification of peptides and stereo-specific synthesis of peptides and peptide derivatives.

Less widespread is the approach to transfer enzyme active sites to stucturally unrelated proteins (Smith *et al.*, 1996) since this implies a heavy work of structural modeling.

Engineering feature to ease protein purification

An important issue for the production of industrial recombinant proteins is the possibility to obtain them in a pure and active form. Two contribution of this volume specifically focus on the introduction of "tags" allowing purification and folding of the protein of interest. **Hober and Uhlen** describe the use of fusion proteins and co-expression strategies to ease folding and recovery of recombinant proteins. **Schmidt and Skerra** developed by molecular repertoire techniques a novel tag able to bind with high affinity to streptavidin from where it can be eluted by competition with biotin analogues.

Engineering Antibodies

Antibodies are important targets for protein engineering that are opening a new era for their application in health care, agriculture and industry. The two major techniques in this category concern reengineering of natural antibodies and catalytic antibodies (reviewed by Schultz and Lerner, 1995). A very remarkable progress in the field was generated by the application of novel techniques, in particular of combinatorial antibody libraries displayed on the surface of phages. This technique, developed in the early 1990s, revolutionises the generation of monoclonal antibodies and their engineering, since the use of animals or hybridomas cultures is not longer necessary, being recombinant antibodies expressed in bacteria, yeast, plants. Antibody phage display is obtained by fusing the sequences coding for the variable region of immunoglobulins to proteins of phage coat. The expressed fusion protein is incorporated in the envelope of the mature phage particle being presented at its surface. This accessibility is the basis for selection procedures consisting, for example, in binding with immobilized antigenes and the enrichment of phage antibodies specific for the desired antigen. In this way large repertoires of antibodies can be created and subjected to subsequent screening. The coding sequence is in addition amenable to any further modification by mutagenesis and protein engineering techniques. One of the main applications of this strategy is the generation of antibodies with high affinity for antigenes that do not necessarily need to have been previously purified, as it was the case of the previously used immunization procedures. On the contrary, antibody's diversity is first randomly generated and specificity is obtained through the subsequent screening against even complex materials containing the antigen of interest. Researcher's ability to control the antibody specificity has increased so to make possible the reational design of tailor-made antibodies against any antigen of choice. In the field of health care, the possibility of producing recombinant human antibodies ev. engineered to increase their specificity, open completely new perspectives as for drug targeting. Immunotoxins, immunoconjugates and bispecific antibodies can be generated for application in therapy and diagnosis. The present state of antibody engineering for health care has been recently reviewed (Carter and Merchant, 1997).

In the field of agriculture, engineered antibodies are studied as substitutes of plant hormones or to decrease the need of pesticides. Finally, the use of engineered antibodies during the removal of organic pollutants have been described.

In the second approach, the ability of the immune system to generate an enormous diversity of binding sites is exploited as a possible source of new catalysts (reviewed in Benkovich, 1992). The concept underlying the generation of antibodies-based catalytst — catalytic antibodies or abzymes — is that, if an antibody is created with an antigen binding site able to bind to analogues of the transition states of chemical reactions, binding will lower the transition energy of the reaction. Through immunization with transition state analogues, monoclonal antibodies have been obtained able to catalyze a wide range of reactions, eventually not performed by natural enzymes (Smithrud and Benkovic, 1997). Catalytic efficiency however, is still lower than with enzymes. Newest developments in abzymes technology exploit combinatorial libraries displayed on the surface of phages, a methods which enormously enlarge the possible size of the library (Janda *et al.*, 1997). Abzymes have been used to activate prodrugs (Wentworth *et al.*, 1996)

Engineering signal molecules (hormones/receptors)

Peptide hormones, receptors and in general component of the cell signaling machinery are all important targets for protein engineering, in view of novel applications in the field of health care. It was demonstrated that hormone/receptor interactions mostly depend on interactions among secondary structure elements. This observation led to the design of polypeptides with reduced size but enhanced biological activity, where secondary structure homology rather than sequence homology was maintained.

PRESENT APPLICATIONS OF PROTEIN ENGINEERING

Protein engineering is a pervasive technology and is a powerful tool to overcome instrinsic limitations weakening industrial applications of available proteins. The properties to introduce in a commercial protein strongly depend on the process where this molecule has to be employed and reather can be a combination of multiple *ad hoc* features.

Industrial Enzymology

Enzymes to be employed for industrial transformations have been the first targets of protein engineering and several engineered proteases, amylases, lipases, cellulases and xylanases have been introduced in the chemical, food and, more recently, environmental protection industry. The market for industrial enzymes in fact is steadily growing and is expected to further increase as scientists will become able to generate novel catalysts with optimized performances for the transformation process.

Most desirable features to be introduced being that molecules are stable and active under unusual conditions of temperature, pH or are active in unusual environments, as for example, in organic solvents. Besides, one should recall another property that may be of importance in industrial processes, i.e. the surface properties of the used catalysers. In fact, several important industrial enzymes, for example cellulases, xylanases and lipases rely for their activity on their ability to bind or adsorb to substrates and interfaces.

Hereafter a brief overview of protein engineering performed on industrial enzymes is presented, where both rational design and directed evolution have been successfully applied (Table 5).

Proteases are the main class of industrial catalyzers, particularly because of their use in detergents. They accounut for over 40% of the global enzyme market. Proteases are added to detergents because of their action on protein stain (spots). Subtilisin in particular has been the target of in-depth engineering: its stability in the chelating environment typical of detergents has been increased of about 1.000 folds by deleting a calcium binding site (Strausberg *et al.*, 1995), whereas the

Table 5 Industrial enzymes targets of protein engineering

Starch processing enzymes: α-amylase, glucoamylase, glucose isomerase

Detergent enzymes: proteases, lipases, amylases

Texiles: cellulases, amilases, lipases, catalases, peroxidases

Leader industry: proteases, lipases

Pulp and paper: xylanases, cellulases

introduction of surface negative charges improved its laundry performances (Rubingh, 1996). More in general serine proteases have been one preferential target for protein engineers, to modulate their specificity and/or regulation (Perona and Craik, 1995)

Several instances of engineered enzymes for food processing have been reported. Among them, **glucose isomerase** used in the conversion of corn starch to high fructose syrup for use in sweeteners, in particular for soft drinks. One of the most desirable technical improvements would be a higher tolerance to temperature. Rational approaches have been applied to stabilize the interactions helding together enzyme monomers in the quaternary structure (Quax *et al.*, 1991). **Calf chymosin** has been subjected to multiple point mutations to made specificity pH-dependent thus allowing modulation of flavor development (Law and Mulholland, 1991). **Papain** has been modified to make it active even at low pH whereas phospholipase A2 termal stability has been increased through mutations in α-helical regions (reviewed in Pickersgill and Goodenough, 1991). **Amylases** are employed to convert starch in simpler sugars. First α-amylase yields maltodextrins which are then processed by glucoamylase giving glucose and β-amylase resulting in maltose. Application of the already mentioned glucose isomerase can finally produce fructose. Starch processing is carried out at high temperature and therefore several protein engineering studies have been aimed at improving the thermal stability of the enzymes involved (Joyet *et al.*, 1992; Decklerc *et al.*, 1995; Gottschalk *et al.*, 1998). Protein engineering of glucoamylase from *Aspergillus niger* is the topic of the contribution by **Frandsen** *et al.* that, based on an extensive mutagenesis work, provides insight in the mechanism of action, stability and specificity of this enzyme.

Cellulases and xylanases are used in the processing of textiles and in detergents. Most commercial cellulases are endoglucanases and contain a catalytic region and a cellulose-binding domain. Efforts have been attempted to increase the binding ability of these enzymes to their substrates as well as enzyme activity (Linder *et al.*, 1996; Koivula *et al.*, 1996). In this book, β-glycosyde hydrolases are the topic of the chapter by **R.A.J. Warren**, focusing on the basis of their thermal stability, pH optimum, catalytic activity and specificity, as elucidated by protein engineering. These enzymes hydrolyse chitin, cellulose, xylan and mannan as well as oligosaccharides used in the detergent, food processing and paper industry.

Pectinolytic enzymes find application in both food and pharmaceutical industry. The state of the knowledge about structure of this enzymes, regions involved in substrate recognition, mechanism of action and structural basis of specificity are reported by **Mayans and colleagues** in their contribution.

Lipases are employed in several sectors, food, chemical transformations and detergents. Several mutagenesis studies have been applied to improve their performances. Fungal lipases used in detergents have been modified as to improve their resistance to proteases also present in cleaning agents and to change surface properties and make them more soluble; in addition several agents present in detergents such as surfactants and calcium sequestering agents reduce lipase activity and have been therefore targets for protein engineering. Site directed mutagenesis has been applied to the control of selectivity for chain length specificity towards triglycerides (Martinelle *et al.*, 1996), towards triradyglycerols (Scheib *et al.*, 1998), whereas directed evolution methods yielded an increased enantioselectivity (Reetz *et al.*, 1997). Lipases are the subject of some contributions in this book. **Lotti and Alberghina** report on studies on the lipases from *Candida rugosa*, underlying the importance of evolutionary analysis for protein engineering of this family of isoenzymes. The paper by Bornscheuer *et al.* illustrates the state of the art on the lipases produced by *Rhizopus species*, enclosing X-ray crystallography, computer modelling and site-directed mutagenesis, opening wide application in lipid modification and organic synthesis. Finally, **Canaan and colleagues** describe the development of gastric acidic lipases for possible applications in substitutive enzyme therapy for exocrine pancreatic insufficiency.

Protein Engineering for Health Care

Only a few engineered proteins are already on the market for health care products but perspectives in this field are so promising that an explosion of application is easily to forsee, following the completion of clinical tests. The efficacy of a drug is dependent not only on its activity and selectivity but also on stability, specific targeting and rate of clearance in the body, all issue that are being addressed through protein engineering (Buckel, 1996).

Peptides and peptidomimetics are probes that mimic the molecular interactions of natural proteins and are therefore useful in the process of drug discovery (reviewed in Kieber-Emmons *et al.*, 1997).

Peptide and polypeptide vaccines, especially for protection against hazardous viruses, since they mimic immunogenic epitopes of the viral coat. Hybrid proteins are constructed also with the purpose of correctly presenting the antigen to the immune system to elicit immune response. Hybrid vaccines already on the market against malaria and hepatitis. The chapter by **Grandi** *et al.* illustrates the role of protein engineering to design vaccine components with enhanced immunogenicity and devoided of any toxicity as well as for mucosal delivery of vaccines.

A big deal of work is devoted to the immune system, especially to the production of **engineered antibodies** for medical therapy and diagnosis: 30% of all proteins under clinical trials are antibodies.

Component of the intracellular **signal transduction machinery** and proteins involved in the regulation of gene transcription. Recently, superactive analogues for human glycoprotein hormones are being developed with potential applications in the therapy of thyroid and reproductive disorders. Knowledge-based mutagenesis was performed on a protein domain rich in basic residues involved in receptor binding and signal transduction, leading to the development of human thyroid stimulating hormone and chorionic gonadotropin analogues with increased receptor binding affinity and bioactivity (Szuklinski *et al.*, 1996). Injection of a variant of the IL-6 receptor proved able to induce a strong anti hIL-6 antibody response; the elicited antibodies bound circulating IL-6 with high affinity thus masking its biological activity, a result very promising for the treatment of immune and neoplastic diseases (Ciapponi *et al.*, 1997). **Vanoni and colleagues** in their contribution propose an approach to disease therapy based on the inhibition of a specific pathway of signal transduction, acting on a ras-specific guanine nucleotide exchange factor.

Protein engineering of **hormone receptors** aims to develop controllable gene expression systems regulated through the administration of small drugs, an experimental system allowing to dissect the mechanism of action. These systems require the expression of genetically engineered chimeric transcription factors that function as molecular or gene switches and regulate the expression of an exogenous target gene in response to administration of a drug. Gene switches are based on the genetic engineering of chimeric intracellular receptors that contain functional domains including a ligand binding domain, a DNA binding domain and a transactivation (transrepression) domain. Gene switches have been developed based on steroid hormone receptors. Estrogen receptors and their inhibitors as targets towards the discovery of new drugs are the topics of the chapter by **Connor and McDonnell**. **Koller and colleagues** deal with surface receptors, suggesting protein engineering techniques to produce soluble receptors to be used in high throughput screens.

Growth factors are also considered as promising targets for mutagenesis, especially towards the development of wound repair drugs. Several companies are being involved in the search of non-peptide analogs of growth factors, i.e. fibroblast growth factor and platelet-derived growth factor.

Generally speaking, **bioscreening** , i.e. methods used to evaluate the molecular interactions of candidate drugs with their targets, is a central issue in drug discovery. Techniques of bioscreening in oncology are outlined in the paper by **Draetta and Boisclair**, with focus on their central role in the process of drug discovery.

Of course several **enzymes** of importance as drugs are possible targets for modifications. Among them we would like to mention recent studies on serine proteases, in particular thrombin. This enzyme is subjected to Na^+-dependent allosteric regulation: suppressing Na^+ binding by mutagenesis shift the balance between the procoagulant and anticoagulant activities of the enzyme, with perspective of application in the field of anticoagulant drugs (Dang *et al.*, 1997).

Finally, **drug delivery** will strongly benefit of the use of fusion proteins designed as to selectively target disease-causing cells and extracellular targets; recombinant fusion proteins are under clinical trial for the treatment of diseases including tumor types (Murphy, 1996).

Enzymes for Environment Diagnosis and Rehabilitation

In the field of bioremediation the major contribution of protein engineering concerns the development of enzymes or modified microbial species able to degrade xenobiotic compounds released in the environment by human activities. We might recall the two previously mentioned examples where molecular evolution methods allowed for the production of enzymes degradating polychorinated biphenyls (Kumamaru *et al.*, 1998) and recombinant bacteria endowed with the ability to detoxify soluble arsenate released during mining (Crameri *et al.*, 1997). This book reports two contribution focussing on environmental issues. The development of biosensors, in particular based on fusion proteins useful for enzyme immunoassay and bioluminescent enzyme immunoassay to detect environmentally hazardous compounds is the topic of the paper by **Aizawa and colleagues**. Enzymes involved in the degradation of chlorinated compounds, one major class of recalcitrant polluting compounds are treated in-depth by the contribution of **Janssen and colleagues**.

REFERENCES

Arnold, F.H. (1996) Directed evolution: creating biocatalysts for the future. *Chem. Eng. Sci*, 51(23), 5091–5102.

Arnold, F.H. (1993) Protein engineering for unusual environments. *Curr. Opin. Biotechnol*, 4(4), 450–455.

Ballesteros, A., Bornscheuer, U., Capewell, A., Combes, D., Condoret, J.S., Koenig, K., Kolisis, F.N., Marty, A., Menge, U., Scheper, T., Stamatis, H. and Xenakis, A. (1995) Enzymes in non-conventional phases. *Biocatalysis & Biotransformation*, 13(1), 1–42.

Ballinger, M.D., Tom, J. and Wells, J.A. (1996) Designing subtilisin BPN' to cleave substrates cotaining dibasic residues. *Biochemistry*, 34, 13312–13319.

Beauregard, M., Dupont, C., Teather, R.M. and Hefford, M.A. (1995) Design, expression and initial characterization of MB1, a *de novo* protein enriched in essential amino acids. *Bio-Technology*, 13, 974–981.

Benkovich, S.J. (1992) Catalytic antibodies. *Ann. Rev. Biochem.*, 61, 29–54.

Benson, D.E., Wisz, M.S. and Hellinga, H.W. (1998) The development of new biotechnologies using metalloprotein design. *Curr. Opin. Biotechnol*, 9(4), 370–376.

Buckel, P. (1996) Recombinant proteins for therapy. *Trends. Pharmacol. Sci.*, 17(12), 450–456.

Carter, P. and Merchant, A.M. (1997) Engineering antibodies for imaging and therapy. *Curr. Opin. Biotechnol.*, 8(4), 449–454.

Cernia, E., Palocci, C. and Soro, S. (1998) The role of the reaction medium in lipase-catalyzed esterifications and transesterifications. *Chem. Phys. of Lipids*, 93(1–2), 157–168.

Chen, K. and Arnold, F. (1993) Tuning the activity of an enzyme for unusual environments: sequential random mutagenesis of subtilisin E for catalysis in dimethylformamide. *Proc. Natl. Acad. Sci. USA*, 90, 5618–5622.

Chen, R., Greer, A. and Dean, A.M. (1995) A highly active decarboxylating dehydrogenase with rationally inverted coenzyme specificity. *Proc. Natl. Acad. Sci. USA*, 92, 11666–11670.

Ciapponi, L., Maione, D., Scoumanne, A., Costa, P., Hansen, M.B., Svenson, M., Bendtzen, K., Alonzi, T., Paonessa, G., Cortese, R., Ciliberto G and Savino, R. (1997) Induction of interleukin-6 (IL-6) autoantibodies through vaccination with an engineered IL-6 receptor antagonist. *Nat. Biotechnol*, 15, 997–1001.

Cleland, J.L. and Craik, C.S. (eds) (1996) *Protein engineering: principles and practice*. Wiley-Liss Inc.

Clore, G.M. and Gronenborn, A.M. (1998) Determining the structures of large proteins and protein complexes by NMR. *Trends Biotechnol*, 16(1), 22–34.

Crameri, A., Dawes, G., Rodriguez, E., Silver, S. and Stemmer, W.P.C. (1997) Molecular evolution of an arsenate detoxification pathway by DNA shuffling. *Nat. Biotechnol*, 15, 436–438.

Crameri, A., Whitehorn, E.A., Tate, E. and Stemmer, W.P.C. (1996) Improved green fluorescent protein by molecular evolution using DNA shuffling. *Nat. Biotechnol*, 3, 315–319.

Dahiyat, B. and Mayo, S. (1997) De novo design: fully automated sequence selection. *Science*, 278, 82–87.

Dalboege, H. and Lange, L. (1998) Using molecular techniques to identify new microbial catalysts. *Trends Biotechnol*, 16, 265–272.

Dang, Q.D., Guinto, E.R. and Dicera, E. (1997) Rational engineering of activity and specificity in a serine protease. *Nat.Biotechnol*, 15(2), 146–149.

Decklerc, N., Joyet, P., Trosset, J.-Y., Garnier, J. and Gaillardin, C. (1995) Hyperthermostable mutants of *Bacillus licheniformis* β-amylase: multiple amino acid replacements and molecular modelling. *Prot.Eng.*, 8, 1029–1037.

20   L. ALBERGHINA and M. LOTTI

Dolgikh, D.A., Gabrielian, A.E., Uversky, V.N. and Kirpichnikov, M.O. (1996a) Protein engineering of *de novo* protein with predesigned structure and activity. *Appl. Biochem. Biotechnol.*, **61**, 85–96.

Dolgikh, D.A., Kirpichnikov, M.P., Ptitsyn, O.B. and Chemeris, V.V. (1996b) Protein engineering of man-made proteins *Mol. Biol.*, **30**(2), 149–156.

Dufour, E., Storer, A.C. and Ménard, R. (1995) Engineering nitrile hydratase activity into a cysteine protease by a single mutation. *Biochemistry*, **34**, 16382–16388.

Fersht, A.R. and Serrano, L. (1993) Principles of protein stability derived from protein engineering experiments. *Curr. Opin. Struct. Biol.*, **3**, 75–83.

Francisco, J.A., Earhart, C.F. and Georgiou, G. (1992) Transport and anchoring of beta-lactamase to the external surface of *Escherichia coli. Proc. Natl. Acad. Sci. USA*, **89**, 2713–2717.

Gottschalk, T.E., Fierobe, H.P., Mirgorodskaya, E., Clarke, A.J., Tull, D., Sigurskjold, B.W., Christensen, T., Payre, N., Frandsen, T.P., Juge, N., Mcguire, K.A., Cottaz, S., Roepstorff, P., Driguez, H., Williamson, G. and Svensson, B. (1998) Structure, function and protein engineering of starch-degrading enzymes. *Biochem. Soc. Transact.*, **26**(2), 198–204.

Halfon, S. and Craik, C.S. (1996) Regulation of proteolytic activity by engineered tridentate metal binding loops. *J. Am. Chem. Soc.*, **118**, 1227–1228.

Hedstrom, L., Szilagyi, L. and Rutter, W. (1992) Converting trypsin to chymotrypsin: the role of surface loops. *Science*, **255**, 1249–1253.

Hellinga, H.W. (1998) Computational protein engineering. *Nat. Struct. Biol.*, **5**(7), 525–527.

Jacobsen, C.S. (1995) Microscale detection of specific bacterial DNA in soil with a magnetic capture-hybridization and PCR amplification assay. *Appl. Environ.Microbiol.*, **61**, 3347–3352.

Janda, K.D., Lo, L.-C., Lo, C.-H.L., Sim, M.-M., Wang, R., Wong, C.-H. and Lerner, R.A. (1997) Chemical selection for catalysis in combinatorial antibody libraries. *Science*, **275**, 945–948.

Joyet, P., Declerck, N. and Gaillardin, C. (1992) Hyperthermostable variants of a highly thermostable α-amylase. *Bio-Technology*, **10**, 1579–1583.

Kieber-Emmons, T., Murali, M. and Green, M.I. (1997) Therapeutic peptides and peptidomimetics. *Curr. Opin. Biotechnol.*, **8**(4), 435–441.

Klibanov, A.M. (1997) Why are enzymes less active in organic solvents than in water. *Trends Biotechnol.*, **15**(3), 97–101.

Koivula, A., Reinikainen, T. Ruohonen, L., Valkeajarvi, A., Claeyssens, M., Teleman, O. Kleywegt, G.J., Szardenings, M., Rouvinen J, Jones, T.A. and Teeri, T.T. (1996) The active site of *Trichoderma reesei* cellobiohydrolase II. the role of tyrosine 169. *Prot. Eng.*, **9**, 691–699.

Kornacker, M.G. and Pugsley, A.P. (1990) The normally periplasmic enzyme beta-lactamase is specifically and efficiently translocated through the *Escherichia coli* outer membrane when it is fused to the cell-surface enzyme pullulanase. *Mol. Microbiol.*, **4**(7), 1101–1109.

Kumamaru, T., Suenaga, H., Mitsuoka, M., Watanabe, T. and Furukawa, K. (1998) Enhanced degradation of polychlorinated biphenyls by directed evolution of biphenyl dioxygenase. *Nat. Biotechnol.*, **16**, 663–666.

Law, B.A. and Mulholland, F. (1991) The influence of biotechnological developments on cheese manufacture. *Biotech. Genet. Eng. Rev.*, **9**, 369–409.

Lee, K.N., Park, S.D. and Yu, M.-H. (1996) Probing the native strain in α1-antitrypsin. *Nat. Struct. Biol.*, **3**, 497–500.

Linder, M., Salovuori, I., Ruohonen, L. and Teeri, T.T. (1996) Characterization of a double cellulose-binding domain. *J. Biol. Chem*, **271**, 21268–21272.

Mainfroid, V., Mande, S.C., Hol, W.G., Martial, J.A. and Gorai, K. (1996) Stabilization of human triosephosphate isomerase by improvement of the stability of individual alpha-helices in dimeric as well as monomeric forms of the protein. *Biochemistry*, **35**, 4110–4117.

Malakauskas, S.M. and Mayo, S.L. (1998) Design, structure and stability of a hyperthermophilic protein variant. *Nat. Struct. Biol.*, **5**(6), 470–475.

Martinelle, M., Holmquist, M., Clause, I.G., Patkar, S., Svendsen, A. and Hult, K. (1996) The role of Glu87 and Trp89 in the lid of *Humicola lanuginosa* lipase. *Prot. Eng.*, **9**, 519–524.

Matthews, B.W. (1995) Studies on protein stability with T4 lysozyme. *Adv. Prot. Chem.*, **46**, 249–278.

Mingarro, I. Abad, C. and Braco, L. (1995) Interfacial activation-based molecular bioimprinting of lipolytic enzymes. *Proc. Natl. Acad. Sci. USA*, **92**(8), 3308–3312.

Moore, J.C. and Arnold, F.H. (1996) Directed evolution of a para-nitrobenzyl esterase for aqueous-organic solvents. *Nat. Biotechnol.*, **14**, 458–467.

Murphy, J.R. (1996) Protein engineering and design for drug delivery. *Curr. Opin. Struct. Biol.*, **6**(4), 541–545.

Nixon, A.E., Ostermeier, M. and Benkovic, S. (1998) Hybrid enzymes: manipulating enzyme design. *Trends Biotechnol.*, **16**, 258–264.

Okkels, J.S., Svendsen, A., Patkar, S.A. and Bork, K. (1996) Protein engineering of microbial lipases with industrial interest. In *Engineering of/with lipases*, edited by F.X. Malcata, pp. 203–217, NATO ASI Series. Kluwer Academic Publisher.

Patkar, S., Vind, J., Kelstrup, E., Christensen, M.W., Svendsen, A., Borch, K. and Kirk, O. (1998) Effect of mutations in *Candida antarctica* B lipase. *Chem. Phys. of Lipids.*, **93**(1–2), 95–101.

Perona, J.J. and Craik, C.S. (1995) Structural basis of substrate specificity in the serine proteases. *Protein Sci.*, **4**(3), 337–360.

Pickersgill, R.R. and Goodenough, P.W. (1991) Enzyme engineering: a review. *Trends Food Sci. Technol.*, May, 122–126.

Price, S.R. and Nagai, K. (1995) Protein engineering as a tool for crystallography. *Curr. Opin. Biotechnol.*, **6**(4), 425–430.

Quax, W.J., Mrabet, N.T., Luiten, R.G.M., Schwerhuizen, P.W., Stanssens, P. and Laters I (1991) Stabilization of xylose isomerase by lysine residue engineering. *Bio/Technology*, **9**, 783–742.

Quinn, T., Tweedy, N., Williams, R., Richardson, J. and Richardson, D. (1994) Betadoublet: *de novo* design, synthesis and caracterization of a β-sandwich protein. *Proc. Natl. Acad. Sci. USA*, **91**, 8747–8751.

Ramstrom, O., Ye, L. and Mosbach, K. (1996) Artificial antibodies to corticosteroids prepared by molecular imprinting. *Chemistry & Biology.*, **3**(6), 471–477.

Reetz, M.T., Zonta, A., Schimossek, K., Liebton, K. and Jaeger K-E (1997) Creation of enantioselective biocatalysts for organic chemistry by *in vitro* evolution. *Angew. Chem. Int. Ed. Engl.*, **36**(24), 2830–2832.

Rich, J.O. and Dordick, J.S. (1997) Controlling subtilisin activity and selectivity in organic media by imprinting with nucleophilic substrates., *J. Am. Chem. Soc.*, **119**(14), 3245–3252.

Riddle, D.S., Santiago, J.V., Brayhall, S.T., Doshi, N., Grantcharova, V.P., Yi, Q. and Baker, D. (1997) Functional rapidly folding proteins from simplified amino acid sequences. *Nat. Struct. Biol.*, **4**(10), 805–809.

Rubingh, D.N. (1996) Engineering proteases with improved properties for detergents. In *Enzyme Technology for Industrial Applications*, edited by S.L. Southborough, pp. 98–123. IBC Biomedical Library Series.

Scheib, H., Pleiss, J., Stadler, P., Kovac, A., Potthoff, A.P., Haalck, L., Paltauf, F. and Schmid, R.D. (1998) Rational design of *Rhizopus oryzae* lipase mith modified stereoselectivity toward triradylglycerols. *Prot. Eng.*, **11**(8), 675–682.

Schreuder, M.P., Mooren, A.T., Toschka, H.Y., Verrips, C.T. and Klis, F.M. (1996) Immobilizing proteins on the surface of yeast cells. *Trends Biotechnol.*, **14**(4), 115–120.

Schultz, P.G. and Lerner, R.A. (1995) From molecular diversity to catalysis: lessons from the immune system. *Science*, **269**, 1835–1842.

Shao, Z. and Arnold, F.H. (1996) Engineering new functions and altering existing functions. *Curr. Opinion. Struct. Biol.*, **6**, 513–517.

Shih, P., Holland, D.R. and Kirsch, J.F. (1995) Thermal stability determinants of chicken egg-white lysozyme core mutants: hydrophobicity, packing volume and conserved buried water molecules. *Prot. Sci.*, **4**, 2050–2062.

Smith, J.W., Tachias, K. and Madison, E.L. (1996) Protein loop grafting to construct a variant of tissue-type plasminogen activator that binds platelet integrinaIIbb3., *J. Biol. Chem.*, **270**, 30486–3049.

Smithrud, D.B. and Benkovic, S.J. (1997) The state of antibody catalysis. *Curr. Opin. Biotechnol.*, **8**(4), 459–466.

Stahl, S. and Uhlen, M. (1997) Bacterial surface display: trends and progress. *Trends Biotechnol.*, **15**, 185–192.

Stemmer, W.P.C. (1994b) Rapid evolution of a protein *in vitro* by DNA shuffling. *Nature*, **340**, 389–391.

Stemmer, W.P.C. (1994a) DNA shuffling by random mutagenesis and reassembly: *in vitro* recombination for molecular evolution. *Proc. Natl. Acad. Sci. USA*, **91**, 10747–10751.

Strausberg, S.L., Alexander, P.A., Gallagher, D.T., Gilliland, G.L., Barnett, B.L. and Bryan, P.N. (1995) Directed evolution of a subtilisin with calcium-independent stability. *Bio-Technology*, **13**, 669–673.

Strauss, A. and Goetz, F. (1996) *In vivo* immobilization of enzymatically active polypeptides on the cell surface of *Staphylococcus carnosus*. *Mol. Microbiol.*, **21**(3), 491–500.

Szkudlinski, M.W., Teh, N.G., Grossmann, M., Tropea, J.E. and Weintraub, B.D. (1996) Engineering human glycoprotein hormone superactive analogues. *Nat. Biotechnol.*, **14**, 1257–1263.

Vita, C. (1997) Engineering novel proteins by transfer of active sites to natural scaffolds. *Curr. Opin. Biotechnol.*, **8**, 429–434.

Wagner, G. (1997) An account of NMR in structural biology. *Nat. Struct. Biol.*, **4**, 841–844.

Wells, J.A., Cunningham, B.C., Graycar, T.P. and Estell, D.A. (1987) Recruitment of substrate-specificity properties from one enzyme into a related one by protein engineering. *Proc. Natl. Acad. Sci. USA*, **84**, 5167–5171.

Wentworth, P., Datta, A., Blakey, D., Boyle, T., Partridge, L.J. and Blackburn, G.M. (1996) Toward antibody-directed "abzyme" prodrug therapy, ADAPT: carbamate prodrug activation by a catalytic antibody and its *in vitro* application to human tumor cell killing. *Proc. Natl. Acad. Sci. USA*, **93**(2), 799–803.

Wescott, C.R. and Klibanov, A.M. (1994) The solvent dependence of enzyme specificity. *Biochim. Biophys. Acta*, **1206**(1), 1–9.

Westhead, D.R. and Thornton, J.M. (1998) Protein structure prediction. *Curr. Opin. Biotechnol.*, **9**(4), 383–389.

Wrede, P. and Schneider, G. (eds) (1994) *Concepts in Protein Engineering and design.* Berlin, New York: Walter de Gruyter.

2. ENHANCED RECOVERY AND FOLDING OF RECOMBINANT PROTEINS USING FUSION PROTEIN STRATEGIES

SOPHIA HOBER and MATHIAS UHLÉN

Department of Biochemistry and Biotechnology, Royal Institute of Technology (KTH), 100 44 Stockholm, Sweden

An important issue in the design of production schemes for recombinant proteins is to minimize the number of recovery steps during the purification. This can be achieved by introducing highly selective unit operations either with or without the use of affinity "tags". Another issue of great importance is to develop improved methods for protein folding based upon the understanding of the mechanisms which govern folding, both *in vivo* and *in vitro*. The pathway of folding *in vivo* is dependent on the host and the growth conditions. Protein folding *in vitro* is affected by the solvent environment and other physical conditions, such as temperature. This chapter will describe different techniques for purification and folding of recombinant proteins expressed in *Escherichia coli (E. coli)*.

PROTEIN FOLDING

The Protein Folding Problem

In protein chemistry, significant efforts have been made trying to solve how information from the amino acid sequence is transferred into a three dimensional structure. The understanding of the protein folding problem has increased in recent years. From the beginning the topic was primarily of academic interest but has now become an issue of industrial importance. Production of recombinant proteins in heterologous hosts often lead to unfolded and inactive proteins, in particular when the production levels are high. In order to generate useful protein products it is necessary to refold the protein *in vitro*. The folding problem was elegantly outlined by Levinthal and is referred to as "the Levinthals paradox" (Levinthal, 1969): Assume a protein with 100 amino acids where every residue can have two different conformations. Despite this very simplified model, the protein can adopt 10^{30} different conformations. Even if the protein is able to sample new configurations at the rate of 10^{13} per second it will need 3×10^9 years to fold, similar to the age of earth. The true allowed number of conformations for an amino acid is of course much higher than two, and the time for a protein to fold completely random to search for the native conformation must exceed life of the earth. In contrast to these statistical figures, folding of an α-helix is estimated to take place in a time scale of 10^{-6} seconds (Creighton, 1993). Consequently, there must be a non-random pathway for a protein to fold into an active structure.

Normally, folding of a protein can be achieved without any extrinsic factors, as first described by Anfinsen (1973). Many different hypothetical models have been proposed for the folding process. It has been compared with a jigsaw puzzle (Harrison & Durbin, 1985). A nucleation model has also been proposed where small "nucleus" form randomly and these serve as templates upon which folding can proceed (Wetlaufer, 1973). The random formation of the "nucleus" is supposed to be the rate limiting step. An alternative hypothesis is the frame work model which is the classical way to look upon folding. Secondary structure elements are supposed to fold first. The tertiary structure is then being formed by packing the secondary elements together (Kim & Baldwin, 1982). This hypothesis presumes a certain stability of secondary structure elements. In the diffusion-collision-adhesion model, micro domains fold up and interact with each other to form a stable three dimensional structure. The rate constant for folding depends on the diffusion rate and the stability of certain structures (Karplus & Weaver, 1976). This model is similar to the frame work model but it does not assume partly already existing secondary structures. Another theory of protein folding proposes that proteins undergo a rapid non-specific hydrophobic collapse at the start of the folding process in order to reach a state close to the molten globule (see below). The hydrophobic effect is thought to be the driving force. As the protein becomes more compact, the number of possible interactions within the protein diminishes. Constraining the polypeptide backbone might increase the probability to reach the native state (Gregoret & Cohen, 1991). From available experimental data it is not possible to build a simple kinetic folding model that is applicable to all studied proteins. When studying proteins with proline residues or disulfide bonds, the rate determining step of the folding process is often that of proline isomerization or disulfide bond rearrangement. The folding of larger multidomain proteins is even more unpredictable. In many different folding systems it has been found that the rate limiting step is late in the folding process. The rate determining step was concluded to go from a compact intermediate to the final packed native conformation. This intermediate was designated "the molten globule" (MG) (Ohgushi & Wada, 1983). There is confusion in the literature about the definition of the molten globule state. One of the main sources for confusion is that the term is used both for the original model (Ptitsyn, 1987, Ptitsyn, 1992) and for isolated intermediates. The original definition of a molten globule is a protein with a high secondary structure content but an absence of unique tight packing of side chains. It is very difficult to study the molten globule state, since the real intermediate molecule is unstable and transient. Therefore, the most frequently studied variant is a partial denatured protein. Even though this is not a true molten globule, it is closely similar.

Protein Folding in the Cell

During *in vitro* refolding proteins aggregate in much lower concentrations than usually found *in vivo*. In biological systems there are several molecules helping newly synthesized proteins to fold. Molecular chaperones and foldases seem to prevent proteins from incorrect folding and aggregation and increase the rate of productive folding. One important function of chaperones is to reduce the concentration of misfolded and unfolded proteins in the cell and thereby decrease the aggregation rate (Hartl & Martin, 1995, Nilsson & Anderson, 1991). Some proteins are initially synthesized as precursors with prosequences that are removed after folding. BPTI (Weissman & Kim, 1992), subtilisin (Eder *et al.*, 1993) and α-lytic protease (Baker *et al.*, 1992) are examples of such proteins. Prosequences have been demonstrated to affect folding at the kinetic level.

Although the presence of thiol/disulfide redox buffers significantly increases the rate of oxidative folding *in vitro* (Saxena & Wetlaufer, 1970), it still takes place on a time-scale ranging from minutes to hours (Gilbert, 1994). If proteins are to fold faster *in vivo* the reaction must be catalyzed. There are a number of characterized proteins that catalyze disulfide formation *in vivo*. One of those is Protein Disulfide Isomerase (PDI) (Freedman, 1989, Noiva, 1994, Noiva & Lennarz, 1992). PDI is resident in the lumen of the endoplasmic reticulum (ER). It has two disulfide active sites, one near the N-terminus and the other near the C-terminus. PDI catalyzes thiol-disulfide exchange by oxidation and reduction and thereby assists in rearranging protein disulfides. These proteins that catalyze disulfide formation are thought not to alter the pathway, but accelerate it. PDI is relatively non-specific, but has slightly higher specificity for cysteine containing stretches. In *E. coli*, several proteins involved in the formation of disulfide bonds have recently been discovered (Freedman, 1995). The periplasmic protein DsbA, discovered in 1991 by Bardwell *et al.* (Bardwell *et al.*, 1991), is assumed to serve as an oxidant of secreted proteins (Wunderlich *et al.*, 1993). The inner membrane protein DsbB is thought to transfer oxidizing equivalents from the cytoplasm to DsbA (Creighton *et al.*, 1995, Missiakas *et al.*, 1993). A more efficient catalyst of disulphide rearrangement than DsbA is DsbC, a soluble protein found in the periplasm (Missiakas *et al.*, 1994, Zapun *et al.*, 1995). Also, a fourth disulphide isomerase-like protein in *E. coli* has been identified as an inner membrane protein suggested to supply the periplasm with reducing equivalents (Missiakas *et al.*, 1995). The isomerization of the proline imide bonds is often a rate-limiting step in protein folding. (Schmidt *et al.*, 1990). Proteins with prolyl peptide isomerase activity have been found in most organisms and in different cellular compartments (Gething & Sambrook, 1992). *In vitro*, they have been shown to catalyze several slow folding reactions.

The high concentration of unfolded and partly folded proteins in the cell should without the presence of molecular chaperones result in aggregation and loss of a large fraction of proteins. Molecular chaperones protect against aggregation by binding and releasing proteins along their pathway to the native structure (reviewed by Hartl and Martin, 1995). They are found in all cell types and are divided into several families of proteins that are structurally unrelated. At first, they were denoted stress or heat shock proteins, since several of them were induced under heat or harsh conditions, although they function also under normal conditions. Molecular chaperones important to protein folding can at present be divided into two main families; the hsp70 family which binds to the nascent polypeptide chain to prevent premature folding, and the hsp60 family or chaperonins which at a later stages promotes the correct folding of the protein. Several investigations have been made to estimate the fraction of proteins that need chaperonins to fold. A genetic study indicated that a minimum of 30% of the soluble proteins in E. coli requires chaperonins to reach the native state (Horwich *et al.*, 1993). Two other investigations base their calculations on the *in vitro* rate of GroEL action and the rate of protein synthesis in *E. coli.*, and reach values of 5% (Lorimer, 1996) and 7.5% (Ellis & Hartl, 1996). It is however questionable to what degree data from *in vitro* experiments can be used when describing the *in vivo* action of chaperonins (Frydman & Hartl, 1996).

Inclusion Bodies

Since prokaryotic organisms do not have the necessary machinery for complete processing and folding of eukaryotic proteins, production of eukaryotic proteins in prokaryotic organisms often results in formation of inactive proteins. One consequence of the genetic modification of microorganisms in order to produce recombinant proteins, is accumulation of insoluble intracellular aggregates, inclusion bodies (Marston, 1986). The aggregation of foreign recombinant proteins has been observed in most cases but high level expression of wild-type protein will under certain circumstances also lead to inclusion body formation. Several factors influence the tendency of recombinant proteins to misfold or aggregate. The recombinant protein use to be produced in a heterologous host or expressed in levels that by far exceed the normally expressed amounts in the organism. Therefore, the environment is often significantly different in both chemical properties and in the availability of "helping proteins" e.g. foldases and chaperones. The formation of inclusion bodies is determined by the rates of two parallel competing reactions, the first order folding and the second order aggregation. Hence, the rate of aggregation is dependent on the rate of translation within the cell and also on the rate of folding for the protein of interest. The insolubility results from the formation of incorrectly or partly folded proteins. Attempts have been made to elucidate if there are particular properties of a protein that are connected with the inclusion body formation, but mostly without success (Wilkinson & Harrison, 1991). Neither the propensity to form precipitate, the size of the expressed protein nor the hydrophobicity are shown to be correlated to the inclusion body formation. Experiences suggest that the host cell physiology is the most important factor in inducing inclusion body formation. Hence, under different fermentation conditions the same protein might be able to form inclusion bodies or stay in the solution (Schein, 1991, Schein, 1989, Strandberg & Enfors, 1991). Analysis of how amino acid substitutions affects the inclusion body formation indicates that the folding pathway is more important than the stability of the fully folded protein itself (King *et al.*, 1996, Mitraki & King, 1989, Mitraki & King, 1992). The effect of other physical conditions, such as temperature and pH, may also be due to an alteration of the folding pathway. There are however, different methods to influence the inclusion body formation. One way is to make point mutations in the particular protein, to increase the solubility during folding (Mitraki & King, 1992, Murby *et al.*, 1995). Large deletions or insertions can also affect the propensity to form inclusion bodies. Solubilizing linkers can affect the inclusion body formation, as shown for thioredoxin by LaVallie and co-workers (1993). Alternatively, one can over express chaperones or other proteins that can help in folding and thereby increase the solubility of the target protein (Roman *et al.*, 1995, Yasukawa *et al.*, 1995). To obtain active proteins from intracellular aggregates, they are to be resolved and renatured (Figure 1).

How to obtain functional protein in adequate quantities for research or commercial purposes is a main issue in both industry and academia. Here, the problems are often related to the production of functional proteins in a foreign host or, if this is not possible, how to perform *in vitro* refolding successfully. The main challenge is to overcome aggregation of unfolded or partially folded intermediates, which compete with the correct folding pathway. Since aggregation is a second-order process and correct folding the result of first order reactions, the concentration of

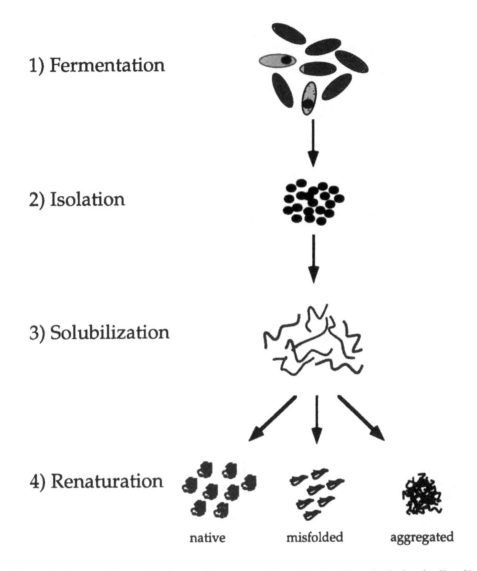

1) Fermentation

2) Isolation

3) Solubilization

4) Renaturation

native   misfolded   aggregated

Figure 1   Strategy for production and recovery of proteins that form inclusion bodies. 1) The extent of inclusion body formation is dependent on the conditions during fermentation, such as temperature and chemical environment. Also the rate of protein synthesis is affecting the solubility of the proteins within the cell. 2) Isolation of the inclusion bodies starts by disrupting the cell membrane by high pressure homogenization or sonication. In order to get rid of the membrane proteins, washing with a low concentration of denaturing chemicals and thereafter centrifugation is recommendable. 3) To solubilize the inclusion bodies, high concentration of denaturing agents are usually needed. When the target protein contains cysteines, a reducing agent has to be added to prevent non-native disulfide formation. 4) The renaturation step is normally performed using stepwise dilution or dialysation. Composition of the solution used during renaturation is dependent on the protein of interest and the conditions have to be experimentally determined for each product. If native disulfides have to form, a low concentration of a reducing agent, such as β-mercaptoethanol, can make the refolding process more efficient

the protein to fold is a critical factor (Kiefhaber *et al.*, 1991). To obtain a functional protein one can either refold the protein produced as inclusion bodies *in vitro* or avoid the formation of inclusion bodies. Some proteins are very sensitive to degradation or deleterious to the host, and therefore it might be an advantage to produce them as inclusion bodies (Kane & Hartley, 1988, Kitano *et al.*, 1987). Fusion to the E. coli TrpE protein can for example increase the propensity of a protein to form inclusion bodies (Yansura & Henner, 1990). Other proteins are very hard to refold *in vitro* into an active form and are therefore preferably produced as soluble and native in the host cell.

Isolation and Solubilization of Inclusion Bodies

Since the inclusion bodies are intracellular or periplasmic structures, cells have to be fractured (Figure 1). The cells can be disrupted through high pressure homogenization or sonication and the inclusion bodies can be isolated (Babbitt *et al.*, 1990, Fischer *et al.*, 1993). As the membranes contain a number of proteolytic enzymes it is of importance to extract these contaminants or inhibit their activity. The most generally applicable method for selective extraction is washing the inclusion bodies with a buffer containing a chelating agent (EDTA), a neutral detergent (Triton X-100) and a chaotrop (Urea) (Babbitt *et al.*, 1990).

When the inclusion bodies are isolated they are to be solubilized (Figure 1). The forces involved in inclusion body formation are mainly non covalent (McCaman, 1989) and the presence of disulfides do not play any major role, even though disulfides in inclusion bodies have been reported (Schoemaker *et al.*, 1985). Solubilization, in most cases, involves total denaturation of the protein. If the protein has substantial amount of native conformation in the inclusion bodies it is possible to choose a weaker solubilizing buffer that do not disrupt the already formed structure. The most commonly used solubilizing agents are chaotropic salts or detergents (Rudolph, 1995). Once solubilized, the thiols are able to form intra- or intermolecular disulfides. Therefore, its essential to include a reducing agent in the solubilizing mixture, in order to keep the thiols in the reduced state, or protect them by forming S-derivatives as have been used in the production of insulin (Frank *et al.*, 1981, Nilsson *et al.*, 1996).

Renaturation

Thereafter, the protein has to be refolded *in vitro* (Figure 1). In some cases it is necessary to purify the protein before the renaturation step. DNA and host proteins might interfere with the refolding step by interaction with partly folded molecules. If purification in denaturating conditions is needed, gene fusion technology is useful. By fusing the protein of interest with an affinity tag that is usable under non-native conditions, the purification step can be both easy and selective. Affinity handles that are frequent used are the poly-histidine tails (Flaschel & Friehs, 1993). The charged histidines residues are used because of their ability to bind to metal ions. It has been shown that using six histidines in the tail makes it possible to purify in denaturating conditions (6M guanidine hydrochloride). However, the purification method requires the imidazole residues in the non-protonated state (Hochuli, 1990, Hochuli *et al.*, 1988). Poly-arginine tails have also been shown to be usable in denaturating conditions since their ability to bind to anionic residues are independent of the three dimensional structure (Stempfer et al., 1996).

One of the main issues in refolding proteins is to inhibit aggregation. The low solubility of unfolded proteins necessitates very dilute protein solutions in the refolding step. An alternative in order to reduce the reaggregation is dialysis or step wise dilution of the protein mixture (Rudolph & Lilie, 1996). The staged dilution approach is critically dependent on the particular protein. Therefore the conditions have to be optimized carefully by experimentation for each protein. It is also possible to increase the solubility of a protein by chemical modification. This strategy could be used either to prevent covalent aggregation by disulfide formation (Frank et al., 1981, Nilsson et al., 1996) or to change the net-charge of the protein by modifying the amides (Light, 1985). Buffer additives which reduces the aggregation are also usable (Fischer et al., 1993, Rozema & Gellman, 1996, Schein, 1990), especially when the particular protein needs a cofactor or a metal ion to achieve the native state. Other frequently used additives are polyethylene glycol (Cleland et al., 1992, Cleland & Wang, 1990), lauryl maltoside (Tandon & Horowitz, 1986, Zardeneta & Horowitz, 1992) and arginine (Buchner et al., 1992b, Winkler & Blaber, 1986). Proteins can also be rendered either less or more soluble by gene fusion technology. There are several protein fusions that have proven to increase the solubility of the product. One example of a fusion protein that can affect the solubility is glutathione-S-transferase (Ray, 1993). Also, by using the highly soluble protein A domain (Z) (Nilsson et al., 1987) as a solubilizer of misfolded and multimeric IGF-I, Samuelsson et al. (1994, 1991) were able to show that high yields of native protein could be recovered. The aggregation and precipitation of misfolded material was reduced. Since the refolding could be performed in high concentrations without any expensive chemical, the approach is well suited for large scale refolding. This method has also been utilized for IGF-II with success (Forsberg et al., 1992).

Creating micelles with the unfolded protein inside has also been tried in order to avoid aggregation (Hagen et al., 1990, Zardeneta & Horowitz, 1992). Another way to prevent aggregation in the refolding step has been elegantly shown by Rudolph and co-workers (Stempfer et al., 1996). They have fused α-glucosidase to a hexa-arginine polypeptide. This construction allows them to attach the fusion protein to a solid support containing polyanionic groups. Upon removal of the denaturant, α-glucosidase are able to refold without aggregation.

Disulfide Formation

Since many of the proteins of interest in biotechnology are extracellular and use disulfides for stabilization, disulfide formation is of great importance. These problems have motivated many researchers to study the periplasmic space of E. coli with the regard to disulfide formation (Missiakas & Raina, 1997, Wulfing & Plüchthun, 1994). Formation of a disulfide is a two electron oxidation reaction and requires an oxidant, for example oxygen with metal ions, oxidized DTT or oxidized β-mercaptoethanol. The most abundant disulfide exchange agent in most cells is glutathione (Gilbert, 1990, Hwang et al., 1992). The reaction starts with an attack of a nucleophilic thiolate ion on the disulfide bond. This reaction gives rise to a symmetrical transition state, with the charge uniformly distributed on the two sulfurs. The rate of disulfide formation is high with a basic attacking group, and a

Disulfide formation:

a) air oxidation, catalyzed by metal ions, irreversible

$+ 2e + 2H^+$

b) disulfide exchange, reversible

$+ GSSG \rightleftarrows \ldots + GSH \rightleftarrows \ldots + 2GSH$

Figure 2 a) Disulfide formation with oxygen as the oxidant. This is an irreversible reaction and therefore molecules might get kinetically trapped with nonnative disulfides. If disulfide formation with oxygen as the oxidation agent is going to be successful, the protein must be able to form the native structure without having the native disulfides. One way to solve this problem, is to add small amounts of a reducing agent in to the refolding buffer
b) A disulfide exchange reaction with glutathione. Since this reaction is reversible, mispaired disulfides can be reduced and new disulfides are able to form. Here the protein is able to break and reform disulfides until the thermodynamically most stable ones are formed

more acidic leaving group. The leaving group will be protonated. Therefore, increasing the pH will increase the rate of disulfide formation until the attacking group is mainly in the thiolate ion form (Gilbert, 1990, Gilbert, 1994). Disulfide formation can proceed by one of two chemical processes, as shown in Figure 2. The cheapest option for disulfide formation is air oxidation catalysed by metal ions (Saxena & Wetlaufer, 1970) (Figure 2a). The reaction is irreversible and therefore most effective when the protein can adopt the native conformation before the disulfides have formed. If the protein is unable to form such stable intermediates, some of the molecules may get kinetically trapped with thermodynamically unstable, mispaired disulfides. Therefore, air oxidation may be more successful if small amounts of a reduced thiol compound, such as ß-mercaptoethanol is present (Epstein & Goldberg, 1963). *In vivo* disulfide formation of a protein is a reversible process (Figure 2b) in the ER with glutathione as a redox buffer (Hwang *et al.*, 1992). The thermodynamic stability for protein disulfides varies greatly, and is dependent on the structure of the protein and the number of amino acids between the cysteines. According to Zhang *et al.* (Zhang & Snyder, 1989) odd numbers of intervening amino acids unfavour disulfides compared to even numbers. To form a disulfide between two adjacent cysteines a significant amount of energy is needed. If there are long stretches with amino acids between the cysteines, the disulfide formation is more dependent on the 3D-structure of the protein and is therefore more unpredictable. Charges around the cysteines can also affect the disulfide formation. Positive charges stabilize the reduced state, consequently retaining free cysteines. Negative charges around the cysteines stabilize the oxidized state, promoting disulfides formation. pKa for a cysteine is approximately 8.5 (Stryer, 1995) but varies in different proteins, depending on the three-dimensional environment.

ENGINEERING TO ASSIST IN PROTEIN FOLDING

Attempts to get a more efficient refolding have been made by adding chaperones and foldases to the folding mixture. Since aggregation during the refolding step is due to a slow folding pathway, it might be successful to add foldases to the refolding mixture. Such experiments have been tried with both disulfide and prolyl isomerases. Lilie et al. have shown that refolding is accelerated when prolyl isomerase is used in the refolding of an antibody. Although the isomerase is shown to act as a true catalyst it does not affect the aggregation since prolyl isomerization is a late step in the folding process (Lilie et al., 1993). On the other hand, when using PDI as a catalyst for disulfide formation, researchers have been able to improve the yields of native protein remarkably (Buchner et al., 1992a, Ryabova et al., 1997). Because of the high amounts of "helping proteins" needed it is not yet feasible to use them in vitro (Mendoza et al., 1991). Another approach to favour correct folding is to alter the environment in host cell by co-expression of mammalian chaperones and catalysts (Bardwell & Beckwith, 1993, Goloubinoff et al., 1989, Knappik et al., 1993, Roman et al., 1995, Yasukawa et al., 1995). However, the results are unpredictable and the effects often minor, as in the antibody fragments expressed in E. coli (Knappik et al., 1993).

The use of selective binders for assisted folding, when the particular protein lacks the necessary conformational information to fold completely, has been successfully tried. Monoclonal antibodies have been used for in vitro refolding of protein S (Carlson & Yarmush, 1992), immobilized trypsin for refolding of Bowman-Birk type proteinase inhibitors (Flecker, 1989) and also simultaneous expression of two receptor partners that are able to form a heterodimer (Chuan et al., 1997). Samuelsson and co-workers showed that co-expression of a specific binding protein in Escherichia coli can significantly improve the relative yields of correctly folded human insulin growth factor I (IGF-I) (Samuelsson et al., 1996) (Figure 3a). The molecule has been shown to possess a thermodynamic folding problem and is unable to quantitatively form the native disulfides (Hober et al., 1992, Hober et al., 1997, Miller et al., 1993). Instead, all attempts to produce IGF-I in heterologous hosts have resulted in misfolded species (Elliott et al., 1990, Forsberg et al., 1990, Meng et al., 1988). Yet, when the protein was allowed to fold in the presence of its binding protein (IGFBP-1) it was able to quantitatively attain the native conformation (Hober et al., 1994). Therefore, IGF-I turned out to be an interesting molecule for affinity-assisted folding. Since both the binding protein and IGF-I were produced as fusion proteins with affinity handles, the purification process was also greatly facilitated (Figure 3b). A requirement for this concept to succeed is that the disulfides of the produced proteins are allowed to break and reform (Saxena & Wetlaufer, 1970, Wetlaufer, 1984). Therefore a redox buffer was added to the cultivation media. In contrast to earlier attempts

a) Co-expression in *E. coli*

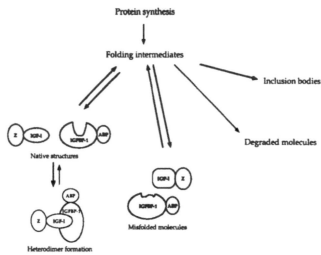

b) Co-purification

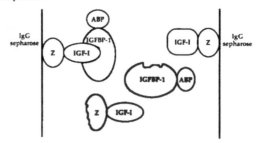

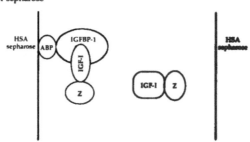

Figure 3   a) Affinity assisted *in vivo* folding by co-expression in *Escherichia coli*. The binding protein (IGFBP1) and native IGF-I form a complex and thereby lower the apparent concentration of the native conformation. The equilibrium between misfolded and native IGF-I is shifted towards the native form. b) Co-purification of IGF-I and IGFBP-1. By using different affinity tags on the target proteins, two separate and sequentially used affinity columns are able allow enrichment of the correctly folded fullength material

to co-express "helping proteins" this system has shown to result in quantitative yields of correctly folded IGF-I.

ENGINEERING PROTEINS TO FACILITATE RECOVERY

Recombinant DNA techniques allow fusion of genes or fragments thereof to alter the properties of recombinant proteins. The most widespread use of this strategy is to enable affinity purification of the product. It is possible to fuse the protein of interest to different extensions, proteins or peptides, in order to facilitate the purification procedure (Nygren *et al.*, 1994, Uhlén *et al.*, 1992). Furthermore, by using DNA-technology it is possible to affect both the amount and the localization of the product. Effective systems for secretion of the protein to the medium or to the periplasm, give us the opportunity of achieving an initial purification step. The expression level of recombinant proteins has in several cases been increased by fusion (Ståhl *et al.*, 1997) and the number of other useful applications is constantly increasing. Gene fusions offer means to improve the proteolytic stability of sensitive recombinant proteins. Ubiquitin, well known for its resistance to proteases, can stabilize proteins in *E. coli* (Butt *et al.*, 1989). Fusion to the IgG-binding tail ZZ has allowed recovery of full-length proinsulin and IGF-I (Nilsson *et al.*, 1996, Samuelsson *et al.*, 1991). Also, a region consisting of several hydrophilic amino acids, fused to the C-terminus has been shown to stabilize a labile form of the phage repressor protein P22 Arc in the E. coli cytoplasm (Bowie & Sauer, 1989). To enable specific recovery of full length protein, a dual affinity system also has been developed by Hammarberg *et al.* (Hammarberg *et al.*, 1989). The protein of interest is fused with two different affinity tags that separately can be used in affinity chromatography. This system has also proven to further stabilize the protein compared to single affinity fusion (Murby *et al.*, 1991). By what mechanism a protein fusion inhibits proteolysis remains unclear in many cases. Suggested mechanisms include (i) protected N- or C-terminus, (ii) sterical hindrance from proteases and (iii) improved folding.

A variety of different gene fusion systems in order to use affinity chromatography have been developed since the first example was presented (Flaschel & Friehs, 1993, LaVallie & McCoy, 1995, Nygren *et al.*, 1994, Uhlén *et al.*, 1983). Different proteins show very divergent characteristics, regarding stability, solubility, size and secretability. Therefore, proteins need individual treatment and no ideal fusion system exists that is applicable on all proteins.

Fusion protein purification systems can be based on several kinds of interactions. Examples of interactions are protein-protein interactions, enzyme-substrate interactions, protein-carbohydrate interactions and protein-metal interactions. Some of the most frequently used fusion protein systems for affinity purification are glutathione-S-transferase (GST) (Smith & Johnson, 1988), protein A (SPA) (Ståhl & Nygren, 1997), maltose binding protein (MBP) (Di Guan *et al.*, 1988) and thioredoxin (TR) (LaVallie *et al.*, 1993) (Table 1). There are also a number of short peptides used for selective purification, such as poly-histidine (Hochuli, 1990, Hochuli *et al.*, 1988), poly-arginine (Sassenfeld & Brewer, 1984), and tryptophan tags (Köhler *et al.*, 1991) (Table 1). When choosing a system for affinity purification there are several factors to consider. The choice of system will influence the conditions

Table 1 Examples on available proteins or peptides for selective purification

Fusion partner	Ligand	Interaction	Elution conditions	Reference
Protein A (SPA)	hIgG/Fc	protein-protein	low pH	(Ståhl and Nygren, 1997)
Z	hIgG/Fc	protein-protein	low pH	(Ståhl and Nygren, 1997)
ProteinG (SPG)	hIgG/Fc	protein-protein	low pH	(Ståhl and Nygren, 1997)
ABP	HSA	protein-protein	low pH	(Ståhl and Nygren, 1997)
Glutathione-S-transferase (GST)	glutathione	protein-substrate	reduced glutathione	(Smith and Johnson, 1988)
Maltose-binding protein (MBP)	amylose	protein-carbohydrate	maltose	(Di Guan et al., 1988)
FLAG peptide	mAb	peptide-protein	low pH/FLAG pept.	(Hopp et al., 1988)
Biotinylated protein	Monoavidin	biotin-avidin	Biotin	(Wilchek and Bayer, 1990)
polyhistidine	Me2+	peptide-IMAC	low pH/imidazole	(Hochuli, 1990, Hochuli et al., 1988)
polyarginine	anion	peptide-ionexchanger	salt	(Sassenfeld and Brewer, 1984)
polytryptophane	–	2-phase system	–	(Köhler et al., 1991)

for purification and elution. Also the size and stability of the fusion partner, the stability and regeneration possibility of the affinity matrix and the cost of the affinity matrix has to be considered. Recently developed fusion partners tend to be smaller in size to improve the ratio target protein/fusion partner, and more attention is being paid to mild elution conditions. Important is also whether the fusion partner has to be removed or not. In many cases this is crucial and there are several cleavage methods, chemical and enzymatic, available.

REFERENCES

Anfinsen, C.B. (1973) Principles that Govern the Folding of Protein Chains. *Science*, 181, 223–230.

Babbitt, P.C., West, B.L., Buechter, D.D., Kuntz, I.D. and Kenyon, G.L. (1990) Removal of a proteolytic activity associated with aggregates formed from expression of ceratine kinase in Escerichia coli leads to improved recovery of active enzyme. *Bio/Technology*, 8, 945–949.

Baker, D., Sohl, J.L. and Agard, D.A. (1992) A protein-folding reaction under kinetic control. *Nature*, 356, 263–265.

Bardwell, J.C.A. and Beckwith, J. (1993) The Bonds That Tie: Catalyzed Disulfide Bond Formation. *Cell*, 74, 769–771.

Bardwell, J.C.A., McGovern, K. and Beckwith, J. (1991) Identification of a Protein Required for Disulfide Bond Formation *In Vivo. Cell*, 67, 581–589.

Bowie, J.U. and Sauer, E.T. (1989 Identification of C-terminal extensions that protect proteins from intracellular proteolysis. *J. Biol. Chem.*, 264, 7596–7602.

Buchner, J., Brinkmann, U. and Pastan, I. (1992a) Renaturation of a single-chain immunotoxin facilitated by chaperones and protein disulfide isomerase. *Bio/Technology*, 10, 682–685.

Buchner, J., Pastan, I. and Brinkmann, U. (1992b) A Method for Increasing the Yield of Properly Folded Recombinant Fusion Proteins: Single-Chain Immunotoxins from Renaturation of Bacterial Inclusion bodies. *Anal. Biochem.*, 205, 263–270.

Butt, T.R., Jonnalagadda, S., Monia, B.P., Sternberg, E.J., Marsh, J.A., Stadel, J.M., Ecker, D.J. and Crooke, S.T. (1989) Ubiquitin fusion augments the yield of cloned gene products in *Escherichia coli. Proc. Natl. Acad. Sci. USA*, 86, 2540–2544.

Carlson, J.D. and Yarmush, M.L. (1992) Antibody assisted refolding. *Bio/Technology*, 10, 86–91.

Chuan, L., Schwabe, J.W.R., Banayo, E. and Evans, R.M. (1997) Coexpression of nuclear receptor partners increase their solubility and biological activities. *Proc. Natl. Acad. Sci. USA*, 94, 2278–2283.

Cleland, J.L., Builder, S.E., Swartz, J.R., Winkler, M., Chang, J.Y. and Wang, D.I.C. (1992) Polyethylene Glycol Enhanced Protein Refolding. *Bio/Technology*, 10, 1013–1019.

Cleland, J.L. and Wang, D.I.C. (1990) Cosolvent assisted protein refolding. *Bio/Technology*, 8, 1274–1278.

Creighton, T.E. (1993) *Proteins: structures and molecular properties*. New York: W.H. Freeman.

Creighton, T.E., Zapun, A. and Darby, N.J. (1995) Mechanisms and catalysts of disulphide bond formation in proteins. *Trends Biotechnol.*, 13, 18–23.

Di Guan, C., Li, P., Riggs, P.D. and Inouye, H. (1988) Vectors that facilitate the expression and purification of forign peptides in *Escherichia coli* by fusion to maltose-binding protein. *Gene*, 67, 21–30.

Eder, J., Rheinnecker, M. and Fersht, A.R. (1993) Folding of Subtilisin BPN': Role of the Pro-sequence. *Journal of Molecular Biology*, 233, 293–304.

Elliott, S., Fagin, K.D., Narhi, L.O., Miller, J.A., Jones, M., Koski, R., Peters, M., Hsieh, P., Sachdev, R., Rosenfeld, R.D., Rohde, M.F. and Arakava, T. (1990) Yeast-derived recombinant human insulin-like growth factor I: production, purification and structural characterization. *J. Prot. Chem.*, 9, 95–104.

Ellis, R.J. and Hartl, F.-U. (1996) Protein folding in the cell: competing models of chaperonin function. *FASEB J.*, 10, 20–26.

Epstein, C.J. and Goldberg, R.A. (1963) A study of factors influencing the reactivation of reduced egg white lysozyme. *J. Biol. Chem.*, 238, 1380–1383.

Fischer, B., Sumner, I. and Goodenough, P. (1993) Isolation, renaturation and formation of disulfide bonds of eukaryotic proteins expressed in *Escherichia coli* as inclusion bodies. *Biotechnol. Bioeng.*, 41, 3–13.

Flaschel, E. and Friehs, K. (1993) Improvement of downstream processing of recombinant proteins by means of genetic engineering methods. *Biotech. Adv.*, 11, 31–78.

Flecker, P. (1989) A new and general procedure for refolding mutant Bowman-Birk-type proteinase inhibitors on trypsin-Sepharose as a matrix with complementary structure. *FEBS Letters*, **252**, 153–157.

Forsberg, G., Palm, G., Ekebacke, A., Josephson, S. and Hartmanis, M. (1990) Separation and characterization of modified variants of recombinant human insulin-like growth factor I derived from a fusion protein secreted from *Escherichia coli. Biochem. J.*, **271**, 357–363.

Forsberg, G., Samuelsson, E., Wadensten, H., Moks, T. and Hartmanis, M. (1992) Refolding of human recombinant insulin-like growth factor II (IGF-II) *in vitro*, using a solubilization handle. In *Techniques in Protein Chemistry*, edited by R.H. Angeletti, pp. 0–0. San Diego: Academic Press.

Frank, B.H., Pettee, J.M. and Zimmermann, R.E. (1981) *The production of human proinsulin and its transformation to human insulin and C-peptide.* In *Peptides: Synthesis, Structure and Function, Proceedings of the Seventh American Peptide Symposium*, edited by D.H. Rich and E. Gross, pp. 729–738. Rockford, IL, USA.

Freedman, R.B. (1989) Protein Disulfide Isomerase: Multiple Roles in the Modification of Nascent Secretory Proteins. *Cell*, **57**, 1069–1072.

Freedman, R.B. (1995) The formation of protein disulphide bonds. *Curr. Op. Struct. Biol.*, **5**, 85–91.

Frydman, J. and Hartl, F.U. (1996) Principles of Chaperone-Assisted Protein Folding: Differences Between *in Vitro* and *in Vivo* Mechanisms. *Science*, **272**, 1497–1502.

Gething, M.-J. and Sambrook, J. (1992) Protein folding in the cell. *Nature*, **355**, 33–45.

Gilbert, H.F. (1990) Molecular and Cellular Aspects of Thiol-Disulfide Exchange. *Advances in Enzymology*, **63**, 69–170.

Gilbert, H.F. (1994) The formation of native disulfide bonds. In *Mechanism of protein folding*, edited by R.H. Pain, pp. 104–136. Oxford: Oxford University Press.

Goloubinoff, P., Gatenby, A.A. and Lorimer, G.H. (1989) GroE heat-shock proteins promote assembly of foreign prokaryotic ribulose bisphosphate carboxylase oligomers in *Escherichia coli. Nature*, **337**, 44–47.

Gregoret, L.M. and Cohen, F.E. (1991) Protein folding. Effect of packing density on chain conformation. *J. Mol. Biol.*, **219**, 109–122.

Hagen, A.J., Hatton, T.A. and Wang, D.I.C. (1990) Protein Refolding in Reverse Micelles. *Biotechn. Bioeng.*, **35**, 955–965.

Hammarberg, B., Nygren, P.-Å., Holmgren, E., Elmblad, A., Tally, M., Hellman, U., Moks, T. and Uhlén, M. (1989) Dual affinity fusion approach and its use to express recombinant human insulin-like growth factor II. *Proc. Natl. Acad. Sci. USA*, **86**, 4367–4371.

Harrison, S.C. and Durbin, R. (1985) Is there a single pathway for the folding of a polypeptide chain. *Proc. Natl. Acad. Sci. USA*, **82**, 4028–4030.

Hartl, F.U. and Martin, J. (1995) Molecular chaperones in cellular protein folding. *Curr Opin Struct Biol*, **5**, 92–102.

Hober, S., Forsberg, G., Palm, G., Hartmanis, M. and Nilsson, B. (1992) Disulfide exchange folding of insulin-like growth factor I. *Biochemistry*, **31**, 1749–56.

Hober, S., Hansson, A., Uhlen, M. and Nilsson, B. (1994) Folding of insulin-like growth factor I is thermodynamically controlled by insulin-like growth factor binding protein. *Biochemistry*, **33**, 6758–61.

Hober, S., Uhlen, M. and Nilsson, B. (1997) Disulfide exchange folding of disulfide mutants of insulin-like growth factor I *in vitro. Biochemistry*, **36**, 4616–4622.

Hochuli, E. (1990) Purification of recombinant proteins with metal chelate adsorbent. *Genet Eng (N Y)*, **12**, 87–98.

Hochuli, E., Bannwarth, W., Döbeli, H., Gentz, R. and Stüber, D. (1988) Genetic approach to facilitate purification of recombinant proteins with a novel metal chelate adsorbent. *Bio/Technology*, **6**, 1321–1325.

Hopp, T.H., Prickett, K.S., Price, V.L., Libby, R.T., March, C.J., Cerretti, D.P., Urdal, D.L. and Conlon, P.J. (1988) A short polypeptide marker sequence useful for recombinant protein identification and purification. *Bio/Technology*, **6**, 1204–1210.

Horwich, A.L., Low, K.B., Fenton, F.A., Hirshfield, I.N. and Furtak, K. (1993) Folding *in vivo* of bacterial cytoplasmic proteins: role of GroEL. *Cell*, **74**, 909–917.

Hwang, C., Sinskey, A.J. and Lodish, H.F. (1992) Oxidized Redox State of Glutathione in the Endoplasmic Reticulum. *Science*, **257**, 1496–1502.

Kane, J.F. and Hartley, D.L. (1988) Formation of recombinant protein inclusions in *E. coli. Trends Biotechnol.*, **6**, 95.

Karplus, M. and Weaver, D.L. (1976) Protein-folding dynamics. *Nature*, **260**, 404–406.

Kiefhaber, T., Rudolph, R., Kohler, H.-H. and Buchner, J. (1991) Protein aggregation *in vitro* and *in vivo*: A quantitative model of the kinetic competition between folding and aggregation. *Bio/Technology*, **9**, 825–829.

Kim, P.S. and Baldwin, R.L. (1982) Specific intermediates in the folding reactions of small proteins and the mechanism of protein folding. *Annu. Rev. Biochem.*, **51**, 459–489.

King, J., Haase-Pettingell, C., Robinson, A.S., Speed, M. and Mitraki, A. (1996) Thermolabile folding intermediates: inclusion body precursors and chaperonin substrates. *FASEB J.*, **10**, 57–66.

Kitano, K., Fujimoto, S., Nakao, M., Watanabe, T. and Nakao, Y. (1987) Intracellular degradation ot recombinant proteins in relation to their location in *Escherichia coli* cells. *J. Biotech.*, **5**, 77–86.

Knappik, A., Krebber, C. and Plückthun, A. (1993) The effect of folding catalysts on the *in vivo* folding process of different antibody fragments expressed in Escherichia coli. *Bio/Technology*, **11**, 77–83.

Köhler, K., Ljungquist, C., Kondo, A., Veide, A. and Nilsson, B. (1991) Engineering proteins for enhanced partitioning properties in aqueos two-phase. *Bio/Technology*, **9**, 642–646.

LaVallie, E.R., DiBlasio, E.A., Kovacic, S., Grant, K.L., Schendel, P.F. and McCoy, J.M. (1993) A Thioredoxin Gene Fusion Expression System That Circumvents Inclusion Body Formation in the *E. coli* Cytoplasm. *Bio/Technology*, **11**, 187–193.

LaVallie, E.R. and McCoy, J.M. (1995) Gene fusion expression systems in *Escherichia coli. Curr. Opin. Biotechnol.*, **6**, 501–505.

Levinthal, C. (1969) *How to fold graciously.* In *Mossbauer Spectroscopy in Biological Systems*, edited by P. Debrunner, J.C.M. Tibris and E. Münck, pp. 22–24. Monticello, IL: University of Illinois Press, Urbana, Allerton House.

Light, A.L. (1985) Protein solubility, protein modifications and protein folding. *BioTechniques*, **3**, 298.

Lilie, H., Lang, K., Rudolph, R. and Buchner, J. (1993) Prolyl isomerases catalyze antibody folding *in vitro. Protein Science*, **2**, 1490–1496.

Lorimer, G.H. (1996) A quantitative assessment of the role of the chaperonin proteins in protein folding *in vivo. FASEB J.*, **10**, 5–9.

Marston, F.A. O. (1986) The purification of eukaryotic polypeptides synthesized in *Escherichia coli. Biochemical Journal*, **240**, 1–12.

McCaman, M.T. (1989) Fragments of prochymosin produced in E. coli form insoluble inclusion bodies. *J. Bacteriol.*, **171**, 1225–.

Mendoza, J.A., Rogers, E., Lorimer, G.H. and Horowitz, P.M. (1991) Chaperonins facilitate *in vitro* refolding of monomeric of monomeric mitochondrial rhodanese. *J. Biol. Chem.*, **266**, 13044–13049.

Meng, H., Burleigh, B.D. and Kelly, G.M. (1988) Reduction studies on bacterial recombinant somatomedin C/insulin-like growth factor-I. *J. Chromatogr.*, **443**, 183–192.

Miller, J.A., Narhi, L.O., Hua, Q.X., Rosenfeld, R., Arakawa, T., Rohde, M., Prestrelski, S., Lauren, S., Stoney, K.S., Tsai, L. and Et, A.L. (1993) Oxidative refolding of insulin-like growth factor 1 yields two products of similar thermodynamic stability: a bifurcating protein-folding pathway. *Biochemistry*, **32**, 5203–13.

Missiakas, D., Georgoploulos, C. and Raina, S. (1994) The Escherichia coli DsbC (xprA) gene encodes a periplasmic protein involved in disulfide bond formation. *EMBO J.*, **13**, 2013–2020.

Missiakas, D., Georgopoulos, C. and Raina, S. (1993) Identification and Characterization of the *Escherichia Coli* gene DsbB, whose Gene Product Is Involved in the Formation of Disulfide Bonds *in Vivo. Proc. Natl. Acad. Sci.*, **90**, 7084–7088.

Missiakas, D. and Raina, S. (1997) Protein Folding in the Bacterial Periplasm. *J. Bacteriol.*, **179**, 2465–2471.

Missiakas, D., Schwager, F. and Raina, S. (1995) Identification and characterization of a new disulfide isomerase-like protein (DsbD) in *Escherichia coli. EMBO J.*, **14**, 3415–3424.

Mitraki, A. and King, J. (1989) Protein Folding Intermediates and Inclusion Body Formation. *Bio/Technology*, **7**, 690–697.

Mitraki, A. and King, J. (1992) Amino acid substitutions influencing intracellular protein folding pathways. *FEBS*, **307**, 20–25.

Murby, M., Cedergren, L., Nilsson, J., Nygren, P.-Å., Hammarberg, B., Nilsson, B., Enfors, S.-O. and Uhlén, M. (1991) Stabilization of recombinant proteins from proteolytic degradation in Escherichia coli using a dual affinity fusion strategy. *Biotechnol. Appl. Biochem.*, **14**, 336–346.

Murby, M., Samuelsson, E., Nguyen, T.N., Mignard, L., Power, U., Binz, H., Uhlen, M. and Stahl, S. (1995) Hydrobicity engineering to increase solubility and stability of a recombinant protein from respiratory syncytial virus. *Eur J Biochem*, **230**, 38–44.

Nilsson, B. and Anderson, S. (1991) Proper and improper folding of proteins in the cellular enviroment. *Annu. Rev. Microbiol.*, **45**, 607–635.

Nilsson, B., Moks, T., Jansson, B., Abrahmsén, L., Elmblad, A., Holmgren, E., Henrichson, C., Jones, T.A. and Uhlén, M. (1987) A synthetic IgG-binding domain based on staphylococcal protein A. *Protein Eng.*, **1**, 107–113.

Nilsson, J., Jonasson, P., Samuelsson, E., Ståhl, S. and Uhlén, M. (1996) Integrated production of human insulin and its C-peptide. *J Biotechnol.*, **48**, 241–250.

Noiva, R. (1994) Enzymatic Catalysis of Disulfide Formation. *Protein Expression and Purification*, **5**, 1–13.

Noiva, R. and Lennarz, W.J. (1992) Protein Disulfide Isomerase, a multifunctional protein resident in the lumen of the endoplasmic reticulum. *J Biol. Chem.*, **267**, 3553–3556.

Nygren, P.-Å., Ståhl, S. and Uhlén, M. (1994) Engineering proteins to facilitate bioprocessing. *Trends Biotechnol.*, **12**, 184–188.

Ohgushi, M. and Wada, A. (1983) "Molten-globule state": a compact form of globular proteins with mobile side-chains. *FEBS Lett.*, **164**, 21–24.

Ptitsyn, O.B. (1987) Protein Folding: Hypotheses and Experiments. *J. Protein. Chem.*, **6**, 272–293.

Ptitsyn, O.B. (1992) The molten globule state. In *Protein folding*, edited by T.E. Creighton, pp. 243–300. New York: W.H. Freeman.

Ray, M.V. L., Van Duyne, P., Jackson-Matthews, D.E., Sturmer, A.M., Merkler, D.J., Consalvo, A.P., Young, S.D., Gilligan, J.P., Shields, P.P. (1993) Production of recombinant Salmon Calcitonin by *in vitro* amidation of an *Escherichia coli* produced precursor peptide. *Bio/Technology*, **11**, 64–70.

Roman, L.J., Shjeta, E.A., Martasek, P., Gross, S.S. and Masters, B.S. S. (1995) High-level expression of functional rat neuronal nitric oxide synthase in *Escherichia coli*. *Proc. Natl. Acad. Sci. USA*, **92**, 8428–8432.

Rozema, D. and Gellman, S.H. (1996) Artificial chaperone-assisted refolding of carbonic anhydrase B. *J. Biol. Chem.*, **271**, 3478–3487.

Rudolph, R. (1995) Successful Protein Folding on an Industrial Scale. In *Principles and Practice of Protein Engineering*, edited by J.L. Cleland and C.S. Craik, pp. 00–00. New York: Wiley.

Rudolph, R. and Lilie, H. (1996) *In vitro* folding of inclusion body proteins. *FASEB J.*, **10**, 49–56.

Ryabova, L.A., Desplancq, D., Spirin, A.S. and Plückthun, A. (1997) Fuctional antibody production using cell-free translation: Effects of protein disulfide isomerase and chaperones. *Nature Biotechn.*, **15**, 79–84.

Samuelsson, E., Jonasson, P., Viklund, F., Nilsson, B. and Uhlén, M. (1996) Affinity assisted *in vivo* folding of a secreted human peptide hormone in *Escherichia coli*. *Nature Biotechnology*, **16**, In press.

Samuelsson, E., Moks, T., Nilsson, B. and Uhlen, M. (1994) Enhanced *in vitro* refolding of insulin-like growth factor I using a solubilizing fusion partner. *Biochemistry*, **33**, 4207–11.

Samuelsson, E., Wadensten, H., Hartmanis, M., Moks, T. and Uhlén, M. (1991) Facilitated *in vitro* refolding of human recombinant insulin-like growth factor I using a solubilizing fusion partner. *Bio/Technology*, **9**, 363–366.

Sassenfeld, H.M. and Brewer, S.J. (1984) A polypeptide fusion designed for the purification of recombinant fusion proteins. *Bio/Technology*, **2**, 76–81.

Saxena, V.P. and Wetlaufer, D. (1970) Formation of Three-Dimensional Structure in Proteins. I. Rapid Nonenzymatic Reactivation of Reduced Lysozyme. *Biochemistry*, **9**, 5015–5021.

Schein, C. (1990) Solubility as a function of protein structure and solvent components. *Bio/Technology*, **8**, 308–317.

Schein, C. (1991) Optimizing protein folding to the native state in bacteria. *Current Opinion in Biotechnology*, **2**, 746–750.

Schein, C.H. (1989) Production of Soluble Recombinant Proteins in Bacteria. *Bio/Technology*, **7**, 1141–1148.

Schmidt, F.X., Lang, K., Kiefhaber, T., Mayer, S. and Schönbrunner, E.R. (1990) *Prolyl isomerase. Its role in protein folding and speculations on its function in the cell.* Washington: AAAS Press.

Schoemaker, J.M., Brasnett, A.H. and Marston, F.A. O. (1985) Examination of calf prochymosin accumulation in Escherichia coli: disulphide linkages are a structural component of prochymosin inclusion bodies. *EMBO J.*, **4**, 775–780.

Smith, D.B. and Johnson, K.S. (1988) Single-step purification of polypeptides expressed in *Escherichia coli* as fusions with glutathione S-transferase. *Gene*, **67**, 31–40.

Ståhl, S. and Nygren, P.-Å. (1997) The use of gene fusions to Protein A and Protein G in immunology and biotechnology. *Pathol. Biol.*, **45**, 66–76.

Ståhl, S., Nygren, P.-Å. and Uhlén, M. (1997) *Strategies for Gene Fusions*. In *Methods Mol. Biol.*, edited by R. Tuan, pp. 37–54. Totowa: Humana Press Inc.

Stempfer, G., Höll-Neugebauer, B. and Rudolph, R. (1996) Improved Refolding of an Immobilized Fusion Protein. *Nature Biotechn.*, **14**, 329–334.

Strandberg, L. and Enfors, S.-O. (1991) Factors influencing inclusion body formation in the production of fused protein in *Escherichia coli. Appl. Env. Microbiol.*, **57**, 1669–1674.

Stryer, L. (1995) *Biochemistry*, 4th ed. New York: W.H. Freeman.

Tandon, S. and Horowitz, P. (1986) Detergent-assisted Refolding of Guanidinium Chloride-denatured Rhodanese, The Effect of Lauryl Maltoside. *J. Biol. Chem.*, **261**, 15615–15618.

Uhlén, M., Forsberg, G., Moks, T., Hartmanis, M. and Nilsson, B. (1992) Fusion proteins in biotechnology. *Curr. Op. Biotechn.*, **3**, 363–369.

Uhlén, M., Nilsson, B., Guss, B., Lindberg, M., Gatenbeck, S. and Philipson, L. (1983) Gene fusion vectors based on staphylococcal protein A. *Gene*, **23**, 369–378.

Weissman, J.S. and Kim, P.S. (1992) The Pro Region of BPTI Facilitates Folding. *Cell*, **71**, 841–851.

Wetlaufer, D.B. (1973) Nucleation, rapid folding and globular intrachain regions in proteins. *Proc. Natl. Acad. Sci. USA*, **70**, 697–701.

Wetlaufer, D.B. (1984) Nonenzymatic Formation and Isomerization of Protein Disulfides. *Methods Enzymol.*, **107**, 301–304.

Wilchek, M. and Bayer, E.A. (1990) Introduction to Avidin-Biotin technology. *Methods Enzymol.*, **184**, 5–13.

Wilkinson, D.L. and Harrison, R.G. (1991) Predicting the solubility of recombinant proteins in Escherichia coli. *Bio/Technology*, **9**, 443–448.

Winkler, M.E. and Blaber, M. (1986) Purification and characterization of recombinant single chain urokinase produced in *E. coli. Biochemistry*, **25**, 4041–.

Wulfing, C. and Plückthun, A. (1994) Protein Folding in the Periplasm of Escherichia coli. *Mol. Microbiol.*, **12**, 685–692.

Wunderlich, M., Otto, A., Seckler, R. and Glockshuber, R. (1993) Bacterial Protein Disulfide Isomerase: Efficient Catalysis of Oxidative Protein Folding at Acidic pH. *Biochemistry*, **32**, 12251–12256.

Yansura, D.G. and Henner, D.J. (1990) Use of *Escherichia coli trp* Promoter for Direct Expression of Proteins. *Methods Enzymol.*, **185**, 54–60.

Yasukawa, T., Kanei-Ishii, C., Maekawa, T., Fujimoto, J., Yamamoto, T. and Ishii, S. (1995) Increase of Solubility of Foreign Proteins in *Escherichia coli* by Coproduction of the Bacterial Thioredoxin. *J. Biol. Chem.*, **1995**, 25328–25331.

Zapun, A., Missiakas, D., Raina, S. and Creighton, T.E. (1995) Structural and Fuctional Characterization of DsbC, a Protein Involved in Disulfide Bond Formation in *Escherichia coli. Biochemistry*, **34**, 5075–5089.

Zardeneta, G. and Horowitz, P.M. (1992) Micelle-assisted Protein Folding. *J. Biol. Chem.*, **267**, 5811–5816.

Zhang, R. and Snyder, G.H. (1989) Dependence of formation of small disulfide loops in two-cysteine peptides on the number and types of intervening amino acids. *J. Biol. Chem.*, **264**, 18472–18479.

3. PROTEIN ENGINEERING FOR AFFINITY PURIFICATION: THE *STREP*-TAG

THOMAS G.M. SCHMIDT[1] and ARNE SKERRA[2]

[1]*Institut für Bioanalytik GmbH, Rudolf-Wissell-Str. 28, D-37079 Göttingen, Germany*
[2]*Lehrstuhl für Biologische Chemie, Technische Universität München,
D-85350 Freising-Weihenstephan, Germany*

INTRODUCTION

Since the early days of protein chemistry the specific detection and the purification to homogeneity were the two critical problems that needed to be solved before the biochemical investigation of a protein could be commenced. Traditionally, a long series of precipitation, adsorption/desorption, and column chromatography steps had to be carried out in order to obtain a pure protein. Although the advent of affinity chromatography made it possible to shorten the laborious procedure, a suitable combination of ligand, matrix, and elution conditions has first to be optimized in each case.

Working out an appropriate purification scheme was not only a common pursuit in classical biotechnology, i.e., when proteins had to be isolated from their natural sources. It remained a challenge after the development of recombinant DNA techniques, even though the problem was often made easier, due to the possible over-production of the protein of interest. Nevertheless, it is often necessary to specifically detect the foreign polypeptide in the course of heterologous protein production in order to optimize expression and subsequent purification yields. Unfortunately, as long as the desired biomolecule is not an enzyme, which can be quantified according to its catalytic activity, or a protein that had been isolated before, so that antibodies could have been obtained by immunization, this is a difficult task.

Fortunately, the concept of affinity tags was invented. The initial idea was to take a short sequence of amino acids that is recognized as an epitope by a monoclonal antibody and to append it to a recombinant protein on the genetic level. Thus, the molecular recognition property of the peptide is transferred to the new protein context and, given the proper choice of the amino acid sequence, it is likely not to interfere with the protein's biological activity. An 11 residue sequence from the c-*myc* proto-oncogene product, which had been fused to GRP78 (Munro and Pelham, 1986) and utilized for studying its intra-cellular targeting by means of the cognate antibody 9E10 (Evan *et al*, 1985), was probably the prototypic example for this approach. Thus, the so-called *myc* tag was created, which served since then for the detection of numerous heterologous proteins and gained particular popularity in the field of antibody engineering (for references see Schiweck *et al*, 1997).

A number of other epitope peptides were, together with their corresponding monoclonal antibodies, subsequently used as affinity tags for over-produced proteins. Some prominent examples are summarized in Table 1. Although these systems clearly facilitate recombinant protein detection, most of them suffer from the

Table 1 Common affinity tags other than the *Strep*-tag

Fusion sequence	Residues	Target	Washing/Elution conditions	Reference
Myc tag	11	Monoclonal antibody	Physiological/Low pH	Evan *et al.*, 1985
Flag peptide	8	Monoclonal antibody	Physiological/ Chelator compounds	Hopp *et al.*, 1988
S-tag	15	S-fragment of RNaseA	Physiological/ Thrombin cleavage	Kim and Raines, 1993
Calmodulin-binding peptide (CBP)	26	Calmodulin	High salt/Chelator compounds	Stofko-Hahn *et al.*, 1992
His-tag	6	Immobilized transition metal ions	High salt/Low pH, chelator compounds, imidazole	Hochuli *et al.*, 1988

disadvantage that they cannot be readily employed for the purification of the same protein since harsh conditions must be applied for elution of the tagged protein from a column with the immobilized antibody. Furthermore, the monoclonal antibodies are difficult to produce and not available as bulk reagents so that their use is restricted to the analytical scale.

In order to circumvent these problems we set out to develop a new affinity tag, termed *Strep*-tag, by inverting the methodological approach. That is, we first selected a target protein that is biochemically well characterized and commercially available from different sources and we then developed an artificial peptide sequence that specifically binds to it with sufficient affinity. Our choice fell on the protein reagent streptavidin, whose properties are briefly described in the next chapter. The idea was to use competition with biotin, a natural ligand of streptavidin, for the disruption of the binding interaction with the peptide under very gentle conditions so that the structure of the purified recombinant protein remained undisturbed. The engineering of the *Strep*-tag sequence by molecular repertoire techniques and the step-wise improvement of the system under practical aspects are described in the following sections. Finally, the currently available vectors for the production of *Strep*-tag fusion proteins are explained and some applications are reviewed.

CHOICE OF STREPTAVIDIN AS A TARGET

Because of its extraordinary affinity for the small ligand biotin (Green, 1975) streptavidin has become a widespread reagent for detection and separation purposes in almost all areas of modern biochemistry (cf. Bayer and Wilchek, 1990). The ligand, also known as vitamin H, can be chemically coupled to a variety of biomolecules in an easy manner. Streptavidin is naturally secreted by *Streptomyces avidinii* as a complex with a small antibiotic molecule (Chaiet and Wolf, 1964). Its three-dimensional structure was first solved for the apo-protein as well as for the biotin complex (Weber *et al.*, 1989). Streptavidin is a homotetramer of 159 amino acids

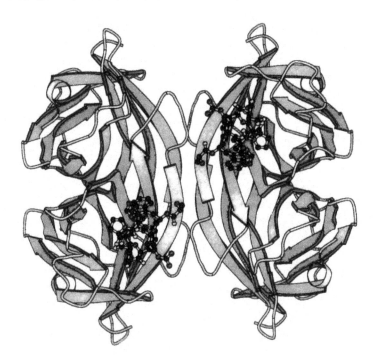

Figure 1 Crystal structure of the complex between streptavidin and the *Strep*-tag (Schmidt *et al.*, 1996). View is along one of the two-fold symmetry axes of the tetramer. Two of the four ligand binding sites can be seen at the front with the bound *Strep*-tag (C: black; N: grey; O: white). The peptides bound to the two other pockets at the back of the tetramer are omitted for clarity. The ribbon representation was generated with the program MolScript (Kraulis, 1991).

per subunit (cf. Figure 1), which possesses four independent biotin binding sites with K_d values as low as 10^{-14} M (Green, 1990; Weber *et al.*, 1992b).

Contrasting with the structurally and functionally related avidin from hen egg white (Green, 1975), streptavidin is not glycosylated, devoid of sulphur-containing amino acids, and slightly acidic (pI = 5–6), which minimizes non-specific adsorption to nucleic acids and negatively charged cell membranes. Therefore, it causes a significantly lower background in many bioanalytical applications and its use is often preferred. A large number of reagents derived from streptavidin are commercially available, including the immobilized protein and its conjugates with different reporter enzymes. It is a further advantage in this respect that streptavidin is extremely stable against proteases (Bayer *et al.*, 1989), detergents, chaotropic salts, and elevated temperature (Bayer *et al.*, 1990).

Consequently, streptavidin appeared to be a promising target for the engineering of an artificial peptide ligand that binds to it without being itself biotinylated. Instead, it was tempting to search for an amino acid sequence that would bind competitively with the natural ligand so that biotin could be used for the mild displacement of a corresponding fusion protein from a streptavidin column.

Soon after the development of the *Strep*-tag sequence (see next chapter) it turned however out that the practical situation was more complicated. Several commercial preparations consist of streptavidin which is proteolytically truncated at both termini of its polypeptide chain. This modification to the primary structure is partially caused by endogenous proteases of the bacterial host cell and brought to completion by a rigorous proteolytic treatment according to undisclosed protocols. As result, so-called core streptavidin is produced, starting with Ala 13 or Glu 14 (Pähler *et al*, 1987; Bayer *et al*, 1989) of the mature polypeptide sequence and ending with Ser 139 (Bayer *et al*, 1989), Ala 138 (Pähler *et al*, 1987), or even with Ser 136 (Chait, 1994), depending on the commercial source. This shortened version of streptavidin has a higher solubility and is less prone to oligomerization. Most importantly, its binding activity for conjugates of biotin is significantly enhanced, probably due to better sterical accessibility of the ligand pocket (Bayer *et al*, 1989).

Streptavidin is usually sold for the binding of biotinylated molecules, rather than for the complexation of a new peptide ligand, and details of the product composition were not available from the manufacturers. Only after testing several different brands we found out that the capability of binding the *Strep*-tag varied considerably, at least during the purification of a recombinant protein (see below). In addition, it turned out that some of the preparations were contaminated with residual proteolytic activity, probably resulting from the production process (Schmidt and Skerra, 1994b). In order to obtain a material with homogeneous amino and carboxy termini, without the necessity for protease treatment, an *E. coli* expression system was established that yielded directly functional core streptavidin (Schmidt and Skerra, 1994b).

For this purpose, the structural gene comprising residues 14 to 139 was amplified from *S. avidinii* chromosomal DNA, equipped with a methionine initiation codon, and placed under transcriptional control of the T7 promoter (Studier *et al*, 1990). Thus the recombinant streptavidin was produced in large amounts as inclusion bodies. After disruption of the cells these were isolated, solubilized in Gdn/HCl, and the protein was refolded upon quick dilution into phosphate-buffered saline. Following a final purification by fractionated ammonium sulphate precipitation, pure and fully functional recombinant core streptavidin was routinely obtained in yields up to 160 mg per liter *E. coli* culture (Schmidt and Skerra, 1994b).

ENGINEERING OF THE *STREP*-TAG PEPTIDE

Earlier random mutagenesis experiments had shown that peptides with streptavidin-binding activity could be found by means of phage display (Devlin *et al*, 1990) or via chemical synthesis on plastic beads (Lam *et al*, 1991). In these studies a His-Pro-Gln consensus sequence was identified and considered as critical for binding. However, those peptides were not directly suited as affinity tags because their affinities were too low (Weber *et al*, 1992a). Therefore, further engineering was needed in order to make such sequences amenable to practical use.

Our criteria for an ideal affinity tag were as follows. It should i) display an easily controllable binding behaviour with sufficient affinity, allowing high purification yields; ii) not interfere with the folding or bioactivity of a recombinant protein; iii) not influence its expression yield; and iv) be useful for detection assays. In order to screen for a streptavidin-binding peptide sequence that possesses most of these features we established a plasmid library which encoded random peptides with a conserved His-Pro-Gln triplet at the centre. These peptides were displayed at the carboxy terminus of the V_H domain of the lysozyme-binding antibody D1.3, which was functionally expressed in *E. coli* as part of an F_v fragment (Schmidt and Skerra, 1993). A filter sandwich colony screening assay was then carried out.

After transformation of *E. coli* with the plasmid library, up to 10 000 cells were plated on a nitrocellulose filter membrane on top of an agar plate and the cells were grown under antibiotic selection. Meanwhile, a second filter membrane was coated with lysozyme, the antigen of the F_v fragment. The first membrane was placed on this second membrane with the *E. coli* colonies on top and the whole stack was incubated on another agar plate in the presence of an inducer for gene expression. Synthesis of the F_v fragment effected partial leakage of the *E. coli* outer membrane during this stage and the released protein diffused to the second membrane with the antigen. There the different F_v fragment-peptide fusions became functionally captured and formed a replica of the still viable *E. coli* colonies from which they originated.

Finally, both membranes were separated and the one with the *E. coli* colonies was placed on a fresh agar plate for later recovery of the clones. The second membrane was probed for streptavidin-binding activity of the peptides with a streptavidin-alkaline phosphatase conjugate, followed by chromogenic staining. Colonies which gave rise to coloured spots were propagated and their plasmid DNA was sequenced so that the peptide composition could be deduced.

The first random library comprised the sequence Xaa-Xaa-His-Pro-Gln-Xaa-Xaa-Xaa-COOH, which was displayed after residue Ser 116 of the V_H domain of the D1.3 F_v fragment. 145 000 clones were screened via the filter sandwich assay. Out of these merely nine clones gave rise to a detectable binding signal for streptavidin. Their subsequent characterization revealed that just one peptide was suitable for the detection in a Western blot and exhibited sufficient affinity for the purification of the F_v fragment on immobilized streptavidin (Schmidt and Skerra, 1993). However, this peptide sequence (-Cys-Trp-His-Pro-Gln-Ala-Gly-Cys-COOH) caused a lytic effect on the *E. coli* cells and a decreased expression yield for the recombinant F_v fragment. This behaviour was probably due to the additional disulfide bond in the hybrid protein, which arose from the two Cys residues in the peptide, and undesirable for general use. In contrast, two rather hydrophilic sequences (Arg-Thr-His-Pro-Gln-Phe-Glu-Arg-COOH and Lys-Thr-His-Pro-Gln-Phe-Glu-Arg-COOH) did not influence *E. coli* viability or expression yields. These peptides did even better perform on the Western blot although their streptavidin-binding activities were too low for efficient protein purification.

Sequence analysis of the nine peptides revealed preference for an amino acid with a bulky hydrophobic side chain — Phe in six cases, Val in two cases — at the position immediately following the His-Pro-Gln moiety. Consequently, another library was constructed with the sequence Xaa-Xaa-Xaa-His-Pro-Gln-Phe-Xaa-Xaa-COOH. The spacer between Ser 116 of the D1.3 V_H domain and the conserved part was extended by an additional random codon in order to permit better sterical accessibility. Again, 145 000 clones were screened for streptavidin binding using the filter sandwich assay as above (Schmidt and Skerra, 1993). In this case a much larger fraction of colonies (ca. 5 %) gave rise to a signal. Altogether 52 clones with the most intense spots were isolated and subjected to another filter sandwich assay for mutual comparison. As result, 15 clones were further characterized in applications like Western blotting, ELISA, and affinity purification experiments on immobilized streptavidin.

Finally, two sequences were identified which exhibited an optimal behaviour in the light of the criteria given above. Their sequences were quite similar: Ala-Trp-Arg-His-Pro-Gln-Phe-Gly-Gly-COOH and Gln-Trp-Leu-His-Pro-Gln-Phe-Gly-Gly-COOH. The more hydrophilic first sequence showed a slightly better signal on the Western blot and in the ELISA. It was therefore chosen for practical use and termed the "*Strep*-tag".

THE STREP-TAG II, A VARIANT WITH IMPROVED VERSATILITY

Although the *Strep*-tag proved to be useful for the detection and purification of a variety of recombinant proteins when attached to their carboxy-terminal end (see below), its affinity for streptavidin was substantially diminished when placed at the amino terminus or in between two protein domains. The key to overcome this restriction came from a comparison of the peptide sequences that were identified in our first random screening experiment. Some of the not yet optimal peptides carried a glutamate residue at the penultimate position, followed by Arg or Lys. The glutamate side chain was assumed to functionally substitute the Gly-Gly-COOH moiety of the *Strep*-tag as a negatively charged group. However, the introduction of this glutamate into the *Strep*-tag sequence necessitated re-optimization of the terminal residue (Gly) and of the residue preceding the His-Pro-Gln triplet (Arg), as it became apparent from a careful analysis of the sequences that had been identified so far (*cf.* Schmidt and Skerra, 1993).

In order to engineer the *Strep*-tag sequence in a systematic manner, a subset of 400 peptides, acetyl-Ser-Asn-Trp-Xaa$_1$-His-Pro-Gln-Phe-Glu-Xaa$_2$, comprising all possible amino acid combinations at the two positions Xaa$_1$ and Xaa$_2$, was generated on a paper filter by the "spot synthesis" technique (Frank, 1992). As part of this approach a chemical neighbourhood as within a polypeptide chain was mimicked for the peptides by covalently coupling their carboxy termini via a spacer to the filter support and by acetylating their free amino groups.

The streptavidin-binding activity of the different peptides in the array was probed with streptavidin-alkaline phosphatase conjugate under competitive conditions, i.e., in the presence of 1 mM diaminobiotin or 1 mM of the synthetic *Strep*-tag peptide (Schmidt *et al.*, 1996). The result was the same in both cases. A clear preference for Lys emerged at position Xaa$_2$ whereas the preference for the amino acid at Xaa$_1$ was not so pronounced. Densitometric analysis revealed however that the peptide with a Ser at position Xaa$_1$, together with Lys at position Xaa$_2$, gave rise to the most intense staining. The signal was by the factor 13 stronger than that of the original *Strep*-tag sequence, which had been spotted on the same filter for comparison. The corresponding peptide, Ser-Asn-Trp-Ser-His-Pro-Gln-Phe-Glu-Lys, was thus named *Strep*-tag II (Schmidt *et al.*, 1996).

CRYSTALLOGRAPHIC ANALYSIS OF THE STREPTAVIDIN/STREP-TAG COMPLEXES

The structural mechanism of the interaction between the *Strep*-tag as well as the *Strep*-tag II and streptavidin was elucidated by X-ray crystallography (Schmidt *et al.*, 1996). Suitable crystals were prepared from recombinant core streptavidin in the presence of the synthesized peptides at physiological pH with ammonium sulphate as the precipitant. The two crystal structures were both refined at a resolution of 1.7 Å. One streptavidin subunit occupied the asymmetric unit so that the streptavidin tetramer emerged as part of the crystal lattice (Figure 1).

The streptavidin monomer exhibited the typical eight-stranded antiparallel β-barrel with the ligand pocket at one of its ends (Figure 2), similar as in the previously described crystal structures of apo-streptavidin and of its biotin complex (Weber *et al.*, 1989). Most part of the polypeptide chain was virtually superimposable between the different structures. However, there was a striking exception regarding residues 44 to 53, which formed an exposed loop. This loop segment was previously described to be flexible and to act as a mobile lid that is structurally disordered in the absence of biotin but shuts down on the pocket upon complexation of this ligand (Weber *et al.*, 1989). However, in the complex with the *Strep*-tag peptides the same loop exhibited a defined but open conformation. Compared with the biotin complex, the loop had flipped into an almost perpendicular orientation, giving access to the pocket for the much bigger peptide ligand (Figure 2).

The actual binding cleft for biotin was obviously too narrow to accomodate the entire peptide, but two of its side chains could penetrate deep enough, occupying at least part of the cavity. Thus, a certain molecular mimicry was observed between the artificial peptide ligands and biotin. The carboxamide group of the Gln side chain was in the approximate position of the biotin sulphur atom and the imidazole group of the His residue replaced the valeryl carboxylate. The latter finding explained why the *Strep*-tag lost its affinity under acidic conditions, when the His side chain was protonated and adopted a positive charge. Interestingly, the space of the biotin imidazolidinone ring, which is responsible for most of the hydrogen bond and Van der Waals interactions in the tight complex between biotin and streptavidin, was not approached by peptidic groups and just occupied by a structurally ordered water molecule (Schmidt *et al.*, 1996).

Most of the observed interactions between the *Strep*-tag or the *Strep*-tag II and streptavidin were of a polar nature. Some of them were mediated by structural water molecules, which also seemed to play an important role in fixing the peculiar loop conformation described above. An exceptional feature in the structure of the *Strep*-tag complex was a salt bridge between the terminal carboxylate group of the peptide and the side chain of streptavidin residue Arg 84. This electrostatic interaction was possible because of the characteristic conformational flexibility of the two Gly residues at the end of the *Strep*-tag. Thus, this structural finding explained not only the repeated occurrence of the Gly-Gly-COOH motif during the screening for the *Strep*-tag but also why the *Strep*-tag had diminished affinity when it was placed at the amino terminus or amidst a recombinant polypeptide so that the free carboxylate group was not present.

In the complex of the *Strep*-tag II a similar salt bridge was formed in a different manner. In this case the side chain of the penultimate Glu residue provided the negative charge. Its carboxylate group was almost situated in the same location as the peptide carboxylate group of the *Strep*-tag. Consequently, a free carboxy terminus was no longer necessary. Apart from the two carboxy-terminal amino acids in the *Strep*-tag and in the *Strep*-tag II the three-dimensional structures of their complexes with streptavidin were very similar. The central part of the peptide adopted the conformation of a 3_{10}-helix. The Ser residue which had replaced an Arg in front of the His residue in the *Strep*-tag II exerted its specific structural role by forming

(A)

(B)

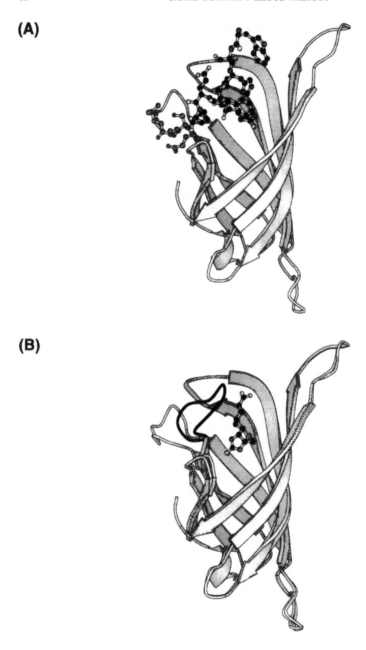

Figure 2 Complexation of the *Strep*-tag II and of biotin by streptavidin and structural properties of the flexible loop. Only one streptavidin subunit from the tetramer is shown (cf. Figure 1). (A) Ribbon representation of the streptavidin complex with the *Strep*-tag II (Schmidt *et al.*, 1996). The side chains of the four amino acids in the flexible loop of streptavidin that were subjected to random mutagenesis are displayed in grey. They are located in close proximity to the carboxy-terminus of the peptide, whose amino-terminal Trp residue can be seen at the top. (B) Superposition between the crystal structure of the streptavidin complex with biotin (Weber *et al.*, 1989) and the streptavidin structure shown in (A). The bound biotin ligand is displayed. The flexible loop in the corresponding holo-streptavidin structure is coloured black.

a hydrogen bond with one of the carboxylate oxygens of the Glu side chain. In this way, the conformation of the bound peptide became stabilized. This observation was in agreement with the preference for Ser and also for Thr at this position, which had been seen in the screening for the *Strep*-tag II.

ENGINEERING OF STREPTAVIDIN

Based on the knowledge about the molecular interaction between streptavidin and the *Strep*-tag peptides it appeared possible to introduce changes into streptavidin itself in order to achieve higher affinity for the artificial ligands (Voss and Skerra, 1997). Emphasis was laid on the *Strep*-tag II in this study because its use was not restricted to a particular location in the recombinant protein. Furthermore, the affinity between the *Strep*-tag II and streptavidin was slightly lower compared with the original *Strep*-tag (Schmidt *et al.*, 1996) so that a gain in binding strength was of special advantage for the application of this peptide.

A mutagenesis of the amino acids 44 to 47 in the flexible loop of streptavidin, which were in close spatial proximity to the carboxy-terminal amino acids of the peptide (Figure 2A), appeared to be promising for two reasons. First, side chain exchanges within this exposed region of streptavidin could influence the binding of the peptide without affecting the structure of the cavity for biotin so that further use of it as a competitor could be made. Second, contrasting with the structurally optimized and rigid β-barrel that constitutes most part of the streptavidin monomer, the flexible loop appeared likely to be more tolerant to amino acid substitutions without loss of protein stability or folding efficiency.

In order to search for streptavidin mutants with enhanced binding activity for the *Strep*-tag II these four amino acid positions were subjected to a targeted random mutagenesis, and then the filter sandwich colony screening assay was applied. For this purpose a secretory expression system for streptavidin was established. The structural gene for core strepavidin was fused with the coding region for the *OmpA* signal sequence and placed under the tight transcriptional control of the chemically inducible *tet* promoter (Skerra, 1994). Consequently, the polypeptide was translocated into the periplasm of *E. coli*, where, after processing, it could fold and assemble into the functional core streptavidin tetramer.

The colony screening assay was in principle performed as described above in the context of the screening for the *Strep*-tag sequence. In the present case the lower membrane was coated with an anti-streptavidin immunoglobulin from rabbits. Thus, the streptavidin mutants that became secreted from the *E. coli* colonies on the upper membrane were functionally captured. In order to probe for tight peptide binding a reporter reagent was produced which consisted of bacterial alkaline phosphatase and the *Strep*-tag II that was fused to its carboxy terminus. The membrane with the immobilized streptavidin mutants was incubated with the *PhoA/ Strep*-tag II enzyme, followed by staining with chromogenic reagents. Already after the first thousand colonies had been screened, two of them gave rise to unambiguous signals. In contrast, neither the recombinant wild-type streptavidin, which was subjected to the same screening assay, nor any other colonies caused coloured spots.

The two colonies were propagated and the plasmid DNA was isolated and sequenced. The deduced amino acid sequences of the mutagenized loop segment

44 to 47 were quite similar: Val-Thr-Ala-Arg and Ile-Gly-Ala-Arg (cf. the wild-type sequence: Glu-Ser-Ala-Val). In order to further characterize the biochemical properties of the streptavidin mutants their genes were subcloned into the plasmid for cytoplasmic expression and produced by refolding from inclusion bodies, in the same manner as the recombinant wild-type core streptavidin (see above). The mutants were investigated in ELISA experiments with the *PhoA/Strep*-tag II enzyme and subjected to fluorescence titration with the synthetic peptide, which had been coupled with a chromophoric group. In both experiments their affinities were found to be approximately 1 μM and, therefore, higher by more than one order of magnitude compared with wild-type streptavidin (Voss and Skerra, 1997). The K_d values of the mutants for the original *Strep*-tag were very similar (Voss and Skerra, unpublished).

The improved binding activity led to significant advantages under practical circumstances. First, the binding interaction was now tight enough to permit the immobilization of a recombinant protein carrying the *Strep*-tag II to an ELISA plate that had been coated with the streptavidin variant. Thus, new sandwich assay formats for the detection and quantification of proteins, e. g. in competition assays, can now be developed. Second, regarding purification purposes, especially the first of the two mutants clearly revealed a better performance than wild-type streptavidin (Voss and Skerra, 1997). Recombinant proteins, either fused with the *Strep*-tag or with the *Strep*-tag II, get more tightly bound to the streptavidin affinity column. This is particularly helpful in critical cases, when sticky components in a cell extract necessitate extended washing, or where the affinity tag is partially obscured by some moiety of the recombinant polypeptide chain.

VECTOR SYSTEMS FOR THE PRODUCTION OF *STREP*-TAG FUSION PROTEINS

The first generic vector that had been constructed for the production of recombinant proteins fused with the *Strep*-tag was pASK60-Strep (Schmidt and Skerra, 1993). This plasmid carried the IPTG-inducible *lac* promoter, followed by a tandem ribosomal binding site and an expression cassette. This cassette encoded the *OmpA* signal sequence and the *Strep*-tag, with a multiple cloning site in between. After proper insertion of a foreign structural gene a polypeptide was thus produced that carried the *Strep*-tag at its carboxy terminus and was secreted to the periplasmic space of *E. coli*. This strategy was successfully employed for the functional production of a number of proteins that possess structural disulfide bonds (Schmidt and Skerra, 1994b; see also next chapter).

Although the *lac* repressor gene had been cloned on pASK60-Strep the *lac* promoter/operator appeared to be somewhat leaky in the absence of the inducer. This behaviour of the *lac* system, which had also been noted by others, was undesirable because heterologous proteins are often toxic to *E. coli*, especially when they are secreted. This problem was solved by recruiting the *tet* promoter for bacterial expression from the vector pASK75 (Skerra, 1994). In this system the *tetA* promoter is tightly shut off by its cognate repressor until low concentrations of a tetracycline derivative are applied for chemical induction (e.g. anhydrotetracycline at 200 μg/ l *E. coli* culture). A further advantage is that *tet* promoter-based expression plasmids are independent of the genotypic background of *E. coli* (there are normally no additional copies of the *tet* repressor gene) and the metabolic state (the $tet^{P/o}$ is not subject to catabolite repression). Thus, their use for bacterial fermentation at higher cell densities is significantly facilitated (Schiweck and Skerra, 1995).

A series of such vectors was constructed for differing purposes in conjunction with the *Strep*-tag method. pASK75, the original $tet^{P/o}$ plasmid, had the same expression cassette as pASK60-Strep so that its use for the periplasmic expression of *Strep*-tag fusion proteins was similarly possible (Skerra, 1994). In another variant, pASK75CA, the signal sequence was deleted in order to facilitate direct cytoplasmic expression. An enhanced set of unique restriction sites in the multiple cloning region is present on the corresponding plasmids pASK-IBA1 and pASK-IBA1CA (Figure 3). The recognition sites of two type IIa endonucleases (*Bsa*I and *Bsm*FI), which generate sticky ends with arbitrary nucleotide overhangs remote from their recognition sequences, were introduced at both sides of the polylinker. Thus, given the proper cloning strategy, a foreign structural gene can be directly fused with the *Omp*A signal sequence and with the *Strep*-tag in a single step. The use of these restriction enzymes has further the advantage that cloning avoids the introduction of extraneous amino acids into the recombinant protein and is independent of the particular gene sequence. In addition, just one restriction enzyme is sufficient for a wide range of cloning purposes, especially when the polymerase chain reaction (PCR) is employed.

Finally, a special set of vectors was developed for the production of recombinant antibody fragments that carry the *Strep*-tag (Table 2). These plasmids provide standardized restriction sites for the cloning of immunoglobulin variable regions by PCR and permit their bacterial secretion and subsequent streptavidin affinity purification in the form of either F_v, scF_v, or F_{ab} fragments (Skerra, 1993).

APPLICATIONS FOR THE *STREP*-TAG

The *Strep*-tag has been successfully used both for the detection and for the purification of recombinant proteins in a variety of examples (Table 3). Detection of recombinant antibody fragments on a Western blot by means of a streptavidin-alkaline phosphatase conjugate (Schmidt and Skerra, 1993; Tsiotis *et al.*, 1995) was employed for guiding the optimization of expression and purification yields. After its fusion to an antibody fragment the *Strep*-tag was also utilized for the indirect detection of the cognate antigen, either in an ELISA with streptavidin-alkaline phosphatase conjugate (Schmidt and Skerra, 1993), by electron microscopy with gold-labelled streptavidin (Kleymann *et al.*, 1995a,b; Tsiotis *et al.*, 1995), or by fluorescence microscopy with fluorescein-labelled streptavidin (Ribrioux *et al.*, 1996). Possible cross reactions in these assays, due to the presence of biotinylated host proteins, were avoided by adding egg-white avidin, which does not recognize the *Strep*-tag but masks the biotin groups (Schmidt and Skerra, 1993).

The oriented immobilization of *Strep*-tag fusion proteins on microtitre plates became possible with the development of the streptavidin mutants that exhibit enhanced affinity for the *Strep*-tag (Voss and Skerra, 1997). Thus, the *Strep*-tag/ streptavidin interaction permits the development of modular setups for the selective capture of recombinant antigens, followed by their quantification with an antibody, which should be advantageous for diagnostic ELISAs in the future (Bouche *et al.*, in preparation).

A

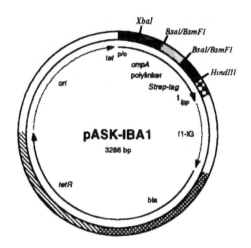

B

```
   RBS                                        XbaI        RBS
TAGAGAAAAGTGAAATGAATAGTTCGACAAAAATCTAGATAACGAGGGCAAAAAATGAAAAAGACAGCTATCGCG
   ***    tetA: MetAsnSerSerThrLysIleEnd       ****         MetLysLysThrAlaIleAla
```

```
                                                     PshAI
                                            BsaI NcoI BamFI EcoRI  SstI   KpnI
ATTGCAGTGGCACTGGCTGGTTTCGCTACCGTAGCGCAGGCCGGAGACCATGGTCCCGAATTCGAGCTCGGTACC
IleAlaValAlaLeuAlaGlyPheAlaThrValAlaGlnAlaGlyAspHisGlyProGluPheGluLeuGlyThr
   OmpA
```

```
                                             PshAI
SmaI BamHI   XhoI   SalI   PstI   BamFI NcoI BsaI Eco47III
CGGGGATCCCTCGAGGTCGACCTGCAGGGGGACCATGGTCTCAGCGCTTGGCGTCACCCGCAGTTCGGTGGTTAA
ArgGlySerLeuGluValAspLeuGlnGlyAspHisGlyLeuSerAlaTrpArgHisProGlnPheGlyGlyEnd
                                              Strep-tag
```

```
HindIII             <<<<<<< <<<<<   >>>>> >>>>>>>
TAAGCTTGACCTGTGAAGTGAAAAATGGCGCACATTGTGCGACATTTTTTTTGTCTGCCGTTTACCGCTACTGCG
                              mRNA End
```

C

```
   RBS                                        XbaI        RBS              PshAI
                                                                   BsaI SacII BamFI
TAGAGAAAAGTGAAATGAATAGTTCGACAAAAATCTAGATAACGAGGGCAAAAAATGGGAGACCGCGGTCCC
   ***    tetA: MetAsnSerSerThrLysIleEnd       ****        MetGlyAspArgGlyPro
```

```
                                                                    PshAI
EcoRI   SstI   KpnI SmaI BamHI   XhoI   SalI   PstI   BamFI NcoI BsaI Eco47III
GAATTCGAGCTCGGTACCCGGGGATCCCTCGAGGTCGACCTGCAGGGGGACCATGGTCTCAGCGCTTGGCGTCAC
GluPheGluLeuGlyThrArgGlySerLeuGluValAspLeuGlnGlyAspHisGlyLeuSerAlaTrpArgHis
                                                                  Strep-tag
```

```
             HindIII              <<<<<<< <<<<<   >>>>> >>>>>>>
CCGCAGTTCGGTGGTTAATAAGCTTGACCTGTGAAGTGAAAAATGGCGCACATTGTGCGACATTTTTTTTGTCTG
ProGlnPheGlyGlyEnd                                        mRNA End
```

Table 2 tet$^{p/o}$ expression vectors for the bacterial production of antibody fragments which carry the *Strep*-tag

Plasmid	Expression cassette[a]	Immunoglobulin fragment
pASK99-D1.3	OmpA-V$_H$-C$_H1_{\gamma1}$-*Strep*II; PhoA-V$_\kappa$-C$_\kappa$	F$_{ab}$
pASK98-D1.3	OmpA-V$_H$-sc-V$_\kappa$-*Strep*	"single chain" F$_v$
pASK90-D1.3	OmpA-V$_H$-*Strep*; PhoA-V$_\kappa$-myc	F$_v$
pASK90	OmpA-(V$_H$)-*Strep*; PhoA-(V$_\kappa$)-myc	F$_v$ (generic vector)

[a]OmpA and PhoA denote the bacterial signal sequences, whereas *Strep* and myc represent the affinity tags. Most of the vectors have the variable domain genes of the anti-lysozyme antibody D1.3 inserted, which can be replaced by those from another antibody via standardized restriction sites (cf. for example Schiweck and Skerra, 1995; Schiweck *et al.*, 1997). In the case of the generic vector pASK90 just the cloning sites are present.

The main application of the *Strep*-tag was so far in the purification of functionally produced recombinant proteins. The fact that the streptavidin affinity chromatography can be performed under gentle physiological conditions is a particular advantage of this method. The cell extract that contains the *Strep*-tag fusion protein is usually prepared in a mild buffer (e. g., phosphate-buffered saline) and applied to the column with the immobilized streptavidin (Schmidt and Skerra, 1994b). Due to the low non-specific binding of streptavidin, host cell proteins and

Figure 3 The generic *tet* promoter vectors pASK-IBA1 for periplasmic secretion in *E. coli* and pASK-IBA1CA for cytosolic expression. A schematic plasmid map of pASK-IBA1 is shown (A) together with the sequence of the corresponding expression cassette (B). The sequence of the expression cassette of pASK-IBA1CA, wherein the *omp*A signal sequence was replaced by an ATG start codon and one of the *Nco*I sites was substituted by a *Sad*I site, is shown as well (C). Both plasmids are based on pASK75, which has been described before (Skerra, 1994), and carry the following elements: *Col*EI origin of plasmid replication (*ori*), f1 intergenic region for the preparation of single-stranded DNA (f1-IG), the β-lactamase gene conferring ampicillin resistance (*bla*), and the *tet* repressor gene (*tet*R). Efficient initiation of translation is ensured by a tandem arrangement of ribosomal binding sites (RBS). Singular restriction sites are shown, whereby recognition sites for endonucleases that cut twice within the cloning region are marked bold. The secondary structure of the lipoprotein transcription terminator (t$_{lpp}$) is indicated (< >). The precise, directed cloning of a foreign structural gene between either the *Omp*A signal sequence (pASK-IBA1) or the ATG start codon (pASK-IBA1CA) and the *Strep*-tag can be achieved using the type IIa endonucleases *Bsa*I or *Bsm*FI. These enzymes cleave the DNA strands at sites which are distant from their recognition sequences. They were placed such that different 5'-overhangs are generated at each end of the cloning region (GGCC and GCGC, respectively, in the case of pASK-IBA1 and AATG and GCGC, respectively, in the case of pASK-IBA1CA; see the underlined stretch in the sequence). For cloning of a foreign gene, compatible protruding ends can be produced for example via PCR by incorporating the corresponding restriction sites into the primers. When the recognition sites are placed at the outward end in respect to the cleavage sites they will be removed in the course of the restriction digest. Thus, no limitations are imposed on the sequence of the structural gene itself. Of course, compatible ends can also be generated with conventional endonucleases, depending on the flanking regions, or a gene can be inserted using one of the several other restriction sites within the cloning region.

Table 3 Applications for the *Strep*-tag

Recombinant protein[a]	Use of the *Strep*-tag	Reference
V_H domain of the F_v fragment from the lysozyme-binding antibody D1.3	Purification of the intact F_v fragment Detection of the F_v fragment via ELISA and Western blotting	Schmidt and Skerra, 1993
V_L domain of the D1.3 F_v fragment	Purification of the intact F_v fragment Detection of the F_v fragment via ELISA and Western blotting Purification of the complex with the antigen	Schmidt and Skerra, 1993
Azurin (*P. aeruginosa*)	Purification of the metallo-protein	Schmidt and Skerra, 1994b
Cytochrome b_{562} (*E. coli*)	Purification of the protein with heme complexed Use as a coloured marker for testing the streptavidin columns	Schmidt and Skerra, 1994b
Mammalian serum retinol-binding protein	Purification of the functional apo-protein	Müller and Skerra, 1994
Chicken egg-white cystatin	Purification of the active inhibitor	Schmidt and Skerra, 1994b
Bilin-binding protein (*Pieris brassicae*)	Purification of the functional apo-protein	Schmidt and Skerra, 1994a
Chicken GPA receptor α-chain	Purification of the soluble protein fragment	Heller *et al.*, 1995
Human CNTF receptor α-chain	Purification of the soluble protein fragment	Heller *et al.*, 1995
V_H domain of the cytochrome c oxidase-binding F_v fragment 7E2	Purification of the intact F_v fragment Purification of the complex with cytochrome c oxidase (4 subunits) Detection of the antigen via the F_v fragment by immunoelectron microscopy and immuno-fluorescence microscopy	Ostermeier *et al.*, 1995 Kleymann *et al.*, 1995c Kleymann *et al.*, 1995b Ribrioux *et al.*, 1996
V_H domains of the ubiquinol-cytochrome c oxidoreductase-binding F_v fragments 7D3 and 2D6	Purification of the intact F_v fragment Purification of the complex with ubiquinol-cytochrome c oxidoreductase (3 subunits) Detection of the antigen via the F_v fragment by immunoelectron microscopy and immuno-fluorescence microscopy	Kleymann *et al.*, 1995b Kleymann *et al.*, 1995a Kleymann *et al.*, 1995b Kleymann *et al.*, 1995c Ribrioux *et al.*, 1996
V_H domain of the photosystem I-binding F_v fragment 1E7	Purification of the intact F_v fragment Purification of the complex with photosystem I (3 subunits) Detection of the antigen via the F_v fragment by Western blotting and immunoelectron microscopy	Tsiotis *et al.*, 1995

Table 3 Continued

Recombinant protein[a]	Use of the Strep-tag	Reference
Oat phytochrome A	Purification of the functional apo-protein as well as complexes with phytochromobilin and similar ligands (expressed in yeast)	Murphy and Lagarias, 1997
Cyanobcterial phytochrome Cph1	Purification of the active, light regulated histidine kinase	Yeh *et al.*, 1997
Akaline phosphatase (*E. coli*)	Purification of the active metalloenzyme	Hengsakul and Cass, 1997 Voss and Skerra, 1997
Glutathione S-transferase and fusion proteins	Purification of the dimeric fusion enzyme Detection in ELISA	Tudyka and Skerra, 1997
Urease subunit B (*H. pylorii*)	Purification of the heterodimeric protein Detection by Western blotting	this contribtion
Curli assembly factor CsgG (*E. coli* outer membrane)	Purification of the native membrane protein	Loferer *et al.*, 1997
Green fluorescent protein (*A. victoria*)	Purification of the functional protein Detection by Western blotting	T.G.M. Schmidt, unpublished

[a]If not otherwise stated the proteins were produced in *E. coli.*

other components are rapidly removed after short washing (Figure 4). Selective elution of the *Strep*-tag fusion protein is then achieved with a low concentration of biotin or, for repeated use of the column, some derivatives of it in the washing buffer. Thus, the method benefits from two steps which confer specificity. This explains the high purification efficiency, permitting in most cases the isolation of a homogeneous protein in a single step.

When working with crude extracts of *E. coli* the biotin carboxyl carrier protein (Sutton *et al.*, 1977) is the only other protein that might be bound to the column, but this step is irreversible owing to the tight interaction between its prosthetic group and streptavidin. This complex formation can however be avoided by incubating the lysate with a small amount of egg-white avidin, which masks both protein-bound and free biotin (the biotin content in an *E. coli* cell extract, which is prepared from a culture in 1 l LB-medium with $OD_{550} = 1$, is in the range of 1 nmol).

A point of practical relevance which should be considered is the choice of the elution reagent. When purification is performed with wild-type streptavidin a solution of 2.5 or 5 mM diaminobiotin has proven to be advantageous over the use of other biotin derivatives that bind to streptavidin in a reversible manner, like iminobiotin or lipoic acid (Schmidt and Skerra, 1993; 1994b). Alternatively, 1 mM dimethyl-hydroxyazobenzoic acid (DMHABA, Weber *et al.*, 1995), a red coloured organic dye, can be used (Figure 4), which makes it possible that both the elution step and the regeneration can be visually followed.

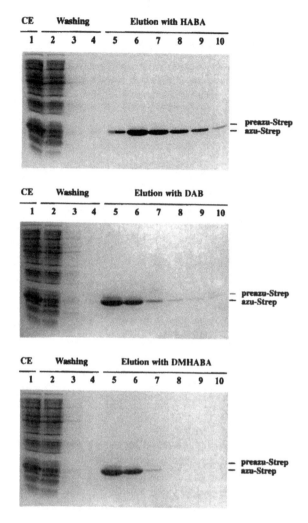

Figure 4   Streptavidin affinity purification of *P. aeruginosa* azurin carrying the *Strep*-tag and influence of different competitors for elution. The azurin gene was fused with the *OmpA* signal sequence and cloned on pASK-IBA1. After expression, the fusion protein was isolated from the soluble part of the total *E. coli* cell extract in the following manner. 1 ml crude extract, which was prepared from a 50 ml *E. coli* culture (OD_{550} = 1.4), was applied to a 1 ml column with immobilized recombinant core streptavidin (5 mg/ml). After washing the column with 3.5 ml of 100 mM Tris/HCl pH 8.0 elution of the bound protein was effected with 4 ml solutions of 5 mM HABA, diaminobiotin (DAB), or DMHABA, respectively, in the same buffer. 0.5 ml fractions were collected during the washing (W1–W7) and elution (E1–E8) steps. 20 μl aliquots from the fractions were applied to a 15% SDS polyacrylamide gel, which was stained with Coomassie brilliant blue. Lane 1: crude extract with the fusion protein; lane 2: pooled fractions W1–W3; lane 3: pooled fractions W4–W6; lane 4: fraction W7; lanes 5 to 10: fractions E3 to E8, respectively. Small amounts of preazurin, which has the signal peptide still attached, were copurified in this case. This can, however, be prevented when the purification is performed from the periplasmic cell extract. Washing away of the host cell proteins was already complete after the addition of just 3 ml washing buffer. Contrasting with the use of HABA for elution DMHABA or diaminobiotin yielded more concentrated solutions of the purified protein. The overall yield of the 15 kDa azurin/*Strep*-tag fusion protein was between 0.6 and 0.7 mg in all cases.

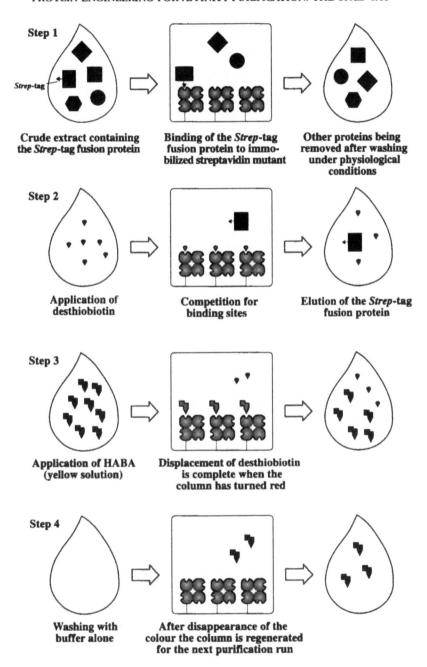

Figure 5 Schematic illustration of the purification of a *Strep*-tag fusion protein with one of the new streptavidin mutants. Desthiobiotin is used for quick competitive elution of the bound recombinant protein. Regeneration of the column can be shortened by using the commercially available dye hydroxyazobenzoic acid (HABA). At pH 8.0 a 1 mM solution of HABA is coloured yellow. Upon binding to streptavidin the colour shifts to red, due to the formation of the hydrazone isomer. Thus, the presence of HABA in the regeneration buffer provides a visual indicator of the column status and HABA competes with desthiobiotin for free binding sites at the same time so that the removal of desthiobiotin is accelerated. Finally, HABA itself is displaced in a brief washing step and the column is ready for next use.

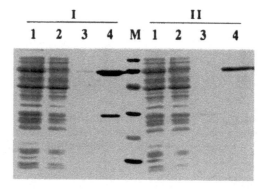

Figure 6 Purification of the *H. pylorii* urease, which is composed of the subunits *Ure*A (26.5 kDa) and *Ure*B (62 kDa). The two structural genes were arranged on pASK-IBA1CA as a dicistronic operon with the *Strep*-tag fused to *ure*B. After cytosolic expression in *E. coli* and preparation of the crude extract streptavidin affinity chromatography was performed with different buffers, I: 100 mM Tris/HCl pH 8.0, and II: 100 mM NaHCO$_3$, 100 mM NaCl, pH 8.1. Samples from different steps of the chromatography were subjected to 15 % SDS-PAGE, followed by staining with Coomassie brilliant blue. Lane 1: crude extract; lane 2: flow through; lane 3: washing fraction; lane 4: elution with 3 mM diaminobiotin. M denotes the molecular size standard (97.4 kDa, 66.2 kDa, 45 kDa, 31 kDa, 21.5 kDa, and 14.4 kDa). While *Ure*A remained associated with *Ure*B in buffer I, so that the intact heterodimer was isolated, it was quickly washed off in the presence of buffer II. This demonstrates that the choice of the buffer may be critical if the purification of proteins with non-covalently associated subunits is intended

In the case of the streptavidin mutants the competitive strength of diaminobiotin is not sufficient for a quick release of the bound *Strep*-tag fusion protein so that desthiobiotin (2.5 mM), which has a higher affinity for streptavidin (Green, 1975), is preferentially used (Voss and Skerra, 1997). This behaviour reflects the tighter interaction between the peptides and the engineered versions of streptavidin. In comparison with diaminobiotin, desthiobiotin has the additional advantage that it is less expensive and chemically more stable so that its use adds to the safe and facile application of the *Strep*-tag protein purification technique (Figure 5).

CONCLUSIONS

In summary, the engineering of the *Strep*-tag and the step-wise improvement of the other necessary components has provided a method which is of general assistance in recombinant protein chemistry. The specific adsorption/desorption events during affinity chromatography, together with the almost ideal biochemical properties of streptavidin, have enabled the efficient purification of many recombinant proteins from crude extracts. Neither high salt concentrations, nor extreme pH, elution gradients, chelator compounds, or a reducing milieu are required. The gentle conditions permit even the purification of non-covalently associated protein complexes which have the *Strep*-tag attached to one of their subunits only. This was first shown for bacterially produced F$_v$ fragments of antibodies (Schmidt and Skerra, 1993, 1994b; Kleymann *et al.*, 1995a,b,c; Ostermeier *et al.*, 1995a,b; Tsiotis *et al.*,

1995). It was also demonstrated for the urease from *Helicobacter pylorii*, which is similarly composed of two different polypeptide chains (Figure 6). Furthermore, F_v fragments fused with the *Strep*-tag were successfully employed for antigen capture and purification of the entire complex (Schmidt and Skerra, 1993). This strategy was particularly useful in the rapid isolation of solubilized membrane proteins (Kleymann *et al.*, 1995c; Tsiotis *et al.*, 1995). Membrane proteins which are directly fused with the *Strep*-tag can be purified as well (Loferer *et al.*, 1997). Thus, the *Strep*-tag saves time and effort even in areas where protein purification has traditionally been regarded a laborious business. It is probably just a matter of a few years until the *Strep*-tag methodology will be extended to applications beyond the research laboratory, possibly even to the production of commercially valuable proteins at an industrial scale.

REFERENCES

Bayer, E.A., Ben-Hur, H., Hiller, Y. and Wilchek, M. (1989) Postsecretory modifications of streptavidin. *Biochem. J.*, **259**, 369–376.

Bayer, E.A., Ben-Hur, H. and Wilchek, M. (1990) Isolation and properties of streptavidin. *Methods Enzymol.*, **184**, 80–89.

Bayer, E.A. and Wilchek, M. (1990) Avidin-biotin technology. *Meth. Enzymol.*, **184**.

Chaiet, L. and Wolf, F.J. (1964) The properties of streptavidin, a biotin-binding protein produced by *Streptomycetes*. *Arch. Biochem. Biophys.*, **106**, 1–5.

Chait, B.T. (1994) Mass spectrometry — a useful tool for the protein X-ray crystallographer and NMR spectroscopist. *Structure*, **2**, 465–467.

Devlin, J.J., Panganiban, L.C. and Devlin, P.E. (1990) Random peptide libraries: A source of specific protein binding molecules. *Science*, **249**, 404–406.

Evan, G.I., Lewis, G.K., Ramsay, G. and Bishop, J.M. (1985) Isolation of monoclonal-antibodies specific for human c-myc proto-oncogene product. *Mol. Cell. Biol.*, **5**, 3610–3616.

Frank, R. (1992) Spot synthesis: An easy technique for the positionally addressable, parallel chemical synthesis on a membrane support. *Tetrahedron*, **48**, 9217–9232.

Green, N.M. (1975) Avidin. *Adv. Protein Chem.*, **29**, 85–133.

Green, N.M. (1990) Avidin and streptavidin. *Methods Enzymol.*, **184**, 51–67.

Heller, S., Finn, T.P., Huber, J., Nishi, R., Geissen, M., Püschel, A.W. and Rohrer, H. (1995) Analysis of function and expression of the chicken GPA receptor (GPARa) suggests multiple roles in neuronal development. *Development*, **121**, 2681–2693.

Hengsakul, M. and Cass, A.E.G. (1997) Alkaline phosphatase-Strep-tag fusion protein binding to streptavidin: Resonant mirror studies. *J. Mol. Biol.*, **266**, 621–632.

Hochuli, E., Bannwarth, W., Döbeli, H., Gentz, R. and Stüber, D. (1988) Genetic approach to facilitate purification of recombinant proteins with a novel metal chelate adsorbent. *Bio/Technology*, **6**, 1321–1325.

Hopp, T.P., Prickett, K.S., Price, V.L., Libby, R.T., March, C.J., Ceretti, D.P., Urdal, D.L. and Conlon, P.J. (1988) A short polypeptide marker sequence useful for recombinant protein identification and purification. *Bio/Technology*, **6**, 1204–1210.

Kim, J.-S. and Raines, R.T. (1993) Ribonuclease S-peptide as a carrier in fusion proteins. *Protein Sci.*, **2**, 348–356.

Kleymann, G., Iwata, S., Wiesmüller, H.-H., Ludwig, B. and Michel, H. (1995a) Immunoelectronic microscopy and epitope mapping with monoclonal antibodies suggests the existence of an additional N-terminal transmembrane helix in the cytochrome b subunit of bacterial ubiquinol:cytochrome-c oxidoreductases. *Eur. J. Biochem.*, **230**, 359–363.

Kleymann, G., Ostermeier, C., Heitmann, K., Haase, W. and Michel, H. (1995b) Use of antibody fragments (Fv) in immunochemistry. *J. Histochem. Cytochem.*, **43**, 607–614.

Kleymann, G., Ostermeier, C., Ludwig, B., Skerra, A. and Michel, H. (1995c) Engineered Fv fragments as a tool for the one-step purification of integral multisubunit membrane protein complexes. *Bio/Technology*, **13**, 155–160.

Kraulis, P.J. (1991) MOLSCRIPT: a program to produce both detailed and schematic plots of protein structures. *J. Appl. Cryst.* 24, 946–950.

Lam, K.S., Salmon, E.S., Hersh, E.M., Hruby, V.J., Kazmierski, W.M. and Knapp, R.J. (1991) A new type of synthetic peptide library for identifying ligand-binding activity. *Nature*, 354, 82–84.

Loferer, H., Hammar, M. and Normark, S. (1997) Availability of the fiber subunit CsgA and the nucleator protein CsgB during assembly of fibronectin-binding *curli* is limited by the intracellular concentration of the novel lipoprotein CsgG. *Mol. Microbiol.*, 26, 11–13.

Müller, H.N. and Skerra, A. (1994) Grafting of a high-affinity Zn(II)-binding site on the β-barrel of retinol-binding protein results in enhanced folding stability and enables simplified purification. *Biochemistry*, 33, 14126–14135.

Munro, S. and Pelham, H.R.B. (1986) An Hsp70-like protein in the ER: Identity with the 78 kd glucose-regulated protein and immunoglobulin heavy chain binding protein. *Cell*, 46, 291–300.

Murphy, J.T. and Lagarias, J.C. (1997) Purification and characterization of recombinant affinity peptide-tagged oat phytochrome A. *Photochem. Photobiol.*, 65, 750–758.

Ostermeier, C., Essen, L.-O. and Michel, H. (1995a) Crystals of an antibody Fv fragment against an integral membrane protein diffracting to 1.28 Å resolution. *Proteins: Struct., Funct., Genet.*, 21, 74–77.

Ostermeier, C., Iwata, S., Ludwig, B. and Michel, H. (1995b) Fv fragment-mediated crystallization of the membrane protein bacterial cytochrome c oxidase. *Nature Struct. Biol.*, 2, 842–846.

Pähler, A., Hendrickson, W.A., Gawinowicz Kolks, M.A., Argarana, C.E. and Cantor, C.R. (1987) Characterization and crystallization of core streptavidin. *J. Biol. Chem.*, 262, 13933–13937.

Ribrioux, S., Kleymann, G., Haase, W., Heitmann, K., Ostermeier, C. and Michel, H. (1996) Use of nanogold- and fluorescent-labeled antibody Fv fragments in immunochemistry. *J. Histochem. Cytochem.*, 44, 207–213.

Schiweck, W. and Skerra, A. (1995) Fermenter production of an artificial Fab fragment, rationally designed for the antigen cystatin, and its optimized crystallization through constant domain shuffling. *Proteins: Struct., Funct., Genet.*, 23, 561–565.

Schiweck, W., Buxbaum, B., Schätzlein, C., Neiss, H. G. and Skerra, A. (1997) Sequence analysis and bacterial production of the anti-c-*myc* antibody 9E10: the V_H domain has an extended CDR-H3 and exhibits unusual solubility. *FEBS Lett.*, 414, 33–38.

Schmidt, F.S. and Skerra, A. (1994a) The bilin-binding protein of *Pieris brassicae*: cDNA sequence and regulation of expression reveal distinct features of this pigment protein. *Eur. J. Biochem.*, 219, 855–863.

Schmidt, T.G.M. and Skerra, A. (1993) The random peptide library-assisted engineering of a C-terminal affinity peptide, useful for the detection and purification of a functional Ig F_v fragment. *Protein Eng.*, 6, 109–122.

Schmidt, T.G.M. and Skerra, A. (1994b) One-step affinity purification of bacterially produced proteins by means of the "Strep tag" and immobilized recombinant core streptavidin. *J. Chromatogr. A*, 676, 337–345.

Schmidt, T.G.M., Koepke, J., Frank, R. and Skerra, A. (1996) Molecular interaction between the *Strep*-tag affinity peptide and its cognate target streptavidin. *J. Mol. Biol.*, 255, 753–766.

Skerra, A. (1993) Bacterial expression of immunoglobulin fragments. *Curr. Opin. Immunol.*, 5, 256–262.

Skerra, A. (1994) Use of the tetracycline promoter for the tightly regulated production of a murine antibody fragment in *Escherichia coli*. *Gene*, 151, 131–135.

Stofko-Hahn, R.E., Carr, D.W. and Scott, J.D. (1992) A single step purification for recombinant proteins. *FEBS Lett.*, 302, 274–278.

Studier, F.W., Rosenberg, A.H., Dunn, J.J. and Dubendorff, J.W. (1990) Use of the T7 RNA polymerase to direct expression of cloned genes. *Methods Enzymol.*, 185, 60–89.

Sutton, M.R., Fall, R.R., Nervi, A.M., Alberts, A.W., Vagelos, P.R. and Bradshaw, R.A. (1977) Amino acid sequence of *Escherichia coli* biotin carboxyl carrier protein (9100). *J. Biol. Chem.*, 252, 3934–3940.

Tsiotis, G., Haase, W., Engel, A. and Michel, H. (1995) Isolation and structural characterization of trimeric cyanobacterial photosystem I complex with the help of recombinant antibody fragments. *Eur. J. Biochem.*, 231, 823–830.

Tudyka, T. and Skerra, A. (1997) Glutathione S-transferase can be used as a C-terminal, enzymatically active dimerization module for a recombinant protease inhibitor, and functionally secreted into the periplasm of *Escherichia coli*. *Protein Sci.*, 6, 2180–2187.

Voss, S. and Skerra, A. (1997) Mutagenesis of a flexible loop in streptavidin leads to higher affinity for the *Strep*-tag II peptide and improved performance in recombinant protein purification. *Protein Eng.*, 10, 975–982.

Weber, P.C., Ohlendorf, D.H., Wendoloski, J.J. and Salemme, F.R. (1989) Structural origins of high-affinity biotin binding to streptavidin. *Science*, **243**, 85–88.

Weber, P.C., Pantoliano, M.W. and Thompson, L.D. (1992a) Crystal structure and ligand-binding studies of a screened peptide complexed with streptavidin. *Biochemistry*, **31**, 9350–9354.

Weber, P.C., Wendoloski, J.J., Pantoliano, M.W. and Salemme, F.R. (1992b) Crystallographic and thermodynamic comparison of natural and synthetic ligands bound to streptavidin. *J. Am. Chem. Soc.*, **114**, 3197–3200.

Weber, P.C., Pantoliano, M.W. and Salemme, F.R. (1995) Crystallographic and thermodynamic comparison of structurally diverse molecules binding to streptavidin. *Acta Cryst.*, **D51**, 590–596.

Yeh, K.-C., Wu, S.-H., Murphy, J. T. and Lagarias, J. C. (1997) A cyanobacterial phytochrome two-component light sensory system. *Science*, **277**, 1505–1508.

4. *CANDIDA RUGOSA* LIPASES: FROM MOLECULAR EVOLUTION ANALYSIS TO THE DESIGN OF A SYNTHETIC GENE

MARINA LOTTI and LILIA ALBERGHINA

Dipartimento Biotechnologie e Bioscienze, Università Degli Studi Milano-Bicocca, via Emanueli 12, 20126 Milano, Italy

INTRODUCTION

The family of lipase enzymes (CRLs) synthesized by the imperfect fungus *Candida rugosa* provides issues of interest for protein engineers because of a few unique features. First of all, *C. rugosa* lipases are members of a large set of proteins closely related in both sequence and function. In an evolutionary perspective, one can consider CRL proteins as the result of a successful process of mutagenesis where a defined group of amino acid changes have been positively selected since they gave rise to functional variants. The study of lipase sequences therefore may provide a good starting point towards the design of functional mutants. On the other hand, *C. rugosa* belongs to a subgroup of *Candida* species which, using a non-conventional decoding of one serine codon, requires the development of novel approaches for the heterologous expression of its genes. Information arising may therefore be of interest for researchers using non conventional expression systems.

Lipases are hydrolytic enzymes of growing potential in industrial processes, especially because of their unique versatility in the use of substrates and in the reactions performed. They are classified as acyl glycerol ester hydrolases (EC 3.1.1.3) their main activity being the hydrolysis of ester bonds in a variety of insoluble lipid substrates. In low water conditions lipases are effective catalysts of synthetic reactions. These properties make them suitable for use in both hydrolysis of fats and triglycerides and in reactions of transesterification were fatty acid chains may be exchanged giving rise to new combinations. Lipases are often employed for asymmetric synthesis based on prochiral or racemic substrates since they are catalytically active in water, in mixtures of water and organic solvents and in organic solvents. Present and potential applications of lipase-based biotransformations have been extensively reviewed elsewhere (Vulfson, 1994; *Methods in Enzymology*, Vol 286, 1998).

Most lipases have molecular masses of about 30–40 kDa, but some microbial enzymes are larger, with Mr up to 60 kDa. Microbial lipases are single domain enzymes, whereas lipases from animals may be more complex in structure related to their interaction with other proteins/substances/cofactors within the source organism (van Tilbeurgh *et al.*, 1992). Despite a somehow surprising variability in their amino acid sequences, all known lipases have a similar molecular architecture consisting of a central hydrophobic β-sheet packed between amphiphilic α-helices. The active site is composed of a triad of serine, histidine and a carboxylic acid (aspartic or glutammic acid), reminiscent of those found in serine proteases but with a different orientation (Ollis *et al.*, 1992). To display their full activity, most lipases require the presence of an oil-water interface, a phenomenon known as "interfacial activation". Accordingly, in most lipases the catalytic centre is accommodated in a crevice or in a depression and is sheltered from the solvent by a "lid" formed by surface helices or loops. In such cases lipase activation has been ascribed to conformational changes in the enzymes upon their adsorption to the interface. Due to the discovery of lipases which are not subjected to interfacial activation, or have water accessible active sites or do not show any clear correlation between these two features, neither substrate-mediated activation nor the presence of the lid are longer considered as appropriate criteria to define lipases, that are now generally referred to as carboxylesterases specific for long chain substrates (Verger, 1997).

Several parameters related to the physicochemical state of the lipid substrate such as surface pressure and "interfacial quality", are known to affect the catalytic behavior of lipases in terms of both activity and selectivity (Rogalska et al., 1993). This flexibility modulated by the reaction environment, together with the already mentioned variability observed in lipase proteins, provide researchers with a large — and to a considerable extent still unexploited — pool of candidate biocatalysts. As a consequence, a big deal of research is being devoted to the development of lipase-based reactions (see as a reference K. Drauz and H. Waldman, 1995). Some dozen of lipases are now commercially available; nevertheless those employed in large-scale industrial processes and products are still limited to a few cases, mainly due to the high price/low availability or non-optimal operational features of the naturally available enzymes. Therefore, perspectives in the use of lipases as industrial catalysts strongly rely on the production of recombinant enzymes with biochemical and catalytic features improved by protein engineering methods.

FUNGAL LIPASES

The search for novel biocatalysts led in recent years to the purification and characterization of dozens of lipases from the most diverse origin including animals, plants, bacteria and fungi. These enzymes are namely widespread in nature were they are involved in the hydrolysis of neutral lipids to generate metabolic energy, whereas in pathogenic bacteria lipases are primary factors of virulence (Jaeger et al., 1994). A special attention has been paid to enzymes of easy and stable working conditions that possibly should be also easily purified from the natural source. In this spirit, the extracellular lipases secreted by several fungi are considered to be good candidates for use in biotransformation. Several out of them are secreted by Ascomycetes and Zygomycetes fungi and have been thoroughly characterized and eventually produced through recombinant DNA technologies. Some three-dimensional structures are also available (Table 1).

Taken as a whole, lipases produced by fungi are not related more closely to each other than to other lipases from completely different origin and this variability concerns biochemical properties, aminoacid sequence and catalytic specificity. Attempts to group lipases in families have been proposed basing on both sequence or structural similarities. A structure-based classification (SCOP, Murzin et al., 1995) divides fungal lipases into two groups, carboxylesterases enclosing the C. rugosa (CRLs) and G. candidum (GCLs) lipases and triacylglycerol lipases enclosing the

Table 1 Examples of fungal lipases characterized at the molecular and structural level

			ASCOMYCOTA			
ARCHAEASCOMYCETES Organism	Mr	aa	localization	3D	isoenzymes	ref
Schizosaccharomyces pombe	n.d.	452	?	–		1
HEMIASCOMYCETES Organism	Mr	aa	localization	3D	isoenzymes	ref
Candida rugosa	60 kDa	534	extracellular	+	>5	2
Candida antarctica	33 kDa	342	"	+	2	3
Candida albicans	38 kDa	351	?	–	?	4
Yarrowia lipolytica	55 kDa	486	?	–	?	5
Geotrichum candidum	60 kDa	544	extracellular	+	2	6
Saccharomyces cerevisiae	63 kDa	548	?	–	n.d.	7
EUASCOMYCETES Organism	Mr	aa	localization	3D	isoenzymes	ref
Penicillium camembertii	33 kDa	276	extracellular	+	n.d.	8
Aspergillus oryzae		306	?	–	n.d.	9
Aspergillus nidulans		286	?	+	n.d.	10
Fusarium heterosporum	32 kDa	301	+	–	n.d.	11
Humicola lanuginosa		270	+	–	n.d.	12
ZYGOMYCOTA Organism	Mr	aa	localization	3D	isoenzymes	ref
Rhizomucor miehei	27 kDa	269	extracellular	+	+	13
Rhizopus oryzae	29 kDa	269	"	+	n.d.	14

[1] Devlin K. and Churcher, C.M., unpublished [2] Kawaguchi et al., 1989; Longhi et al., 1992; Lotti et al., 1993; Grochulski et al., 1993 and 1994. [3] Uppenberg et al., 1994. [4] Fu et al., 1997. [5] Gonzales and Dominguez, in preparation. [6] Shimada et al., 1989 and 1990; Bertolini et al., 1994; Phillips et al., 1995; Schrag et al., 1991. [7] Abraham et al., 1992. [8] Yamaguchi et al., 1991. [9] Tsuchiya et al., 1996. [10] Brown et al., 1996. [11] Nagao et al., 1994. [12] Derewenda et al., 1994. [13] Boel et al., 1988. [14] Haas et al., 1991

enzymes produced by Mucorales and similar fungi. Recently, novel sequences become available in databases allowing to implement the accuracy of this classification (Figure 1). One subfamily encloses the enzymes from the Hemiascomycetes *C. rugosa* and *Geotrichum candidum* (GCL) both secreting lipase isoforms which have been shown by Cygler and colleagues (1993) to belong to a family of lipases, esterases and other related proteins distinct from the other known lipases. Recently, another lipase isolated from the yeast *Yarrowia lipolytica* (YARLIP) has been proposed to belong to this group, based on a 30% identity with both CRL and GCL (F. Gonzales and A. Dominguez, manuscript in preparation). Another two enzymes synthesized by other species of *Candida* can be now taken into consideration. *C. antarctica* produces a lipase of 342 residues (CANAR) that has been already industrially developed as a recombinant enzyme for detergent formulation (Uppenberg et al.,

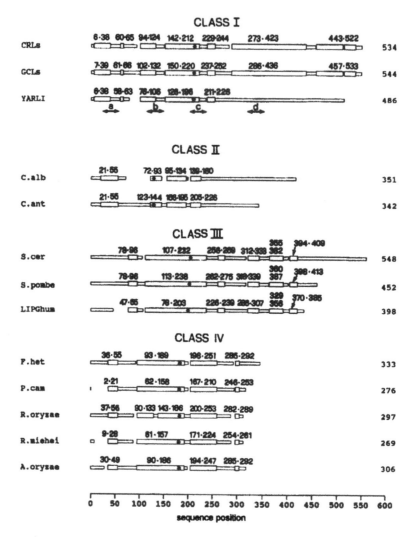

Figure 1 Homology blocks in fungal lipases families. Sequences available in databases were first aligned by CLUSTALW (Higgins *et al.*, 1994) and divided in four classes according to sequence similarity. Stretches of residues conserved within each subfamily were extracted by MACAW (Schuler *et al.*, 1991) **Class I**: CRLs represents the consensus sequence for five *C. rugosa* isoenzymes (AC: P20261; P32946; P32947; P32948; P32942); GCLs the consensus sequence for 9 lipases from *Geotrichum* strains (AB000260; S41096; S41095; S41093; S41092; S41091; S41090; P22394; P17573); YARLI is the *Yarrowia lipolytica* lipase (Q99156). Lines marked with **a**, **b** and **c**, represent regions of homology to the *S. cerevisiae* lipase whose aminoacids 96–109 (**a**), 130–152 (**b**); and 194–204 (**c**) can be aligned with the corresponding regions of CRLs. Line **d** is a region of homology between CRLs and the *C. albicans* lipase (residues 381–400 and 304–320, respectively). **Class II**. CALB = *C. albicans* lipase (U34807); CANAR = *C. antarctica* lipase B (P41365). **Class III**: S. cer = *S. cerevisiae* lipase (P34163); S. pombe = putative lipase from *Schizosaccharomyces pombe* (Z98887); LIPGhum represents the consensus of 3 mammalian lipases: human gastric lipase (P07098), rat gastric lipase (P04634) and human lysosomal lipase (P38571). **Class IV**: F. het = *Fusarium heterosporum* (Q02351); P. cam = Penicillium camembertii (P25234); R. oryzae = *Rhizopus oryzae* (P21811); R. miehei = *Rhizomucor miehei* lipase (P19515); A. oryzae = *Aspergillus oryzae* (P78583). Boxes enclosing the catalytic serine are marked by a star.

1994). Even the pathogen *C. albicans* has been recently found to possess an endogenous lipase built up of 351 aminoacids (CALB). Since this enzyme has been cloned by functional complementation and never purified from the producing strain, its cellular localization is not clear, although a stretch of hydrophobic amino acids at its amino terminus is consistent with the presence of a signal for secretion (Fu *et al.*, 1997). Surprisingly, the two latter proteins are only very distantly related to both the previous group and to each other, neither significant homology to other lipolytic enzymes is evidenced by algorithms for sequence similarity search such as FASTA and BLAST (Pearson and Lipman, 1988; Altschul *et al.*, 1990). Nevertheless, local similarities relating CALB and CANAR can be detected using the MACAW program (Schuler *et al.*, 1991). A more detailed analysis of multiple alignments within this family is reported in the following paragraph.

Lipases from the budding yeast *Saccharomyces cerevisiae* (Hemyascomycetes) and the fission yeast *Schizosaccharomyces pombe* (Archiascomycetes) share 50% identity in 362 residues overlap and in addition display an astonishing homology (33–34%) to three mammalian lipases: a lysosomal acid human lipase, a gastric human lipase and a gastric lipase from rat. This similarity excludes the C-terminal part of the yeast protein that seems to build an additional domain. Short stretches of aminoacids can be aligned with the corresponding regions of CRL and CANAR, as reported below.

A remarkable homology in both sequence and structure relates also a fourth group of fungal lipases composed by enzymes from *Rhizopus, Rhizomucor, Humicola, Penicillium, Aspergillus* with pairwise identities up to 35–55% (Derewenda *et al.*, 1994). To this family a further enzyme synthesized by *Fusarium heterosporum* can be added, with 40% aminoacid identity to the *Penicillium* and *Aspergillus* enzymes (Table 1).

In conclusion, despite in some cases it is possible to relate sequence homology with phylogeny, the availability of more and more sequences obtained from direct cloning or from the genome sequencing projects, is ruling out a simple equation between evolutionary relationships and fungal lipase sequences. Clear examples of this are the lipases from *Candida* species or from *S. cerevisiae*, which displays a more remarkable similarity to mammalian lipases than to most fungal enzymes. Thus the pattern followed by lipase sequences during evolution still need to be unraveled and even close evolutionary relations cannot be used in the absence of other information, as an analytical tool, for example, in designing primers for gene cloning.

A Closer Look: Sequence Analysis of CRLs and Related Lipases

The family of lipases composed by CRLs, GCLs and the *Yarrowia* lipase at present encloses at least 15 sequences. Available CRL sequences correspond to isoenzymes all derived from the strain ATCC14830, whereas GCLs are produced by different strains each of them can synthesize 1 or 2 protein isoforms. Whether *Yarrowia* does also produce multiple lipases has been not yet investigated. Identity is very high among homogeneous groups of isoenzymes (70–90%), about 40% between CRLs and GCLs and about 30% between the *Yarrowia* and the other two fungal lipases. A detailed comparison between CRLs, GCLs and other related esterases has been reported previously (Cygler *et al.*, 1993). This work pointed out to a high degree of homology in the N-terminal half of this family of proteins. In the following, we will extend sequence analysis to lipases characterized in the subsequent years but excluding non-lipase enzyme.

Alignment of all available CRLs and GCLs detects high identity through the whole length of the sequences indicating a very close relationship (Figure 1). Introducing the *Yarrowia* lipase lowers the overall similarity which become restricted to a highly conserved core encompassing the central to N-terminal regions of the sequences. Taking as the reference the CRLs numbering, this region extends from Leu6 to Arg 324. Compact blocks of conservation can be identified in the stretch 23–44 which is also one of the few regions of homology to the *Saccharomyces cerevisiae* enzyme. Conservation is lower in the correspondence of the lid regions with the exception of the pattern CxQxxP and the N-terminal edge of this loop (60–65). A large deletion is observed in this part of the *Yarrowia* sequence. A very high sequence conservation encompasses the large block extending from E95 to S214, where the aminoacids from 114 to 142 can be aligned also with the corresponding region of the *S. cerevisiae* lipase and, what can be more surprising, of the human hormone-sensitive lipase. This latter homology among unrelated enzymes could be recently detected by a methodology based on the alignment of 2D structure elements (Contreras *et al.*, 1996). Attempts to locate blocks conserved also in the other two *Candida* lipases failed with the notable exceptions of the aminoacids surrounding the active serine and of the region from Ala 380 to Thr 400 which is maintained in CRLs, GCLs, and the *C. antarctica* lipases but is completely missing in the *Yarrowia* sequence. Interestingly the two enzymes from *C. antarctica* and *C. albicans*, which appeared to be unrelated in the previous analysis, when scanned for local homologies reveal the presence of regions of conservation as shown in Figure 1.

REDUNDANCY OF LIPASE PROTEINS IN *C. RUGOSA*

C. rugosa is a major source of a commercial lipase produced by several enzyme companies and classified as non-specific with respect to the position of the fatty acid chain released from the glycerol molecule. CRL has a molecular mass of ca. 60 kDa being composed of 534 amino acids and moderately glycosylated. In recent years it was discovered that apparent inconsistencies in the catalytic properties described in different laboratories were due to the fact that CRL contains different lipase proteins whose composition and ratio may be different in preparations distributed by different producers, possibly depending on the strain or the composition of the growth medium (Chang *et al.*, 1994). Multiple lipase isoenzymes have been proposed to differ in biochemical properties and in some cases in substrate specificity (Rua *et al.*, 1993; Hernaiz *et al.*, 1995; Linko and Wu, 1996). The production of lipase isoforms is common also to other fungi, as it is the case of the mold *G. candidum* with two separate lipase genes (Shimada *et al.*, 1989 and 1990; Bertolini *et al.*, 1994). Interestingly, the lipase isoforms from G. candidum ATCC34614 have 86% identical primary structure and yet they differ as catalysts, being only isoform I specific for *cis* (D-9) unsaturated fatty acyl moieties in the ester substrates (Sidebottom *et al.*, 1991). What is exceptional in *C. rugosa* is the number of the encoding genes. In fact we cloned from strain ATCC 14830, 7 lipase sequences, five of which as whole coding sequences (Longhi *et al.*, 1992; Lotti *et al.*, 1993; Brocca *et al.*, 1995). Recently, Fu and colleagues reported about the presence in *C. albicans* of a family of lipase genes, one of them was cloned and characterized in detail and others were identified by Southern hybridization (Fu *et al.*, 1997). In a way the existence of very large gene families seems to be a recurrent feature within the genus *Candida*. Thus seven genes encoding secreted aspartic proteinases have been found in *C. albicans* (Monod *et al.*, 1994) whereas seven and nine genes respectively code for cytochromes P450 in *C. tropicalis* (Seghezzi *et al.*, 1992) and *C. maltosa* (Ohkuma *et al.*, 1991). Whatever their origin and physiological significance, the secretion by *C. rugosa* of multiple lipase enzymes is of interest for their application in biocatalysis, particularly in the case CRL proteins should turn out to possess slightly different catalytic properties, as it has been suggested by several authors working on subsets of CRL isoenzymes separated from the total pool by chromatographic approaches (Rua *et al.*, 1993; Hernaiz *et al.*, 1996; Linko and Wu, 1996). The uniqueness of CRL among lipases lies just in the availability of several similar but not identical enzymes. While initially at least some members of the gene family were thought to be silent (pseudogenes), it is now generally accepted that all lipase-encoding genes may be expressed by the fungus. Whether all enzymes are expressed at the same levels and are therefore equally represented in the CRL is not clear. We recently obtained indication that CRL-encoding genes are subjected to regulation at the level of transcription depending on the composition of the culture medium and on the physiology of culture growth (unpublished). This observation may add a further element of interest to the study of this complex family of enzymes that apparently provide *Candida* cells with a large and versatile pool of catalysts to be produced and used according to the composition of the environment. However, the significance of lipases for *Candida rugosa* growth is still far to be understood.

THE UNUSUAL GENETIC CODE IN *CANDIDA*: ENGINEERING CRL TOWARDS EXPRESSION IN YEAST CELLS

The heterologous expression of cloned genes in host cells is particularly desirable in the case of CRLs not only because of the well known advantages in making available large amounts of purified enzyme but also because this seems to be the only feasible approach to obtain pure CRL isoforms without making use of very sophisticated purification techniques as required to separate proteins which are extremely similar in their features.

For this reason, extensive work was devoted to the development of appropriate expression systems able to by-pass the hurdles posed by genes cloned from the so-called "CUG-serine *Candidas*" (Table 2). In fact *C. rugosa* belongs to a group of *Candida* species, recently identified as using the universal leucine codon CUG as the seventh triplet for serine (Kawaguchi *et al.*, 1989; Ohama *et al.*, 1993). The reassignment of CUG from leucine to serine in *Candida* sp. is so far the only known change concerning an aminoacid codon, whereas several instances of reassignment of stop codons have been reported both in mithocondral and nuclear genes (Osawa, 1995). This phenomenon may be disregarded during heterologous expression of most genes cloned from *Candida* sp. and in fact a number of genes isolated from *C. albicans* and from *C. maltosa* have been produced in a functional form in *Saccharomyces* as well as in other conventional expression systems (Mauersberger *et al.*, 1996; Kawachi *et al.*, 1996; Kargel *et al.*, 1996; Fu *et al.*, 1997; Smolenski *et al.*, 1997). This is possible due to the very low occurrence of the unusual codon in most *Candidas* where it accounts only for 1–2% of all serine codons and is often completely

Table 2 Candida species decoding CUG as serine

Species	ref
Candida parapsilosis	Ohama et al., 1993
Candida zeylanoides	"
Candida albicans	"
Candida cylindracea JMC1613	"
Candida melibiosica	"
Candida rugosa ATCC 14830	Kawaguchi et al., 1991
Candida maltosa	Sugiyama et al., 1995
Candida guillermondii	Ueda et al., 1994
Candida lusitaniae	"
Candida tropicalis	"
Candida wisvananthii	Pesole et al., 1995

neglected in highly expressed genes (Pesole et al., 1995). The introduction of a few wrong residues in the protein sequence may be tolerated if substitutions do not occur at positions essential for the function or correct folding of the proteins. When this is not the case, expression is hindered to different extents. As an example, the expression of C. maltosa cytochrome P450 genes in S. cerevisiae resulted in the production of active but unstable enzymes, unless CUG triplets were replaced by other serine codons (Zimmer and Schunck, 1995).

Candida rugosa is unique in the group of CUG-ser Candidas for its extreme bias for this codon. In the five genes coding for lipase, CUG is the major codon for serine and it is used for about 40% of all serine residues, in good agreement with the presence in the genome of multiple copies of seryl-tRNACAG genes instead of the single copy found in the genomes of other CUG-Candidas (Suzuki et al., 1994). Not only the frequency but also the positioning of serine coded by the anomalous codon at locations of crucial importance for the structure and function of lipase proteins — as for example their catalytic Ser209 — might suggest that C. rugosa was subjected to some kind of positive selection for the use of CUG that did not act on other species belonging to the same group (Alberghina and Lotti, 1997). Serine is inserted at 13 conserved CUG sites in the CRL proteins and at another few positions that differ from an isoenzyme to the other; thus it can be hardly imagined that expression in a universal-code host would not affect lipase function. How this dramatic change in codon usage may be tolerated and of which kind of advantage it provided C. rugosa cells, is still matter of speculation.

Expression in Saccharomyces cerevisiae of a CRL-encoding gene was attempted relying on the site directed mutagenesis of CTGs with universal codons for serine. In preliminary studies, the natural LIP1 gene was expressed under the control of the strong inducible GAL1-GAL10 promoter. Interestingly, we were not able to obtain any recombinant protein, unless the leader sequence for secretion was replaced with a yeast signal peptide, in spite of an effective synthesis of specific mRNA in both kind of constructs (Fusetti et al., 1996). This provided evidence about the role of the signal sequence in the expression and stabilization of foreign proteins within the host cells. It is very likely that this effect become dramatically evident because of an intrisinc instability of the recombinant protein due to the many leucine residues erroneously inserted by the expression system. In fact, a codon-optimized *LIP1* gene preceded by its own signal could be expressed, although in low amounts in *Pichia pastoris* cells (Brocca *et al.*, 1997). Since the *LIP1* gene contains 20 anomalous codons, 13 of which are conserved in the five CRL genes, we tried to define a hierarchy of the functionally and structurally most important Ser residues to be mutated. Based on the alignment with other homologous enzymes and on the analysis of the three-dimensional structure, we selected a first set of 8 triplets located at the most important structural positions. Replacement of the target codons to TCC codons resulted in high level of a recombinant inactive protein accumulated within the yeast cells, as we already observed during expression of the non-mutated gene. Defects in the secretion as well as the results of native electrophoresis suggested a misfolding of the recombinant peptide and revealed the existence of unpredicted constraints on the folding of this lipase (Fusetti *et al.*, 1996, Brocca *et al.*, 1997).

DESIGN OF A CRL SYNTHETIC GENE

In a completely different approach, a synthetic *LIP1* gene was obtained by the method of mutually priming long overlapping oligonucleotides (Ausubel *et al.*, 1994). Four segments of 400bps were sythesized separately as to replace all CTGs with universal serine codons and to optimize codon usage for expression in yeasts (Brocca *et al.*, 1997). The importance of taking into account also the codon bias of cells used as the host, can be better appreciated if one considers that the *C. rugosa* genome has an exceptional high content in G+C (63%) and is therefore expected to prefer codons carrying a purine at the third codon position. This strong bias may cause a shortage of the correspondent tRNAs in the host cells, hampering therefore heterologous expression. The synthetic gene was fused with the appropriate leader sequences for secretion and expressed in both *S. cerevisiae* and *Pichia pastoris* cells. Though an active recombinant protein was secreted in the culture medium by both organisms, *Pichia pastoris* was found to support a higher expression with the synthetic gene cloned under the control of the strong alcohol oxidase promoter. The recombinant protein was glycosylated, according to the presence of three potential N-glycosylation sites in the LIP1 sequence. When compared with the commercial lipase preparation, the recombinant LIP1 showed a comparable activity towards triacylglycerides different in acyl group chain length and various methyl esters differing in acyl group chain length (Brocca *et al.*, 1997).

Having available a CRL-encoding gene amenable of high level expression in conventional hosts provides us with a functional scaffold where to study the effect of amino acid changes on both the folding and the function of lipase enzymes. In a previous study, differences among CRL isoenzymes where identified which are localized in regions of interaction with the substrate molecules (Lotti *et al.*, 1994). Thus information generated from the comparative analysis of lipase isoforms provides the basis for trying to gain some insight in the patterns of lipase evolution and for the production by mutagenesis of functional mutants.

ACKNOWLEDGEMENTS

The work described in this paper was supported by the BRIDGE and BIOTECHNOLOGY programmes of the European Commission. The authors wish to thank Stefania Brocca, Fabrizia Fusetti, Rita Grandori and Sonia Longhi for their precious participation to this research. We are especially gratefull to Rolf D. Schmid and Claudia Schmidt-Dannert, University of Stuttgart, for their major contribution to work on the synthetic sequence.

REFERENCES

Abraham, P.R., Mulder, A., van't Riet, J., Planta, R.J. and Rave, H.A. (1992) Molecular cloning and physical analysis of an 8.2 Kb segment of chromosome XI of *Saccharomyces cerevisiae* reveals five tightly linked genes. *Yeast*, 8(3), 227–38.

Alberghina, L. and Lotti, M. (1997) The evolution of a non universal codon as detected in *Candida rugosa* lipase. *J. Mol. Cat. B: Enzymatic*, 3, 37–41.

Altschul, S.F., Gish, W., Miller, W., Myers, E.W. and Lipman, D.J. (1990). Basic local alignment search tool. *J. Mol. Biol.*, 215, 403–410.

Ausubel, F.M., Brent, R., Kingston, R.E., Moore, D.D., Seidman, J.G., Smith, J.A. and Struhl, K. (1994) Current Protocols in Molecular Biology Vol. 1. New York: Greene Publishing Associates and Wiley-Interscience.

Bertolini, M.C., Laramee, L., Thomas, D.Y., Cygler, M., Schrag, J.D. and Vernet T. (1994) Polymorphism in the lipase genes of *Geotrichum candidum* strains. *Eur. J. Biochem.*, 219, 119–125.

Boel, E., Huge-Jensen, B, Christensen, M., Thim, L. and Fiil, N.P. (1988) *Rhizomucor miehei* triglyceride lipase is synthesized as a precursor. *Lipids*, 23, 701–706.

Brocca, S., Grandori, R., Breviario, D. and Lotti, M. (1995) Localization of lipase genes on *Candida rugosa* chromosomes. *Curr. Genet.*, 28, 454–457.

Brocca, S., Schmidt-Dannert, C., Lotti, M., Alberghina, L. and Schmid, R.D. (1998) Design, total synthesis and functional overexpression of the *Candida rugosa lip1* gene coding for a major industrial lipase. *Protein Science*, 7, 1415–1422.

Brown, D.W, Yu, J.H., Kelkar, H.S., Fernandes, M., Nesbitt, T.C., Keller, N.P., Adams, T.H. and Leonard, T.J. (1996) Twenty-five coregulated transcripts define a sterigmatocystin gene cluster in *Aspergillus nidulans*. *Proc. Natl. Acad. Sci. USA*, 93(4), 1418–1422.

Chang, R.C., Chou, S.J. and Shaw, J.F. (1994) Multiple forms and functions of *Candida rugosa* lipase. *Biotechnol. Appl. Biochem.*, 19, 93–97.

Contreras, J.A. Karlsson, M., Osterlund, T., Laurell, H., Svensson, A. and Holm, C. (1996) Hormone-sensitive lipase is structurally related to acetylcholinesterase, bile salt-stimulated lipase and several fungal lipases. Building of a three-dimensional model for the catalytic domain of hormone-sensitive lipase. *J. Biol. Chem.*, 271(49), 31426–31430.

Cygler, M., Schrag, J.D., Sussman, J.L., Harel, M., Silman, I., Gentry, M.K. and Doctor, B.P. (1993) Relationship between sequence conservation and three-dimensional structure in a large family of esterases, lipases and related proteins. *Protein Science*, 2, 366–382.

Derewenda, U., Swenson, L., Green, R., Wei, Y., Dodson, G.G., Yamaguchi S., Haas, M.J. and Derewenda Z.S. (1994) An unusual buried polar cluster in a family of fungal lipases. *Structural biology*, 1, 36–47.

Drauz, K. and Waldmann, H. (eds) (1995) *Enzyme catalysis in organic synthesis. A comprehensive handbook.* Weinheim, Germany: VCH Verlagsgesellschaft mbH.

Fu, Y., Ibrahim, A.S., Fonzi, W., Zhou, X., Rmos, C.F. and Ghannoum, A. (1997) Cloning and characterization of a gene (LIP1) which encodes a lipase from the pathogenic yeast *Candida albicans*. *Microbiology*, 143, 331–340.

Fusetti, F., Brocca, S., Porro, D. and Lotti, M. (1996) Effect of the leader sequence on the expression of recombinant *C. rugosa* lipase by *S. cerevisiae* cells. *Biotechnol. Lett.*, 18(3), 281–286.

Grochulski, P. Li, Y., Schrag, J.D., Bouthillier, F., Smith, P., Harrison, D., Rubin, B. and Cygler, M. (1993) Insights into interfacial activation from an open structure of *Candida rugosa* lipase. *J. Biol. Chem.*, 268, 12843–12847.

Grochulski, P., Li, Y., Schrag, J.D. and Cygler, M. (1994) Two conformational states of *Candida rugosa* lipase. *Protein Science*, **3**, 82–91.

Haas, M.J., Allen, J. and Berka, T.R. (1991) Cloning, expression and characterization of a cDNA encoding a lipase from *Rhizopus delemar*. *Gene*, **109**, 107–113.

Hernaiz, M.J., Sanchez-Montero, J.M. and Sinisterra, J.V. (1995) Hydrolysis of (R,S)2-aryl propionic esters by pure lipase B from *Candida cylindracea*. *J. Mol. Cat. A: Chemical*, **96**, 317–327.

Higgins, D., Thompson, J., Gibson, T., Thompson, J.D., Higgins, D.G. and Gibson T.J. (1994). CLUSTAL W: improving the sensitivity of progressive multiple sequence alignment through sequence weighting, position-specific gap penalties and weight matrix choice. *Nuc. Acids Res.*, **22**, 4673–4680.

Jaeger, K.E., Ransac, S., Dijkstra, B.W., Colson, C., van Heuvel, M. and Misset, O., (1994) Bacterial lipases. *FEMS Microbiology Reviews*, **15**(1), 29–63.

Kargel, E., Menzel, R., Honeck, H., Vogel, F., Bohner, A. and Schunck, W.H. (1996) *Candida maltosa* NADPH-cytochrome P450 reductase: cloning of a full-length cDNA, heterologous expression in *Saccharomyces cerevisiae* and function of the N-terminal region for membrane anchoring and proliferation of the endoplasmic reticulum. *Yeast*, **12**(4), 333–348.

Kawachi, H., Atomi, H., Ueda M., Hashimoto, N., Kobayashi, K., Yoshida, T., Kamasawa, N., Osumi, M. and Tanaka, A. (1996) Individual expression of *Candida tropicalis* peroxisomal and mithocondrial carinitine acetyltransferase-encoding genes and subcellular localization of the products in *Saccharomyces cerevisiae*. *J. Biochem.*, **120**, 731–735.

Kawaguchi, Y., Honda, H., Taniguchi-Morimura, J. and Iwasaki, S. (1989) The codon CUG is read as serine in an asporogenic yeast *Candida cylindracea*. *Nature*, **341**, 164–166.

Linko, Y.Y. and Wu, X.Y. (1996) Biocatalytic production of useful esters by two forms of lipase from *Candida rugosa*. *J. Chem. Tech. Biotechnol.*, **65**, 163–170.

Longhi, S., Fusetti, F., Grandori, R., Lotti, M., Vanoni, M. and Alberghina, L. (1992) Cloning and nucleotide sequences of two lipase genes from *Candida cylindracea*. *Biochim. Biophys. Acta*, **1131**, 227–232.

Lotti, M., Grandori, R., Fusetti, F., Longhi, S., Brocca, S., Tramontano, A. and Alberghina, L. (1993) Cloning and analysis of *Candida cylindracea* lipase sequences. *Gene*, **125**, 45–55.

Lotti, M., Tramontano, A., Longhi, S., Fusetti, F., Brocca, S., Pizzi, E. and Alberghina, L. (1994) Variability within the *Candida rugosa* lipases family. *Prot. Eng.*, **7**, 531–535.

Mauersberger, S., Ohkuma, M., Schunck, W-H and Takagi, M. (1996) *Candida maltosa*. In *Non conventional yeasts in biotechnology*, edited by Klaus Wolf. Berlin, Heidelberg, Germany: Springer Verlag.

Methods in Enzymology Vol. 286. *Lipases, Part B: Enzyme characterization and utilization*, edited by B. Rubin and E.A. Dennis, 1998, Academic Press.

Monod, M., Togni, G., Hube, B. and Sanglard, D. (1994) Multiplicity of genes encoding secreted aspartic proteinases in *Candida* species. *Mol. Microbiol.*, **13**, 357–368.

Murzin, A.G., Brenner, S.E., Hubbard, T. and Chothia, C. (1995) Scop: a structural classification of proteins database for the investigation of sequences and structures. *J. Mol. Biol.*, **247**, 536–540.

Nagao, T., Shimada, Y., Sugihara, A. and Tominaga, Y. (1994) Cloning and nucleotide sequence of cDNA encoding a lipase from *Fusarium heterosporum*. *J. Biochem.*, **116**(3), 536–540.

Ohama, T., Suzuki, T., Mori, M., Osawa, S., Ueda, T., Watanabe, K. and Nakase, T. (1993) Non-universal decoding of the leucine codon CUG in several *Candida* species. *Nucl. Acid Res.*, **21**(17), 4039–4045.

Ohkuma, M., Tanimoto, T., Yano, K. and Takagi, M. (1991) CYP52 (cytochrome P450alk) multigene family in *Candida maltosa*: molecular cloning and nucleotide sequence of the two tandemly arranged genes. *DNA Cell Biol*, **10**, 271–282.

Ollis, D.L., Cheah, E., Cygler, M., Dijkstra, B., Frolow, F., Franken, S.M., Harel, M., Remington, S.J., Silman, I., Schrag, J., Sussman, J.L., Verschueren, K.H.G. and Goldman A. (1992) The α–β hydrolase fold. *Prot. Eng.*, **5**, 197–211.

Osawa, S. (1995) *Evolution of the genetic code*. Oxford, UK: Oxford University Press.

Pesole, G., Lotti, M., Alberghina, L. and Saccone, C. (1995) Evolutionary origin of nonuniversal CUGSer codon in some Candida species as inferred from a molecular phylogeny. *Genetics*, **141**, 903–907.

Pearson, W.J. and Lipman, D.J. (1988) Improved tools for biological sequence comparison. *Proc. Natl. Acad. Sci. USA*, **85**, 2444–2448.

Phillips, A., Pretorius, G.H. and van Resenburg, H.G. (1995) Molecular characterization of a *Galactomyces geotrichum* lipase, another member of the cholinesterase/lipase family. *Biochim. Biophys. Acta*, **1252**(2), 305–311.

Rogalska, E., Ransac, S. and Verger, R. (1993). Controlling lipase stereoselectivity via the surface pressure. *J. Biol. Chem.*, **268**(2), 792–794.

Rua, M.L., Diaz-Maurino, T., Fernandez, V.M., Otero, C. and Ballesteros, A. (1993) Purification and characterization of two distinct lipases from *Candida cylindracea. Biochim. Biophys. Acta,* 1156, 181–189.

Schrag, J.D., Li, Y., Wu, S. and Cygler, M. (1991) Ser-His-Glu triad forms the catalytic site of the lipase from *Geotrichum candidum. Nature,* 351, 761–764.

Schuler, G.D., Altschul, S.F. and Lipman, D.J. (1991) A workbench for multiple alignment construction and analysis. *Proteins Struct. Funct. Genet.,* 9, 180–190.

Seghezzi, W., Meili, C., Ruffiner, R. Kuenzi, R., Sanglard, D. and Fiechter, A. (1992) Identification and characterization of additional members of the cytochrome P450 multigene family CYP52 of *Candida tropicalis. DNA & Cell Biology,* 11(10), 767–780.

Shimada, Y., Sugihara, A., Iizumi, T. and Tominaga, Y. (1990) cDNA cloning and characterization of *Geotrichum candidum* lipase II. *J. Biochem.,* 107, 703–707.

Shimada, Y., Sugihara, A., Tominaga, Y., Iizumi, T. and Tsunasawa, S. (1989) cDNA molecular cloning of *Geotrichum candidum* lipase. *J. Biochem.,* 106, 383–38.

Sidebottom, C.M., Charton, E., Dunn, P.P., Mycock, G., Davies, C., Sutton J., Macrae, A.R. and Slabas, A.R. (1991) *Geotrichum candidum* produces several lipases with markedly different substrate specificities. *Eur. J. Biochem.,* 202, 485–491.

Smolenski, G., Sullivan, P.A., Cutfield, S.H. and Cutfield, J.F. (1997) Analysis of secreted aspartic proteinases from *Candida albicans*: purification and characterization of individual Sap1, Sap2 and Sap3 isoenzymes. *Microbiology,* 143, 349–356.

Sugiyama, H., Ohkuma, M., Masuda, Y., Park, S.-M., Ohta A. and Takagi M. (1995) *In vivo* evidence for non-universal usage of the codon CUG in *Candida maltosa. Yeast,* 11, 43–52.

Suzuki, T., Ueda, T., Yokogawa, T., Nishikawa, K. and Watanabe, K. (1994) Characterization of serine and leucine tRNAs in an asporogenic yeast *Candida cylindracea* and evolutionary implications of genes for tRNA(Ser)CAG responsible for translation of a non-universal genetic code. *Nucl. Acids Res.,* 22(2), 115–123.

Tsuchiya, A., Nakazawa, H., Toida, J., Ohnishi, K. and Sekiguchi, J. (1996) Genome structure and nucleotide sequence of the mono- and diacylglycerol lipase gene (mdlB) of *Aspergillus oryzae. FEMS Microbiol. Lett,* 143(1), 63–67.

Ueda, T., Suzuki, T., Yokogawa, T., Nishikawa, K. and Watanabe, K. (1994) Unique structure of new serine tRNAs responsible for decoding leucine codon CUG in various *Candida* species and their putative ancestral tRNA genes. *Biochimie,* 76, 1217–1222.

Uppenberg, J., Patkar, S., Bergfors, T. and Jones, T.A. (1994) Crystallization and preliminary X-ray studies of lipase B from *Candida antarctica. J. Mol. Biol.,* 235(2), 790–792.

Van Tilbeurgh, H., Sarda, L., Verger, R. and Cambillau C. (1992) Structure of the pancreatic lipase-procolipase complex. *Nature,* 359, 159–162.

Verger, R. (1997) "Interfacial activation" of lipases: facts and artifacts. *Tibtech,* 15, 32–38.

Vulfson, E.N. (1994) Industrial applications of lipases. In *Lipases. Their structure, biochemistry and application,* edited by P. Woolley and S.B. Petersen. UK: Cambridge University Press.

Yamaguchi, S., Mase, T. and Takeuchi, K. (1991) Cloning and structure of the mono- and diacylglycerol lipase-encoding gene from *Penicillium camembertii* U-150. *Gene,* 103, 61–67.

Zimmer, T. and Schunck, W.H. (1995) A deviation from the universal genetic code in *Candida maltosa* and consequences for heterologous expression of cytochromes P450 52A4 and 52A5 in *Saccharomyces cerevisiae. Yeast* 11(1), 33–41.

5. STRUCTURE-FUNCTION STUDIES ON CUTINASE, A SMALL LIPOLYTIC ENZYME WITH A WATER ACCESSIBLE ACTIVE SITE

SONIA LONGHI[1*], ANNE NICOLAS[1,2], CHRISTIAN JELSCH[1,3] and
CHRISTIAN CAMBILLAU[1]

[1] *Architecture et Fonction des Macromolécules Biologiques, UPR 9039, CNRS, IFR1,
31 Chemin Joseph Aiguier, 13402 Marseille CEDEX 20, France*
[2] *Actual address Dept. of Crystal and Structural Chemistry.
Bijvoet Center for Biomolecular Research. Utrecht University, Padualaan 8,
3584 CH Utrecht, The Netherlands.*
[3] *Actual address Laboratoire de Cristallographie et Modelisation des Materiaux
Mineraux et Biologiques. LCM3B-CNRS, Bvd des Aiguillettes, B.P. 239,
54506 Vandoeuvre les Nancy, CEDEX, France.*

Cutinases are enzymes of 22–26 KDa which are able to hydrolyse ester bonds in the cutin polymer (Kolattukudy, 1984). Cutinases are produced by several phytopathogenic fungi and pollen (Ettinger *et al.*, 1987). For many years, cutinases have been thought to play a crucial role in the fungal invasion of host plants: the hydrolysis of cutin, an insoluble monocarboxylic acid polyester representing the main component of the plant cuticle, has been thought to enable the penetration of the fungus in the plant. However, interpretations of recent data on fungal cutinase activity and pathogenicity are contradictory, and range from cutinase having little or no influence at all on pathogenicity, to it enhancing adhesion of fungal spores to the plant surface (Schäfer, 1993). In addition to cutin polymer degradation, cutinases are also able to hydrolyse a large variety of synthetic esters and show activity on short and long chains of emulsified triacylglycerols as efficiently as pancreatic lipases (Table 1a). Contrary to lipases, the activity of which is greatly enhanced in the presence of a lipid/water interface, cutinases do not display, or display little, interfacial activation, being active on both soluble and emulsified triglycerides (Sarda & Desnuelle, 1958; Verger & de Haas, 1976) (Table 1b).

Cutinase from the fungus *Fusarium solani* is a 22 KDa enzyme which has been extensively investigated from both biochemical (Kolattukudy, 1984; Lauwereys *et al.* 1991, Mannesse *et al.*, 1995 a,b; Mannesse, 1997; Mannesse *et al.*, 1997) and structural standpoints. Since its 3D structure was solved at 1.6 Å (R-factor 15.4%) (Martinez *et al.*, 1992 & 1993), considerable efforts have been devoted to the elucidation of the catalytic mechanism and of the involvement of specific residues in substrate binding and catalysis, as well as to the rational design of mutants endowed with a better catalytic performance.

---

* Corresponding author. Architecture et Fonction des Macromolécules Biologiques, UPR 9039, CNRS, IFR1, 31 Chemin Joseph Aiguier, 13402 Marseille CEDEX 20, France. Phone: (33) 4 91 16 45 04; Fax: (33) 4 91 16 45 36; E-mail: longhi@afmb.cnrs-mrs.fr

Table 1   (a) Specific activity of cutinase, hpL and RmL as a function of the chain length of the substrate. (b) Specific activity of cutinase and of RmL on tributyrin below and beyond the CMC (compiled in martinez, 1992; Verger, personal communication)

a U/mg	Tributyrin (C4)		Triolein (C18:1)
Cutinase	3 200 (40%)		670 (14%)
hpL	8 000 (100%)		4 600 (96%)
RmL	8 000 (100%)		4 800 (100%)
b U/mg (tributyrin)	Below CMC	Beyond CMC	Ratio
Cutinase	1 280	3 200	2.5
RmL	53	8 000	150

During the last six years, the structures of 35 cutinase variants, including covalently inhibited complexes (Martinez et al., 1994; Longhi et al., 1997a), as well as rationally designed point mutants (Nicolas et al., 1996; Longhi et al., 1996), have been determined, and the resolution of native cutinase was extended to 1.15 Å (R-factor 16.7%) (Nicolas, 1996). During all these investigations, striking evidences on the stability of cutinase were obtained, thus making this enzyme an excellent candidate to collect data on high energy X-ray sources. Accordingly, monochromatic (Longhi et al., 1997b; Prangé et al., 1998) and Laue data (Bourgeois et al., 1997) were collected on different synchrotron sources.

In this review we describe the most relevant information arising from this considerable body of structural information.

THREE-DIMENSIONAL STRUCTURE OF NATIVE CUTINASE

The Overall Structure

Crystals of recombinant *Fusarium solani* native cutinase (Lauwereys et al., 1991) crystallize in space group $P2_1$ (Abergel et al., 1990). No density could be observed for 16 amino acids at the N terminus and two at the C terminus. Residues 1 to 15 are part of a propeptide which is not present in the natural mature enzyme, which could explain their high extent of disorder. Cutinase is a compact one domain molecule (45 * 30 * 30 Å³) consisting of 197 residues. It has a central β-sheet consisting of 5 parallel β-strands covered by 2 α-helices on each of the two sides of the sheet (Figure 1a). The active site — consisting of the catalytic triad Ser120, Asp 175 and His188 — is reminiscent of that found in serine proteases, but with an opposite handedness. Unlike RmL (Brady et al., 1990), hpL (Winkler et al., 1990; van Tilbeurgh et al., 1992) and GcL (Schrag et al., 1991), the catalytic serine is not buried under an amphipathic loop but is accessible to the solvent: it is located at one extremity of the protein ellipsoid at the bottom of a crevice delimited by the loops 80–87 and 180–188 (Figure 1a) (Martinez et al., 1992 & 1993). These loops, which bear hydrophobic amino acids, constitute the lipid binding site (LBS) of cutinase. The extended active site crevice is partly covered by two thin bridges formed by the side

a)

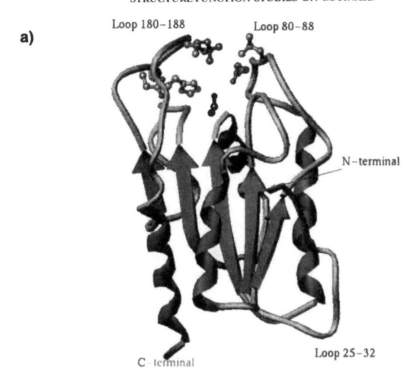

b)

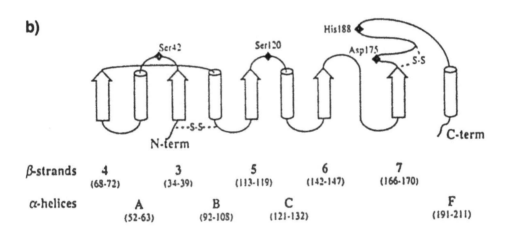

Figure 1 (a) Ribbon representation of cutinase. α-helices (red), β-strands (blue) and coils (yellow) are shown. The residues of the catalytic triad (Ser120, His188, Asp175), as well as the residues constituting the two thin bridges over the catalytic crevice (Leu 81, Val 184, Leu 182 and Asn 84) are represented in balls and sticks (atom type colouring). The loops delimiting the catalytic crevice (residues 80–88 and 180–188) as well as the loop 25–32 are highlighted. (b) Topology representation of cutinase. Disulfide bridges are represented by dashed lines, strands by arrows and helices by cylinders. The catalytic residues are represented by diamonds

chains of residues Leu 81 and Val 184, and Leu 182 and Asn 84, respectively (Figure 1a). The presence of a catalytic site directly exposed to the solvent may provide a structural explanation for the absence of interfacial activation. The side chains of the hydrophobic amino acids belonging to the LBS display B-factor values higher than the average (Martinez *et al.*, 1992 & 1993). This may be an indication that these side chains are mobile and could move upon interface binding. Based on the 3D structure of the native, uncomplexed enzyme it was postulated that the binding of cutinase to the interface would not require major main chain rearrangements but only the re-orientation of a few lipolytic side chains that would play the role of a mini-lid. The accessibility of the catalytic serine would provide a structural explanation for the esterase activity on short triglycerides (tributyrin, C4) at substrate concentration below the critical micellar concentration (CMC) (see Table 1b).

Besides the two loops delimiting the catalytic crevice, the loop 25–32, which is located far away and opposite to the LBS (Figure 1a), was found to diplay the highest temperature factors. This region does not participate in the crystal packing and is located in front of the largest solvent channel. The first N-terminal residue and the last two C-terminal residues are disordered. Atoms in the central β-sheet show the lowest temperature B-factors. A small number of residues, which are characterized by high B-values, are all located on the surface of the molecule. Two disulfide bonds are present in cutinase. The first, Cys31–Cys109, links the N-terminal end to a β-turn and participates to the stabilization of the global molecular folding (Figure 1b). The second disulfide bridge, Cys171–Cys178, can be assumed to play an important role in the stabilization of the two consecutive β-turns on which is located the catalytic Asp175 (Figure 1b). The two disulfide bonds adopts a right-handed hook conformation (Ridchardson, 1981).

The Tertiary Fold

Cutinase is the smallest lipase/esterase structure solved so far (Brady *et al.*, 1990; Winkler *et al.*, 1990; van Tilbeurgh *et al.*, 1992; Sussmann *et al.* 1991; Schrag *et al.*, 1991; Schrag & Cygler 1993; Grochulsky *et al.*, 1993 & 1994b; Bourne *et al.*, 1993; Bott, 1993; Noble *et al.*, 1994; Uppenberg *et al.*, 1994; Derewenda *et al.*, 1994a, b,c; Lang *et al.*, 1994 & 1996; Ghosh *et al.*, 1995; Hermoso *et al.*, 1996). Like all the lipolytic enzymes investigated so far, cutinase belongs to the class of serine esterases and to the superfamily of the α/β hydrolases (Ollis *et al.*, 1992). According to Ollis *et al.* (1992), the members belonging to this protein family share the following structural features: *i)* their core consists of a non-barrel α/β sheet *ii)* their active site is composed of a catalytic triad *iii)* the triad residues, nucleophile-histidine-acid, are located on loops which are the best conserved structural elements of the overall fold; *iv)* the nucleophile is invariably located at the center of an extremely sharp turn between a β-strand and an α-helix. The sharpness of the turn results in ϕ and ψ angles which lie in a very unfavourable region of the plot of Ramachandran (ε conformation). The occurrence of an α/β hydrolase fold and of a catalytic machinery composed by a nucleophile, an acid and an histidine, seems to be a common feature to all the esterases described so far, with the only exception represented by *Streptomyces scabiaes* esterase (Wei *et al.*, 1995).

Cutinase, which is the smallest member of the α/β hydrolase superfamily (Table 2), is an α/β protein with an hydrophobic core comprising a slightly twisted five-parallel-stranded β-sheet, surrounded by four α-helices (Figure 1b). The catalytic

Table 2 Structures of the α/β hydrolase fold family members

Enzyme source-complexed with	Abbreviation	Res. (Å)	Reference	PDB code
Lipase				
human pancreatic	hpL	2.3	Winkler et al., 1990	n.d.*
-porc colipase-C11 phosphonate	hpL-COL-C11	2.45	Egloff et al., 1995	1LPB$
horse pancreatic	hopL	2.3	Bourne et al., 1994	1HPL*
guinea pig (N-term)	gpL		Martinez et al., 1996	*
Rhizomucor miehei	RmL	1.9	Brady et al., 1990	3TGL*
-n-hexylphosphonate ethyl ester		1.9	Brzozowski et al., 1991	5TGL$
Penicillium camembertii	PcL	2.0	Derewenda et al, 1994a-c	1TIA*
Humicola lanuginosa	HlL	1.84	Derewenda et al., 1994a-b-c	1TIB*
Rhizopus delemar	RdL	2.6	Derewenda et al., 1994a-b-c	1TIC*
Geotrichum candidum	GcL	1.8	Schrag & Cygler, 1993	1THG*
Candida rugosa	CrL	2.06	Grochulsky et al., 1993	1CrL*
-(1S)-menthyl hexyl phosphonate	(1S)-CrL-MPC	2.2	Grochulsky et al., 1994a	1LPS$
Pseudomonas mendocina			Bott et al., 1993	n.d.
Pseudomonas glumae	PGL	3.0	Noble et al., 1994	1TAH*
Candida antarctica B	CaBL	1.55	Uppenberg et al., 1994	1TCA$
Chromobacterium viscosum		3.0	Lang et al.,1994 & 1996	n.d.
Cutinase				
Fusarium solani pisi	FsC	1.6	Martinez et al., 1992	1CUS*
-diethylphosphate	FsC-E600	1.9	Martinez et al., 1994	2CUT$
Esterase				
Torpedo californica acetylcholinesterase	AChE	2.8	Sussman et al., 1991	1ACE*
-tacrine	AChE-THA	2.8	Harel et al., 1993	1ACJ$
Pseudomonas fluorescens esterase	PfE		Se Won Suh, unpublished	n.d.*
Vibrio harveyi thioesterase	TE	2.1	Lawson et al., 1994	1THT*
Bromoperoxydase				
Streptomyces aureofaciens A2		2.05	Hecht et al., 1994	
Serine carboxypeptidase				
Triticum vulgaris II	CPD-WII	2.2	Liao & Remington, 1992	3SC2*
-L-benzylsuccinate	CPD-WII-BZS	2.0	Bullock et al., 1994	1WHT$
Saccharomyces cerevisiae	CPD-Y	2.8	Endrizzi et al., 1994	1YSC*
Dienelactone hydrolase				
Pseudomonas sp. B13	DHL	2.8	Pathak et al., 1988	n.d.
Haloalkane dehalogenase				
Xanthobacter autotrophicus	HAL	1.9	Verschueren et al., 1993a	2HAD*
-2-chloroethane, pH = 5.0, 22° C	HAL-DCE	2.0	Verschueren et al., 1993b	2DHD$
Protective protein				
Human	HPP	2.2	Rudenko et al., 1995	

Are reported all the uncomplexed forms of the α/β hydrolase fold members and one of their complexed forms. *, $: atomic co-ordinates used in the structural comparison of the open and closed forms, respectively. n.d.: atomic co-ordinates are not deposited with the Brookhaven Protein Data Bank.

serine (Ser120) is located at the top of a very sharp γ-like turn between the strand
β5 and the helix αC (Figure 1b). The sharpness of this turn, which is a type II'
turn, results in the energetically unfavourable ε conformation of Ser 120 ($\phi = 59°$,
$\psi = -122°$) typical of the nucleophilic residue of all the members of the α/β
hydrolase fold (Ollis *et al.*, 1992).

Secondary Structure

The topology of the molecule is shown schematically in Figure 1b. The secondary
structure elements based on the definition of Kabsch & Sander (1983) have been
numbered to emphasize the similarity in the secondary structures of the α/β
hydrolase fold family. The five β-strands of the central β-sheet represent 15% of
the total residues. 23 out of the 25 residues of the β-sheet, are hydrophobic and
contribute to mantain a central water-excluded environment in the core of the
protein. 6 out of the 12 residues located on the concave side of the β-sheet are
aromatic or long side chain amino acids. On the convex side, there are 13 residues
with shorter side chains. In the central β-sheet, the hydrogen bond pattern typical
of parallel strands is observed, all the 17 hydrogen bonds occur between backbone
atoms and fall within the expected range (Baker & Hubbart, 1984). There is no
internal β-turn in the cutinase model.

The four crossover connections between the parallel strands are either of the
common right-handed β/α/β or β/loop/β type (Richardson, 1981). In the proximity
of the active site region, 57 residues form five α/β loops. The loop connecting the
strand β-7 and the helix α F (residues 171–190) contains the catalytic Asp 175 and
His 188 residues. On the opposite side of the active site, 31 residues form four
α/β loops.

The four helices of the α/β core represent 31% of the total residues. Their length
ranges from 11 to 21 residues. The four helices have the same orientation and are
located on the surface of the molecule on each side of the central β-sheet. Helices
α A and α F pack against the concave side of the β-sheet. They are parallel to strand
β-3 and β-7, respectively. Helices α B and α C pack against the convex side of the
β-sheet. The packing of the four α-helices against the β-sheet is mediated by a
network of hydrophobic interactions involving 14 aromatic residues. No hydrogen
bonds between either the β-sheet and the α-helices or two α-helices are observed.

Helices α A (11 residues), α B (17 residues), and α C (12 residues) are classical
right-handed 3.6_{13} α-helices. The 3.6_{13} helix αF (21 residues) presents a pronounced
kink with a distorted α-helix hydrogen bond pattern, caused by the occurence of
an additional residue. As a result, this helix accepts three proline residues. Pro193
is in the turn preceeding the N-terminal part of the kink, while Pro198 and Pro200
are distributed in the two first turns following the C-terminal of the kink. As a result,
the α-helix hydrogen bond pattern is not classical, with the Asp194 backbone
oxygen atom of the kink being hydrogen bonded with the (n+5)th residue amide
group. The four residues (194–197) of the kink resemble to a type III turn except
for the turn hydrogen bond. As Asp194 O is involved in the α-helix hydrogen
bonding, its hydrogen bond with Gly197 NH is in an unvafourable conformation
and very weak (3.42 Å). The backbone oxygens of Ala195 and Arg196 do not
participate in the α-helix hydrogen bonding pattern.

Table 3 Ionic interactions and side chain to side chain hydrogen bonds in cutinase

Residue 1		Distance (Å)		Residue 2
Arg20	NH2	2.80	Oδ1	Asp22
	NH2	2.77	Oε1	Glu60
	NH1	2.86	Oε1	
Arg78	NH1	2.88	Oδ1	Asp83
Arg88	NH2	2.70	Oδ1	Asp132
	NH1	2.88	Oδ2	
	NH1	2.95	Oε2	Glu131
Arg20	Nε	3.13	Oδ2	Asp22
Asp21	Oδ2	3.17	Nδ2	Asn47
	Oδ2	2.92	Nε2	Gln71
Glu44	Oε2	2.72	Oγ1	Thr80
Arg78	Nε	2.92	Oγ1	Thr90
Asn106	Oδ1	2.74	NZ	Lys140
Thr113	Oγ1	2.87	NH1	Arg211
Thr144	Oγ1	3.18	OH	Tyr162
Asn155	Nδ2	2.99	OH	Tyr149
* Ser42	Oγ	3.12	Nε2	Gnl121
* Gln121	Oδ1	2.86	Oγ1	Thr90
* His188	Nε2	2.76	Oγ	Ser120
* His188	Nδ1	2.86	Oδ1	Asp175
* His188	Nδ1	3.14	Oδ2	

The upper and lower part of the table list the salt-bridges and the side chain hydrogen bonds, respectively. * indicates active site residues, including the three catalytic residues and the residues involved in the stabilization of the substrate oxyanion.

Among the three one-turn helices external to the α/β core, the helices α A (4 residues) and α C (6 residues) are 3.6_{13} α-helices, while the helix α B (5 residues) is the only 3_{10} helix of the molecule. The helix α A, located on the surface of the molecule, belongs to the N-terminal loop preceeding the first strand (β-3) of the central β-sheet. The helix α B connects the strand β-4 and the helix α B. The helix α C, located in the loop region opposite side to the active site, connects the helix α C and the strand β-7.

Hydrogen Bonds and Salt Bridges

In Table 3 are listed the side chain to side chain hydrogen bonds and salt-bridge interactions between polar and charged residues. All these interactions are found on the surface of the protein, except the hydrogen bond cluster of the active cavity. The 7 salt bridges form three clusters of ionic interactions, namely Arg20 with

Glu60, Arg78 with Asp83, and Arg88 with Asp132 and Glu131. Contrary to the RmL (Brady *et al.*, 1990) and hPL (Winkler *et al.*, 1990; van Tilbeurgh *et al.*, 1992), cutinase has no internal charged residues. There are 183 main chain to main chain hydrogen bonds. None of the 253 solvent molecules present in the 1.15 Å model establish contact with the central β-sheet. Most of them are located on the protein surface near α-helices. The active site cavity contains two water molecules. There are five internal water molecules stabilized by hydrogen bonds with carboxyl and amide main chain groups of hydrophobic residues.

The Catalytic Triad

The active site of cutinase is formed by the catalytic triad (Ser 120, Asp 175 and His 188), which is reminiscent of that found in serine proteases. However, in the α/β hydrolase fold enzymes, the twist of the β-sheet imposes a handedness on the catalytic triad which is approximately a mirror-image of that observed in the serine proteases (Ollis *et al.*, 1992). The amino acid sequence encompassing Ser 120 (Gly-Tyr-Ser-Gln-Gly) matches the consensus sequence Gly-His(Tyr)-Ser-X-Gly commonly present in lipase sequences. The catalytic residues are in the same sequence order observed in all the members of the α/β hydrolase fold family (Ollis *et al.*, 1992). These residues are located in turns (Figure 1b) that project their side chain away from the polypeptide backbone into accessible positions. Cutinase Ser 120 is located in a tight turn, classified as type II', between the C-terminal end of the strand β-5 and the helix α C. The sharpness of the turn forces the side chain od Ser 120 to point into the active site cavity. Asp 175 and His 188 are very close in sequence and are located on the same loop between the strand β-7 and the helix α F. His 188 is located at position 4 of a type I β-turn. As usually observed in this type of reverse turns, the residue preceeding the histidine is a proline. Asp 175 is located at position 1 of a type I turn. The carboxylate of Asp175 is stabilized by means of two hydrogen bonds to the backbone nitrogens of the residues Cys 178 and Val 177. All hydrogen bonds identified in the active site are characteristic of the serine proteases: Ser 12 Oγ is bonded to His 188 Nε2 (2.76 Å) while Nδ1 of the imidazole ring is hydrogen bonded to Asp 175 Oδ2 (3.14 Å) and Oδ1 (2.86 Å).

STRUCTURAL COMPARISON BETWEEN CUTINASE AND THE OTHER α/β HYDROLASE FOLD MEMBERS

Even if the members of the α/β hydrolases fold family vary widely in their sizes, amino acid sequences, substrates and physical properties, they present an appreciable structural similarity. As already mentioned, among the common features defining the α/β hydrolase fold, the key points are the central β-sheet, the presence of a catalytic triad and the location of the nucleophile active site residue in a β-hairpin. Despite these common features, the α/β hydrolase fold members present significant differences. The differences include variations of *i)* the number of strands in the central β-sheet, *ii)* the nucleophilic and acidic residue type of the catalytic triad, and *iii)* the number of external central core motifs.

   Figure 2 shows the topology of the 21 α/β hydrolase fold structures known so far (Nicolas, 1996). The secondary structure labels have been chosen to emphasize the similarity in the secondary structures of the different members of the α/β

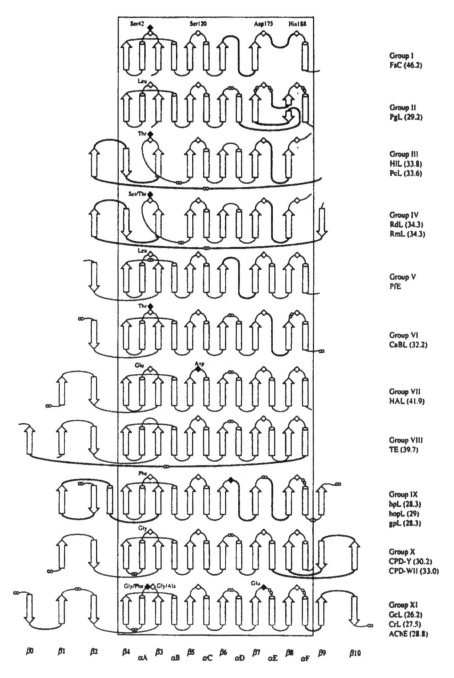

Figure 2 The eleven topologic groups of the α/β hydrolase fold proteins. The percentages of the residues belonging to the consensus α/β hydrolase fold (out of the total residues) are shown in brackets. The $\alpha6/\beta6$ consensus motif is contoured. Consensus and extra loops are shown in thin and in thick, respectively. Extra β-strands or α-helices are represented as two small circles. Catalytic residues are represented by empty diamonds, except for the acidic member of groups IX and XI which are represented by a solid diamond. The second oxyanion group (main chain) is represented by an empty diamond. A solid diamond is added when the residue serves also as the third component of the oxyanion hole (side chain)

hydrolase fold family. The structural elements of the central core have been used to define eleven topological groups. The groups are classified based on the number of α/β hydrolase motifs; the group I is the smallest member of the family with five strands and four helices while the group XI is the largest one with eleven strands and six helices. Five groups (VII to XI) contain the central eight stranded motif defined by Ollis *et al.* (1992). HlL, PcL, RdL, and RmL which belong to a same enzymatic family are distributed into the topological groups II and IV. The consensus topology (Richardson, 1981) of the central β-sheet is: $-1x, +2x, +1x, +1x, +1x$. The groups V, VI, VII, VIII, IX, and XI show a complete agreement with this consensus topology, while the groups I, II, III, IV, and V present small differences in the external part of the central core. The best representant of the consensus α/β hydrolase fold topology is the group VII which does not present extra motifs. Ollis *et al.* (1992) have defined it as the prototypic fold of the α/β hydrolase fold family. Cutinase matches the consensus fold even if it has the smallest number of secondary elements of the family; the missing elements, with respect to the consensus fold, are the strand β-8, and the helices α D and α E. The helix α D is replaced by two β-turns and the motif from the helix α E to the strand β-8 is replaced by a disulfide bond. Consequently, the strand β-6 is directly connected to the strand β-7 which is in its turn directly connected to the helix α F.

The strands β-3, β-4, β-5, β-6, and β-7, and the helices α B and α C are always presents and the motif from strand β-5 to helix αC is the only one which is conserved among all the groups. The superimposition of the β-sheets of all the α/β hydrolase fold members is shown in Figure 3a.

The six loops of the consensus topology play an important role in enzymatic activity. The loops L5C, L7E, and L8F contains the nucleophilic, acidic, and histidine residues, respectively. The loops L4B and L6D are involved in the binding of the substrate. The loop L3A contains the second oxyanion hole group. The first oxyanion group is provided by the amide of the residue following the nucleophile. In all the α/β hydrolase fold groups the nucleophilic and histidine residues occur always at the same topological location, whereas location of acidic residue accepts one exception (hpL) as compared to the consensus fold.

The β-5 — L5C — α C motif, called the nucleophile elbow, is highly conserved in the family (Figure 3b). The position 2 in the loop L5C is invariably occupied by the nucleophile with its typical ε conformation (Derewenda & Derewenda, 1991). In all α/β hydrolase fold proteins, the nucleophilic residue is a serine except in the case of HAL and of dienelactone hydrolase in which it is replaced by an aspartic and a cysteine, respectively.

The acidic residue is located in the loop L7E, except for the the group IX in which it has a different topological origin, being located in the loop L6D. In all groups, the acidic residue is an aspartic, with the only exception being provided by the group XI in which it is replaced by a glutamic. The best conserved catalytic residue is the histidine which is conserved in all the members and which is invariably located in the loop L8F.

Cutinase is presently the smallest α/β hydrolase member, with only five strands in the central β-sheet surrounded by four helices. The "minimal topology" of cutinase suggests that the essential structural requirements for enzymatic activity are provided by six loops located on the active side of the molecule, two of which serve for the binding of the substrate, one for the stabilization of the tetraedral intermediate and three for the catalytic activity.

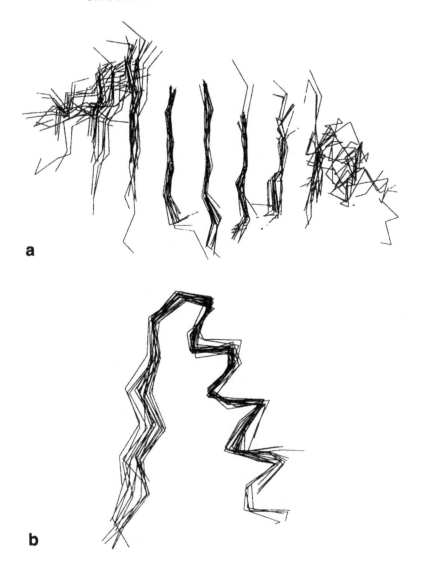

Figure 3 Superimposition (Cα tracing) of (a) the β-sheet and of (b) the nucleophile elbow of all the α/β hydrolase fold members

THE COVALENTLY INHIBITED COMPLEXES AND THE OXYANIONE HOLE

Based on the comparison between the native and the complexed structures of RmL (Brozozowski *et al.*, 1991, Derewenda *et al.*, 1992) and of hpL (van Tilbeurgh *et al.*, 1993; Egloff *et al.*, 1995), it has been suggested that interfacial activation, can be explained by an essential structural rearrangement when lipases bind to a lipid-water interface (see also Smith *et al.*, 1992; Lawson *et al.*, 1992; Cambillau & van Tilbeurgh, 1993; Cambillau *et al.*, 1996). In solution, the active site of lipases was shown to be covered by an amphypatic "lid". The opening of this lid upon interaction with an hydrophobic interface, renders the catalytic site accessible to the substrate and increases the rate of hydrolysis. The conformational change in the lid is thought to be closely related to the interfacial activation phenomenon. The rearrangement of the lid results in a large increase in the lipolytic surface, stabilized by interactions with the interface. The exposure to solvent of these wide hydrophobic surfaces would probably be thermodynamically unfavourable in the absence of a lipid-water interface. Open forms in the absence of a bound inhibitor could be obtained in the case of *Pseudomonas cepacia* lipase (Kim *et al.*, 1997; Schrag *et al.*, 1997) and CrL (Grochulski *et al.*, 1993), however, owing to the rather high hydrophobic nature of the solutions used in crystallization.

In some lipases, a second consequence of the movement of the lid is the formation of the oxyanion hole (Brozozowski *et al.*, 1991; Derewenda *et al.*, 1992; van Tilbeurgh *et al.*, 1993; Egloff *et al.*, 1995). As in serine proteases (Kraut *et al.*, 1977), the lipase oxyanion hole provides an electrophilic environment that stabilizes the negative charge generated during the nucleophilic attack on the sissile bond of the substrate. The tetrahedral intermediate is hydrogen bonded to two main chain nitrogens and possibly stabilized by further determinants. The role played by the amide nitrogen of the active site serine in the serine proteases is played by the amide nitrogen of the residue following the nucleophile in the α/β hydrolase superfamily. The relative position of the oxyanion hole residues is highly conserved in the α/β hydrolase superfamily (Figure 2): whereas the histidine and the acidic member of the catalytic triad are located on one side of the nucleophile (ser), the residues forming the oxyanion hole lie on the other side.

Phosphorylating agents, which block the serine protease reaction at the acylation step via the formation of a tetrahedral intermediate (Kraut, 1977), provide a powerful tool to study one of the activated states of lipases/esterases and to characterise their oxyanion hole. Covalent complexes with organophosphate and organophosphonate inhibitors of four lipases (Brzozowski *et al.*, 1991; Derewenda *et al.*, 1992; Grochulski *et al.*, 1994a; Cygler *et al.*, 1994; Egloff *et al.*, 1995; Uppenberg *et al.*, 1995) and of cutinase (Martinez *et al.*, 1994, Longhi *et al.*, 1997a) were obtained and their structures were solved. These structures allowed to identify precisely the residues of the oxyanion binding site and in addition they suggested a relation between the position of the lid and interfacial activation. In most cases, the optimal oxyanion hole geometry is achieved only upon the conformational change which takes place after adsorption to the interface.

The cut-E600 Complex

The comparison between the structures of native enzyme and of a covalently inhibited complex with *n* diethyl-*p*-nitrophenyl phosphate (E600) (1.9 Å) (Martinez *et al.*, 1994) revealed that cutinase has a preformed oxyanion hole. The overall structure of the cut-E600 complex (crystallizing in a different space group) is very similar to that of the native enzyme (Figure 4a). The only significant differences concern the loop 25–32 which is located opposite to the active site (Figure 4a) and which displays the highest B-factor values even in the native structure (Martinez *et al.*, 1992 & 1993; Nicolas, 1996; Longhi *et al.*, 1996 & 1997a,b). The Ramachandran plots of native and complexed cutinase are almost identical indicating that no conformational changes occur upon inhibitor binding, contrary to what has been observed in the case of RmL (Brzozowski *et al.*, 1991; Derewenda *et al.*, 1992) and hpL (van Tilbeurgh *et al.*, 1993; Egloff *et al.*, 1995). The E600 accomodation in the

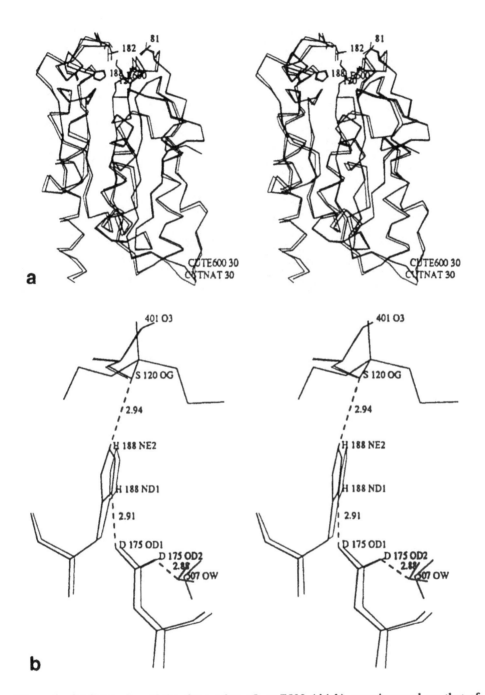

Figure 4 (a) Stereoview of the Cα tracing of cut-E600 (thick) superimposed on that of native cutinase (thin). The catalytic triad and the inhibitor are shown in balls and sticks. The loop 25–32 is highlighted. (b) Stereoview of the catalytic triad of cut-E600 (thick) superimposed on that of native cutinae (thin). The inhibitor, as well as the triad bound water molecule are shown

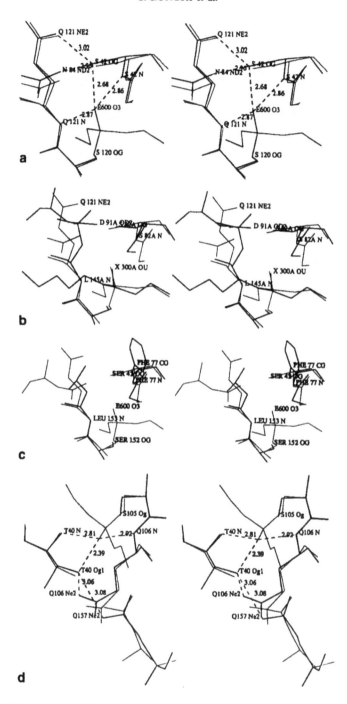

Figure 5   (a) Stereoview of the oxyanion hole of cut-E600 (thick) superimposed on that of native cutinase (thin). The E600 inhibitor as well as the hydrogen bonds involved in the oxyanion hole stabilization are shown. (b, c, d). Stereoview of the oxyanion hole of the inhibited RmL (b) (thick), that of the inhibited hpL (c) (thick) and that of the CaBL (d) (thick) superimposed on that of cut-E600 (thin). The residues involved in the oxyanion hole formation are shown

active site crevice takes place with only minor changes in the backbone and side chain positions of the loops 80–88 and 180–188 delimiting the active site crevice. The active site contains a water molecule which is bound with three main chain nitrogens (Leu 176 N, Ala 185 N and Ile 183 N) and to the catalytic Asp 175 Oδ1 atom, thus maintainig this residue in the proper orientation. The structure of the catalytic triad is basically the same in the native and inhibited enzyme, except for a water molecule, located in front of Ser 120 in the native structure, which has been replaced by the inhibitor (Figure 4b). After the central β-sheets of RmL, hpL, and native and complexed cutinase were superimposed, the catalytic triads of these three enzymes have been found to almost overlap, expecially the atoms involved in the catalytic process (Ser 120 Oγ, His 188 Nε2, Asp 175 Oδ1). The analysis of the environment of the phosphate oxygen in the cut-E600 structure allows to identify the oxyanion hole. The main chain nitrogens of Ser 42 and Gln 121, the residue following the catalytic serine, are within hydrogen bonding distance of the E600 O3 atom (Figure 5a). The Ser 42 Oγ atom is located at 2.68 Å. The position of the Ser 42 side chain is further stabilized by hydrogen bonds with Asn 84 Nδ2 and Gln 121 Nε2 (Figure 5a). In the case of RmL (Brzozowski et al., 1991; Derewenda et al., 1992) the two main chain nitrogen stabilizing the oxyanion come from the residues Leu 145 and Ser 82 (Figure 5b). The Ser 82 Oγ, which the equivalent of Ser 42 Oγ of cutinase, may also be hydrogen bonded to the phosphate oxygen and may therefore contribute to the oxyanion stabilization. The Ser 82 Oγ atom is stabilized by only one interaction with Asp 81 (Figure 5b). The role of the RmL Ser 82 Oγ in the stabilization of the oxyanion hole is further supported by the results obtained with mutants RdL Thr 83 (Joerger & Haas, 1994) and mutants PcL Ser 84 (Yamaguchi et al., 1992) since these residues are homologous to RmL Ser 82 (Derewenda et al., 1994a). Winkler et al. (1990) hypothesized that the oxyanion hole in hpL may be formed by the main chain nitrogens of Phe 77 and Leu 153 (the residue following the catalytic Ser 152). This was confirmed by the analysis of ternary lipase-colipase-phospholipid complex (van Tilbeurgh et al., 1993). In this case however, the first member of the oxyanion hole would fail to provide an additional stabilization by means of its side chain (Figure 5c). The position of the cutinase Ser 42 side chain is also conserved in CaBL with Thr 40 Oγl (see Figure 5d) (Uppenberg et al., 1995).

The atoms involved in the oxyanion hole of cutinase do not undergo conformational changes upon inhibitor binding (see Figure 5a), contrary to what has been observed in the case of RmL (Brzozowski et al., 1991; Derewenda et al., 1992) and of hpL (van Tilbeurgh et al., 1993; Egloff et al., 1995), thus indicating that in cutinase a preformed oxyanion hole is present.

The Oxyanion Hole Mutants

Based on the structure of the cut-E600 complex (Martinez et al., 1994), it has been suggested that the cutinase oxyanion hole may consist not only of two main chain nitrogen atoms (Ser 42 N and Gln 121N) but also of the Ser 42 Oγ side chain. Moreover, Asn 84 Nδ2 and Gln 121 Nε2 maintain Ser 42 Oγ in the proper orientation for the hydrogen bonding of the oxyanion to occur. Point mutants that selectively disrupt each of these interactions were designed in order to assess the potential function of the Ser42 side chain as part of the cutinase oxyanion hole (Nicolas et al., 1996). To evaluate the extent of the tetrahedral intermediate stabilization provided

Table 4 Steady state kinetic parameters of native and mutants cutinase during the hydrolysis of pnp-butyrate

	k_{cat} s^{-1}	k_{cat}/K_M $s^{-1}.mM^{-1}$	% residual activity
WT	1800	2647	100.00
S42A	4	12	0.22
N84A	478	1062	26.5
N84L	55	190	3.00
N84W	2	6	0.11
N84D	2.5	19	0.16

by Ser 42 Oγ, Ser 42 has been mutated into alanine. Since the proper orientation of Ser 42 Oγ is directed by Asn 84, the mutants Asn 84 to Ala, Leu, Asp and Trp, respectively, were produced to investigate the importance of this indirect interaction in the stabilization of the oxyanion hole. Biochemical modifications of cutinase mutants have been evaluated with regard to their ability to hydrolyze *para*-nitrophenyl-butyrate (pnp-butyrate) in solution, since cutinase is able to hydrolyze this soluble substrate (Table 4). Any secondary effects originating from interfacial activation can therefore be dissociated from the primary effects on the integrity of the oxyanion hole.

The S42A mutation resulted in a drastic decrease in the activity (450 fold, see Table 4) without significantly perturbing the three dimensional structure and the active site geometry (Figure 6a). Mutations N84W, N84L and N84A disrupt the hydrogen bond stabilization of Asn84 Nδ2 with Ser 42 Oγ and modify the steric environment in the active site pocket. The activity is reduced to 0.11% in the Trp mutant and remains at 3% and 26.5% with Leu and Ala mutants, respectively (see Table 4). As the result of these three mutations with non polar residues, the catalytic triad and the oxyanion hole geometry remain unchanged (Figure 6b,c,d). The introduction of a net negative charge into the oxyanion hole pocket resulting from the N84D mutation modifies the hydrogen bonding network in the catalytic triad and the oxyanion hole (Figure 6e) and the activity is reduced to 0.13% (see Table 4). Although Asn84 appears to play a significant role in the electrostatic stabilization of the active site pocket, it does not seem to be absolutely essential, since mutants Asn84Leu and Asn84Ala retain some esterase activity. Our kinetic experiments and structural analysis of these cutinase mutants provide new evidence supporting the idea that Ser42 Oγ is an extra component of the cutinase oxyanion hole, acting in combination with the two main chain nitrogen atoms of residues Ser 42 and Gln 121. However, the reason(s) why cutinase and some lipases need three hydrogen bonds to stabilize the oxyanion, and subtilisin and other lipases need only two (Table 5), still remains to be elucidated.

The cut-TC4 Complex

Although studies based on the use of small organophosphate and organophosphonate inhibitors have widely contributed to the elucidation of the catalytic mechanism of lipases, these compounds bear little resemblance to a natural triglyceride substrate.

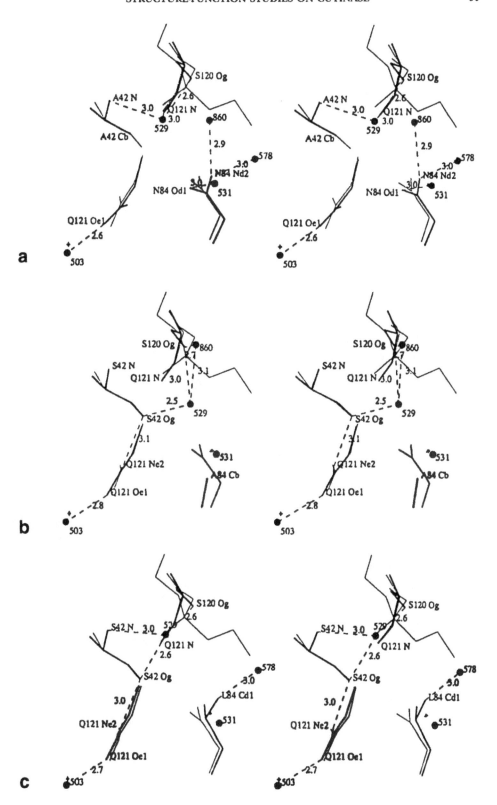

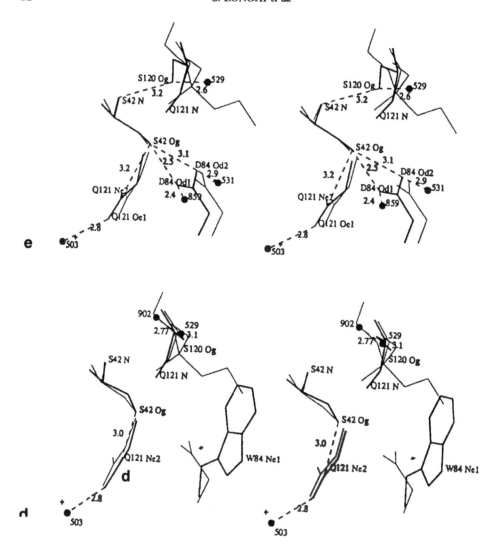

Figure 6 Stereoview of the oxyanion hole of E600 inhibited cutinase (thin) superimposed on that (thick) of S42A (**a**), and those of mutants N84A (**b**), N84L (**c**), N84W (**d**), and N84D (**e**)

In this context, the resolution of RmL (Brzozowski *et al.*, 1991; Derewenda *et al.*, 1992) complexed to a C6-phosphonate (*n*-hexyl *o*-ethyl chloro phosphonate, EECP), of CrL complexed with long alkyl sulfonyl chains (Grochulski *et al.*, 1994a) and of a human pancreatic lipase-colipase complex covalently inhibited by a long (C11) alkyl chain phosphonate (Egloff *et al.*, 1995) represent important attempts to mimic more closely and more realistically tetrahedral intermediates.

Taking into account the marked preference of cutinase for short chain (C4) triglycerides (Verger, personal communication; Mannesse *et al.*, 1995a), a four carbon phosphonate inhibitor (TC4) that mimics 1-2-trihexanoyl-3-pentanoyl glycerol (Mannesse *et al.*, 1995b) has been selected for co-crystallization experiments. The

Table 5 Examples of oxyanion stabilization components

Nucleophile-His-Asp (Glu) hydrolase with its nucleophilic residue		Main chain group			Side chain group	
		1st NH	2nd NH	3rd NH	NH	OH
Serine protease Trypsin	(S195)	S195	G193			
Chymotrypsin	(S195)	S195	G193			
Proteinase A	(S195)	S195	G193			
Subtilisin	(S221)	S221			N155 Nδ2	
Carboxypeptidase II	(S146)	S147	G53			
Cysteine protease Papain	(C25)	C25			Q19 Nε2	
Esterase *Streptomyces scabies* esterase	(S14)	S14	G66		N106 Nδ2	
Acetylcholinesterase	(S200)	A201	G118	G119		
Cholesterol esterase	(S209)	A210	G124	G123		
Lipase *Candida rugosa* lipase	(S209)	A210	G124	G123		
Human pancreatic lipase	(S152)	L153	F77			
Horse pancreatic lipase	(S152)	L153	F77			
Pseudomonas glumae lipase	(S87)	Q88	L17			
Mucor miehei lipase	(S144)	L145	S82			S82 Oγ
Penicillium camembertii lipase	(S145)	L146	S81			S81 Oγ
Humicola lanuginosa lipase	(S146)	L147	S83			S83 Oγ
Rhizopus delemar lipase	(S145)	L146	T83			T83 Oγl
Candida antartica B lipase	(S107)	Q106	T40			T40 Oγl
Cutinase	(S120)	Q121	S42			S42 Oγ

structure of cutinase covalently inhibited by (R)-1,2-dibutyl-carbamoylglycero-3-O-*p*-nitrophenylbutyl-phosphonate (TC4) provides the first reported case of a complex with a triglyceride analogue (Longhi *et al.*, 1997a). Besides the obvious difference concerning the presence of a phosphorous atom as target of the nucleophilic attack, this inhibitor differs from a triglyceride in that the ester bonds have been replaced by the non-hydrolysable carbamates.

The cut-TC4 complex crystallises in space group $P2_1$, with two protein molecules (Mol A and Mol B) in the asymmetric unit. The overall structure of both molecules of the cut-TC4 complex is very similar to that of the native enzyme (Figure 7a). As in the case of the cut-E600 complex (Martinez *et al.*, 1994), the central β-sheet is the best conserved part and the most important Cα deviations occur in the C-terminus and in the loop 25–32 opposite to the substrate binding site (Figure 7a). As already observed with the cut-E600 complex (Martinez *et al.*, 1994), only minor differences could be observed between the lipid binding sites of both molecules and that of the native cutinase. The absence of large conformational changes in the substrate binding region upon binding to a non-hydrolysable analogue, represents to date a feature unique to cutinase (Martinez *et al.*, 1994; Longhi *et al.*, 1997a), contrary to the large movements that have been reported for covalently inhibited

a)

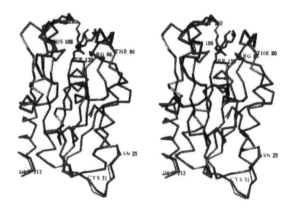

b)

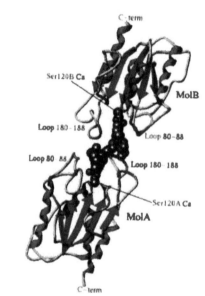

c)

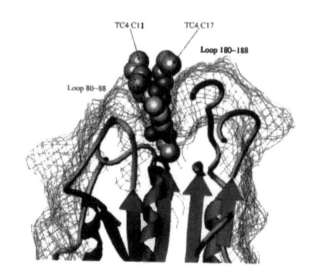

d)

e)

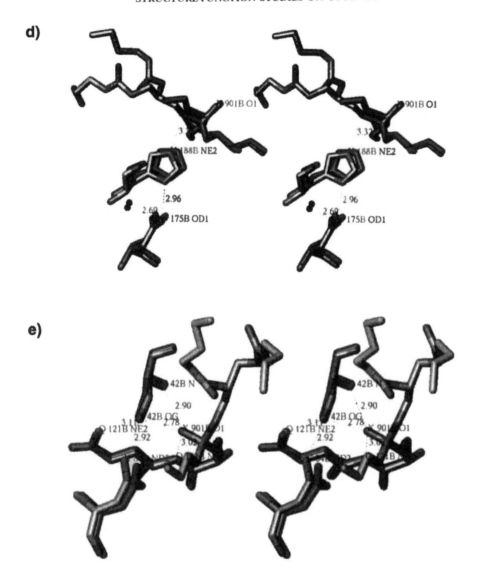

Figure 7 (a) Stereo view of the superimposition of the Cα tracing of Mol B (yellow) on the native structure (cyan). The TC4 B molecule is also shown (atom type colouring). (b) Secondary structure ribbon representation of the cut-TC4 complex, showing the two molecules in the asymmetric unit (Mol A and Mol B). The two inhibitor molecules are shown in CPK. (c) The lipid binding site region of Mol B. The solvent accessible surface (blue), the inhibitor (cpk), the loops 80–87 and 180–188 surrounding the catalytic crevice (yellow), the central β-sheet (blue) and the α-helices (red) are shown. (d) Stereo view of the catalytic triad of Mol B (atom type colouring) superimposed on that of the native enzyme (blue). The TC4 B (atom type colouring) and the E600 (magenta) inhibitors are displayed. The triad bound water molecule of the TC4 inhibited cutinase (red) and of the native enzyme (blue), together with the hydrogen bond network are also shown. (e) Stereo view of the oxyanion hole of Mol B (atom type colouring) superimposed on that of the native enzyme (blue). The TC4 inhibitor (atom type colouring) as well as the hydrogen bonds involved in the oxyanion hole stabilization are also shown

complexes of other lipases (Brzozowski *et al.*, 1991; Derewenda *et al.*, 1992; van Tilbeurgh *et al.*, 1993; Grochulski *et al.*, 1994a; Cygler *et al.*, 1994; Egloff *et al.*, 1995; Hermoso *et al.*, 1996).

The two protein molecules of the asymmetric unit are oriented head-to-head, with their substrate binding sites facing and interacting each other (Figure 7b). This structural arrangement allows the large hydrophobic surface, formed by the substrate binding region itself and by the inhibitor molecule, to be shielded from the solvent. The burying of hydrophobic surfaces in the crystal lattice by means of dimerization has been also observed in other lipase/esterase complexes (van Tilbeurgh *et al.*, 1993; Ghosh *et al.*, 1995) as well as in an open form of CrL (Grochulski *et al.*, 1993).

Only the S_P enantiomer of the inhibitor could be fitted into the electron density map of both sites. This observation suggests that only one enantiomer at phosphorous of the TC4 inhibitor had reacted, in agreement with the kinetic data showing that one of the stereoisomers at phosphorous reacts 60-fold faster than the other one. In both molecules, the overall shape of the inhibitor is that of a fork: the two dibutyl-carbamoyl chains point toward the surface of the protein (Figure 7c), while the butyl chain, linked to the phosphorus atom, is roughly perpendicular to the two dibutyl-carbamoyl chains. The butyl chain is embedded in a buried position at the bottom of the active site crevice, while the dibutyl-carbamoyl chains occupy a more exposed position (Figure 7c). The whole inhibitor molecule is almost entirely embedded in the active site crevice, with only the C10, C11, C16 and C17 atoms protruding from the surface of the protein (Figure 7c).

The presence of a rather small pocket which accomodates the butyl chain, provides a structural explanation for the preference of cutinase for substrates with short acyl chains in the *sn*-3 position (Mannesse *et al.*, 1995a). The limited number of atoms which can be embedded in this pocket would be responsible for the drop in the activity when using longer ester chain substrates. Conversely, no structural explanation can be found for the different chain-length selectivity of cutinase for position *sn*-1 and *sn*-2 (Mannesse *et al.*, 1995a). The drop in the activity observed when increasing the *sn*-1 chain length would suggest the existence in the structure of a site specific for the *sn*-1 chain and able to accomodate only short acyl chain substrates. Actually, both the dibutyl-carbamoyl chains point away from the catalytic crevice and no such a site is found. It should be noticed however, that the actual inhibitor conformation may be the result of the packing constraints: the crystal lattice contacts may stabilise the inhibitor in an extended conformation which might be unfavourable in solution. On the other hand, the intermolecular contacts between the inhibitor and the hydrophobic region from the facing molecule in the crystal lattice (see Figure 7b), mimic the interactions of a natural substrate with the lipidic matrix. Accordingly, the observed conformation might indeed occur in physiological conditions, with the C11 and C17 atoms of the inhibitor contacting the interface. This model, although possibly able to explain a general preference for short chain fatty acid esters, fails to account for the drop in the cutinase activity observed when increasing the length of the *only* *sn*-1 chain.

The total contact surface between the inhibitor and the protein is about 300 Å² in both molecules of the asymmetric unit. Most of the contribution is provided by the hydrophobic residues Val 184 and Leu 189 of the lipid binding site, followed by Thr 43, Ser 42 and Leu 81. Further residues in contact are Asn 84, Tyr 119, and most of the residues of the loop 180–188. In both molecules of the asymmetric unit, there are very few contacts between the residues of the substrate binding site and the region of the inhibitor mimicking the diacyl-glycerol moiety of a natural substrate. As already discussed, this lack of interactions may be related to the crystal lattice constraints which select an extended conformation of the inhibitor. As a result, the inhibitor points away from the active site cleft and establishes contacts with the facing hydrophobic surface. This interaction might nevertheless be reminiscent of the physiological interaction with the lipidic interface.

The structure of the catalytic triad (Figure 7d) and of the oxyanione hole (Figure 7e) of the cut-TC4 complex is basically unchanged with respect to that of the native enzyme. As in the case of the cut-E600 complex (Martinez et al., 1994) a water molecule located close to Ser 120 in the native enzyme, is replaced by the inhibitor. The active site water molecule which favours the proper position of the Oδ2 atom of Asp 175 in the catalytic network, is conserved in spatially equivalent positions in the native and in both TC4 inhibited molecules (Figure 7d). In both molecules of the asymmetric unit, the His 188 Nε2 atom is located 3.3 Å away from the O2 atom of the inhibitor, that is the proper spatial position which would allow the release of the alcohol and the formation of the acyl-enzyme for a natural substrate (Figure 7d). The superimposition of the cut-TC4 (Mol B) structure onto that of the cut-E600 complex shows that the overall position of the two inhibitors in the active site is the same (Figure 7d), thus suggesting the existence of a well-determined and favoured mode of interaction and accomodation in the active site.

The crystal structure of the cut-TC4 complex, although hardly explaining the different sn-1 and sn-2 chain length stereoselectivity of cutinase, provides a possible structural interpretation for its preference for short chain fatty acid esters. While cutinase activity on tributyrin is 40% of that of human pancreatic or RmL, on long chain triglycerides (triolein) the activity drops to 15% of the activities observed for these two lipases (Table 1a) (Kolattukudy, 1984; Lauwereys et al., 1991). The TC4 inhibitor is almost completely embedded in the active site crevice, as would be also a tributyrin molecule. Conversely, the human pancreatic lipase, which has a high activity on triolein, can accomodate a C11 mono-alkyl phosphonate inhibitor (Egloff et al., 1995).

Cutinase activity on soluble tributyrin (below the CMC) is accounted for by the accessibility of its active site and by the deep embedding of the substrate in the hydrophobic catalytic crevice. This embedding reduces the unfavourable interactions between the acyl chains of the substrate and the surrounding bulk water. At concentrations exceeding the CMC, the binding of the substrate at the lipid/water interface and its burying in the active site cleft would result in an almost complete removal of the tributyrin from the lipidic phase. Conversely, a triolein molecule, which would protrude from the cutinase active site crevice, would still continue interacting with the lipidic phase. Which situation is the most favourable? Cutinase has a lower penetrating power than pancreatic lipase or RmL in lipidic phases, as evaluated by monolayer studies (Rogalska et al., 1995). We can assume that its ability to extract, even partially, long triglycerides from the lipidic phase is therefore reduced, thus leading to a lower activity. This would not be the case with true lipases which have deep and broad active sites formed upon interaction with the lipid/water interface. It is therefore tempting to establish a correlation between the selectivity of lipases for triglycerides of certain length and the depth of their active site.

DYNAMICS OF CUTINASE AND PACKING FORCES

During the past few years, an increasing interest has been payed to protein dynamics as the result of the elucidation of its crucial role in protein function and enzyme catalysis. X-ray 3D structure determinations give an accurate but rather static description of crystalline proteins. Motion is deeply implicated in biological function, and the crystal structure of a protein constitutes only an average conformation. In addition, when comparing proteins in solution and in a crystalline environment, significant local differences have been found to occur, which suggests that packing forces may affect the dynamics of proteins. Crystal packing provides a stabilising environment which prevents the protein from undergoing substantial conformational changes. The X-ray resolution of various crystal forms folded in different ways makes it possible to describe protein motion by combining several static pictures (Vonrhein *et al.*, 1995). Large conformational changes have been described in this way in several protein families, such as the hexokinase (Bennett & Steitz, 1980), the T4 lysozyme (Dixon *et al.*, 1992; Zhang *et al.*, 1995), the transferrins (Anderson *et al.*, 1990) and the lipases (Brzozowski *et al.* 1991; Derewenda *et al.*, 1992; van Tilbeurgh *et al.*, 1993; Egloff *et al.*, 1995; Grochulski *et al.*, 1994a,b).

Although in the case of cutinase it has been demonstrated that the binding of a non sissile substrate analogue does not induce any significant structural rearrangements (Martinez *et al.*, 1994; Longhi *et al.*, 1997a), minor motions can nevertheless be expected to occur in the substrate-binding region during the interaction with the substrate. No evidence in this regard can be provided by the analysis of the native and complexed forms, which constitute only the starting and final points of the interaction. The availability of several cutinase variant structures in different crystal forms (Table 6) makes cutinase an ideal system for studying protein dynamics by means of a structural comparison among several closely related proteins in different crystal-packing environments.

Structural Comparison Among Different Crystal Forms of its Variants

A comparative structural analysis among several cutinase variants crystallising into 8 different crystal forms, has been carried out (Longhi *et al.*, 1996). Cutinase variants crystallise either in the wild-type form (type I, which is the most represented one with 26 variants) or in a form that is characteristic of the particular variant (see Table 6). Accordingly, two separate structural comparative analyses have been carried out, which included either the type I variants (isomorphous analysis), or one variant for each crystal form (heteromorphous analysis). For each set, the RMSD of the $C\alpha$ distances ($C\alpha$ RMSD) from the average structure was computed for each $C\alpha$ position. The residue-by-residue plots of the $C\alpha$ fluctuations are shown in Figure 8a,b. In both sets, the possibility that the observed $C\alpha$ fluctuations might be related either to differences in the quality of the model, or to eventual local distortions introduced by the mutation(s) or by the accomodation of the inhibitor, was checked and ruled out. Despite the overall weak flexibility detected within the set of isomorphous cutinase variants, a few discrete regions displaying a significantly higher flexibility were identified (Figure 8a). The comparison between the results of the heteromorphous analysis (Figure 8b) with those of the isomorphous one (Figure 8a), points out a much greater overall flexibility in the former case. When

Table 6 List of solved cutinase variant structures

Cutinase variant	Crystal form	Space group	Nb mol per AU	Res. (Å)	PDB code
Native (wt)	I	$P2_1$	1	1.60	1CUS
Native (wt)*	I	$P2_1$	1	1.15	1AGY
Native (wt)*	I	$P2_1$	1	1.00	1CEX
wt-EECP	I	$P2_1$	1	1.69	1XZL
wt-MUCP	I	$P2_1$	1	1.75	1XZM
S120C	I	$P2_1$	1	1.60	1CUJ
S92C	I	$P2_1$	1	1.75	1XZE
S213C*	I	$P2_1$	1	2.70	1XZD
S129C	I	$P2_1$	1	1.80	1XZA
T144C	I	$P2_1$	1	1.69	1XZF
A199C	I	$P2_1$	1	1.69	1CUU
L114Y	I	$P2_1$	1	1.75	1CUX
L189F	I	$P2_1$	1	1.69	1CUY
L81G,L182G	I	$P2_1$	1	2.10	1CUZ
N172K,R196E (I)	I	$P2_1$	1	1.75	1CUB
R17E,N172K	I	$P2_1$	1	1.75	1CUG
T45A	I	$P2_1$	1	1.69	1XZG
Y38F	I	$P2_1$	1	1.69	1XZJ
T80P	I	$P2_1$	1	1.69	1XZH
N172K	I	$P2_1$	1	1.80	1CUA
R156L	I	$P2_1$	1	1.75	1CUF
S42A	I	$P2_1$	1	1.69	1FFE
N84A	I	$P2_1$	1	1.69	1FFA
N84L	I	$P2_1$	1	1.75	1FFC
N84W	I	$P2_1$	1	1.69	1FFD
Q121L*	I	$P2_1$	1	2.70	1CUE
S120A*	I	$P2_1$	1	2.70	1CUI
Y119H	I	$P2_1$	1	1.69	1XZI
R196E*	II	$P2_12_12_1$	1	1.75	1CUH
N172K,R196E (II)	II	$P2_12_12_1$	1	1.75	1CUC
N172K,R196E (III)	III	$P2_1$	3	2.70	1CUD
N84D	IV	$P2_1$	1	1.75	1FFB
A85F	V	I222	1	2.01	1CUV
wt-DFP	VI	$P2_1$	2	2.01	1XZK
wt-E600	VII	$P2_12_12_1$	1	1.90	2CUT
G82A,A85F,V184I,A185L,L189F	VIII	C2	2	2.70	1CUW
wt-TC4*	IX	$P2_1$	2	2.30	1OXM

* not included in the structural comparison; EECP: n-hexyl o-ethyl chloro phosphonate; MUCP: n-uncedyl o-methyl chloro phosphonate; DFP: di-isopropyl fluoro phosphate; E600: n diethyl-p-nitrophenyl phosphate ; TC4: (R)-1,2-dibutyl-carbamoylglycero-3-O-p-nitrophenylbutyl-phosphonate

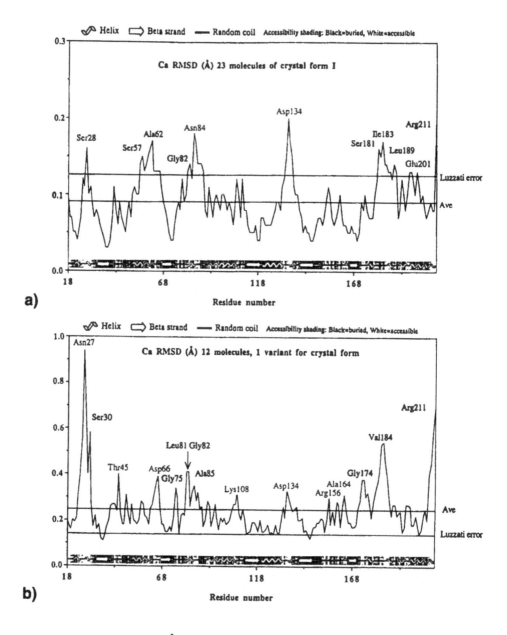

Figure 8 (**a, b**) Cα RMSD (Å) as a function of residue number, as obtained by including in the structural comparison (**a**) the 23 isomorphous variants belonging to crystal form I (isomorphous set) or (**b**) a single variant for each crystal form (heteromorphous set). (**c, d**) Secondary structure ribbon model of the Cα trace of cutinase in two different orientations. The most flexible regions detected upon heteromorphous structural analysis are highlighted. (**e**) Cα RMSD (Å) from the original native structure, as obtained from an 800 ps MD simulation trajectory (solid line) and as observed within the heteromorphous crystallographic data set (dotted line)

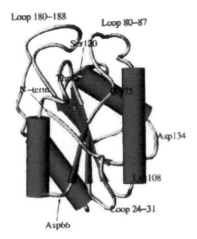

c)

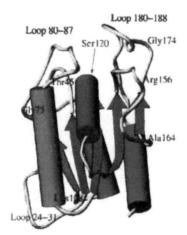

d)

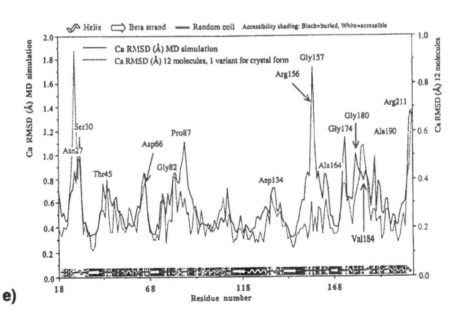

e)

comparing different crystal environments, the Cα fluctuations turn out to be amplified and a few additional peaks become detectable. The stretches 24–30, 64–66, 81–87, 173–176, 180–187 and 208–211 were identified as flexible domains (Figure 8b). Asn 27 and Ser 30 map on an exposed loop (residues 25–32) that is located far from and opposite to the substrate-binding region (Figure 8c,d). Arg 211, the second RMSD peak, corresponds to the C-terminal residue (Figure 8c,d), whereas Leu 81 and Val 184 fall into the two loops (residues 80–88 and 180–188) that delimit the active site crevice (Figure 8c,d) The identification of these two stretches of residues as flexible domains provides the first direct structural evidence of the expected dynamic behavior of the surface loops surrounding the catalytic crevice. This mobility had been previously suggested and predicted to occur in order to allow a productive interaction with the substrate (Martinez *et al.*, 1992). The occurrence of an appreciable flexibility in the case of Thr45 (Figure 8b), which maps on an exposed loop in the proximity of the catalytic crevice (Figure 8c,d), further confirms that the structural elements that belong to the substrate-binding region have a greater mobility. As regards the location in the secondary structure elements, not surprisingly, all the flexible points fall into loops (Figure 8c,d), thus indicating that the structural elements building up the protein scaffold are endowed with a greater rigidity.

In both sets, the flexibility has been shown to correlate with thermal motion. In order to evaluate whether residue mobility is affected by crystal packing, the involvement of the residues in intermolecular interactions has been analysed. When considering the same crystal packing environment, a reduced flexibility is observed in the case of residues/regions greately involved in packing contacts, and *vice versa*. However, when different crystal packing environments are concerned, a greater degree of participation in crystal packing contacts may even result in a greater flexibility. Let us assume, as an example, that a given loop displaying high B-factor values and which is not involved in crystal packing contacts in crystal form I, becomes involved in intermolecular interactions in a different crystal form. The setting up of these packing contacts may stabilise the loop in a different conformation, thus resulting in large deviations with respect to crystal form I. An increased participation in crystal packing contacts would therefore be correlated with greater positional fluctuations. In the same way, a reduced participation in packing contacts may cause a given loop to occur in a different spatial position. When several crystal forms are concerned, a qualitative difference in the environment surrounding a certain loop would therefore result in appreciable flexibility. The availability of several crystal forms makes it possible to detect different conformations with similar energy in each crystal form, thus allowing to identify motions which would have remained unidentified if only a single crystal form had been available.

By combining several static pictures in different crystal environments, the dynamics of the protein can therefore be inferred and described much more realistically than what can be achieved on the basis of static pictures alone. A crystallographic approach is therefore a quite suitable tool for investigating protein dynamics, providing that crystal polymorphism is available.

Molecular Dynamics Simulation

The application of such a crystallographic approach for studying protein dynamics is still limited by the rather small number of suitable protein families. The fundamental prerequisite, that is, the availability of 3D structures of a set of closely related proteins, is still the "rate-limiting step" in this approach. In most cases, no such information is available, which makes molecular dynamics (MD) simulation the most frequently used method of analysing protein motion as a function of time and obtaining a physical description of protein behavior. The abundance of experimental crystallographic data on cutinase provides an excellent means of testing the reliability of MD simulation approaches.

Accordingly, the information about protein motion deduced from the comparative structural analysis have been compared with that obtained by subjecting native cutinase to a MD simulation (Longhi et al., 1996). The Cα fluctuations calculated from the MD simulation of cutinase have been compared to the experimental mobilities observed within the heteromorphous crystallographic data set (Figure 8e). Even if an overall greater flexibility is observed upon MD simulation (as a result of the much greater degree of freedom of a protein in water rather than in a crystal environment), a fairly good agreement has been found between the experimental data (Figure 8e, dotted line) and the MD simulation (Figure 8e, solid line). The quite good correlation observed between the crystallographic mobilities and those inferred from a MD simulation approach is particularly relevant, since it confirms the validity of the latter approach as a means of describing protein motion.

Packing Forces

At present, much still remains to be elucidated about the underlying mechanism driving protein-protein interactions in the crystallization process. Only a few studies have focused on the relationships between the crystal growth conditions, the packing and the diffraction power of individual crystals. Mittl et al. (1994) have examined the effect of enlarging the packing contacts of crystals of glutathione reductase. They mutated two residues and thus introduced an additional crystal contact, consisting of two intermolecular hydrogen bonds. The mutations had no effect on the diffraction power or on the stability of the crystal, but the nucleation process was greatly accelerated and the crystallization time reduced. Similarly the intermolecular cross-linking of the T4 lysozyme resulted in a more rapid crystallization of the protein (Heinz & Mathews, 1994). Crosio et al. (1992) analysed the molecular packing of six crystal forms of bovine pancreatic ribonuclease A obtained from different crystallization conditions. They found that the six packings differ in the number of molecules in contact and in the size of pairwise interfaces. However, they found no systematic differences in either the total area of the protein surface buried in crystal contacts or the total number of polar interactions. The protein-protein interactions in the crystal lattice turned out to be non-specific and to involve essentially all regions of the protein surface (Crosio et al., 1992).

In the case of cutinase, most of the various crystal forms have been obtained under crystallization conditions similar to those used with the native enzyme (Abergel et al., 1990; Longhi et al., 1996). The availability of various crystal forms (Longhi et al., 1996 & 1997a), resulting from either mutation(s) or complexation with inhibitors, provides a powerful tool to get insights into the nature of the packing forces as well as on the effect of various mutations on the crystal packing. Accordingly, the lattice contacts in the 9 different crystal forms of cutinase have been analyzed. The relationships between hydrophobicity and accessibility from one hand, and propensity to participate in crystal packing contacts on the other hand, have been investigated. With a view of determining the nature of the crystal lattice contacts, the hydrophobicities/polarities of surfaces in mutual contact have been analyzed as well (Jelsch et al., 1998).

Protein mutants with a different charge distribution form new intermolecular salt bridges or long range electrostatic interactions which are accompanied by a change in the crystal packing. The introduction or the modification of charged residues at the protein surface leads to a change in the crystal packing, so that the protein molecules become oriented in such a way as to compensate for their opposite charges. Likewise, increasing the hydrophobicity at the protein surface by means of amino acid substitutions or by complexation with hydrophobic ligands result in the establishment of new crystal packings (see Table 6).

The whole protein surface is involved in packing contacts and the hydrophobicities of the protein surfaces in mutual contact have been shown to be non-correlated, which indicates that the packing interactions are essentially non specific. In the case of the hydrophobic variants however, the packing contacts exhibit a certain degree of specificity, as the protein in the crystal tends to form either crystallographic or non-crystallographic dimers, which shield the hydrophobic surface from the solvent.

The likelyhood of surface atoms to be involved in a crystal contact is the same for both polar and non-polar atoms. However, when considering areas in the 200–600 Å^2 range instead of individual atoms, the surface regions either highly hydrophobic or highly polar have been found to have an increased probability to establish crystal lattice contacts. The protein surface surrounding the active site crevice of cutinase constitutes a large hydrophobic area which is involved in packing contacts in all the various crystalline contexts.

One of the most relevant implications arising from this study, is the possibility of using site-directed mutagenesis for "crystal engineering" purposes. The introduction of either charged or hydrophobic residues at the protein surface, seems to be a very promising tool to crystallize a given protein in a new form. Such an approach has been already successfully used in the case of the human H ferritin (Lawson et al., 1991) and apoferritin (Takeda et al., 1995).

THE ATOMIC STRUCTURE (1.0 Å) OF NATIVE CUTINASE

Implications for Protein Sterochemistry

One of the most relevant implications of atomic resolution is the possibility of a better definition of the stereochemical parameter dictionaries. To date, even though about 50% of the crystal protein structures in the Protein Data Bank (Bernstein et al., 1977) are quoted as high resolution, only few of them have been solved at truly atomic resolution. Moreover, despite the increasing number of proteins for which data have been recorded at atomic resolution (for a comprehensive review see Dauter et al., 1995), the available data in the literature are still scarce. In most cases, small protein molecules ($M_r < 10$ KDa) are concerned. Although, in the next future the number of proteins at atomic resolution is expected to exponentially increase, at present the resolution at 1.0 Å of cutinase, a 22 KDa protein, represents the largest structure ever refined to such an extent of resolution (Longhi et al., 1997b). The data were collected on the EMBL beamline X11 at the DORIS strorage ring,

DESY, Hamburg. The structure has been refined anisotropically to a final R-factor of 9.4%. The overall accuracy of the final model is very high due to an observation to parameter ratio of 5.5. The mean coordinate error for protein atoms is 0.013 Å.

The availability of an accurate cutinase structure at atomic resolution allowed us to carry out an extensive analysis of several stereochemical parameters, such as peptide planarity, main chain and some side chain bond distances. The results support a need for revision of some well-established parameters in protein stereochemistry (Engh & Huber, 1991). Some of the interatomic distances in existing libraries are not accurate enough and significant discrepancies from planarity were found for the ω angles. The reliability of the observed deviations is validated by means of the associated estimated error. Large size of the proteins provides larger sample and thus more reliable estimates. The atomic structure of cutinase provides a further example which will contribute to the challenge of finding new and more appropriate parameters for protein stereochemistry.

Beyond this aspect, which is of quite general interest for structural biologists, the atomic resolution of cutinase provided insights in the catalytic mechanism and in the dynamics of this enzyme.

The Active Site Region

The X-ray data allowed the identification of most hydrogen atoms in the electron density. Accordingly, a precise and detailed description of the network of hydrogen bonds in the catalytic site has been provided. Unambiguous evidences on the protonation state of the catalytic histidine have been obtained. When an omit map was computed with the His 188 Hδ1 atom removed from the model, a difference Fourier peak ($\sigma = 3$) close to the Nδ1 atom could be observed (Figure 9a). The possibility of identifying His hydrogens in the electron density has a relevant implication concerning the catalytic mechanism of cutinase: even though the occurrence of the Hδ1 atom with the concomitant absence of the Hϵ2 one on the catalytic histidine, had been postulated for a long time, our data represent the first direct structural evidence on the protonation state of the catalytic His of cutinase.

In the 1.0 Å structure, the Ser 120 Oγ as well as the Asn 84 Cβ, Cγ, Nδ2 and Oδ1 atoms are in double conformation. The more and less occupied components of the Ser 120 Oγ are located at 2.71 Å and at 3.71 Å from the His 188 Nϵ2 atom, respectively (Figure 9b). The less occupied position, being located too far away from the catalytic histidine, is not likely to play a relevant role in catalysis. The active site water molecule has been modelled as two sites, in which the occupancy of the less occupied member was tied to that of the less occupied member of the Ser 120 Oγ. The two Ser 120 side chain conformations are hydrogen bonded to the two components of the solvent molecule in double conformation in an arrangement scheme in which the less occupied member of the solvent molecule faces the more occupied Ser 120 Oγ conformation (Figure 9b). The third component of the oxyanion hole, i.e. the Ser 42 Oγ atom, is stabilised by two hydrogen bonds with the Gln 121 Nϵ2 atom (3.15 Å) and with the more occupied Nδ2 site of the Asn 84 side chain (3.06 Å) (Figure 9c). The less occupied Asn 84 Nδ2 site, occurring at 3.96 Å from the Ser 42 Oγ, does not seem to contribute to the stabilisation of the Ser 42 hydroxyl group position. The less occupied alternate conformation of Asn 84 is stabilised by the solvent site in double conformation (Figure 9c). The Ser 42

a)

b)

c)

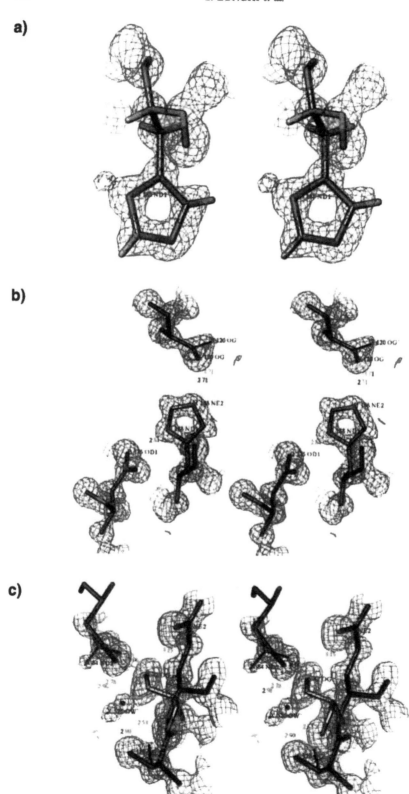

d)

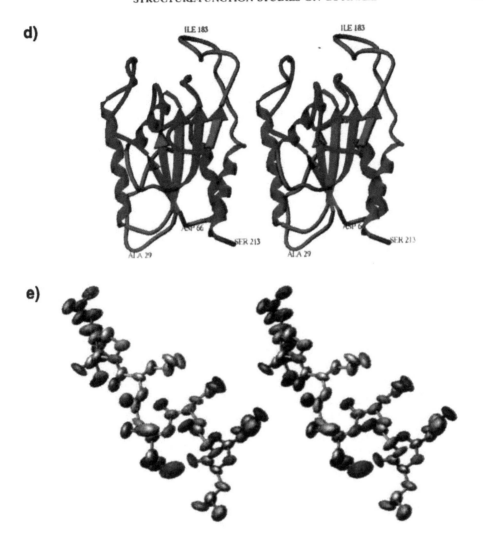

e)

Figure 9 (a) Stereo view of the omit map around His 188 as obtained after removal of the His 188 Hδ1 atom. The 3Fo-2Fc (1σ) and Fo-Fc (3σ) electron density maps are shown in blue and in red, respectively. (b, c) Stereo view of the electron density 3Fo-2Fc (1σ) around (b) the catalytic triad and (c) the oxyanion hole. The less occupied conformations of the split positions are shown in thin. The solvent site in double conformation coupled to Ser 120 Oγ is also shown. The relevant hydrogen bonds are displayed. (d) Stereo view of the Cα tracing (ribbon representation) of cutinase. Colours are according to anisotropy, ranging from blue to red with increasing anisotropy. (e) TURBO-FRODO stereo view of the 20% thermal ellipsoids of the loop 80–88

side chain displays an appreciable extent of flexibility, as indicated by the electron density map which points out a high extent of anisotropy (Figure 9c).

The atomic resolution allowed to detect an appreciable extent of flexibility in the active site of cutinase. Although the less occupied alternate conformations do not seem to play a role in catalysis, they clearly indicated that the active site is less rigid than suggested by previous crystallographic studies (Martinez *et al.*, 1992 & 1993; Nicolas *et al.*, 1996, Nicolas, 1996; Longhi *et al.* 1996). These results are in agreement with those recently obtained by an NMR study of cutinase which showed that the residues in, or near, the active site display functional motions in the ms-ms time scale (Prompers *et al.*, 1996). The biological relevance of this flexibility might reside in a possible adaptation to different substrates which might be amplified in the course of evolution.

The Analysis of the Anisotropy

The refinement of cutinase at atomic resolution, beyond the possibility of detecting more sensibly atomic motions, provided information into their anisotropic nature. In order to get some quantitative insights into the distribution of anisotropy within the protein, the anisotropy (A) of each atom was expressed as

$$A = \sigma(\lambda_i)/\langle\lambda_i\rangle$$

where λ_1, λ_2 and λ_3 are the principle components of the U_{ij} matrix, $\langle\lambda_i\rangle$ and $\sigma(\lambda_i)$ are the mean and the standard deviation of λ_i, respectively. The computation method chosen has the advantage of rendering A independent from thermal motion, thus providing a convenient measure of the "ellipsoidicity". The average anisotropy of each residue has been computed. A few highly anisotropic regions (25–31, 62–67 and 134–136) and residues can be observed (Figure 9d). The region exhibiting the greatest anisotropy is the loop 24–31 (Figure 9d). As already mentioned, this loop, which is exposed to the solvent and opposite to the substrate binding site, displays a significantly greater extent of thermal motion in all the different crystal forms of cutinase variants solved so far (Martinez *et al.*, 1994; Nicolas *et al.*, 1996; Longhi *et al.*, 1996 & 1997a). Besides this disordered loop, the loops 80–88 and 180–188 exhibit an extent of anisotropy greater than the average (Figure 9d). In Figure 9e are shown the 25% thermal ellipsoids of the loop 80–88 which, together with the loop 180–188, builds up the catalytic crevice. Based on the visual inspection, a main coherent direction for motion can be identified which runs from the left bottom to the right top corner of the figure. This motion, bringing the loop 80–88 close to the facing loop 180–188, produces a sort of breath-like movement in the substrate binding region. The high extent of thermal motion and of anisotropy of this loop is an agreement with what expected for an efficient substrate binding region, which should display a certain flexibility in order to allow productive interaction with the substrate. The analysis of the anisotropic motions confirms the indications previously obtained (Longhi *et al.*, 1996) on the dynamic behaviour of the substrate binding site of cutinase.

CONCLUSIONS AND PERSPECTIVES

The availability of an atomic model of cutinase and of the 3D structure of several cutinase mutants and covalently inhibited complexes, have greatly contributed to elucidate the structure-function relationships of cutinase. Despite this enormous body of information, the precise behaviour of cutinase *during* the interaction with the substrate, i.e. the mode of approach to the interface, and the release of the substrate still remain to be elucidated. The structure of cutinase covalently complexed with an analogue of triglyceride has revealed that the acyl moiety of the ligand establish very few interactions with the substrate binding region, thus failing to provide a convenient explanation for the stereoselectivity of cutinase. Some elucidations in this regard are likely to be provided by time-resolved crystallographic approaches, which would allow to follow the release of the substrate. In this frameship, the accuracy of structural information obtained from very rapid Laue data collection (X-ray pulses of 150 psec and total exposure time shorter than 10 nsec) on native cutinase, assumes a particular relevance and is very encouraging (Bourgeois *et al*, 1997). Since tertiary structural changes in proteins generally occur in the msec to nsec timescale at physiological temperatures, new opportunities are opened in the field of time resolved crystallography. Short laser pulses could be used to trigger a photoreaction in cutinase crystals by using suitable caged compounds. The structural modifications intervening during the release of the substrate could be probed at various time delays (in the msec to nsec time scale) by recording Laue data from single X-ray flashes.

From an applicative standpoint, the identification of the structural determinants responsible for the specificity and stereoselectivity of cutinase may open the way to the rational design of point mutants endowed with a different substrate preference. New variant forms could therefore be obtained by site directed mutagenesis approaches, thus allowing the use of cutinase in the catalysis of new substrates. Cutinase variants with an increased resistence to detergents and high temperatures have been already obtained and are presently used in the detergent industry.

ACKNOWLEDGEMENTS

We thank the Unilever Research Laboratorium for providing us with purified recombinant cutinase (native and mutant forms), Philippe H. Hüenenberg for providing us with the MD simulation data and Lucia Creveld for the computation and analysis of the MD trajectories. We are also grateful to Maurice Mannesse and Hubertus M. Verheij for synthesizing the TC4 inhibitor and for kinetically characterizing it. We thank Keith Wilson for assigning us synchrotron beam-time at the EMBL Hamburg outstation and Victor Lamzin for his precious assistance. We thank Anne Gael Inisan and Alain Roussel for implementing in TURBO-FRODO the option for the representation of the anisotropic ellipsoids. We thank Michael Wulff for assigning us synchrotron beam-time at the ESRF Grenoble outstation and Dominique Bourgeois for his kind assistance. Finally we wish to thank Maarten Edmond, Jakob de Vlieg, Hubertus M. Verheij and Robert Verger for stimulating discussions and critical reading of some papers.

These studies were supported by the EC BRIDGE-T Lipase Project (BIOT CT91-023), the EC BIOTECH Programs (BIO2 CT94-3016 and BIO2 CT94-3013), the EC HCM Large-Scale Facilities Program (CHGE CT93-0040), the CNRS-IMABIO Program and the PACA region.

REFERENCES

Abergel, C., Martinez, C., Fontecilla-Camps, J., Cambillau, C., de Geus, P. and Lauwereys, M. (1990) Crystallization and preliminary X-ray study of a recombinant cutinase from *Fusarium solani pisi*. *J. Mol. Biol.*, **215**, 215–216.

Anderson, B.F., Baker, H.M., Norris, G.E., Rumball, S.V. and Baker, E.N. (1990) Apolactoferrin structure demonstrates ligand-induced conformational change in transferrins. *Nature*, **344**, 784–787.

Baker, E.N. and Hubbart, R.E. (1984) Hydrogen bonding in globular proteins. *Prog. Biophys. Mol. Biol.*, **44**, 97–179.

Barlow, D.J. and Thornton, J.M. (1988) Helix geometry in proteins. *J. Mol. Biol.*, **201**, 601–619.

Bennett, W.S. Jr. and Steitz, T.A. (1980) Structure of a complex between yeast hexokinase A and glucose. Detailed comparisons of conformation and active site conformation with the native hexokinase B monomer and dimer. *J. Mol. Biol.*, **140**, 211–230.

Bernstein, F.C., Koetzle, T.G., Williams, G.J.B., Meyer, E.F. Jr., Brice, M.D., Rogers, J.R., Kennard, O., Shimanouchi, T. and Tasumi, M. (1977) The Protein Data Bank: a computer-based archival file for macromolecule structures. *J. Mol. Biol.*, **112**, 535–542.

Bott, R., Wu, S., Dauberman, J., Power, S., van Kimmenade, A., Jarnigan, A., Fallon, E., Boston, M., Poulose, A.J. and Raghupathy, S. (1993) In *Lipases: Structure, function and protein engineering*, p. 33, program abstract. Denmark: Elsinore.

Bourgeois, D., Longhi, S., Wulff, M. and Cambillau, C. (1997) Accuracy of structural information obtained at ESRF from very fast Laue data collection on macromolecules. *J. Applied Crystallogr.*, **30**, 153–163.

Bourne, Y., Martinez, C., Kerfelec, B., Lombardo, D., Chapus, C. and Cambillau, C. (1994) Horse pancreatic lipase, the crystal structure refined at 2.3 Å resolution. *J. Mol. Biol.*, **238**, 709–732.

Brady, L., Brzozowski, A., Derewenda, Z.S., Dodson, R., Dodson, G., Tolley, S., Turkenburg, J.P., Christiansen, L., Huge-Jensen, B., Norskov, L., Thim, L. and Menge, U. (1990) Serine protease triad forms the catalytic centre of a triacylglycerol lipase. *Nature (London)*, **343**, 767–770.

Brzozowski, A.M., Derewenda, U., Derewenda, Z.S., Dodson, G., Lawson, D. and Turkenburg, J.P. (1991) A model for interfacial activation in lipases from the structure of a fungal lipase-inhibitor complex. *Nature (London)*, **351**, 491–494.

Bullock, T.L., Branchaud, B. and Remington, S.J. (1994) Structure of the complex of L-benzylsuccinate with wheat serine carboxypeptidase II at 2.0-Å resolution. *Biochemistry*, **33**, 11127–11134.

Cambillau, C. and van Tilbeurgh, H. (1993) Structure of hydrolases: lipases and cellulases. *Curr. Opinion Struct. Biol.*, **3**, 885–895.

Cambillau C., Longhi S., Nicolas A. and Martinez C. (1996) Acyl glycerol hydrolases: inhibitors, interface and catalysis. *Curr. Opinion Struct. Biol.*, **6**, 449–455.

Crosio, M.P., Janin, J., Jullien, M. (1992) Crystal packing in six crystal forms of pancreatic ribonuclease. *J. Mol. Biol.*, **228**, 243–251.

Cygler, M., Grochulski, P., Kazlauskas, R.J., Schrag, J.D., Bouthillier, F., Rubin, B., Serregi, A.N. and Gupta, A.K. (1994) A structural basis for the chiral preferences of lipases. *J. Am. Chem. Soc.*, **116**, 3180–3186.

Dauter, Z., Lamzin, V.S. and Wilson, K.S. (1995) Proteins at atomic resolution. *Curr. Op. Struct. Biol.*, **5**, 784–790.

de Geus, P., Lauwereys, M. and Matthyssens, G. (1989) European Patent Application Nr. PCT 89.400. 462.1.

Derewenda, U. and Derewenda, Z.S. (1991) Relationships among serine hydrolases: evidence for a common structural motif in triacylglyceride lipases and esterases. *Biochem. Cell Biol.*, **69**, 842–851.

Derewenda, U., Brzozowski, A.M., Lawson, D.M. and Derewanda, Z.S. (1992) Catalysis at the interface: the anatomy of a conformational change in a triglyceride lipase. *Biochemistry*, **31**, 1532–1541.

Derewenda, U., Swenson, L., Green, R., Wei, Y., Yamaguchi, S., Joerger, R., Haas, M.J. and Derewenda, Z.S (1994a) Current progress in crystallographic studies of new lipases from filamentous fungi. *Protein Eng.*, **7**, 551–557.

Derewenda, U., Swenson, L., Green, R., Wei, Y., Kobos, M.P., Joerger, R., Haas, M.J. and Derewenda, Z.S. (1994b) Conformational lability of lipases observed in the absence of an oil-water interface: crystallographic studies of enzymes from the fungi *Humicola lanuginosa* and *Rhizopus delemar*. *J. Lipid Res.*, **35**, 524–534.

Derewenda, U., Swenson, L., Green, R., Wei, Y., Dodson, G.G., Yamaguchi, S., Haas, M.J. and Derewenda, Z.S. (1994c) An unusual buried polar cluster in a family of fungal lipases. *Struct. Biol.*, **1**, 36–47.

Dixon, M.M., Nicholson, H., Shewchuk, L., Baase, W.A. and Matthews, B.W. (1992) Structure of a hinge-bending bacteriophage T4 lysozyme mutant, Ile → Pro. *J. Mol. Biol.*, **227**, 917–933.

Egloff, M.-P., Marguet, F., Buono, G., Verger, R., Cambillau, C. and van Tilbeurgh, H. (1995) The 2.46 Å resolution structure of the pancreatic lipase colipase complex inhibited by a C11 alkyl phosphonate. *Biochemistry*, **34**, 2751–2762.

Endrizzi, J.A., Breddam, K. and Remington, S.J. (1994) 2.8-Å structure of yeast serine carboxypeptidase. *Biochemistry*, **33**, 11106–11120.

Engh, R.A. and Hubert, R. (1991) *Acta Crystallogr.*, A47, 392–400.

Ettinger, W.F., Thukral, S.K. and Kolattukudy, P.E. (1987) Structure of cutinase gene, cDNA, and the derived amino acid sequence from phytopathogenic fungi. *Biochemistry*, **26**, 7883–7892.

Ghosh, D., Wawrzak, Z., Pletnev, V.Z., Li, N., Kaiser, R., Pangborn, W., Jörnvall, H., Erman, M. and Duax, W.L. (1995) Structure of uncomplexed and linoleate-bound Candida cylindracea cholesterol esterase. *Structure*, **3**, 279–288.

Grochulski, P., Li, Y., Schrag, J.D., Bouthillier, F., Smith, P., Harrison, D., Rubin, B. and Cygler, M. (1993) Insights into interfacial activation from an "open" structure of *Candida rugosa* lipase. *J. Biol. Chem.*, **268**, 12843–12847.

Grochulski, P., Bouthillier, F., Kazlauskas, R.J., Serregi, A.N., Schrag, J.D., Ziomek, M. and Cygler, M. (1994a) Analogs of reaction intermediates identify a unique substrate binding site in *Candida rugosa* lipase. *Biochemistry*, **33**, 3494–3500.

Grochulski, P., Li, Y., Schrag, J.D., and Cygler, M. (1994b) Two conformational states of *Candida rugosa* lipase. *Protein Sci.*, **3**, 82–91.

Harel, M., Schalk, I., Ehret-Sabattier, L., Bouet, F., Goeldner, M., Hirth, C., Axelsen, P.H., Silman, I. and Sussman, J.L. (1993) Quaternary ligand binding to aromatic residues in the active-site gorge of acetylcholinesterase. *Proc. Nat. Acad. Sci. (USA)*, **90**, 9031–9035.

Hecht, H.J., Sobek, H., Haag, T., Pfeifer, O. and van Pée, K.-H. (1994) The metal-ion-free oxidoreductase from *Streptomyces aureofaciens* has an a/b hydrolase fold. *Struct. Biol.*, **1**, 532–537.

Heinz, D.W. and Matthews, B.W. (1994) Rapid crystallization of T4 lysozyme by intermolecular disulfide cross-linking. *Prot. Eng.*, **7**, 301–307.

Hermoso, J., Pignol, D., Kerfelec, B., Crenon, I., Chapus, C. and Fontecilla-Camps, J.C. (1996) Lipase activation by non-ionic detergents. The crystal structure of porcine lipase-colipase-tetraethylene glycol monooctyl ether complex. *J. Biol. Chem.*, **271**, 18007–18016.

Jelsch, C., Longhi, S. and Cambillau, C. (1998) Packing forces in nine crystal forms of cutinase. *Proteins: Structure, Function and Genetics*, **31**, 320–333.

Joerger, R.J. and Haas, M.J. (1994) Alteration of chain length selectivity of a *Rhizopus delemar* lipase through site-directed mutagenesis. *Lipids*, **29**, 377–384.

Kolattukudy, P.E. (1984) In *Cutinases from fungi and pollen*, edited by B. Borgström and H. Brockman, pp. 471–504. Amsterdam, The Netherlands: Elsevier.

Kabsch, W. and Sander, C. (1983) Dictionnary of protein secondary structure: pattern recognition of hydrogen-bonded and geometrical features. *Biopolymers*, **22**, 2577–2637.

Kim, K.K., Song, H.K., Shin, D.H., Hwang, K.Y. and Suh, S.W. (1997) The crystal structure of a triacylglycerol lipase from *Pseudomonas cepacia* reveals a highly open conformation in the absence of a bound inhibitor. *Structure*, **5**, 173–185.

Kraut, J. (1977) Serine proteases: structure and mechanism of catalysis. *Annu. Rev. Biochem.*, **46**, 331–358.

Lang, D., Hofmann, B., Burgdorf, T., Haalck, L., Hecht, H.-J., Spener, F., Schmid, R.D. and Schomburg, D. (1994) Structure elucidation of two lipases of the *Pseudomonas* family. In *Characterisation of lipases for industrial applications: 3D structure and catalytic mechanism*, Abstract Book, p. 26. France: Bendor Island.

Lang, D., Hofmann, B., Haalck, L., Hecht, H.J., Spener, F., Schmid, R.D. and Schoburg, D. (1996) Crystal structure of a bacterial lipase from *Chromobacterium viscosum* ATCC 6918 refined at 1.6 Å resolution. *J. Mol. Biol.*, **259**, 704–717.

Lauwereys, M., de Geus, P., de Meutter, J., Stanssens, P. and Matthyssens, G. (1991) In *GBF Monographs*, edited by L. Alberghina, R.D. Schmid and R. Verger, vol. 16, pp. 243–251. Weinheim, Germany: VCH Verlagsgesellschaft mbH.

Lawson, D.M., Artymiuk, P.J., Yewdall, S.J., Smith, J.M., Livingstone, J.C., Treffry, A., Luzzago, A., Levi, S., Arosio, P., Cesareni, G., Thomas, C.D., Shaw, W.V. and Harisson, P.M. (1991) Solving the structure of human H ferritin by genetically engineering intermolecular crystal contacts. *Nature*, **349**, 541–544.

Lawson, D.M., Brozozowski, A.M. and Dodson, G.G. (1992) Lifting the lid off lipases. *Curr. Opinion Struct. Biol.*, **9**, 473–475.

Lawson, D.M., Derewenda, U., Serre, L., Ferri, C., Szittner, R., Wei, Y., Meighen, E.A. and Derewenda, Z.S. (1994) Structure of a myristoyl-ACP-specific thioesterase from *Vibrio harveyi. Biochemistry*, **33**, 9382–9388.

Liao, D.-L. and Remington, S.J. (1992) Structure of wheat serine carboxypeptidase II at 3.5 Å resolution. *J. Biol. Chem.*, **265**, 6528–6531.

Longhi, S., Nicolas, A., Creveld, L., Egmond, M., Verrips, C.T., de Vlieg, J., Martinez, C. and Cambillau, C. (1996) Dynamics of *Fusarium solani* cutinase investigated through structural comparison among different crystal forms of its variants. *Proteins: Struct. Funct. Genet.*, **26**, 442–458.

Longhi, S., Mannesse, M.L.M., Verheij, H.M., de Haas, G.H., Egmond, M., Knoops-Mouthuy, E. and Cambillau C. (1997a) Crystal structure of cutinase covalently inhibited by a triglyceride analogue. *Protein Science*, **6**, 1–12.

Longhi, S., Czjzek, M., Lamzin, V, Nicolas, A. and Cambillau, C. (1997b) Atomic resolution (1.0 Å) crystal structure of *Fusarium solani* cutinase: stereochemical analysis. *J. Mol. Biol.*, **268**, 779–799.

Mannesse, M.L.M., Cox, R.C., Koops, B.C., Verheij, H.M., de Haas, G.H., Egmond, M., van der Hijden, H.T.W.M. and de Vlieg, J. (1995a) Cutinase from *Fusarium solani pisi* hydrolyzing triglyceride analogues. Effect of acyl chain length and position in the substrate molecule on activity and enantioselectivity. *Biochemistry*, **34**, 6400–6407.

Mannesse, M.L.M., Boots, J.W.P., Dijkman, R., Slotboom, A.J., van der Hijden, H.T.W.M., Egmond, M., Verheij, H.M. and de Haas, G.H. (1995b) Phosphonate analogues of triacylglycerols are potent inhibitors of lipase. *Biochim Biophys Acta*, **1259**, 56–64.

Mannesse, M.L.M., de Haas, G.H., van der Hijden, H.T.W.M., Egmond, M.R. and Verheij, H.M. (1997) Chiral preference of cutinase in the reaction with phosphonate inhibitors. *Biochem. Soc. Trans.*, **25**, 165–170.

Mannesse, M.L.M. (1997) Enantioselectivity of cutinase from *Fusarium solani*. The Netherlands: PhD Thesis, University of Utrecht.

Martinez, C. (1992) Crystal structure at 1.6 Å resolution of a recombinant cutinase from *Fusarium solani pisi*. PhD Thesis. Orsay, France: University of Paris XI.

Martinez, C., de Geus, P., Lauwereys, M., Matthyssens, G. and Cambillau, C. (1992) *Fusarium solani* cutinase is a lipolytic enzyme with a catalytic serine accessible to solvent. *Nature (London)*, **356**, 615–618.

Martinez, C., Stanssens, P., de Geus, P., Lauwereys, M. and Cambillau, C. (1993) Engineering cysteine mutants to obtain crystallographic phases with a cutinase from *Fusarium solani pisi. Protein Eng.*, **6**(2), 157–165.

Martinez, C., Nicolas, A., van Tilbeurgh, H., Egloff, M.-P., Cudrey, C., Verger, R. and Cambillau, C. (1994) Cutinase, a lipolytic enzyme with a preformed oxyanion hole. *Biochemistry*, **33**, 83–89.

Martinez Withers, C., Carrière, F., Verger, R., Bourgeois, D. and Cambillau, C. (1996) A pancreatic lipase with a phospholipase A1 activity: crystal structure of a chimeric pancreatic lipase-related protein 2 from guinea pig. *Structure*, **4**, 1363–1374.

Mittl, P.R.E., Berry, A., Scrutton, N.S., Perham, R.N. and Schultz, G.E. (1994) A designed mutant of the enzyme glutathione reductase shortens the crystallization time by a factor of forty. *Acta Cryst.*, **D50**, 228–231.

Nicolas, A., Egmond, M., Verrips, C.T., de Vlieg, J., Longhi, S., Cambillau, C. and Martinez, C. (1996) Contribution of cutinase serine 42 side chain to the stabilization of the oxyanion transition state. *Biochemistry*, **35**, 398–410.

Nicolas, A. (1996) Structure-Function studies of a recombinant *Fusarium solani pisi* cutinase. PhD Thesis. Luminy, France: Mediterranean University of Aix-Marseille II.

Noble, M.E.M., Cleasby, A., Johnson, L.N., Egmond, M.R. and Frenken, L.G.J. (1994) The crystal structure of triacylglycerol lipase from *Pseudomonas glumae* reveals a partially redundant catalytic aspartate. *Protein Eng.*, **7**, 559–562.

Ollis, D.L., Cheah, E., Cygler, M., Dijkstra, B., Frolow, F., Franken, S.M., Harel, M., Remington, S.J., Silman, I., Schrag, J., Sussman, J.L., Verschueren, K.H.G. and Goldman, A. (1992) The *a/b* hydrolase fold. *Protein Eng.*, **5**, 197–211.

Pathak, D., Ngai, K.L. and Ollis, D. (1988) X-ray crystallographic structure of dienelactone hydrolase at 2.8 Å. *J. Mol. Biol.*, **204**, 435–445.

Prangé, T., Pernot, L., Colloc'h, N., Longhi, S., Bourguet, W., Fourme, R. and Schiltz, M. (1998) Exploring hydrophobic sites in proteins with xenon or krypton. *Proteins: Structure Function and Genetics*, **30**, 61–73.

Prompers, J.J., Groenewegen, A., de Jong, S., Pepermans, H.A.M. and Hilbers, C.W. (1996) Backbone dynamics of *Fusarium solani pisi* cutinase on four different time scales as determined by NMR. In *Lipases: Structure, Function and Applications*, Abstract Book, p. 45. Umeå, Swede.

Ridchardson, J.S. (1981) The anatomy and taxonomy of protein structure. *Advan. Protein Chem.*, **34**, 167.

Rogalska, E., Nury, S., Douchet, I. and Verger, R. (1995) Lipase stereoselectivity and regioselectivity toward three isomers of dicaprin: a kinetic study by the monomolecular film technique. *Chirality*, **7**, 505–515.

Rudenko, G., Bonten, E., d'Azzo, A. and Hol, W.G.L. (1995) Three-dimensional structure of the human 'protective protein': structure of the precursor form suggests a complex activation mechanism. *Struct.* **3**, 1249–1259.

Sarda, L. and Desnuelle, P. (1958) *Biochem. Biophys. Acta*, **30**, 513–521.

Schäfer, W. (1993) The role of cutinase in fungal pathogenicity. *Trends in Microbiology*, 1, 69–71.

Schrag, J.D., Li, Y., Wu, S. and Cygler, M. (1991) Ser-His-Glu triad forms the catalytic site of the lipase from *Geotrichum Candidum. Nature (London)*, **351**, 761–764.

Schrag, J.D. and Cygler, M. (1993) 1.8 Å refined structure of the lipase from *Geotrichum candidum. J. Mol. Biol.*, **230**, 575–591.

Schrag, J.D., Li, Y., Cygler, M., Lang, D., Burgdorf, T., Hecht, H.J., Schmid, R., Schhmburg, D., Rydel, T.J., Oliver, J.D., Strickland L.C., Dunaway, C.M., Larson, S.B., Day, J. and McPherson, A. (1997) The open conformation of a *Pseudomonas* lipase. *Structure*, 5, 187–202.

Smith, L.C., Faustinella, F. and Lawrence, C. (1992) Lipases: three-dimensional structure and mechanism of action. *Curr Opinion Struct Biol*, 2, 490–496.

Sussman, J.L., Harel, M., Frolow, F., Oefner, C., Goldman, A., Toker, L. and Silman, I. (1991) Atomic structure of acetylcholinesterase from *Torpedo californica*: a prototypic acetylcholine-binding protein. *Science*, **253**, 872–879.

Takeda, S., Yoshimura, H., Endo, S., Takahashi, T. and Nagayama, K. (1995) Control of crystal forms of apoferritin by site directed mutagenesis. *Proteins: Struc. Funct. and Genet.*, **23**, 548–556.

Uppenberg, J., Hansen, M.T., Patkar, S. and Jones, T.A. (1994) The sequence, crystal structure determination and refinement of two crystal forms of lipase B from *Candida antarctica. Structure*, **2**, 293–308.

Uppenberg, J., Öhrner, N., Norin, M., Hult, K., Kleywegt, G.J., Patkar, S., Waagen, V., Anthonsen, T. and Jones, T.A. (1995) Crystallographic and molecular modeling studies of lipase B from Candida antarctica reveal a stereospecificity pocket for secondary alcohols. *Biochemistry*, **34**, 16838–16851.

van Tilbeurgh, H., Sarda, L., Verger, R. and Cambillau, C. (1992) Structure of the pancreatic lipase-procolipase complex. *Nature (London)*, **359**, 159–162.

van Tilbeurgh, H., Egloff, M.-P., Martinez, C., Rugani, N., Verger, R. and Cambillau, C. (1993) Interfacial activation of the lipase-procolipase complex by mixed micelles revealed by X-ray crystallography. *Nature (London)*, **362**, 814–820.

Verger, R. and de Haas, G.H. (1976) Interfacial enzyme kinetics of lipolysis. *Ann. Rev. Biophys. Bioeng.*, **5**, 77–117.

Verschueren, K.H.G., Kingma, J., Rozeboom, H.J., Kalk, K.H., Janssen, D.B. and Dijkstra, B.W. (1993a) Crystallographic and fluorescence studies of the interaction of haloalkane dehalogenase with halide ions: studies with halide compounds reveal a halide binding site in the active site. *Biochemistry*, **32**, 9031–9037.

Verschueren, K.H.G., Seljee, F., Rozeboom, H.J., Kalk, K.H. and Dijkstra, B.W. (1993b) Crystallographic analysis of the catalytic mechanism of haloalkane dehalogenase. *Nature (London)*, **363**, 693–698.

Vonrhein, C., Schlauderer, G.J. and Schulz, G.E. (1995) Movie of the structural changes during a catalytic cycle of nucleoside monophosphate kinases. *Structure*, **3**, 483–490.

Wei, Y., Schottel, J.L., Derewenda, U., Swenson, L., Patkar, S. and Derewenda, Z.S. (1995) A novel variant of the catalytic triad in the *Streptomyces scabies* esterase. *Nature Struct. Biol.*, **2**, 218–223.

Winkler, F.K., D'Arcy, A. and Hunziker, W. (1990) Structure of human pancreatic lipase. *Nature (London)*, **343**, 771–774.

Yamaguchi, S., Mase, T. and Takeuchi, K. (1992) *Biosci. Biotech. Biochem.* **56**, 315–319.

Zhang, X., Wozniak, J.A. and Matthews, B.W. (1995) Protein flexibility and adaptability seen in 25 crystal forms of T4 lysozyme. *J. Mol. Biol.*, **250**, 527–552.

.

6. LIPASES FROM *RHIZOPUS* SPECIES: GENETICS, STRUCTURE AND APPLICATIONS

UWE T. BORNSCHEUER, JÜRGEN PLEISS, CLAUDIA SCHMIDT-DANNERT
and ROLF D. SCHMID*

*Institut für Technische Biochemie, Universität Stuttgart, Allmandring 31,
70569 Stuttgart, Germany*

INTRODUCTION

Fungi of the genus *Mucorales* are well-established producers of useful enzymes, notably proteases and lipases. Some of these enzymes have found important commercial applications, e.g. the protease from *Rhizomucor miehei* which is used as a microbial rennet in milk clotting. The lipase from *Rhizomucor miehei*, which is commercially available in immobilized form as LIPOZYME™, has also been studied in a wide range of technical applications. Compared to this commercial success, lipases of the genus *Rhizopus* have been less successful, though they are also commercially available. For instance, the Japanese enzyme producer Amano Co. in Nagoya provides lipase N, D and J which are derived from *R. niveus*, *R. delemar* and *R. javanicus*, respectively. Lipases from *Rhizomucor miehei* and *Rhizopus* show 1,3-regiospecificity toward triacylglycerols but exhibit slightly different stereoselectivity and chain length specificity (Rogalska, 1993). A recent reclassification of these fungi had *R. niveus*, *R. delemar* and *R. javanicus* all renamed as *Rhizopus oryzae* (Schipper, 1984, review: Haas and Joerger, 1995) and, consistent with this taxonomic evaluation, the lipases isolated from *R. delemar* (RDL), *R. javanicus* (RJL) and *R. niveus* (RNL) were found to have identical amino acid sequences; the lipase from *R. oryzae* (ROL) differs by only two conservative substitutions (His134 is Asn and Ile234 is Leu in ROL). Thus, the different properties of these lipases observed in commercial preparations may be due to proteolytic cleavage of the same prolipase, which contains extra amino acid residues to guide folding and secretion, at different amino acid residues (Uyttenbroeck *et al.*, 1993).

The structure and function of both *Rhizomucor miehei* lipase and several *Rhizopus* lipases has been investigated on a molecular basis using x-ray crystallography, computer-modelling, site-directed mutagenesis and kinetic experiments with monolayers and pseudosubstrates. In the framework of these and other studies, these enzymes have been cloned, expressed in heterologous hosts and genetically modified. As a result, *Rhizopus* lipases today are a class of enzymes whose function is reasonably well understood. As they were found already to have a strong potential in a wide range of reactions relevant to industry (as summarized in the final paragraph of this review), the better understanding of their structure and function now achieved should be considered as a promising step towards their improvement *via* protein engineering and their wide application in industrial enzyme technology.

Table 1 Some properties of *Mucorales* lipases

Lipase		pI	pH optimum	Temperature optimum (°C)	Specific activity (U/mg)*
Rhizopus oryzae (ROL)	recombinant[1]	8.3[2]	8.0–8.5	30	10000
	native[3]	n.d.	8.5	40	10000
Rhizopus delemar(RDL)	native[4]	8.6	8.0–8.5	30	5094
	recombinant[5]	8.1[2]	8.0	30	n.d.
Rhizopus niveus (RNL)	native[6]	n.d.	6.0	40	n.d.
Rhizopus javanicus (RJL)	native[7]	7.1, 7.7, 7.8	6.0–8.0	n.d.	9260
Rhizomucor miehei (RML)	native[8]	3.9	7.0	n.d.	7500
	recombinant[9]	n.d.	n.d.	70	8810

1) Beer *et al.* (1997), 2) deduced from amino acid sequence, 3) Salah *et al.* (1994), 4) Haas *et al.* (1991), 5) Joerger and Haas (1993), 6) Kohno *et al.* (1994), 7) Uyttenbroeck *et al.* (1993), 8) Huge-Jensen *et al.* (1987), 9) product sheet, Novo Nordisk; *under different assay conditions

BIOCHEMICAL PROPERTIES

The regio- and stereoselectivity of pure *Rhizopus* lipases, their pH- and temperature optimum are summarized in Table 1 and compared to the lipase from *Rhizomucor miehei*. All enzymes are highly 1,3-specific in the hydrolysis or esterification of glycerides. They are most active at neutral or slightly basic pH and at 30°C and exhibit a moderate pH- and temperature stability. If immobilized, the lipase of RML became stable up to 70°C (product sheet, Novo Nordisk).

The native enzymes are only moderately glycosylated, e.g. 11% carbohydrate content of native RML (Boel *et al.*, 1988; Haas *et al.*, 1991). As fully active recombinant lipases can be obtained from heterologous expression in *E. coli*, glycosylation does not seem to be necessary for lipase activity.

MOLECULAR GENETICS

Cloning and Nucleotide Sequence

Up to now, the genes of four *Mucorales* lipases have been cloned, and three have been overexpressed and purified (Table 1). Lipase genes from *Rhizopus delemar* ATCC34612 (RDL) (Haas *et al.*, 1991), *Rhizopus niveus* IFO 9759 (RNL) (Kugimiya *et al.*, 1992), *Rhizopus oryzae* ATCC 853 (ROL) (Beer *et al.*, 1998) and *Rhizomucor miehei* (RML) (Boel *et al.*, 1988) were cloned from cDNA libraries screened either by probes derived from the amino acid sequences or by detecting lipase-producing *E. coli* transformants on tributyrin plates. Analysis of the nucleotide sequence and comparison with the N-terminal sequences derived from the native lipase revealed that all *Mucorales* lipases studied so far are organized as preproenzymes (Figure 1). From the amino acid sequence, a molecular weight of 42 kDa and 39.5 kDa for the

Figure 1 Organization of the lipases from *Rhizopus oryzae* (ROL) and *Rhizomucor miehei* (RML) as preproenzymes. The signal sequence (white box) is followed by the propeptide (gray box) and the mature lipase (black box). The numbers on top display the cleavage sites and the total size of the protein. Numbers in brackets indicate an alternative cleavage site observed with the native lipase of *Rhizopus niveus* (Kugimiya *et al.*, 1992)

lipase precursors of ROL and RML was deduced, and 29.5 kDa for the mature ROL and RML lipases.

At the amino acid level, RNL and RDL are identical, whereas ROL differs in two amino acids in the mature lipase and in six of the prepropeptide. The amino acid sequence of RML, if compared to ROL, shows 54% identity of the mature lipases and 29% identity of the prepropeptides. While *Rhizopus delemar* and *Rhizopus niveus* have been reclassified as *Rhizopus oryzae* (Schipper, 1984), comparison of the nucleotide sequence of the three lipases genes indicates, that RDL and RNL are identical, but differ slightly from ROL (see below).

Overexpression and Purification

cDNA-derived clones of *E. coli* expressing RML, RNL and ROL showed low lipolytic activity on tributyrin plates. Under the control of the lac-promotor on the pUC8-2.14 cloning vector, a low-level expression of RDL in *E. coli* was achieved (Haas *et al.*, 1991). Visualization of the expressed lipase on SDS-gels by immunoblotting resulted in bands with sizes of 45 and 39 kDa, corresponding to the prepro- and proenzyme, respectively. Similar results were found for recombinant ROL (Beer *et al.*, 1998), indicating that *E. coli* is unable to fully process the preproenzyme to its mature product.

The genes encoding the ROL and RDL were therefore cloned into high-level expression vectors, in order to allow for the production of larger amounts of recombinant lipase in *E. coli*. It turned out, however, that the expression of even low amounts of active mature lipase upon translocation into the periplasmic space is toxic for the *E. coli* cells. Expression of mature lipase was possible only if a tightly controlled promoter was used, or cytoplasmically, which resulted in expression of mature inactive lipase as inclusion bodies (Joerger and Haas, 1993). However, expression of the active prolipase into the periplasm was not toxic for *E. coli*. Using a pET-vector (Studier *et al.*, 1990), both the mature lipase and prolipase of *Rhizopus delemar* were cytoplasmatically expressed in *E. coli* under the control of the T7-promotor at a level of 15–20% as inactive inclusion bodies (Joerger and Haas, 1993).

Instead of the less tightly controlled pET-vector system, in our group the *E. coli* expression vector pCYTEXP1 (Belev *et al.*, 1991) for the expression of ROL under the control of the strictly regulated 1 P_RP_L promoter was chosen. Both mature and precursor lipase were fused to an ompA-leader sequence, which directs the expression of the lipase into the periplasm. Using this system, expression levels of 20–30% and 10–20% were obtained for the mature and prolipase, respectively. The mature lipase, which was expressed to a higher level, formed inactive inclusion bodies in the periplasm, whereas the prolipase was partly active.

For the inclusion bodies containing mature and precursor ROL and RDL, a refolding and purification protocol was established, leading to specific activities of 10000 U/mg (ROL) and 5094 U/mg (RDL) for the mature lipases, and of 12800 U/mg (ROL) and 5800 U/mg (RML) for the prolipases (Joerger and Haas, 1993). Thus, the specific activity obtained with the recombinant mature ROL was almost twice as high compared to the recombinant RDL, and 20% higher compared to the native lipase of *Rhizopus delemar* (Haas *et al.*, 1991), but similar to that of the native ROL (Salah *et al.*, 1994). Recently, expression of functional ROL in high yield and purity could be achieved using *Pichia pastoris* (Minning *et al.*, 1998) or *Saccharomyces cerevisiae* (Takahashi *et al.*, 1998) as a host. RML could be functionally expressed at high level by cloning the gene encoding for the lipase precursor under the control of the α-amylase promoter of *Aspergillus oryzae* and transforming *Aspergillus oryzae* with the resulting plasmid. The secreted lipase was correctly processed in *Aspergillus oryzae*, and purification led to a recombinant lipase with a specific activity of 8810 U/mg as compared to 7500 U/mg of the native lipase (Huge-Jensen *et al.*, 1989). It thus appears that fungal expression systems are better suited than *E. coli* to express functional lipases of the genus *Mucorales* at a high level.

The Function of the ROL Prosequence

All *Mucorales* lipases investigated so far have a prosequence, which is unusual for microbial lipases. Genetic studies on the role of this 97 amino acids prosequence suggest at least two functions. First, it allows secretion of ROL without damaging the cells. Thus, expression studies of ROL and ProROL in *E. coli* showed that even low levels of actively expressed ROL in the periplasm of *E. coli* is toxic for the cells, whereas ProROL is not. The main reason for the toxicity might be the fact, that ROL, in contrast to ProROL, possesses phospholipase activity. Thus, the function of the prosequence is analogous to that of other zymogen sequences (Beer *et al.*, 1996a).

Secondly, it seems to play an important role in the folding of the lipase, particulary in the correct formation of disulphide bridges. Both ProROL and ROL could be refolded and purified to a similar specific activity of 12800 and 10000 U/mg, respectively, demonstrating that ProROL is fully active and thus does not behave like a typical zymogen. However, the enzymatic properties of both lipase forms differed markedly. ROL had a significantly broader and higher pH optimum (pH 8.5) in comparison to ProROL (pH 7.5). Additionally, ProROL was thermostable up to 50°C, whereas ROL was stable only up to 30°C (Figure 2). A difference between ROL and ProROL was also observed in terms of substrate affinity, suggesting that structural elements other than the lid are involved in interfacial activation. By systematic mutation analysis of the prosequence and comparison of the *in vitro* folding behaviour of the resulting ProROL and ROL mutants under different conditions, it was shown that the prosequence facilitates folding via a pathway

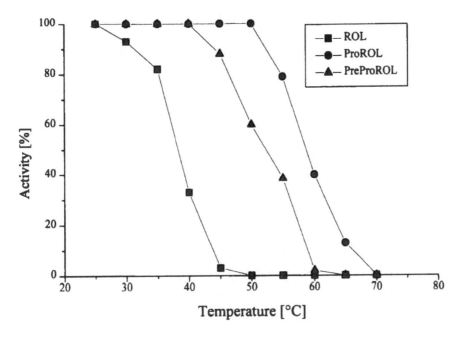

Figure 2 Activity of ROL, ProROL and PreProROL as a function of temperature

including Cys-68 as the key residue, acting as an intramolecular thiodisulphide reagent during folding (Beer *et al.*, 1996a). Furthermore, it was shown that residues in the neighborhood of Cys-68 play an important structural role, in particular Arg-69 which enhances the leaving group character of Cys-68. At low temperatures, native ROL and ProROL fold with similar kinetics, but already at temperatures around 25°C, correct folding of ROL and ProROL mutants is significantly reduced if compared to native ProROL.

STRUCTURE

Primary Structure

Lipases of *Rhizopus delemar, Rhizopus sp., R. niveus, R. javanicus* and *R. arrhizus* are identical and differ from *R. oryzae* lipase (Beer *et al.*, 1996b) by two conservative mutations (H134N, I254L) (for a review: Haas and Joerger, 1995). *Rhizopus* lipases are highly homologous to the lipase from *Rhizomucor miehei* (54% identity) and moderately homologous to lipases from *Humicola lanuginosa, Penicillium camembertii* and *Aspergillus oryzae* (Figure 3).

Secondary and Tertiary Structure

For four *Mucorales* lipases the structure has been determined in closed and open form including two complexes with substrate analogous inhibitors (Table 2).

```
R.delemar      -122 MVSFISISQGVSLCLLVSSMMLGSSAVPVSGKSGSSNTAVSASDNAALPPLISSR
R.miehei        -93 --MVLKQRANYLGFLIVFFTAFLVEAVPIKRQSNS--------TVDSLPPLIPSR
P.camembertii   -25 ------MRLS--FFTALSAVASLGYALPGKLQSR--------------------
H.lanuginosa         -----------------------------------------------------
A.oryzae        -27 ------MRFLSGFVSVLSSVALLGYAYPTAIDVR--------------------

R.delemar       -67 CAPPSNKGSKSDLQAEPYNMQKNTEWYESHGGNLTSIGKRDDNLVGGMTLDLPSD
R.miehei        -48 TSAPSSSPSTTDPEAPAMSRNG-------------PLPSDVETKYGMALNATSY
P.camembertii        -----------------------------------------------------
H.lanuginosa         -----------------------------------------------------
A.oryzae             -----------------------------------------------------

R.delemar       -12 APPISLSSSTNSASDGGKVVAATTAQIQEFTKYAGIAATAyCRSVVPGNKWDCVQ
    A                             EEE   HHHHHHHHHHHHHHHHH
    B                                   HHHHHHHHHHHHHHHH       HH
  R.niveus                        EEE   HHHHHHHHHHHHHHH        HH
R.miehei         -7 PDSVVQAM-----SIDGGIRAATSQEINELTYYTTLSANSYCRTVIPGATWDCIH
  closed                          EEE   HHHHHHHHHHHHHHH        HH
  open                            EEE   HHHHHHHHHHHHHHHH
P.camembertii     1 ---------------------DVSTSELDQFEFWVQYAAASYYEADYTAQVGDKLS
                                   HHHHHHHHHHHHHHHH   HHHHH
H.lanuginosa      1 ---------------------EVSQDLFNQFNLFAQYSAAAYCGKNNDAPAGTNIT
                                   HHHHHHHHHHHH   HHH
A.oryzae          1 ---------------------DIPTTQLEDFKFWVQYAAATYCPNNYVAKDGEKLN
                                                  .        .* .*

R.delemar        43 CQK------WVPDGKIITTFT-SLLSDTNGYVLRSDKQKTIYLVFRGtNSFRSaI
    A                          EEE          EEEEE          EEEE    HHHHH
    B                 HHH      EEE          EE            EE      HHHH
  R.niveus            HHH    H EEEEEEE E   EEEEEEEE    EEEEEEE     HHHH
R.miehei         43 CD-------ATEDLKIIKTWS-TLIYDTNAMVARGDSEKTIYIVFRGSSSIRNWI
  closed             HH       EEEEEEE      EEEEEEE       EEEEE     HHHH
  open                        EEEEEEE      EEEEEEE       EEEEE     HHHHH
P.camembertii    36 CSKGNCPEVEATGATVSYDFSDSTITDTAGYIAVDHTNSAVVLAFRGSYSVRNWV
                     HHHHH    EEEEE              EEEEE       EEEEE     HHHHH
H.lanuginosa     36 CTGNACPEVEKADATFLYSFEDSGVGDVTGFLALDNTNKLIVLSFRGSRSIeNwI
                     HHH  HHHHH   EEEEEE        EEEEEEEE     EEEEEE    HHHH
A.oryzae         36 CSVGNCPDVEAAGSTVKLSFSDDTITDTAGFVAVDNTNKAIVVAFRGSYSIRNWV
                     *                          .  *   . .       .. ***. *

R.delemar        91 TdIvfNFSDY-KPVKGAKVHAGfLSsYEQVVNDYFPVVQEQLTAHPTYKVIVTGh
    A                     EEE        EEEHHHHHHHHHHHHHHHHHHHHHH          EEEEEE
    B                                HHHHHHHHHHHHHHHHHHHHHHHHH         EEEEE
  R.niveus           H   EEE        EEEHHHHHHHHHHHHHHHHHHHHHHHHH        EEEEEEE
R.miehei         90 ADLTFVPVSY-PPVSGTKVHKGFLDSYGEVQNELVATVLDQFKQYPSYKVAVTGH
  closed             HH   EEE        EEEHHHHHHHHHHHHHHHHHHHHHHHHH       EEEEEEE
  open               HHH  EEE        EEEHHHHHHHH  HHHHHHHHHHHH         EEEEEEE
P.camembertii    91 ADATFVHTNP-GLCDGCLAELGFWSSWKLVRDDIIKELKEVVAQNPNYELVVVGH
                     HH   EE         EEHHHHHHHH   HHHHHHHHHHHHH         EEEEEE
H.lanuginosa     91 GNLNFDLKEINDICSGCRGHDGFTSSWRSVADTLRQKVEDAVREHPDYRVVFTGH
                          EEE        EEEHHHHHHHHHHHHHHHHHHHHHHH         EEEEEE
A.oryzae         91 TDATFPQTDP-GLCDGCKAELGFWTAWKVVRDRIIKTLDELKPEHSDYKIVVVGH
                     ..  *        *      **  . * .           *  .    **
```

```
R.delemar      145 SLGGAQALLAGMDLYQREPRLSPKNLSIFTVGGPRVGNPTFAYYVEST--GIPFQ
A                  E HHHHHHHHHHHHHH         EEEEEE      HHHHHHHHH        EE
B                  HHHHHHHHHHHH         HHH EEEEE       HHHHHHHHH        EE
R.niveus           HHHHHHHHHHHHHHHH         EEEEEE       HHHHHHHHH        EE
R.miehei       144 SLGGATALLCALDLYQREEGLSSSNLFLYTQGQPRVGDPAFANYVVST--GIPYR
closed             HHHHHHHHHHHHHHH         EEEEEE       HHHHHHHHH        EE
open               HHHHHHHHHHHHHHHH        EEEEEE       HHHHHHHHH        EE
P.camembertii  145 SLGAAVATLAATDLRG--KGYP--SAKLYAYASPRVGNAALAKYITAQ--GNNFR
                   HHHHHHHHHHH         EEE         HHHHHHHHH
H.lanuginosa   146 SLGGALATVAGADLRG--NGY---DIDVFSYGAPRVGNRAFAEFLTVQTGGTLYR
                   HHHHHHHHHHHHH         EEEEEE       HHHHHHHH        EEE
A.oryzae       145 SLGAAIASLAAADLRT--KNY---DAILYAYAAPRVANKPLAEFITNQ--GNNYR
                   ***.* *  . ** 		    ...  ***..  *  ..       *  .
```

```
R.delemar      198 RTVHKRdIVPHvPPQSfGFLHPGVESWIKSGTS-NVQICTSEIETKDCSNSIVP-
A                  EEEE   HHHH         EEEE       EEEE        HHHH
B                  EEE    HHH   HHH    EEEE       EE
R.niveus           EEEE   HHH   HHH    EEEEEEE    EEEEE       HHH
R.miehei       197 RTVNERDIVPHLPPAAFGFLHAGEEYWITDNSPETVQVCTSDLETSDCSNSIVP-
closed             EEEE   HHH   HHH    EEEEEEE    EEEEE       HHHH
open               EEEE   HHH   HHH    EEEEEEE    EEEEE       HHHH
P.camembertii  194 FTHTN-DPVPKLPLLSMGYVHVSPEYWITSPNN--ATVSTSDIKVIDGDVSFDGN
                          HHH   HHH               HHH              HH
H.lanuginosa   196 ITHTN-DIVPRLPPREFGYSHSSPEYWIKSGTL--VPVTRNDIVKIEGIDATGGN
                   EEE    HHH   HHH    EEEE       HHHHEEE
A.oryzae       193 FTHND-DPVPKLPLLTMGYVHISPEYYITAPDN--TTVTDNQVTVLDGYVNFKGN
                    *   * **..*   *. *   *        .     .
```

```
R.delemar      251 ---------FTSILDHLSYFdINeGSCL---------
A                            HHHEE  EE
B                            HHHH
R.niveus                     HHHH    EE
R.miehei       251 ---------FTSVLDHLSYFGINTGLCT---------
closed                       HHHH
open                         HHHH
P.camembertii  246 TGT--GLPLLTDFEAHIWYFVQ-VDAGKGPGLPFKRV
                   H            HHH
H.lanuginosa   248 NQP-----NIPDIPAHLWYFGL-IGTCL---------
                             HHHH
A.oryzae       245 TGTSGGLPDLLAFHSHVWYFIH-ADACKGPGLPLR--
                             *. **
```

Figure 3 Sequence alignment of lipases from *R. delemar* (LIP_RHIDL in Swiss-Prot (Bairoch and Boeckmann, 1991)), *R. miehei* (LIP_RHIMI in Swiss-Prot), *P.camembertii* (MDLA_PENCA in Swiss-Prot), *H. lanuginosa* (1TIB in Protein Data Bank (Bernstein *et al.*, 1977)) and *A. oryzae* (D85895 in GenBank (Bilofsky *et al.*, 1986)). The start of the mature lipase is indicated by a vertical bar (signal and propeptide left of the bar). *R. oryzae* lipase is nearly identical to *R. delemar* lipase (H134N and I254L). In bold: catalytic triad, oxyanion hole, mutation sites (in lower case letters); the lid helix is I85-D91 in RML; secondary structure analysis by DSSP (Kabsch and Sander, 1983), alignment by CLUSTAL W (Thompson *et al.*, 1994), visualization by BOXSHADE (shareware program by Kay Hofmann, Lausanne)

Table 2 *Mucorales* lipase structures published in Protein Data Bank (PDB) or modelled by homology

Lipase source	PDB entry	Conformation	Inhibitor	Resolution	Reference
R.delemar	1TIC A,B	semi-open	none	2.6 Å	Derewenda *et al.* (1994b)
R.niveus	1LGY A,B,C	closed	none	2.2 Å	Kohno *et al.* (1996)
R.miehei	1TGL	closed	none	1.9 Å	Brady *et al.* (1990)
"	3TGL	closed	none	1.9 Å	Derewenda *et al.* (1992a)
"	4TGL	open	diethylphosphate	2.6 Å	Derewenda *et al.* (1992b)
"	5TGL	open	hexylphosphonate-ethylester	3.0 Å	Brzozowski *et al.* (1991)
R. oryzae	model	open			Beer *et al.* (1996b)
P.camembertii	1TIA	closed	none	2.0 Å	Derewenda *et al.* (1994a)
H.lanuginosa	1TIB	closed	none	1.8 Å	Derewenda *et al.* (1994b)

Lipases are members of the α/β hydrolase fold family (Ollis *et al.*, 1992): a central hydrophobic eight-stranded mixed β-pleated sheet with the β-strands connected by loops and amphiphilic α-helices (Figure 4). Secondary structure is well conserved among the mucorales group and homologous lipases, but is slightly mediated by crystallization conditions. Upon opening of the lid, the helix (I85-D91 in *R. miehei*) increases in length (Figure 3).

In the lipases of the *Mucorales* group, the tertiary structure is further stabilized by three disulfide bridges. The catalytic triad is placed on top of a β-sheet, with the catalytic serine and aspartic acid in two separate loops between a β-strand and an α-helix and the catalytic histidine in a short α-helix near the C-terminus of the lipases (Figure 3). The position of these residues and the oxyanion hole are highly conserved. As in most lipases, a mobile lid covers the catalytic site in the catalytically inactive state. In *Mucorales* lipases, the lid consists of a short α-helix (4–8 amino acids) linked to the body of the lipase by flexible structure elements. In the open,

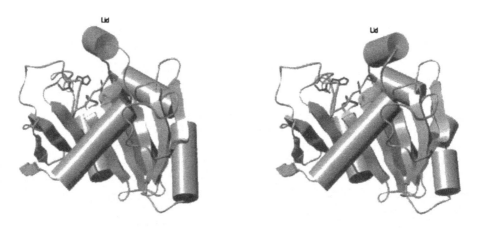

Figure 4   Secondary structure of *R. miehei* lipase in closed (left) and open (right) form

active form of the lipase, the lid moves away and renders the substrate binding site accessible (Figure 4). It has been suggested, however, that lid opening in *Mucorales* lipases does not follow a simple two-state model, but more conformations are possible, some of whom have been observed depending on the crystallization conditions (Derewenda *et al.*, 1994b).

The hydrolytic reaction pathway of lipases is similar to that of serine proteases and consists of 5 subsequent steps (Beer *et al.*, 1996b): i) the ester substrate binds, ii) a first tetrahedral intermediate is formed by nucleophilic attack of the catalytic serine, with the oxyanion stabilized by two or three hydrogen bonds, the so-called oxyanion hole, iii) the ester bond is cleaved, and the alcohol moiety leaves the enzyme, iv) water adds to the acyl enzyme forming a second tetrahedral intermediate, v) the acyl enzyme is cleaved.

Binding of phosphate and phosphonate inhibitors mimicks the first tetrahedral intermediate state of substrates. From the x-ray structure of complexes of *R. miehei* lipase with diethylphosphate and hexylphosphonate ethylester, the binding sites for the alcohol and the fatty acid moiety were identified. Like in other lipases, the alcohol binding site consists of two binding pockets for the large and the medium-sized substituent of a secondary alcohol (Kazlauskas, 1994), which mediate stereoselectivity towards esters of secondary alcohols (Cygler *et al.*, 1994). The residues which mediate binding of the scissile fatty acids can be extracted from the experimentally determined structure. The fatty acid binding site is formed by a hydrophobic crevice, which extends beyond the inhibitor (Norin *et al.*, 1994, Holzwarth *et al.*, 1997). The crevice is long enough to allow binding of fatty acids up to a length of C18. While the scissile fatty acid binding site is easily identified, no x-ray structure is available where the binding site of the second and third fatty acid of a triglyceride substrate is indicated. From an analysis of surface hydrophobicity a second, shallow hydrophobic binding site was postulated (Lawson *et al.*, 1994, Holzwarth *et al.*, 1997) which might be able to bind a fatty acid. Modelling of substrate binding to these binding sites has been used to understand fatty acid chain length (Haas *et al.*, 1996, Klein *et al.*, 1997) and stereoselectivity towards triglycerides and analogs (Holzwarth *et al.*, 1997, Haalck *et al.*, 1997).

Although the five lipases described here have only moderate overall homology, most of the amino acids involved in catalysis are conserved, with two exceptions:

(1) in *Rhizopus* lipases, the oxyanion hole is formed by Thr83, and by serine in the other lipases, and the binding pocket for medium-sized substituent of secondary alcohol, Leu258 of *R. delemar* lipase which is conserved in the lipases of *R. miehei* and *H. lanuginosa*, and Ile and Val in *P. camembertii* and *A. oryzae* lipases, respectively;

(2) the sequence of the lid helix differs considerably.

The residues mediating substrate specificity differ to a bigger extent. It has been suggested to relate these differences to the different biochemical properties of these lipases concerning their activity, stereoselectivity, stability and substrate specificity (Menge *et al.*, 1995). Lipases from *R. delemar*, *R. miehei* and *H. lanuginosa* are triacylglycerol hydrolases, *A. oryzae* lipase is a diacylglycerol hydrolase and *P. camembertii* lipases is a mono- and diacylglycerol hydrolase.

Table 3 Mutants of functionally relevant amino acids

Lipase source	Mutation	Role of the mutated residue	Effect[a] (%)	References
R.oryzae	Y28F	anchor between active site serine and oxyanion hole; or: stabilization of tetrahedral substrate intermediate	0.02	Beer *et al.* (1996a)
R.oryzae	T83A	stabilization of tetrahedral substrate intermediate	0.04	Beer *et al.* (1996a)
R.oryzae	T83V	"	0.14	Beer *et al.* (1996a)
R.oryzae	T83S	"	12	Beer *et al.* (1996a)
H.lanuginosa	E87A	defines lid position	35–70	Martinelle *et al.* (1996)
R.oryzae	A89W	A in RDL, W in RML	56	Beer *et al.* (1996a)
R.oryzae	D92N	hydrogen bond to T83	7	Beer *et al.* (1996a)
R.delemar	S115D,L146K	near the active site	0	Klein *et al.* 1997
R.oryzae	H144F	maintains active site geometry by H-bridging between Y28 and H257 carbonyl	0.04	Beer *et al.* (1996a)
R.oryzae	H144S	"	2.2	Beer *et al.* (1996a)
R.oryzae	D204A	catalytic residue	0.06	Beer *et al.* (1996a)
R.oryzae	D262G	not involved in hydrolysis	96	Beer *et al.* (1996a)
R.oryzae	E265Q	not a catalytic residue, but maintains active site geometry	98	Beer *et al.* (1996a)
R.oryzae	E265D	maintains active site geometry	0.14	Beer *et al.* (1996a)

[a]Change of activity relative to wild-type

PROTEIN ENGINEERING OF *RHIZOPUS* LIPASES

Site-directed mutagenesis has been applied to identify the function of individual amino acids in *Rhizopus* lipases. Modelling of substrate binding and enzymatic catalysis helped to understand their function and to predict mutants with defined properties. Two groups of mutants have been predicted and characterized: mutants related to the enzymatic function (Table 3) and mutants with altered substrate specificity (Table 4).

In addition to the catalytic triad, a network of other amino acids near the active site are necessary to maintain the proper orientation of catalytic residues (Beer *et al.* 1996) in *R.oryzae* lipase. Furthermore, mutation of Thr83 to Ser, Ala and Val indicated that the side chain of this residue is involved in stabilization of the oxyanion, and that formation of the acyl enzyme is the rate limiting step in ester hydrolysis.

Numerous experiments have been performed to probe the fatty acid binding site and to change fatty acid specificity both near and far away from the active site (Table 4). The effect of these latter mutations can be easily understood: increasing the size or the polarity of side chains lining the scissile binding site prevents long-chain fatty

Table 4 Mutants which change fatty acid specificity

Lipase source	Mutation	Role of the mutated residue	Effect of the mutation on the hydrolytic activity	Reference
R.delemar	T83S	oxyanion hole; binding of acyl chain	4 fold decrease of tributyrin, 20% decrease of tricaprylin relative to triolein	Joerger and Haas (1994)
H.lanuginosa	W89 F,L,G,E	binding of acyl chain	Increase of triacetin and trioctanoin relative to tributyrin	Martinelle *et al.* (1996)
R.delemar	V94W	binding of fatty acids >C8	1.4-fold increase of tricaprylin relative to tributyrin	Klein *et al.* 1997
R.oryzae	F95Y	binding of fatty acids >C12	60% (30%) increase of caproic acid methyl ester relative to oleic (stearic) acid	Atomi *et al.* (1996)
R.delemar	F95D	binding of fatty acids	2-fold decrease of hydrolysis	Joerger and Haas (1994)
R.delemar	F95D F214R	distal end of acyl binding groove	3-fold increase of tricaprylin relative to triolein	Klein *et al.* (1997)
R.delemar	F112W	fatty acid binding	50% increase of tributyrin relative to triolein	Joerger and Haas (1994)
R.delemar	F112Q	fatty acid binding	no activity	Klein *et al.* (1997)
R.delemar	V206T	fatty acid binding	10–20% activity	Joerger and Haas (1994)
R.delemar	V209W	binding of fatty acids >C4	2 fold increase of tributyrin relative to triolein	Joerger and Haas (1994)
R.delemar	V209W F112W	binding of fatty acids >C4	80 fold increase of tributyrin relative to tricaprylin; no triolein hydrolysis	Klein *et al.* (1997)
R.oryzae	F214Y	binding of fatty acids >C12	20% increase of caproic acid methyl ester relative to oleic an stearic acid	Atomi *et al.* (1996)
R.oryzae	F216R	mutant should stabilize Δ9 double bond	90% increase of oleic acid methyl ester relative to stearic acid	Atomi *et al.* (1996)

acid chains reaching the mutated site from binding, while binding of shorter fatty acid chains is unimpaired. The observed role of the side chain of Thr83, forming the oxyanion hole, for fatty acid chain length specificity is, however, not fully understood.

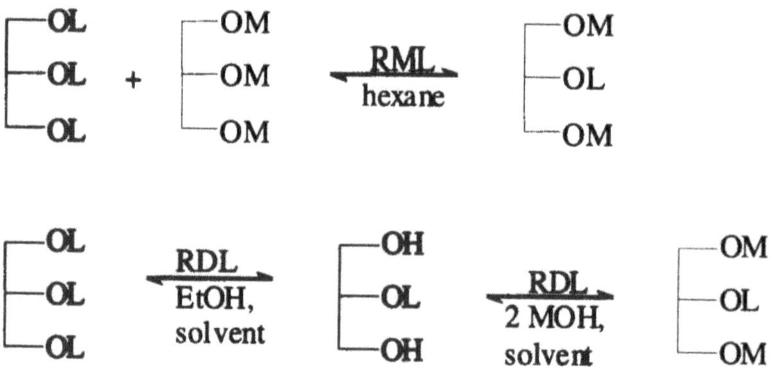

$$\begin{array}{c}\text{OL}\\\text{OL}\\\text{OL}\end{array} + \begin{array}{c}\text{OM}\\\text{OM}\\\text{OM}\end{array} \xrightarrow[\text{hexane}]{\text{RML}} \begin{array}{c}\text{OM}\\\text{OL}\\\text{OM}\end{array} \qquad (1)$$

$$\begin{array}{c}\text{OL}\\\text{OL}\\\text{OL}\end{array} \xrightarrow[\substack{\text{EtOH,}\\\text{solvent}}]{\text{RDL}} \begin{array}{c}\text{OH}\\\text{OL}\\\text{OH}\end{array} \xrightarrow[\substack{\text{2 MOH,}\\\text{solvent}}]{\text{RDL}} \begin{array}{c}\text{OM}\\\text{OL}\\\text{OM}\end{array} \qquad (2)$$

Figure 5 Production of structured lipids by direct interesterification (1) or a combination of alcoholysis and esterification (2) (Soumanou *et al.*, 1997, 1998)

APPLICATIONS

Lipid Modification

The natural substrates of lipases are triacylglycerols and, as a consequence, most applications of lipases deal with the modification of lipids. The low reaction enthalpy of lipase hydrolysis allows to use these enzymes for hydrolysis as well as for the synthesis or interesterification of triacylglycerols. Taking advantage of the mild reaction conditions (permitting the use of labile fatty acids as substrates), lipase regioselectivity, chain-length and fatty acid specificity, the production of structured lipids with a defined distribution of fatty acids along the glycerol backbone (e.g cocoa-butter equivalent, Betapol™) and the synthesis of emulsifiers like mono- or diglycerides (MAG or DAG) were extensively investigated.

As shown by Berger and Schneider (1991) and Waldinger and Schneider (1996), lipases from *Rhizopus sp.* have a strict 1,3-regioselectivity in nonpolar organic solvents where acyl migration is slow. The highly 1,3-selective RDL (Amano D) reacted at the primary position 76 times faster than at the secondary position, while the moderately selective CVL and RML reacted 26 and 11 times faster, respectively.

Interesterification between triglycerides

RDL was used to produce a cocoa-butter like fat by interesterification between olive oil and stearic acid in hexane. The highest activity was found for RDL adsorbed on Celite followed by entrapment with photo-crosslinkable resin prepolymer resulting in up to 40% incorporation of stearic acid (Yokozeki *et al.*, 1982). In similar experiments, Shimada *et al.* (1996) used RDL immobilized on a ceramic carrier to produce structured lipids of the type MLM (Figure 5) by interesterification of safflower or linseed oil with caprylic acid. Confirming the strict 1,3-regiospecifity of RDL, MLM's were the only product after three repeated batch reactions.

A pilot-scale interesterification between shea oleine and stearic acid or shea oil and myristic acid in hexane catalyzed by RAL immobilized on Hyflo Super Cel was performed in a column reactor. Up to 100% conversion of fatty acid was observed and the enzyme lost almost no activity even after 166 h of operation (Wisdom *et al.*, 1987). The interesterification of palm oil midfraction with stearic acid using Celite-RAL in a lecithin-based micellar system in a gas-lift reactor resulted in a 2.8 fold higher productivity and reduced reaction times, if compared to shake flask experiments (Mojovic *et al.*, 1994).

Butter fat was investigated as another suitable starting material for interesterification. Using RNL in a cosurfactant free microemulsion system made up from Span 60, Tween 60 or lecithin, a high proportion of C18:1 at the *sn*2-position was found (Kermasha *et al.*, 1995). ROL immobilized on controlled pore glass was used in the esterification of milk fat with oleic acid in isooctane resulting in a lowered concentration of palmitic acid in milk fat and, consequently, in a lower transition temperature (Oba and Witholt, 1994).

In order to increase the activity of the lipase in the presence of an organic solvent, the enzyme was coated with a lipid or surfactant. For example, a water-insoluble complex containing approximately 10 wt% of protein formed upon mixing aqueous solutions of lipases from RDL or RNL and a nonionic amphiphile (didodecyl *N*-D-glucono-L-glutamate (Okahata and Ijiro, 1988, 1992). The modified lipase was soluble in most organic solvents and was >100 times more active than suspended enzyme in the synthesis of di- and triglycerides from lauric acid and monolaurin in benzene. Similar observations were made by Goto *et al.* (1994), who coated a lipase from *Rhizopus sp.* with the nonionic surfactant glutamic acid dioleyl ester ribitol amide and investigated the interesterification between lauric acid and benzyl alcohol in organic solvents.

RJL was coated with sorbitan monostearate and the lipid-coated enzyme was employed in a hollow fiber membrane reactor and the interesterification between tripalmitin and stearic acid in hexane was investigated. After reaching steady state (4 h), no loss in enzyme activity was observed within 76 h, and PPS and SPS were the major products (Basheer *et al.*, 1995, see also Mogi and Nakajima, 1996).

Recently, structured lipids containing essential fatty acids, especially polyunsaturated fatty acids (PUFA) like docosahexaenoic or eicosapentaenoic acid, have been targets of lipase research. Fish oils contain up to 30% PUFA, mainly in the form of triacylglycerols, and are sensitive to elevated temperatures, extreme pH etc. resulting in side reactions such as oxidation, *cis-trans* isomerization or double-bond migrations. Yadwad *et al.* (1991) described the RNL-catalyzed glycerolysis of cod liver oil (9.6% PUFA) to yield 1(3)-monoacylglycerides with 29% PUFA. Shimada *et al.* (1996) interesterified tuna oil with caprylic acid with RDL, immobilized on a ceramic carrier, and found that after 2 d at 30°C approx. 65% of the fatty acids at 1- and 3-position in tuna oil were replaced by caprylic acid. Moreover, the enzyme could be reused 14 times without significant loss of activity.

Recently, we described the synthesis of highly pure triglycerides of the MLM type, which are important in nutrition (e.g. for patients with pancreatic insufficiency). Initially, we investigated the direct interesterification between a triglyceride bearing long-chain fatty acids (e.g. triolein) with a short-chain triglyceride (e.g tricaprylin) in *n*-hexane (Figure 5, eq. 1). RML was found most suitable, but the highest yield of MLM was only 31% (Soumanou *et al.*, 1997). Significantly higher yields and purities were obtained by a combination of alcoholysis and esterification (Figure 5). In the first step, a triglyceride (e.g. triolein, peanut oil) is subjected to alcoholysis with ethanol at controlled water activity in methyl-*t*-butyl ether yielding

Table 5 Synthesis of di- or monoacylglycerides using lipase from *Rhizopus species*

Acyl acceptor	Acyl donor	Reaction system	Yield (%)	Lipase	Reference
glycerol	palmitic acid	solid-phase	80 (1,3)[b]	RJL	Weiss (1990)
glycerol	vinyl caprylate	MTBE[a]	80 (1,3)[b]	RDL	Berger *et al.* (1992)
glycerol	ethyl caproate	free evaporation/ precipitate	88 (1,3)[b]	RAL	Millqvist-Fureby *et al.* (1996a)
water[1]	palm oil	reverse micelles	(26) 78[c,d]	RDL	Holmberg and Österberg (1988)
water[1]	castor oil	buffer, pH 7.5	(8) 23[c]	RAL	Flenker and Spener (1990)
ethanol[1]	tripalmitin	MTBE	(32) 97[c,d]	RAL	Millqvist *et al.* (1994)
glycerol[2]	palm oil	reverse micelles	(30) 30[c]	RAL	Holmberg *et al.* (1989)
glycerol[3]	lauric acid/ oleic acid	reverse micelles	(20) 62[c]	RDL	Hayes and Gulari (1994, 1992, 1991)
water	crambe oil	reverse micelles	?	RJL	Derksen and Cuperus (1992)
glycerol[3]	oleic acid	water	(20) 60[c]	RDL	Tsujisaka *et al.* (1977)
glycerol[3]	oleic acid	org. solvent, SCCO$_2$	(6–15) 18–44[c]	RAL	Gancet (1990)
glycerol[3]	palmitic acid	hexane	(20) 61[c]	RDL	Kwon *et al.* (1995)

[1]: hydrolysis or alcoholysis of triacylglycerides to 2-MAG's, [2]: glycerolysis of triacylglycerides yielding mixtures of 1(3)-MAG's and 2-MAG's, [3]: esterification or transesterification of glycerol with fatty acids or esters yielding 1(3)-MAG's; [a]methyl *t*-butyl ether; [b]1,3: 1,3-DAG; [c]overall yields of MAG relative to glycerolysis (per mol TAG), hydrolysis (per mol TAG) or esterification (per mol glycerol) are given in brackets; [d]2-MAG

2-monoglycerides (2-MAG). RDL was found most suitable and after crystallization up to 72% yield of 2-MAG were achieved. The subsequent esterification of this 2-MAG with caprylic acid in an organic solvent catalyzed by RDL gave a highly pure MLM-triglyceride, which contained more than 90% caprylic acid in *sn*1- and *sn*3-position and 98.5% unsaturated fatty acids in *sn*2-position (Soumanou *et al.*, 1998). Beside RDL, RJL was also suitable in this reaction.

Rhizopus sp. lipase were also found active in supercritical carbon dioxide, as shown by Nakamura *et al.* (1986) for the RDL-catalyzed interesterification of triolein and stearic acid.

Preparation of MAG or DAG

Monoacylglycerides (MAG) as well as diacylglycerides (DAG) are widely used emulsifiers in the food and pharmaceutical area. Although numerous articles are dealing with the lipase-catalyzed synthesis of these compounds, most MAG and DAG are still chemically produced. Examples for the lipase-catalyzed synthesis of MAG have recently been reviewed (Bornscheuer, 1995), and in several cases lipases from *Rhizopus species* have been used (Table 5).

DAG have been synthesized in high yields (up to 80%) by esterifications between glycerol and fatty acids (or their derivatives) using lipases from *Rhizopus delemar, arrhizus* or *javanicus*, and 1,3-DAG was the major product. For example, Berger *et al.* (1992) adsorbed glycerol onto silica gel to create a large surface area. RML-catalyzed esterification with fatty acid methyl esters or RDL-catalyzed esterification with fatty acid vinyl esters provided 1,3-DAG in good yields (>70%) and high regioisomeric purity (>98%). Vinyl ester reacted faster, but are also more expensive. Although RDL was more regioselective, RML was preferred because it was more active. Several unsaturated or hydroxy fatty acids were also converted to MAG or DAG in high yields with RNL using the same method (Waldinger and Schneider, 1996).

MAG were prepared by lipase hydrolysis of oil in reverse micelles or buffer, but yields were not satisfactory, mainly due to rapid acyl migration from 2-MAG to 1- or 3-MAG, which is further hydrolyzed by the lipase. This problem was overcome by performing an alcoholysis of triglycerides at controlled water activity (Millqvist-Fureby *et al.*, 1996b). Thus, the reaction between tripalmitin with ethanol using a lipase from *Rhizopus arrhizus* immobilized on Celite under optimized conditions gave 97% yield of 2-MAG (Millqvist *et al.*, 1994). Alternatively, esterification between glycerol and fatty acid in various reaction systems with *Rhizopus* lipase was studied by several groups (Table 5). Hayes and Gulari (1991) employed a 1,3-specific lipase from *Rhizopus delemar*, but mainly 1 (3)-MAG was formed, which again was related to acyl migration. Conversion reached 50 to 60%, and the reaction rate decreased in the order oleic acid > caprylic acid > myristic acid = lauric acid > stearic acid = palmitic acid. A third general approach is based on the glycerolysis of fats or oils. Gancet (1990) developed a two fixed-bed segment for the continuous glycerolysis of beef tallow. At a residence time of 75 min, a yield of 38.8 g MAG per 100 g beef tallow was reported. The reactor system was run for several months with good operational stability of the mycelial lipase from *Rhizopus arrhizus*.

Triacylglycerol hydrolysis

A comparison of several lipases in the hydrolysis of crambe oil (rich in erucic acid) in reverse micelles revealed that most lipases followed Michaelis-Menten kinetics, and RJL was found to be the most active enzyme (Derksen and Cuperus, 1992). A similar system was studied by Kim and Chung (1989) using RAL and palm kernel olein in AOT/isooctane reversed micelles. In another study it was found that a cationic surfactant (CTAB) decreased the maximal rate of a RDL-catalyzed hydrolysis of triolein by a factor of 50, if compared to AOT (Valis *et al.*, 1992).

Organic Synthesis

Most lipases exhibit high stability in organic solvents, and their ability to discriminate between stereoisomers has lead to a large number of applications for the resolution of a wide variety of chiral alcohols or acids. For these investigations, mainly lipases from *Pseudomonas sp.*, *Candida sp.*, *Rhizomucor sp.* and from Porcine pancreas were used. Only a few examples deal with lipases from *Rhizopus sp.* — despite the fact that these lipases accept vinyl acetate as an irreversible acylating agent and are stable toward the acetaldehyde concomitantly formed (Weber *et al.*, 1995).

Goto *et al.* (1996) studied the kinetic resolution of ibuprofen using surfactant-coated lipases. They found that surfactant-coated RJL catalyzed efficiently the acylation of (*S*)-ibuprofen with hexadecanol in isooctane, and up to 70%ee were achieved. Moreover, the coated lipase was approx. 100 times more active than the powdered enzyme.

In another example, RJL was used to prepare a key precursor in the synthesis of tautomycin (Nagumo *et al.*, 1996). RJL allowed for high conversion and reaction rates in the acylation of a *meso*-triol with vinyl acetate (53%ee, 1 d) or the hydrolysis of the corresponding triester (93%ee, 10 d).

Excellent optical purities (>95%ee) and good yields (64–95%) were reported in the RDL-catalyzed hydrolysis of *meso*-diacetates employed in the synthesis of the antiviral carbocyclic nucleosides (−)-carbovir and (−)-BCA (Tanaka *et al.*, 1996). Here, RDL was superior to lipases from *Pseudomonas sp.* or Porcine pancreas.

RJL was used in the resolution of *cis*-3-acetyloxy-4-phenyl-2-azetidinone, a compound important in the semi-synthesis of taxol. Optical purities were excellent (98.5%ee), but yields were lower compared to other lipases (Patel *et al.*, 1996).

Lipases can be used to deprotect carboxyl groups in peptides. Braun *et al.* (1991, 1992) cleaved the heptyl ester carboxyl protective group from a wide range of dipeptides using RNL (Amano N). This lipase did not cleave the amino protective groups Cbz, Boc, Aloc, or Fmoc and the heptyl protective group survived conditions used to remove these amino protective groups (hydrogenation, HCl/ether, or Pd(0)/C-nucleophile). Although the crude lipase (Amano N) also hydrolyzed the peptide link, pretreatment with PMSF, a serine protease inhibitor, eliminated this side reaction. Hydrolysis of the heptyl group slowed and sometimes did not proceed when the peptide was hindered and/or hydrophobic.

OUTLOOK

Lipases obtained from the genus *Mucorales* have been explored in a wide range of synthetic reactions in oleochemistry and organic synthesis. *Rhizopus* lipases have found somewhat less applications, than the lipase from *Rhizomucor miehei* which is used, in immobilized form, in some oleochemical biotransformations. However, *Mucorales* lipases can be heterologously expressed in active form, without the need to apply cumbersome refolding procedures. In combination with the better understanding of the structure and function of these lipases, the way is open to taylor these stable and active enzymes to the needs of the organic chemist, and the applications given above already show that *Rhizopus* lipases are versatile catalysts. It is to be expected that *Rhizopus* lipases may be among those enzymes who will find increasing applications in industrial enzyme technology.

ACKNOWLEDGEMENTS

Financial support by the European Community (BRIDGE Lipase-T-project and EU-BIO4-CT96-0005), the German Research Foundation (DFG, SCHM 1240/1-1) and the German Academic Exchange Service (DAAD) is gratefully acknowledged.

REFERENCES

Atomi, H., Bornscheuer, U., Soumanou, M.M., Beer, H.D., Wohlfahrt, G. and Schmid, R.D. (1996) Microbial Lipases — from Screening to Design. In *Oils-Fats-Lipids, Proc. 21st World Congr. Int. Soc. Fat Res.*, Vol. 1, pp. 49–50. Bridgwater: PJ Barnes & Associates.

Bairoch, A. and Boeckmann, B. (1991) The SWISS-PROT protein sequence data bank. *Nucleic Acids Res.*, 19, 2247–2249.

Basheer, S., Mogi, K.-I. and Nakajima, M. (1995) Development of a novel hollow-fibre membrane reactor for the interesterification of triglycerides and fatty acids using modified lipase. *Process Biochem.*, 30, 531–536.

Beer, H.D., Wohlfahrt, G., Schmid, R.D. and McCarthy, J.E.G. (1996a) The folding activity of the extracellular lipase of *Rhizopus oryzae* are modulated by a prosequence. *Biochem. J.*, 319, 351–359.

Beer, H.D., Wohlfahrt, G., McCarthy, J.E.G., Schomburg, D. and Schmid, R.D. (1996b) Analysis of the catalytic mechanism of a fungal lipase using computer-aided design and structural mutants. *Protein Eng.*, 9, 507–517.

Belev, T., Singh, N.M. and McCarthy, J.E.G. (1991) A fully modular vector system for the optimization of gene expression in *Escherichia coli*, *Plasmid*, 26, 147–150.

Berger, M., Laumen, K. and Schneider, M.P. (1992) Enzymatic Esterification of Glycerol I. Lipase-Catalyzed Synthesis of Regioisomerically Pure 1,3-sn-Diacylglycerols. *J. Am. Oil Chem. Soc.*, 69, 955–960.

Berger M. and Schneider, M.P. (1991) Regioselectivity of lipases in organic solvents. *Biotechnol. Lett.*, 13, 333–338.

Bernstein, F.C., Koetzle, T.F., Williams, G.J., Meyer, E.E. Jr., Brice, M.D., Rodgers, J.R., Kennard, O., Shimanouchi, T. and Tasumi, M. (1977) The Protein Data Bank: a computer-based archival file for macromolecular structures. *J. Mol. Biol.*, 112, 535–542.

Bilofsky, H.S., Burks, C., Fickett, J.W., Goad, W.B., Lewitter, F.I., Rindone, W.P., Swindell, C.D. and Tung, C.S. (1986) The GenBank genetic sequence databank. *Nucleic Acids Res.*, 14, 1–4.

Boel, E., Huge-Jensen, B., Christensen, M., Thim, L. and Fili, N.P. (1988) *Rhizomucor miehei* triglyceride lipase is synthezised as a precursor. *Lipids*, 23, 701–706.

Bornscheuer, U.T. (1995) Lipase-catalyzed synthesis of monoacylglycerols. *Enzyme Microb. Technol.*, 17, 578–586.

Brady, L., Brzozowski, A.M., Derewenda, Z.S., Dodson, E., Dodson, G., Tolley, S., Turkenburg, J.P., Christiansen, L., Huge-Jensen, B., Norskov, L., Thim, L. and Menge, U. (1990) A serine protease triad forms the catalytic centre of a triacylglycerol lipase. *Nature*, 343, 767–770.

Braun, P., Waldmann, H., Vogt, W. and Kunz, H. (1991) Selective enzymatic removal of protecting functions: heptyl esters as carboxy protecting groups in glycopeptide synthesis. *Liebigs Ann. Chem.*, 165–170.

Braun, P., Waldmann, H. and Kunz, H. (1992) elective enzymatic removal of protecting functions: heptyl esters as carboxy protecting groups in glycopeptide synthesis. *Synlett*, 39–40.

Brzozowski, A.M., Derewenda, U., Derewenda, Z.S., Dodson, G.G., Lawson, D.M., Turkenburg, J.P., Bjorkling, F., Huge-Jensen, B., Patkar, S.A. and Thim, L. (1991) A model for interfacial activation in lipases from the structure of a fungal lipase-inhibitor complex. *Nature*, 351, 491–494.

Cygler, M., Grochulski, P., Kazlauskas, R.J., Schrag, J. and Joseph, D.A. (1994) Structural basis for the chiral preferences of lipases. *J. Am. Chem. Soc.*, 116, 3180–3186.

Derewenda, Z.S., Derewenda, U. and Dodson, G.G. (1992a) The crystal and molecular structure of the *Rhizomucor miehei* triacylglyceride lipase at 1.9 A resolution. *J. Mol. Biol.*, 227, 818–839.

Derewenda, U., Brzozowski, A.M., Lawson, D.M. and Derewenda, Z.S. (1992b) Catalysis at the interface: the anatomy of a conformational change in a triglyceride lipase. *Biochemistry*, 31, 1532–1541.

Derewenda, U., Swenson, L., Green, R., Wei, Y., Dodson, G.G., Yamaguchi, S., Haas, M.J. and Derewenda, Z.S. (1994a) An unusual buried polar cluster in a family of fungal lipases. *Nat. Struct. Biol.*, 1, 36–47.

Derewenda, U., Swenson, L., Wei, Y., Green, R., Kobos, P.M., Joerger, R., Haas, M.J. and Derewenda, Z.S. (1994b) Conformational lability of lipases observed in the absence of an oil-water interface: crystallographic studies of enzymes from the fungi *Humicola lanuginosa* and *Rhizopus delemar*. *J. Lipid Res.*, 35, 524–534.

Derksen, J.T.P. and Cuperus, F.P. (1992) Lipase-catalyzed hydrolysis of crambe oil in AOT-isooctane reversed micelles. *Biotechnol. Lett.*, 14, 937–940.

Flenker, J. and Spener, F. (1990) Hydroxy-monoglycerides by lipases-catalyzed partial hydrolysis of castor oil. *DECHEMA Biotechnol. Conf.*, **4**, Pt. A, 139–142.

Gancet, G. (1990) Catalysis by *Rhizopus arrhizus* mycelial lipase. *Ann. N.Y. Acad. Sci.*, **613**, 600–604.

Goto, M., Noda, S., Kamiya, N. and Nakashio, F. (1996) Enzymatic resolution of racemic ibuprofen by surfactant-cotaed lipases in organic media. *Biotechnol. Lett*, **18**, 839–844.

Goto, M., Kamiya, N., Miyata, M. and Nakashio, F. (1994) Enzymatic Esterification by Surfactant-Coated Lipase in Organic Media. *Biotechnol. Prog.*, **10**, 263–268.

Haalck, L., Paltauf, F., Pleiss, J., Schmid, R.D., Spener, F. and Stadler, P. (1997) Stereoselectivity of lipase from *Rhizopus oryzae* towards triacylglycerols and analogs: computer aided modeling and experimental validation. *Methods Enzymol.*, **284**, 353–376.

Haas, M.J., Allen, J. and Berka, T.R. (1991) Cloning, expression and characterization of a cDNA encoding a lipase from *Rhizopus delemar*. *Gene*, **109**, 107–113.

Haas, M.J., Joerger, R.D., King, G. and Klein, R.R. (1996) The use of rational mutagenesis to modify the chain length specificity of a *Rhizopus delemar* lipase. *Ann. N.Y. Acad. Sci.*, **799**, 115–128.

Haas, M.J. and Joerger, R.D. (1995) Lipases of the Genera *Rhizopus* and *Rhizomucor*: Versatile catalysts in nature and the laboratory. *Food Biotechnology: Microorganisms*, edited by G.G. Khachatourians and Y.H. Hui, Chap. 15, pp. 549–588. Weinheim: VCH.

Hayes, D.G. and Gulari, E. (1994) Improvement of enzyme activity and stability for reverse micellar-encapsulated lipases in the presence of short-chain and polar alcohols. *Biocatalysis*, **11**, 223–231.

Hayes, D.G. and Gulari, E. (1992) Formation of polyol-fatty acid esters by lipases in reverse micellar media. *Biotechnol. Bioeng.*, **40**, 110–118.

Hayes, D.G. and Gulari, E. (1991) 1-Monoglyceride production from lipase-catalyzed esterification of glycerol and fatty acid in reverse micelles. *Biotechnol. Bioeng.*, **38**, 507–517.

Holmberg, K., Lassen, B. and Stark, M.B. (1989) Enzymatic glycerolysis of a triglyceride in aqueous and nonaqueous microemulsions. *J. Am. Oil Chem. Soc.*, **66**, 1796–1800.

Holmberg, K. and Österberg, E. (1988) Enzymatic preparation of monoglycerides in microemulsion. *J. Am. Oil Chem. Soc.*, **65**, 1544–1548.

Holzwarth, H.-C., Pleiss, J. and Schmid, R.D. (1997) Computer-aided modelling of stereoselective triglyceride hydrolysis catalyzed by *Rhizopus oryzae* lipase. *J. Mol. Catal. B. Enzymatic*, **3**, 73–82.

Huge-Jensen, B., Andreasen, F., Christensen, T., Christensen, M., Thim, L. and Boel, E. (1989) *Rhizomucor miehei* triglyceride lipase is processed and secreted from transformed *Aspergillus oryzae*. *Lipids*, **24**, 781–785.

Huge-Jensen, B., Galluzzo, D.R. and Jensen, R.G. (1987) Partial purification and characterization of free and immobilized lipase from *Mucor miehei*. *Lipids*, **22**, 559–565.

Joerger, R.D. and Haas, M.J. (1994) Alteration of chain length selectivity of a *Rhizopus delemar* lipase through site-directed mutagenesis. *Lipids*, **29**, 377–384.

Joerger, R.D. and Haas, M.J. (1993) Overexpression of a *Rhizopus delemar* lipase gene in *Escherichia coli*. *Lipids*, **28**, 81–88.

Kabsch, W. and Sander, C. (1983) Dictionary of protein secondary structure: pattern recognition of hydrogen-bonded and geometrical features. *Biopolymers*, **22**, 2577–2637.

Kazlauskas, R.J. (1994) Elucidating structure-mechanism relationships in lipases: prospects for predicting and engineering catalytic properties. *Trends Biotechnol.*, **12**, 464–472.

Kermasha, S., Safari, M. and Goetghebeur, M. (1995) Interesterification of butter fat by lipase from *Rhizopus niveus* in cosurfactant-free microemulsion system. *Appl. Biochem. Biotechnol.*, **53**, 229–244.

Kim, T. and Chang, K. (1989) Some characteristics of palm kernel olein hydrolysis by Rhizopus arrhizus lipase in reversed micelle of AOT in isooctane and additive effects. *Enzyme Microb. Technol.*, **11**, 528–532.

Klein, R.R., King, G., Moreau, R.A. and Haas, M.J. (1997) Altered acyl chain length specificity of *Rhizopus delemar* lipase through mutagenesis and molecular modeling. *Lipids*, **32**, 123–130.

Kohno, M., Kugimiya, W., Hashimoto, Y. and Morita, Y. (1994) Purification, characterization and crystallization of two types of lipase from *Rhizopus niveus*. *Biosci. Biotech. Biochem.*, **58**, 1007–1012.

Kohno, M., Funatsu, J., Mikami, B., Kugimiya, W., Matsuo, T. and Morita, Y. (1996) The crystal structure of lipase II from *Rhizopus niveus* at 2.2 A resolution. *J. Biochem.*, **120**, 505–510.

Kwon, S.J., Han, J.J. and Rhee, J.S. (1995) Production and in situ preparation of mono- or diacylglycerol catalyzed by lipases in n-hexane. *Enzyme Microb. Technol.*, **17**, 700–704.

Kugimiya, W., Otani, Y., Kohno, M. and Hashimoto, Y. (1992) Cloning and sequence analysis of cDNA encoding *Rhizopus niveus* lipase. *Biosci. Biotech. Biochem.*, **56**, 716–719.

Lawson, D.M., Brzozowski, A.M., Rety, S., Vernag, C. and Dodson, G.G. (1994) Probing the nature of substrate binding in *Humicola lanuginosa* lipase through X-ray crystallography and intuitive modeling. *Protein Eng.*, 7, 543–550.

Martinelle, M., Holmquist, M., Clausen, I.G., Patkar, S., Svendsen, A. and Hult, K. (1996) The role of Glu87 and Trp89 in the lid of *Humicola lanuginosa* lipase. *Protein Eng.*, 9, 519–524.

Menge, U., Schomburg, D., Spener, F. and Schmid, R.D. (1995) Towards novel biocatalysts via protein design: the case of lipases. *FEMS Microbiol. Rev*, 16, 253–257.

Millqvist, A., Adlercreutz, P. and Mattiasson, B. (1994) Lipase-catalyzed alcoholysis of triglycerides for the preparation of 2-monoglycerides. *Enzyme Microb. Technol.*, 16, 1042–1047.

Millqvist-Fureby, A., Adlercreutz, P. and Mattiasson, B. (1996a) Glyceride synthesis in a solvent-free system. *J. Am. Oil Chem. Soc.*, 73, 1489–1495.

Millqvist-Fureby, A., Virto, C., Adlercreutz, P. and Mattiasson, B. (1996b) Acyl migration in 2-monoolein. *Biocatal. Biotransform.*, 14, 89–111.

Minning, S., Schmidt-Dannert, C., Schmid, R.D. Overexpression of functional *Rhizopus oryzae* lipase in *Pichia pastoris*: high-level production and some properties, in preparation.

Mojovic, L., Siler-Marinkovic, S., Kukic, G., Bugarski, B. and Vunjak-Novakovic, G. (1994) *Rhizopus arrhizus* lipase-catalyzed interesterification of palm oil midfraction in a gas-lift reactor. *Enzyme Microb. Technol.*, 16, 159–162.

Mogi, K.-I. and Nakajima, M. (1996) Selection of surfactant-modified lipases for interesterification of triglyceride and fatty acid. *J. Am. Oil Chem. Soc.*, 73, 1505–1512.

Nagumo, S., Arai, T. and Akita, H. (1996) Enzymatic hydrolysis of *meso*(*syn-syn*)-1,3,5-triacetoxy-2,4-dimethylpentane and acetylation of *meso*(*syn-syn*)-3-benzyloxy-2,4-dimethylpentane-1,5-diol by lipase. *Chem. Pharm. Bull.*, 44, 1391–1394.

Nakamura, K., Chi, Y.M., Yamada, Y. and Yano, T. (1986) Lipase activity and stability in supercritical carbon dioxide. *Chem. Eng. Commun.*, 45, 207–212.

Norin, M., Haeffner, F., Achour, A., Norin, T. and Hult, K. (1994) Computer modeling of substrate binding to lipases from *Rhizomucor miehei*. *Humicola lanuginosa* and *Candida rugosa*. *Protein Sci.*, 3, 1493–1503.

Oba, T. and Witholt, B. (1994) Interesterification of milk fat with oleic acid catalyzed by immobilized *Rhizopus oryzae* lipase. *J. Dairy Sci*, 77, 1790–1797.

Okahata, Y. and Ijiro, K. (1992) Preparation of a lipid-coated lipase and catalysis of glyceride ester syntheses in homogeneous organic solvents. *Bull. Chem. Soc. Jpn.*, 65, 2411–2420.

Okahata, Y. and Ijiro, K. (1988) A lipid-coated lipase as a new catalyst for triglyceride synthesis in organic solvents. *J. Chem. Soc. Chem. Commun.*, 20, 1392–1394.

Ollis, D.L., Cheah, E., Cygler, M., Dijkstra, B., Frolow, F., Franken, S.M., Harel, M., Remington, S.J., Silman, I., Schrag, J., *et al.* (1992) The alpha/beta hydrolase fold. *Protein Eng.*, 5, 197–211.

Patel, R.N., Banerjee, A. and Szarka, L.J. (1996) Biocatalytic synthesis of some chiral pharmaceutical intermediates by lipases. *J. Am. Oil Chem. Soc.*, 73, 1363–1375.

Rogalska, E., Cudrey, C., Ferrato, F. and Verger, R. (1993) Stereoselective hydrolysis of triglycerides by animal and microbial lipases. *Chirality*, 5, 24–30.

Salah, B., Fendri, K. and Gargaoury, Y. (1994) La lipase de *Rhizopus oryzae*: production, purification et charactéristique biochimiques. *Rev. franc. corps gras*, 41, 133–137.

Schipper, M.A.A. (1984) A revision of the genus *Rhizopus* II. The *Rhizopus stolonifer*-group and *Rhizopus oryzae*. *Stud. Mycol.*, 25, 1–19.

Shimada, Y., Sugihara, A., Nakano, H., Yokota, T., Nagao, T., Komemushi, S. and Tominaga, Y. (1996) Production of structured lipids containing essential fatty acids by immobilized *Rhizopus delemar* lipase. *J. Am. Oil Chem. Soc.*, 73, 1415–1420.

Soumanou, M.M., Bornscheuer, U.T., Menge, U. and Schmid, R.D. (1997) Synthesis of structured triglycerides from peanut oil with immobilized lipase. *J. Am. Oil Chem. Soc.*, 74, 427–433.

Soumanou, M.M., Bornscheuer, U.T. and Schmid, R.D. (1998) Two-step enzymatic reaction for the synthesis of pure structured triacylglycerides. *J. Am. Oil Chem. Soc.*, 75, 703–710.

Tanaka, M., Norimine, Y., Fujita, T., Suemune, H. and Sakai, K. (1996) Chemoenzymatic synthesis of antiviral carbocyclic nucleosides: Asymmetric hydrolysis of *meso*-3,5-bis(acetoxymethyl)cyclopentenes using *Rhizopus delemar* lipase. *J. Org. Chem.*, 61, 6952–6957.

Thompson, J.D., Higgins, D.G. and Gibson, T.J. (1994) CLUSTAL W: improving the sensitivity of progressive multiple sequence alignment through sequence weighting, positions-specific gap penalties and weight matrix choice. *Nucleic Acids Res.*, 22, 4673–4680.

Tsujisaka, Y., Okumura, S. and Iwai, M. (1977) Glyceride synthesis by four kinds of microbial lipases. *Biochim. Biophys. Acta*, **489**, 415–422.

Uyttenbroeck, W., Hendriks, D., Vriend, G., de Baere, I., Moens, L. and Scharpé, S. (1993) Molecular characterization of an extracellular acid-resistant lipase produced by *Rhizopus javanicus. Biol. Chem. Hoppe-Seyler*, **374**, 245–254.

Valis, T., Xenakis, P. and Kolisis, F.N. (1992) Comparative studies of lipase from *Rhizopus delemar* in various microemulsion systems. *Biocatalysis*, **6**, 267–279.

Waldinger, C. and Schneider, M.P. (1996) Enzymatic esterification of glycerol III. Lipase-catalyzed synthesis of regioisomerically pure 1,3-sn-diacylglycerols and 1(3)-*rac*-monoacylglycerols derived from unsaturated fatty acids. *J. Am. Oil Chem. Soc.*, **73**, 1513–1519.

Weber, H.K., Stecher, H. and Faber, K. (1995) Sensitivity of microbial lipases to acetaldehyde formed by acyl-transfer reactions from vinyl esters. *Biotechnol. Lett.*, **17**, 803–808.

Weiss, A. (1990) Enzymic preparation of solid fatty acid monoglycerides. *Fat Sci. Technol.*, **92**, 392–396.

Wisdom, R.A., Dunnill, P. and Lilly, M.D. (1987) Enzymic interesterification of fats: Laboratory and pilot-scale studies with immobilized lipase from *Rhizopus arrhizus. Biotechnol. Bioeng.*, **29**, 1081–1085.

Yadwad, V.B., Ward, O.P. and Noronha, L.C. (1991) Application of lipase to concentrate the docosahexaenoic acid (DHA) fraction of fish oil. *Biotechnol. Bioeng.*, **38**, 956–959.

Yokozeki, K., Yamanaka, S., Takinami, K., Hirose, Y., Tanaka, A., Sonomoto, K. and Fukui, S. (1982) Application of immobilized lipase to regio-specific interesterification of triglyceride in organic solvent. *Eur. J. Appl. Microbiol. Biotechnol.*, **14**, 1–5.

7. SOLVENT ENGINEERING MODULATES STEREOSELECTIVITY OF MICROBIAL LIPASES

ENRICO CERNIA, CLEOFE PALOCCI and SIMONETTA SORO

Department of Chemistry, University of Rome " La Sapienza", P.le Aldo Moro, 5, 00185 Rome Italy

INTRODUCTION

Lipases, beside their use as drugs in disorders of the digestive system (Gargouri *et al.*, 1997), have proved to be useful catalysts in biotechnology.

In the last years lipases of different sources have been recognised as valuable catalysts for organic transformations both on the laboratory and on the industrial scale (Zacks *et al.*, 1984; Dordick, 1991). While bulk application already enjoyed a long standing tradition in areas of food processing and detergent formulations, new examples of the beneficial use of lipases have now emerged in the production of fine chemicals. In particular the stereoselectivity of lipases has been employed to manufacture enantiomerically pure pharmaceuticals, agrochemicals and food additives.

Their broad synthetic potential is largely due to the fact that lipases can accommodate a wide range of substrates other than triglycerides, such as, esters, thioesters and amides.

The stability in non-aqueous organic solvents, expanding the synthetic potential of lipases, can be applied to hydrolysis, ester synthesis and transesterifications.

LIPASES APPLICATIONS

Flavour and Feed Industry

The majority of flavours and fragrances used in foods and perfumery are esters, some of which are fatty acid esters. Many of the naturally occurring esters are now synthesised, and lipase biocatalysis could be advantageous due to specific and mild reactions. Although many flavour esters are composed of acids and alcohols very different from fatty acids and glycerol (branched, cyclic, aromatic, very short chain) lipases may still accept these. Synthesis of terpene alcohol esters by various lipases has been reported, as well as the continuous synthesis of ethyl butyrate (pineapple/ banana flavour) by immobilised *Candida rugosa* lipase. A number of other flavour esters were made. Biochemical production of butter flavour can be made by alcoholysis of butter fat and ethanol, e.g. with *Candida rugosa* lipase. This type of reaction provides a mixture of fatty acid ethyl esters, which may be combined with natural components in foods, e.g. margarine and cookies (Harwood *et al.*, 1989). The oil industry is widely interested in the application of lipase to the neutralisation of oil acidity. The acidity of tropical oils is due to the presence of free fatty acids. The content of these acids can be depressed by partial esterifications by lipases (Graille *et al.*, 1988). The Soya-bean oil changes its flavour during storage and during the frying process giving rise to a typical fish smell. The Soya-bean oil flavour instability is due to the linoleic acid. At low temperature *Rhyzomucor miehei* lipase has shown specificity towards polyinsaturated fatty acid. It is possible to decrease the linoleic acid to 3% content by biocatalised transesterification at 10°C (Karmal *et al.*, 1988). Otherwise γ-linoleic acid has an high dietetic and commercial value. It is in the prostaglandin path-synthesis and it is rare in edible oils. At the moment the natural source of this fatty acid is evening primrose seeds. In order to obtain a fraction with an higher content of γ-linoleic acid, from evening primrose oil as starter, turnip lipase was employed (Hills *et al.*, 1989). Using similar techniques it is possible to get hypocaloric products enriching oil with short and medium chain-length fatty acids. The properties of fat depend on the fatty acids content and the commercial value of one fat compared to another is based on its fatty acid structure. Traditionally, fat and oil processors have changed the fatty acid structure of their materials by blending different triglyceride mixtures, by chemical modification of the fatty acids or by re-arrangement of the fatty acids on the glyceride backbone of the fat (interesterification) (Harwood *et al.*, 1989). Specific enzyme catalysts may be used for interesterification. In this field, another example of lipase application in the cookies industry is cocoa butter production. This lipid contains an high amount of stearic acid and is composed of 1-palmitoyl-2-oleyl-3-stearoyl glycerol. An alternative source of it at cheaper cost is palm oil. In fact palm oil is composed of 1,3-dipalmitoyl-2-oleyl glycerol. The 1,3 regioselective transesterification, catalized by *Rhyzomucor miehei* lipase, of palm oil yields cocoa butter (Macrae *et al.*, 1985). *Rhizopus delemar* lipase shows the same regioselectivity, it is able to biotransform olive oil into 1,3-distearyl-2-oleyl glycerol. In the same way the immobilised lipase from *Humicola lanuginosa*, a thermostable enzyme, catalyses the transesterification of olive oil into 1,3-distearyl-2-oleyl glycerol at 50°C in 30 h with 65% yield in stearic acid (Omar *et al.*, 1988).

Flavour development of cheeses is to a great extent due to the ability of lipases to modify milk fat by partial hydrolysis. Specific free fatty acid profiles are generated by naturally occurring microbial lipases or enzymes added for flavour enhancement. In the first case, moulds involved in the ripening of certain varieties of cheeses, e.g. *Penicillium sp.*, produce lipases. *Penicillium roqueforti* lipases have a short-chain fatty acid specificity and the liberation of a suitable balance of mainly short-chain, volatile fatty acids seems to be the general effect behind a characteristic flavour. Probably some of the fatty acids may be released as ethyl- or other alcohol esters with flavour potential, by lipase catalysed alcoholysis or esterification. Pre-gastric esterases are used in the manufacture of Italian cheeses to produce the characteristic piquant flavour.

Flavour development is usually the result of a very small degree of lipid hydrolysis (few percent) whereas a higher degree may lead also to a change in the physical properties of the fat. Such partial hydrolysis of oils and fats can improve the palatability and digestibility of the lipid component in feeds and pet foods. Hard fats may be softened due to mono- and di-glycerides and more energy supplied to the animals if the lipid is "pre-digested".

The production of emulsifiers is another important aspect of the food industry. At the present lecithin, monoglycerides and sugar esterified with fatty acids are employed as emulsifiers in foods. Lecithin is a by-product of seed oil refining. Enzymes are used to produce lyso-lecithin which has better emulsifying properties than normal lecithin. Lyso-lecithin is used in margarine and cosmetics, for instance.

Detergent industry

Fatty acids and oil soiling are difficult to remove from fabrics at low washing temperatures. Spillage of lipid containing foods and accumulation of sebum from skin are common problems. Therefore, intensive research efforts have been directed towards development not only of new detergent compounds and surfactants but also toward enzymes which could facilitate removal of this type of soiling under wash conditions (Newmark, 1988). Hydrolysis of triglycerides and other fatty esters does not increase the water solubility of the lipids as readily as in the case of protein and starch hydrolysis. It has been shown that a mixture with partial glycerides and free fatty acids is more effectively removed from the fabric than free fatty acids alone are, so that a non-specific lipase is not necessary or even preferable in washing. A detergent lipases require an high alkaline stability (common powder detergents pH is 9–10), are thermostable enzymes, stable in the presence of protease and compatible with common detergents. The first detergent lipase on the market was Lipolase™, introduced early in 1988 by Novo. It was also the first industrial enzyme produced by recombinant technology.

Pulp and Paper

Enzyme treatments, including the hydrolytic application of lipases, have recently been taken up by the paper and pulp industry. Although lipids constitute a very small part of wood, they may have important effects. In spruce, about 2% resin is found. The resin is a mixture of triglycerides, free fatty acids and other components. The resin content is reduced to about 0.5% or less in existing pulping processes. However a more complete resin removal is possible by the addition of lipase in the process, with improved water absorption characteristics of cellulose fibres as the result. The resin content may also influence other properties, e.g. paper strength or colour and result in paper making processes (the formation of suspension of wood fibres). A lipase preparation, Resinase™A is produced by Novo and used in pulping processes. Printed paper can be treated by lipase in order to facilitate removal of colour from the printing ink, containing vegetable oil base. This application can be useful for the improvement of return paper.

Chiral Switch

There is an increasing trend towards the use of optically pure stereoisomers with pharmaceutical activity, which are more target-specific and show fewer side-effects than racemic mixtures of isomers. This leads to an increasing demand for efficient processes for the industrial-scale synthesis of optically active compounds and to an increased research and production of chiral drugs via, for instance, lipases or esterases employment (Iriuchijma et al, 1982) The biocatalytic strategy to the resolution of racemic mixture can be applied to compounds containing alcohols, esters and acid moieties. If chiral or prochiral substrates are being used, it is usually just one enantiomer which undergoes reaction, thus leading to the chiral resolution of racemates (Sonomoto and Tanaka, 1988).

DIFFERENT APPROACHES TO THE IMPROVEMENT OF LIPASE CATALYTIC ACTIVITY AND SELECTIVITY

In organic solvents, lipases catalyse the transfer of acyl groups from suitable donors to a wide array of acceptors other than water. Depending on the type of acyl donors and acceptors, lipase-catalysed reactions include esterifications, transesterifications, amidations, peptide synthesis and macrocyclic lactone formations.

As well known, the quantification of enzyme selectivity can be determined using the so-called "Enantiomeric Ratio", or E-value (Shih *et al.*, 1982), and it serves this purpose as it summarises the enantioselectivity of an enzyme in a single number. The E-value has found its way both as a process parameter for the enzyme-catalysed kinetic resolutions, as well as for the evaluation of the intrinsic structure-function relations underlying enzymatic performance.

The presence of competing enzymes in the crude lipase preparations should also be considered. In general, lipases from microbial sources are produced extracellularly and are virtually homogeneous in terms of lipolytic activity. In contrast, crude mammalian, fungal and plant lipases preparations can contain several other interfering enzymes, including proteases and esterases, which may possess opposite or poor streoselectivity as compared with the lipases. As a consequence, low optical yields or unreproducible results are often encountered in using these crude lipases, depending on the state of the enzyme preparations. Therefore, to enhance the optical yield, a number of precautions can be taken. These include treatment of the crude enzyme preparation with chemical reagents (e.g. serine protease inhibitors) to inactivate interfering enzymes; treatment of the crude enzyme preparation by physical means (e.g. partial protein purification; lyophilization) to remove or selectively inactivate competing enzymes; selection of a suitable acyl donor which serves as a poorer substrate for interfering enzymes and/or as a more efficient substrate for the target enzymes.

Although the great potentials of many lipases in stereoselective transformations, their performance under conditions prevailing in e.g. industrial reactors, in general, is far from optimisation; in only a few cases this potential is reflected in an industrial patent ownership. It is therefore important to improve the properties, stability and selectivity, of lipases by using different strategies such as, for example site-directed mutagenesis or protein engineering, or trying to modulate enzyme selectivity and activity employing as reaction media non conventional solvents such as organic solvents or supercritical fluids (Mc Hugh and Krukonis, 1985; Ikushima *et al.*, 1992; Chi *et al.*, 1986).

Within the past ten years, lipase-encoding mutated genes from several organisms have been cloned. Some of these have been expressed in their new genetic background and recently X-ray diffraction studies have produced models of the three dimensional structure of several lipases. Also within this frame, biochemical and molecular genetic data have been gathered that allow the identification, in some lipases, of those amino acids that play a crucial role in the enzyme activity. These areas of research have converged to the extent that a co-ordinated approach, using computer-assisted molecular modelling and recombinant DNA technology, to examine structure-function relationships in these enzymes is now possible.

As far as the "solvent approach" has been concerned during the last decade, the interest for enzymatic reactions in organic media has been intense (Wescott and Klibanow, 1984; Cygler et al., 1994). Many fundamental studies have been carried out and a few industrial processes have been developed. The composition of the reaction medium is decisive with respect to the equilibrium position and substrate specificity. The degree of hydration of the medium and the biocatalyst is acknowledged to be of extreme importance. The thermodynamic water activity (aw) is used to describe the hydration level of these enzymatic systems. Minute amounts of water are necessary for the enzyme to maintain an active conformation. Any further increase of water, results in the enhancement of protein flexibility thus affecting enzyme stability. Recent studies (Hirata et al., 1990) have elucidated some effects of the hydration level on both activity and stability of enzymes in dehydrated environments. The interaction between a protein molecule and the surrounding water leads to a greater conformational mobility of enzymes. Consequently the hydration level is a critical parameter to take into account in order to optimise non-aqueous enzymatic processes (Halling, 1990).

As well known, lipases have proven to be versatile and selective catalysts for the hydrolysis of a large range of interesting esters. In organic media, they catalyse the reverse reaction leading to various combinations of chiral alcohols and/or chiral acids.

The effect of solvent choice has been addressed as "medium engineering" since this might afford a relatively simple way for the fine tuning of enzyme activity, chemioselectivity, regioselectivity and enantioselectivity. The possibility to vary the activity and the selectivity of the lipase biocatalysed reactions, by changing the nature of the reaction medium, can be considered a useful tool to modulate the chemical interactions between enzyme, substrate and reaction media with different physico-chemical properties.

Very recently, the use of microbial lipases in non conventional solvents, such as supercritical fluids, has been proposed as a means of improving the activity and utility of such enzymes in anhydrous environments.

The potential use of supercritical fluids (SCFs) in chemical separation processes has been of considerable research interest for the past decade (McHugh and Krukonis, 1985). The fundamentals of SCF extraction technology and a number of potential applications have been described in several review papers (Ikushima et al., 1992). One very interesting and, as yet, not fully tested offshoot of SCF extraction technology is the use of an SCF solvent as a reaction medium in which an SCF either actively participates in the reaction or functions only as the solvent medium for reactants, or catalysts and/or products (Chi et al., 1986).

By exploiting the unique solvent properties of SCFs (wide variations in density and viscosity are possible with small changes in pressure and/or temperature; Table 1), it may be possible to enhance reaction rates while maintaining or improving selectivity. Also, separating products from reactants can be greatly facilitated by the ease with which the solvent power of the SCF can be adjusted.

The present work reports some recent results and some general observations on parameters which can enhance the reactivity and the stereoselectivity in raw and recombinant lipase catalysed transesterifications of some model compounds,

Table 1 Physical properties of supercritical fluids

	Gas	Supercritical fluid	Liquid
Density (g/cm^3)	10^{-3}	0.1–1	1
Diffusion (cm^2/s)	10^{-1}	10^{-3}–10^{-4}	$<10^{-5}$
Viscosity (g/cm•s)	10^{-4}	10^{-3}–10^{-4}	10^{-2}

Gas Like:
High Diffusivity (mass transport in complex matrices)
Low Viscosity (favourable flow characteristics)

Liquid Like:
High Solvatating Power (dependent on density)

e.g. l-phenylethanol, and some acyl donors, e.g. vinylacetate, in organic solvents and supercritical carbon dioxide.

ENZYME ACTIVITY IN CONVENTIONAL ORGANIC SOLVENTS AND SUPERCRITICAL FLUIDS

Before optimising the environment of an enzyme, we screened a number of wild-type and recombinant lipases for the ability to catalyse the transesterification reactions between a secondary alcohol and vinylacetate, Figure 1, in some organic solvents (heptane, toluene and *tert*-buthylether) and supercritical carbon dioxide with the aim to describe the effect of the solvent on lipase activity.

We have chosen commercial lipases: four wild-type and recombinant from fungal source, and two recombinant from thermiphile microorganisms. These enzymes are usually employed as catalysts in organic synthesis and, in the case of RML even cheese manufactor. The 3D structures are available for only 3 of them (CRL, CAL A, CAL B, RML) in open and closed lid form. While CRL and RML show the "interfacial activation" phenomenon when employed in water systems for hydrolysis reactions, CAL B is active even with the substrate at lower concentration than CMC (Critical Micellar Concentration). The enzymes have different regioselectivity toward *sn*-position in triglyceride hydrolysis.

The isoform composition of the employed enzymes was first checked by gel electrophoresis in non denaturing conditions (Braimi-Horn *et al.*, 1990). The results pointed out the presence of several isoforms with esterasic activity in wild-type enzyme preparations and in recombinant ones.

In our recent research work we studied esterification reactions of model substrates (phenylethanol derivatives) with acetic anhydride catalysed by a bacterial lipase

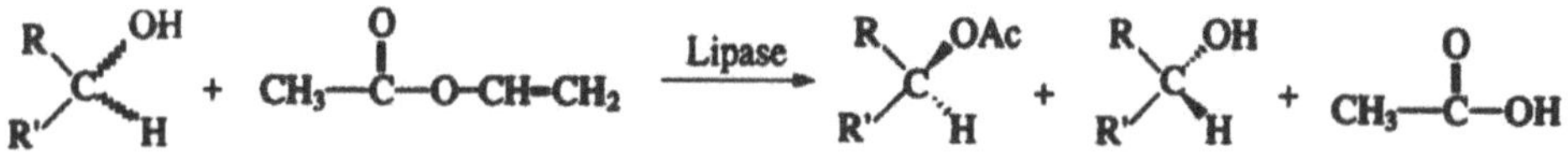

Figure 1 General scheme of transesterification reactions of secondary alchohols and vinylacetate biocatalyzed by lipase

Table 2 Transesterification reactions of (±) 1-phenylethanol with acetic anhydride employing immobilised *Pseudomonas cepacea* lipase in organic solvents and in supercritical CO₂

Substrate	SC-CO₂		Hexane		CCl₄		Toluene		Benzene		CHCl₃		NEt₃			
	C^a	$e.e^b$	C^a	$e.e^b$	C^a	$e.e^b$	C^a	$e.e^b$	C^a	$e.e^b$	C^a	$e.e^b$	C^a	$e.e^b$	C^a	$e.e^b$
1-phenyl-ethanol	46	97	32	42	28	88	28	88	21	87	36	93	20	67	14	85
1-(4-fluorophenyl)-ethanol	52	96	48	69	34	76	42	80	28	82	28	85	25	7	16	75
1-(4-chlorophenyl)-ethanol	52	80	28	58	27	69	23	70	18	67	28	90	24	12	18	85
1-(4-bromophenyl)-ethanol	55	96	21	88	25	70	24	89	20	73	27	80	18	4	11	84
1-(4-methoxyphenyl)-ethanol	29	100	17	60	28	49	26	88	25	86	29	55	19	20	29	39
1-(4-*tert*-buthylphenyl)-ethanol	52	99	32	93	28	62	24	70	16	75	24	80	20	3	10	78
1-(2-bromophenyl)-ethanol	2	100	5	71	3	13	2	100	2	100	1	100	10	4	1	31
1-(2,4-dichlorophenyl)-ethanol	10	90	9	76	4	75	8	75	7	60	2	60	14	88	2	43
1-(2-methoxyphenyl)-ethanol	7	95	5	10	5	13	1	54	2	14	3	7	15	1	3	16

[a] Conversion (%)
[b] Enantiomeric excess of product (%)

Reaction conditions: Phenylethanol = 0.5 mM, acetic anhydride = 0.5 mM, reaction volume = 1.75 mL, T = 40°C, enzyme preparation = 10 mg, 6 h reaction time

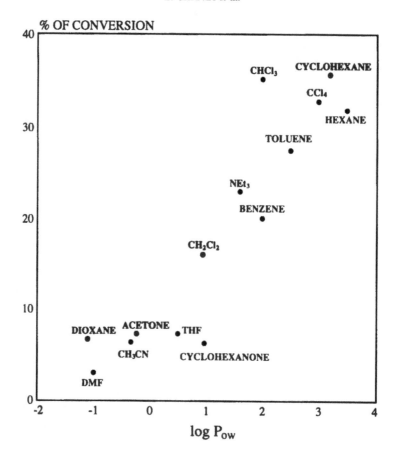

Figure 2 Esterification reactions of 1-phenylethanol and acetic anhydride catalyzed by *Pseudomonas cepacea* lipase in organic solvents. Conversion (%) versus solvent characteristics expressed as logPow. (DMF: dimethylformamide, THF: tetrahydrofuran, NEt₃: triethylamine, CCl₄: carbon tetrachloride, CH₂Cl₂: dichlorometane, CHCl₃: chloroform, CH₃CN, acetonitrile)

from *Pseudomonas cepacia* (Cernia *et al.*, 1996). The results pointed out the percentage of conversion and enatioselectivity dramatically depend on the utilised solvent, while the use of differently activated substrates did not substantially improve the reaction rates in the employed conditions, Table 2. Products were analysed by gas chromatography (Fisons 5300) using two serial capillary columns: OV1: 25m × 0.32 mm ID and DMePeβCDX, Mega 25m × 0.25mm ID. Data proved that there is a correlation between solvent hydrophobicity, expressed in terms of logPow values, and percentage of conversion, Figure 2. Moreover the results obtained in supercritical fluids showed an higher extent of conversion and enantiomeric excess values compared to those obtained in organic solvents. The apparatus employed to carry out the reactions in supercritical medium (SFE 30 Fison Instruments) has been specially designed to investigate various enzymatic reactions in supercritical carbon dioxide in a reactor cell of 0.9 mL, Figure 3. After sealing, pressurisation is achieved by pumping liquid carbon dioxide to the desired final pressure (20 MPa), and the reactor is thermostated at 40°C and 70°C. During the reaction time a 6-way valve

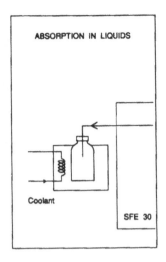
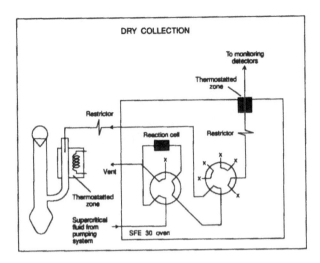

Figure 3 Schematic diagram of supercritical fluid biorecator

(Rheodyne 7125) permits the withdrawal of samples for analysis without depressurisation.

On this basis, we carried out alcoholysis reactions with a broad number of recombinant and wild-type lipases as catalysts in some organic solvents and supercritical fluids.

As expected, the use of hydrophylic organic solvents (*tert*-butylether and toluene) decrease the percent of conversion in comparison with the hydrophobic one (heptane). In some cases (wild-type and recombinant RML), Table 3, there is no detectable reaction product, showing remarkably different suitability of the substrate for the biocatalysts. As far as enzyme selectivity, the reaction product was proved to be the R(+)1-phenylethyl acetate in all the reaction medium employed (e.e. 100%) unless in the case of recombinant CAL A, carried out in heptane, in which an

Table 3 Transesterification reactions of (±) 1-phenylethanol with vinyl acetate employng wild-type and recombinant *Rhyzomucor miehei* lipases

Enzyme	Solvent	Conversion (%)	Stereochemistry	e.e.$_p$ (%)
RML wt	SCCO$_2$	0.8	R(+) acetic ester	100
RML r	SCCO$_2$	1.27	R(+) acetic ester	100
RML wt	heptane	0	/	/
RML r	heptane	7,5	R(+) acetic ester	100
RML wt	toluene	0	/	/
RML r	toluene	0	/	/
RML wt	*tert*-buthylether	0	/	/
RML. r	*tert*-buthylether	0	/	/

Reaction conditions: Phenylethanol = 1mM, vinylacetate = 10mM, reaction volume = 1mL, T = 40°C, enzyme = 5 mg, 24 h reaction time

Table 4 Transesterification reactions of (±) 1-phenylethanol with vinyl acetate employng recombinant *Aspergillus oryzae*, *Candida antartica* form A and B lipases and wild-type *Candida rugosa* lipase

Enzyme	Solvent	Conversion (%)	Stereochemistry	e.e.$_p$ (%)
AOL r	SCCO$_2$	1.39	R(+) acetic ester	100
AOL r	heptane	6	"	
AOL r	toluene	/	/	/
AOL r	*tert*-buthylether	/	/	/
CAL A r	SCCO$_2$	32.2	R(+) acetic ester	100
CAL A r	heptane	75.3	S(−) acetic ester	16
CAL A r	toluene	0.32	R(+) acetic ester	/
CAL A r	*tert*-buthylether	5,8	R(+) acetic ester	9,3
CAL B r	SCCO$_2$	/	/	/
CAL B r	heptane	10	R(+) acetic ester	100
CAL B r	toluene	12,4	R(+) acetic ester	100
CAL B r	*tert*-buthylether	/	/	/
CRL wt	eptano	19,2	R(+)acetic ester	40
CRL wt	toluene	/	/	/
CRL wt	*tert*-buthylether	/	/	/

Reaction conditions: Phenylethanol = 1mM, vinylacetate = 10mM, reaction volume = 1mL, T = 40°C, enzyme = 5 mg, 24 h reaction time

opposite stereopreference was achieved, Table 4. In fact as far as the recombinant CAL A the reaction medium is able to modulate not only the extent of conversion but even the enzyme stereoselectivity.

The use of recombinant thermostable lipases derived from thermophiles are likely to be more thermally robust and more stable in organic solvents than their mesophilic counter parts currently used in organic synthesis. In our experiments the recombinant thermostable lipase ESL-001-01 and ESL-001-02, supplied in partially purified form (60–95% pure) by Recombinant Biocatalysis Inc., showed in heptane a conversion value, after 24 hours reaction time, from 5 to 40% about, but a very poor selectivity. In *tert*-buthylether very low conversion product was achieved, Table 5.

On this basis the relationship between enzyme activity and solvent physical properties cannot be a simple correlation, and much will be learned from future experiments.

Carbon dioxide is the most commonly used supercritical fluid, and as such, we started our investigations of the lipase-catalysed reactions in supercritical fluids in this solvent. Given the results in Table 4 and 5 we can observe that only with recombinant termostable ESL-001-01 and ESL-001-02 lipases the percent of reaction product and the selectivity of biocatalytic reactions in supercritical fluids is higher than in conventional organic solvents. These are the first data reported for biocatalysed reactions in SCCO$_2$ employing recombinant lipases.

As far the results obtained, the solvent hydrophobicity (carbon dioxide is almost non polar as hexane), the unusually high density (0.754g/cm^3) and the high diffu-

Table 5 Transesterification reactions of (±) 1-phenylethanol with vinyl acetate employng recombinant thermophilic lipases

Enzyme	Solvent	Conversion (%)	Stereochemistry	e.e.$_p$ (%)
ESL-001-02	SCCO$_2$	12.26	R(+) acetic ester	100
ESL-001-02	heptane	5,4	R(+) acetic ester	68
ESL-001-02	toluene	0	R(+) acetic ester	/
ESL-001-01	SCCO$_2$	38,6	R(+) acetic ester	67.5
ESL-001-01	heptane	40	R(+) acetic ester	65
ESL-001-01	toluene	9.4	R(+) acetic ester	60

Reaction conditions: Phenylethanol = 1mM, vinylacetate = 10mM, reaction volume = 1mL, T = 70°C, enzyme = 1mg, 24 h reaction time

sivity of supercritical fluid (enzymes suspended in these media are subject to less internal diffusion limitations than those in conventional solvents) are probably the main reasons for the improvement of ester production.

CONCLUSIONS

The use of enzymes in organic solvents is now commonplace, and analysis of the dependence of enzyme activity as a function of solvent will enhance our understanding of protein-environment interactions.

Supercritical fluids can be considered just as another nonaqueous media in which to attempt biocatalytic reactions. Since the physical properties of supercritical fluids are determined by temperature and pressure, these dispersants offer a very real opportunity to investigate the effect of solvent physical properties on an enzyme reaction in a controlled manner.

The work presented in this chapter was performed as first approach to investigate on supercritical fluid and conventional organic solvents suitability as reaction medium in transesterification reactions employing recombinant lipases of different sources.

Although the experimental data do not supply a predictive model, because solvent-enzyme interactions can be different according to the different substrates used, this approach could give rise to a wide range of industrial applications through a careful optimisation of the reaction conditions and screening of more suitable enzymes. Moreover all recombinant enzymes used show an higher activity than wild-type ones. These data could be useful to protein engineering in order to optimise the enzyme performance through side-direct mutagenesis.

REFERENCES

Gargouri, Y., Ransae, S. and Verger, R. (1997) *Biochim. Biophys. Acta*, **1344**, 6–37.

Braimi-Horn, M.C., Guglielmino, M. L., Elling L. and Sparrow, L.G. (1990) *Bioch. Biophys. Acta*, **1042**, 51–59.

Cernia, E., Palocci, C. and Catoni, E. (1996) *J. Mol. Cat.*, **107**, 79–89.

Chi, Y. M., Nakamura, K. and Yano, T. (1989) *Agric. Biol. Chem.*, **52**, 1541–1546.

Cygler, M., Grouchulski, P. Kaslauskas, R.J., Schrag J.D., Boutillier, F., Rubin, B., Serreqi, A.N. and Gupta, A.K. (1984) *J. Am. Chem. Soc.*, **116**, 3180–3186.

Dordick, J.S. (1991) *"Biocatalysis for industry"*, pp. 567–580. New York: Plenum Press.

Graille, J. (1988) *La Rivista delle sostanze grasse*, **65**, 423–428.

Halling, P.J. (1990) *Biotech. Bioeng.*, **35**, 691–701.

Harwood, J. (1989) *TIBS* **14**, 125–126.

Hills, M.J. (1989) *Biotech. Lett.*, **11**, 629–632.

Hirata, H., Higuchi, K. and Yamashina, T. (1990) *J. Biotech.*, **14**, 157–167.

Ikushima, Y., Saito, N. and Arai, M. (1992), *J. Phys. Chem.* **96**, 2293–2297.

Iriuchijma, S., Keiyu, A. and Kojima N. (1982) *Agric. Biol. Chem.*, **46**, 1593–1597.

Karmal, T.N.B. and Saroja, M. (1988) *Biotech. Lett.*, **10**, 337–340.

Macrae, A.R. (1985) *"Biocatalyst in Organic synthesis"*, edited by J. Tramper, H.C. van der plas and P. Link, pp. 195–208. Elsevier.

McHugh, M.A. and Krukonis, K.J. (1985) *"Supercritical Fluid Extraction: Principle and practice"*, pp. 2567–2574. Stoneham, MA: Butterworth.

Newmark, P. (1988) *Biotechnology*, **6**, 369–353.

Omar, I.C. (1988) *Agric. Biol. Chem.*, **52**, 2923–2925.

PhastSystem User Manual, Pharmacia Biotech (1990) Sweden: Uppsala.

Sih, C.C., Fujimoto, Y., Girdaukas, G. and Sih, C.J. (1982) *J. Am. Chem. Soc.*, **104**, 7294–7299.

Sonomoto, K. and Tanaka, J. (1988) *Enz. Eng.*, **65**, 235–239.

Wescott, C.R. and Klibanov, A.M. (1984) *Biochim. Biophys. Acta* **1206**, 1–9.

Zacks, A. and Klibanov, A.M. (1985) *Proc. Natl. Acad. Sci USA*, **82**, 3192–3196.

8. STABILIZATION OF INDUSTRIAL ENZYMES BY PROTEIN ENGINEERING

GUIDO GRANDI*[,†], IMMA MARGARIT, FRANCESCO FRIGERIO,
RENZO NOGAROTTO*, GIOVANNA CARPANI, RENATA GRIFANTINI*
and GIULIANO GALLI*

*Genetic Engineering and Microbiology Laboratories — ENIRICERCHE S.p.A.,
Via F. Maritano 26, San Donato Milanese 20097 Milan, Italy*

INTRODUCTION

One of the major problems encountered in enzyme technology is the tendency of enzymes to get inactivated when exposed to non-physiological conditions. In particular, the fact that generally enzymes denature rapidly at temperatures above 50°C prevents them from being used in many industrial applications.

In the attempt to solve the problem of thermal inactivation of enzymes the industrial research is following two lines of activity. Firstly the search for thermostable enzymes from both thermophilic and hyperthermophilic bacteria has become a very popular and fruitful approach. Indeed, extremely robust enzymes able to survive at temperatures well above 70°C have already found their way to industrial processes. The second approach consists in the application of powerful protein engineering techniques which allow selective substitutions of amino acid residues that are crucial for thermostability. This approach, which is widely utilized nowadays, is well applicable when the 3-D structure of the protein to be stabilized is known. However, although the number of proteins studied at structural level is rapidly increasing, the structures of many industrially useful enzymes remain to be elucidated.

Can enzymes be stabilized by using novel molecular biology and protein engineering approaches even when their 3-D structure is unknown?

In the attempt to answer this question, three industrial enzymes, the *Bacillus subtilis* neutral protease, the *Pseudomonas* isoamylase and carbamylase from *Agrobacterium radiobacter* were taken as model systems and were subjected to protein engineering following two different experimental strategies. From our data, as well as from the data of other laboratories, it appears that protein engineering can indeed become the strategy of choice whenever the stabilization of a protein is required.

STABILIZATION OF THE *BACILLUS SUBTILIS* NEUTRAL PROTEASE

The *Bacillus subtilis* neutral protease is a zinc metalloendopeptidase extensively studied at both biochemical and genetic levels (Yang *et al.*, 1984; Toma *et al.*, 1986;

---

*: Present address: Chiron S.p.A., Via Fiorentina 1, 53100 Siena, Italy
†: Corresponding author

Table 1 Rational approach to protein stabilization

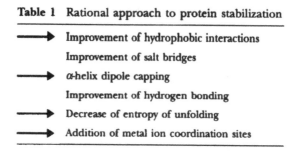

⟶	Improvement of hydrophobic interactions
	Improvement of salt bridges
⟶	α-helix dipole capping
	Improvement of hydrogen bonding
⟶	Decrease of entropy of unfolding
⟶	Addition of metal ion coordination sites

Signor *et al.*, 1990). Although its structure is not known, the enzyme belongs to a highly homologous family of proteases, three of which, the enzymes from *B. thermoproteolyticus* (Holmes *et al.*, 1982), *B. cereus* (Paupit *et al.*, 1988) and *Pseudomonas aeruginosa* (Holland *et al.*, 1992), have been crystallized and structurally characterized. The knowledge of the structures of these three enzymes prompted us to use computer-assisted techniques to model a 3-D structure of the *B. subtilis* neutral protease (Signor *et al.*, 1990). The model was then carefully scanned to localize possible residues which, if properly replaced, might contribute to enzyme stabilization. The rationale behind the selection of the residues to be replaced and the nature of the replacing amino acids is based on the well-known criteria of protein stabilization listed in Table 1. According to such criteria, any protein can be stabilized whenever site-directed mutations are created which 1) improve the hydrophobic interactions, 2) improve salt bridges, 3) stabilize α-helices by dipole capping, 4) decrease the entropy of unfolding and 5) create metal ion binding sites.

Our experimental approach was to produce a series of mutant enzymes, each carrying a mutation designed to satisfy one of the criteria listed in Table 1 (those indicated by the arrows). Therefore, a multiple mutant was obtained in which all the stabilizing mutations were added one after the other.

We started our stabilizing work from a neutral protease mutant selected in our laboratories which presents a Phenylalanine in a region (position 63) exposed to the solvent. We hypothesized that the replacement of the surrounding residues Ser9, Thr11, Thr17 and Thr61 with amino acids whose side chains could create a hydrophobic pocket around the aromatic ring of Phe63 would have stabilized the enzyme. The replacing amino acids were Val for Ser9, Arg for Thr11 and Gln for Thr17 and Thr61. The computer analysis predicted that the methylene groups of the side chains of the new residues would establish favourable hydrophobic interactions with Phe63. An additional mutation (Arg68>Leu) was added to remove the destabilizing electrostatic interaction between the guanidinium groups of Arg68 and Arg11. The stabilizing effect of the hydrophobic pocket modelled around Phe63 is given in Table 2 which shows that the T50 (the temperature at which 50% of the activity is lost after 30-minute incubation) of the (Phe63) neutral protease is enhanced by 3°C after the replacement of the five amino acids.

It has been demonstrated that the interactions between the extremities of the α-helix dipole and point charges favourably contribute to the entalpy of the protein folded state. After scanning the seven α-helices of the neutral protease we found that the most C-terminal one, extending from residue 305 to residue 315, has a

Table 2 Analysis of the neutral protease mutants

Mechanism	Mutant	$\Delta T50$
Hydrophobic interactions	"Hydrophobic pocket"	+ 3.0°C
α-helix capping	K305D	+ 0.4°C
Entropic effect	A4T	+ 0.6°C
	G147A	+ 2.0°C
	G189A	+ 1.0°C
Metal ion coordination	"Ca loop"	+ 3.0°C
	MULTIPLE MUTANT	+ 9.9°C

destabilizing positive charge at its N-terminus (Lys305). The replacement of Lys305 with Asp increases the T50 of the wild type enzyme by about 0.5°C (Table 2).

One of the mechanisms by which proteins can be stabilized is by decreasing their entropy of unfolding. According to the thermodinamic principles, a folded protein (F) finds itself in equilibrium with its unfolded state (U).

$$U \longleftrightarrow F$$

The thermodynamic equation which governs such equilibrium is:

$$\Delta G = G_F - G_U \tag{1}$$

or:

$$\Delta G = (H_F - H_U) - T(S_F - S_U) \tag{2}$$

Equation (2) shows that any mutation able to decrease the entropy of the unfolded state (S_U) makes the ΔG value more negative and therefore stabilizes the protein in its folded conformation.

One way to decrease the entropy of unfolding is to replace amino acids having flexible backbones (e.g. glycine) with more rigid residues, provided that the substitutions do not create an unfavourable strain within the molecule.

With this in mind we selected three substitutions, Ala4>Thr, Gly147>Ala and Gly189>Ala, which turned out to increase the T50 of the wild type neutral protease by 0.5, 2 and 1°C, respectively (Margarit et al., 1992).

In the B. subtilis neutral protease, the residues from 188 to 194 constitute a seven amino acid loop located in proximity of the catalytic site. In thermolysin, the highly stable enzyme belonging to the same protease family, the corresponding region is occupied by a 10-amino acid loop which defines a calcium coordination site (Toma et al., 1991).

After the replacement of the native 7-amino acid loop by the thermolysin loop, we have been able to prove that the mutant neutral protease acquires the ability to bind a third calcium ion (two calcium atoms are naturally bound to the enzyme and turn out to be essential for enzyme stability). The binding energy enhances the T50 value of the mutant enzyme with respect to the wild type by about 3°C, when the protein is exposed to calcium containing buffers.

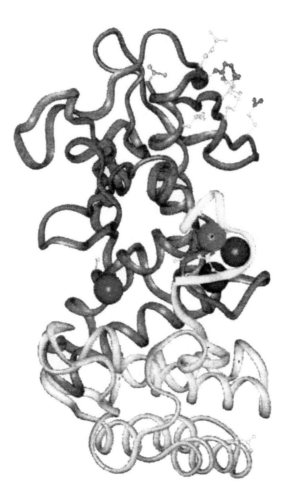

Figure 1 3-D model structure of the multiple mutant of *B. subtilis* neutral protease obtained by side-directed mutagenesis. Indicated are the zinc atom essential for catalytic activity (fuchsia), the two calcium atoms required for stability (violet) and the third calcium atom (blue) bound to the mutant molecule after loop replacement (shown in yellow). The side chains of all amino acids which have been replaced are also shown in yellow color

Table 2 also shows the stabilizing effect of all mutations described above accumulated in a single mutant (Figure 1). Interestingly, a cumulative effect can be observed which results in a net increase in thermostability of the multiple mutant with respect to the (Phe63) neutral protease by about 10°C.

In conclusion, the data described above indicate that 1) a reliable 3-D protein structure can be predicted by computer-assisted molecular modeling whenever a homologous protein, the structure of which is known, does exist. 2) The computer-derived model can be used to select specific amino acid substitutions able to enhance the protein stability on the basis of well defined general rules. 3) Once single stabilizing substitutions have been identified, multiple mutants can be created in which a cumulative stabilizing effect is generally obtained.

STABILIZATION OF THE *PSEUDOMONAS* ISOAMYLASE

Isoamylases, enzymes which hydrolyse the α-1,6 glucosidic inter-chain linkages present in starch, are used industrially for the production of amylose and maltose. In the industrial production of maltose, the starch solution is first partially hydrolysed at high temperatures and thereafter the hydrolysis is brought to near completion by adding thermostable amylases. Finally, 97–98% conversion of starch into maltose is achieved by further addition of pullulanase and/or isoamylase.

Among the different isoamylases isolated so far, one of the best characterized both at biochemical and genetic levels is the isoamylase from *Pseudomonas amyloderamosa* (Harada *et al.*, 1972; Tognoni *et al.*, 1989). This enzyme is currently used by the starch industry but it has the disadvantage of being thermolabile and also of having a optimum pH for stability around 4. This implies that for the enzyme to be used in the industrial process the temperature of the reactor has to be cooled down considerably and the pH of the solution has to be adjusted (in the maltose production process amylases operate at pH 5.5–6). Therefore, for the improvement of the process, the availablity of a thermostable isoamylase with an optimum of pH for stability around 6 would be particularly useful.

Since neither the structure of the enzyme is known nor a homologous enzyme the structure of which has been resolved is available, in order to isolate a pH-dependent thermostable isoamylase mutant we adopted a random-site-directed mutagenesis strategy. Briefly, the experimental approach consists of the following steps: 1) PCR amplification of the isoamylase gene under conditions in which the polymerase makes an average of 1–2 errors per 100–200 nucleotide elongation, 2) preparation of a "library" of recombinant *B. subtilis* clones, each expressing a different isoamylase mutant, 3) screening of the library to identify the clones that produce isoamylase mutants with higher thermostability at pH 6. The library screening was carried out growing the recombinant clones on plates buffered at pH 6 and exposing the colonies to iodine vapours after their incubation for sixteen hours at temperatures between 37 to 45°C. Only the colonies producing thermostable isoamylase mutants were surrounded by blue halos when incubated at temperatures higher than 37°C.

To facilitate the subsequent characterization of the mutants, the isoamylase gene was first mutagenized to create unique restriction sites without changing the protein sequence, allowing the splitting of the gene into 5 restriction fragments. Therefore the PCR-mediated mutation was carried out on each fragment so as to restrict the region to be sequenced for identifying the stabilizing mutation (Figure 2). Table 3 summarizes the results of the screening procedure. One mutant of fragment 2 out of 85,000 analyzed carried a thermostable mutation at position 355 (Thr>Ser) whereas 30 independent fragment 4 mutants were selected. The latter carried, together with different silent mutations, the stabilizing substitutions of Ile367 with either Thr or Ser. Asp507>Val was the relevant mutation present in the 9 mutants of fragment 4 identified by the screening of 90,000 clones. Finally, no thermostable mutant was selected out of 300,000 fragment 1 mutants and out of 12,000 fragment 5 mutants analyzed. To verify whether or not the stabilizing effect of each mutant was additive, double and triple mutants were obtained by site-directed mutagenesis and the thermostability of the purified enzymes was compared with respect to the wild type enzyme. As shown in Figure 3, the triple mutant Ile367>Thr/Asp507>Val/

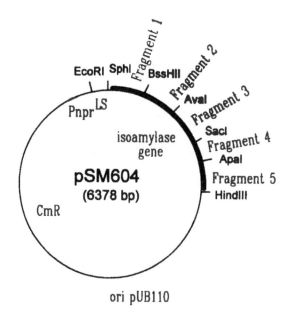

ori pUB110

Figure 2 Functional and restriction map of plasmid pSM604. Pnpr: promoter of the B. subtilis neutral protease gene under which the isoamylase gene is expressed. LS: leader sequence for secretion of subtilisin. The fragments of the isoamylase gene which have been mutagenized by PCR are indicated. For details see text

Thr355>Ser turned out to be the most stable enzyme, the double mutants Ile367>Thr/Asp507>Val and Thr355>Ser/Ile367>Thr being more stable than the wild type but less stable than the triple mutant. Interestingly, when the stability assay was carried out at pH 4, no difference in thermostability between the mutants and the wild type was observed (data not shown).

In conclusion, the random site-directed mutagenesis approach appears to be a powerful general strategy to isolate stability and/or activity mutants of any enzyme, provided that a rapid and reliable screening procedure is available. As in the case of neutral protease, the isoamylase stabilizing mutations turned out to exert a cumulative effect, the double mutants being more stable than both the wild type enzyme and the single mutants but less stable than the triple mutant.

Table 3 Screening of the isoamylase mutants

Fragment	Number of colonies screened	Number of positive clones	Mutation
1	300,00	0	–
2	85,00	1	Thr355Ser
3	60,00	30	Ile367Thr
			Ile367Ser
4	90,00	9	Asp507Val
5	12,00	0	–

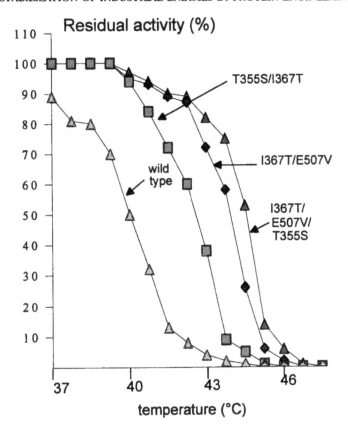

Figure 3 Thermal stability of isoamylase mutants. Residual activity was measured after 3-hour incubation at different temperatures

STABILIZATION OF CARBAMYLASE FROM *AGROBACTERIUM RADIOBACTER*

The D-amino acids D-phenylglycine and D-p-hydroxyphenylglycine are among the most important optically active compounds for their use in the production of semisynthetic penicillins and cephalosporins (Runser *et al.*, 1990).

These amino acids are currently synthesized in a two-step reaction process using as substrates D,L-5 monosubstituted hydantoins (Olivieri *et al.*, 1981). According to this two-step process, the substrate is first hydrolysed to the D-carbamyl derivative by a D-specific hydantoinase and then further converted into the corresponding D-amino acid by an N-carbamyl-D-amino acid amidohydrolase (hereinafter carbamylase). Since several micro-organims expressing both hydantoinase and carbamylase have been isolated, the D-amino acid production process simply involves the incubation, in an oxygen-free environment, of the bacterial biomass with the substrate for a period long enough to allow the quantitative conversion of the D,L hydantoin into the corresponding D-amino acid.

Despite its semplicity, the process has margins for further optimization. In particular, carbamylase is sensitive to thermal inactivation (Ogawa *et al.*, 1994), thus limiting the life span of the biocatalyst.

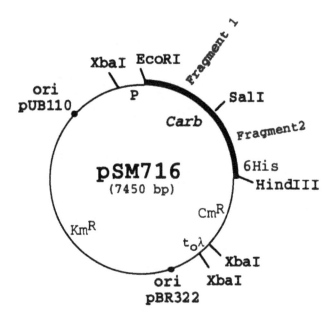

Figure 4 Functional and restriction map of plasmid pSM716. P: costitutive promoter under which the carbamylase gene is expressed. 6 His: histidine tail added to the carboxy-terminal of the protein. The fragments of the carbamylase gene mutagenized by PCR are indicated. For details see text

In the attempt to select thermostable carbamylase mutants, we developed a random site-directed mutagenesis approach similar to the one described for the isoamylase.

The carbamylase gene was divided into two fragments and each fragment was amplified by PCR under conditions which favoured approximately 1–2 errors per 100 base pairs. Each amplified reaction was then ligated to plasmid pSM716 (Figure 4) and the two ligation mixtures used to transform *E. coli* cells. Two libraries were generated, each carrying random mutations at the 5' portion and at the 3' portion of the carbamylase gene, respectively.

The two libraries were replicated on nitrocellulose filters and after lysis of the colonies by freeze-thawing, the filters were incubated at 48°C for 16 hours, conditions that inactivate the wild type carbamylase. When the filters were laid on a medium containing the substrate N-carbamyl-D-phelylglycine and the pH indicator Phenol Red, violet-red halos appeared around a few colonies, suggesting the presence of thermostable carbamylase mutants which, having survived the temperature treatment, converted the substrate to D-phenyl glycine and ammonia (Figure 5).

Two colonies were selected for further study. The sequence analysis of the carbamylase gene from these colonies revealed that a single point mutation had occurred in each mutant, both leading to the replacement of threonine with alanine one at position 212 and the other at position 262.

PCR RANDOM MUTAGENESYS OF D-N-α-CARBAMYLASE

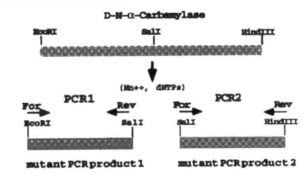

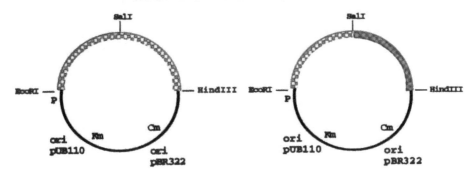

CLONING IN THE EXPRESSION VECTOR

TRANSFORMATION AND PLATING OF THE MUTANT LIBRARIES

REPLICA ONTO NITROCELLULOSE FILTERS, LYSIS OF THE COLONIES BY FREEZE - THAWING, HEAT INACTIVATION AND ENZYME ACTIVITY ASSAY ON PLATES CONTAINING CARBAMYL - AMINOACID AND PHENOL - RED

POSITIVE COLONIES ARE VISIBLE AS RED-VIOLET SPOTS

Figure 5 Scheme of the procedure used for the random mutagenesis of the carbamylase gene and for the screening of the mutant expression libraries. For details see text.

Table 4 Thermal stability of carbamylase mutants

Mutation	$\Delta T50$
Thr212Ala	3.3
Thr262Ala	2.3
Thr212Ala-Thr262-Ala	7.1

The carbamylase mutants were purified from *E. coli* cells as already described (Grifantini *et al.*, 1996) and the ΔT_{50} values determined using the purified proteins (T_{50} is the temperature at which the enzyme loses 50% of its activity after 30-minute incubation and ΔT_{50} is the difference between the T_{50} value of the mutant and the T_{50} value of wild type enzyme). As shown in Table 4, the Thr212>Ala mutant was 3.3°C more stable than the wild type carbamylase whereas Thr262>Ala mutant had a ΔT_{50} value of 2.3. When the $\Delta T50$ value was measured on the Thr212>Ala/Thr262>Ala double mutant, an increased thermal stability of 7.1°C was found indicating that, as in the case of neutral protease and isoamylase, stabilizing mutations have a cumulative effect when present on the same molecule.

REFERENCES

Grifantini, R., Pratesi, C., Galli, G. and Grandi, G. (1996) Topological mapping of cysteine residues of N-carbamyl-D-amino-acid amidohydrolase and their role in enzymatic activity. *J. Biol. Chem.*, **271**, 9326–9371.

Harada, T., Misaki, A., Akai, M., Yokobayashi, K. and Sugimoto, K. (1972) Characterization of *Pseudomonas* isoamylase by its action on amylopectin and glycogen: comparison with aerogenes. *Biochim. Biophys. Acta*, **268**, 497–505.

Holland, D.R., Tronrud, D.E., Pley, M.W., Flaherty, K.M., Stark, W., Jansonius, D.B., McKay, D.B. and Matthews, B.W. (1992) Structural comparison suggests that thermolysin and related neutral proteases undergo hinge-bending motion during catalysis. *Biochemistry*, **31**, 11310–11316.

Holmes, M.A. and Matthews, B.W. (1982) Structure of thermolysin refined at 1.6 A resolution. *J. Mol. Biol.*, **160**, 623–639.

Margarit, I., Campagnoli, S., Frigerio, F., Grandi, G., De Filippis, V. and Fontana, A. (1992) Cumulative stabilizing effects of glycine to alanine substitutions in *Bacillus subtilis* neutral protease. *Protein Eng.*, **5**, 543–550.

Ogawa, J., Ching-Ming Chung, M., Hida, S., Yamada, H. and Shimizu, S. (1994) *J. Biotechnol.*, **38**, 11–19.

Olivieri, R., Fascetti, E., Angelini, L. and Degen, L. (1981) Microbial transformation of racemic hydantoins to D-amino acids. *Biotechnol. Bioeng.*, **23**, 2173–2183.

Paupit, R.A., Karlsson, R., Picot, D., Jenkins, J.A., Niklaus-Reimer A.-S. and Jansonius, J.N. (1988) Crystal structure of neutral protease from *B. cereus* refined at 3 A resolution and comparison with the homologous and more thermostable enzyme thermolysin. *J. Mol. Biol.*, **199**, 525–537.

Runser, S., Chinski, N. and Ohleyer, E. (1990) D-p-hdroxyphenylglycine production from DL-5-p-hydroxyphenylhydantoin by *Agrobacterium* sp. *Appl. Microbiol. Biotechnol.*, **33**, 382–388.

Signor, G., Vita, C., Fontana, A., Frigerio, F., Bolognesi, M., Toma, S., Gianna, R., De Gregoriis, E. and Grandi, G. (1990) Structural features of neutral protease from *B. subtilis* deduced from model-building and limited proteolysis experiments. *Eur. J. Biochem.*, **189**, 221–227.

Tognoni, A., Carrera, P., Galli, G., Lucchese, G., Camerini, B. and Grandi, G. (1989) Cloning and nucleotide sequence of the isoamylase gene for a strain of *Pseudomonas* sp. *J. Gen. Microbiol.*, **135**, 37–45.

Toma, S., Del Bue, M., Pirola, A. and Grandi, G. (1986) nprR1 and nprR2 regulatory regions for neutral protease expression in *B. subtilis. J. Bacteriol.*, **163**, 824–831.

Toma, S., Campagnoli, S., Margarit, I., Gianna, R., Grandi, G, Bolognesi, M., De Filippis, V. and Fontana, A. (1991) Grafting of a calcium-binding loop of thermolysin to *Bacillus subtilis* neutral protease. *Biochemistry*, **30**, 97–106.

Yang, M.V., Ferrari, E. and Henner, D. (1984) Cloning of neutral protease gene of *B. subtilis* and the use of the cloned gene to create an *in vitro* derived deletion mutation. *J.Bacteriol.*, **160**, 15–21.

9. DESIGN OF BIOTECHNOLOGICALLY APPLICABLE YEAST CARBOXYPEPTIDASE Y MUTANTS WITH INCREASED PREFERENCE FOR CHARGED P_1 RESIDUES[†]

KJELD OLESEN and KLAUS BREDDAM*

Department of Chemistry, Carlsberg Laboratory, Gamle Carlsberg Vej 10, DK-2500 Copenhagen Valby, Denmark

CONSTRUCTION AND CHARACTERISATION OF CPD-Y MUTANTS

An inherent property of serine carboxypeptidases is that they specifically act on the C-terminal peptide bond of their substrate. Thus, as opposed to endopeptidases only one position of a peptide sequence may act as a substrate at a time. This makes serine carboxypeptidases useful tools for a number of applications, such as C-terminal sequencing (Breddam and Ottesen, 1987), and C-terminal modification of peptides (Breddam et al., 1981a, 1981b; Stennicke et al., 1997) and for stereo specific synthesis of peptide and peptide derivatives (Breddam and Johansen, 1984). However, due to substrate side chain preferences of any serine carboxypeptidase not all peptides will function equally well as substrate precursors in such reactions. Serine carboxypeptidases display the most pronounced substrate preference with respect to P_1. Among the described enzymes only a few catalyse the hydrolysis of substrates with a basic P_1 residue, i.e. CPD-S1, KEX1, and CPD-WII (Breddam et al., 1981; Latchiniak-Sadek and Thomas, 1994; Olesen and Breddam, 1995), while no enzymes have been found that efficiently catalyse hydrolysis of substrates with an acidic P_1 residue. Thus, it was of interest to produce variants of CPD-Y that efficiently hydrolyse substrates having basic or acidic P_1 residues, also because this enzyme is readily purified in gram quantities.

The process of creating CPD-Y mutants with increased activity toward substrates with charged P_1 residues has gone through three phases. In phase one the tertiary structure of the enzyme was not known but some information on the structure of the enzyme could be gained from the tertiary structure of the homologous enzyme CPD-WII from wheat (Liao and Remington, 1990; Liao et al., 1992). Based on the structure of this enzyme and an alignment of 4 serine carboxypeptidases (Sørensen et al., 1987) the S_1 pocket of CPD-Y was predicted to be comprised of residues Tyr147, Leu178, Glu215, Arg216, Ile340, and Cys341 (Olesen and Kielland-Brandt,

[†]This work was supported by funds from BioNebraska Inc., Lincoln, NE., USA.

*Corresponding author. Abbreviations: CPD, carboxypeptidase; CBZ, carbobenzoxyl, benzyloxycarbonyl; FA, 3-(2-furylacryloyl). Mutant enzyme names, e.g. L178S, indicate that residue 178 has been changed from Leu to Ser. The binding site notation is that of Schechter and Berger (1967). Accordingly, the C-terminal amino acid of the substrate is denoted P_1' and those in the amino terminal direction from the scissile bond are denoted $P_1, P_2, ..., P_n$. In analogy, binding sites are denoted S_1' and $S_1, S_2, ..., S_n$.

1993). These residues were chosen as targets for degenerated oligonucleotide directed mutagenesis using a mutagenesis strategy that eliminates any unmutagenized wild type background (Olesen and Kielland-Brandt, 1993). Eight pools of mutant plasmid DNA were produced targeting from only one to all of the selected residues. The mutant pools of DNA were then introduced into a yeast strain lacking the *PRC1* gene producing more than 10^5 independent yeast transformants. These were screened by a chromogenic overlay assay (modified from Jones, 1977) to differentiate colonies expressing active CPD-Y from those expressing inactive CPD-Y. Depending on how many residues had been targeted the number of colonies expressing active CPD-Y varied from 0 to 50%. Then, the P_1 substrate preference of the CPD-Y expressed by these colonies for Trp, Phe, Gly, Pro, Ser, Glu, His and Lys was estimated using a semiquantitative chromogenic microtiter dish assay. This assay, involving L-amino acid oxidase, peroxidase and o-dianisidine, has previously been used to monitor carboxypeptidase S activity (modified from Lewis and Harris, 1967). Nine mutant enzymes were found that exhibited from 3 to 10-fold increased activity with Lys in P_1 in this particular assay. Plasmid DNA was then recovered from these colonies and sequenced to determine which amino acids had been substituted. All nine mutants contained substitutions of Leu178 while six contained substitutions at position 215 and 216. The two mutants displaying the highest activity in the microtiter assay, L178S and L178S+E215A+R216P, were purified by affinity chromatography and kinetically characterised using CBZ-Xaa-Leu-OH substrates with Phe, Ala, Ser, Lys, and Glu at the P_1 position. The k_{cat}/K_M values for the hydrolysis of CBZ-Lys-Leu-OH and CBZ-Glu-Leu-OH with L178S was found to have increased 150 and 2-fold relative to wild type CPD-Y, respectively. The activities with the triple mutant enzyme were about 50% that of L178S with all substrates. This, together with results from the semiquantitative assay, indicated that it was the substitution of Leu178 rather than those at positions 215+216 that was responsible for the increased activity with basic P_1 residues. Thus, to further investigate the influence of residue 178 on the P_1 substrate preference another 6 mutants were produced by site directed mutagenesis, such that Leu178 had been substituted by Trp, Phe, Ala, Ser, Cys, Asn, Asp, and Lys. These enzymes were kinetically characterised using FA-Xaa-Ala-OH substrates with Phe, Leu, Val, Ala, Ser, Glu, Arg, and Lys in the P_1 position (Olesen et al., 1994). Except for L178K all of these enzymes exhibited increased activity with basic P_1 residues. L178S was the most active enzyme with Arg in P_1 (increased 40-fold) while with Lys in P_1 the most active enzyme was L178D (increased 85-fold) (Table 1). Only 3 enzymes were found to exhibit increased activity with Glu in P_1, i.e. with L178A, L178C and L178K k_{cat}/K_M had increased 2.3, 1.9, and 1.5-fold, respectively (Table 2).

In the next phase the crystal structure of CPD-Y had also been determined (Endrizzi et al., 1994). Comparing this to that of CPD-WII revealed that structural differences exist between the two enzymes giving rise to errors in the alignment of primary structures used to predict the constitution of the S_1 subsite of CPD-Y. Thus, the S_1 pocket of CPD-Y turned out not to be constituted of residues Tyr147, Leu178, Glu215, Arg216, Ile340, and Cys341 as predicted but rather of the residues Tyr147, Leu178, Tyr185, Tyr188, Trp312, Ile340, and Cys341 (Figure 1). Thus, using the two structures as a guide a new and improved alignment of 30 primary sequences of serine carboxypeptidases was constructed (Olesen and Breddam, 1995). Comparing this to the substrate preferences of kinetically characterised enzymes it was found

Table 1 Enzymes engineered for activity with basic P_1 residues

Enzyme	FA-Leu-Ala-OH			FA-Arg-Ala-OH			FA-Lys-Ala-OH		
	k_{cat} (min^{-1})	K_M (mM)	k_{cat}/K_M (min^{-1} mM^{-1})	k_{cat} (min^{-1})	K_M (mM)	k_{cat}/K_M (min^{-1} mM^{-1})	k_{cat} (min^{-1})	K_M (mM)	k_{cat}/K_M (min^{-1} mM^{-1})
Wild type	3800[a]	0.11[b]	35000[b]	nd	nd	14[a]	9[c]	5[c]	2.0[a]
L178S	1300[a]	0.41[a]	3300[a]	5100[a]	9.2[b]	550[a]	nd	nd	140[a]
L178D	320[b]	8.4[b]	38[a]	1500[a]	5.6[b]	260[a]	nd	nd	173[a]
Y185E	3500[a]	0.27[a]	13000[a]	45[b]	1.7[b]	26[a]	16[a]	1.3[a]	12[a]
W312L	7200[a]	0.39[b]	18000[a]	nd	nd	410[a]	nd	nd	16[a]
W312F	5000[a]	0.26[a]	19000[a]	nd	nd	190[a]	nd	nd	8.4[a]
W312D	3400[a]	0.43[a]	8000[a]	nd	nd	830[a]	5700[b]	6.6[b]	860[a]
W312E	3900[a]	0.28[a]	14000[a]	nd	nd	680[a]	3000[a]	1.3[a]	2300[a]
L178D+W312D	70[a]	1.4[a]	51[a]	280[a]	0.39[a]	710[a]	330[a]	0.32[a]	1100[a]
N241D	2300[a]	0.36[a]	6500[a]	nd	nd	370[a]	nd	nd	100[a]
N241D+W312L	4600[a]	0.23[b]	20000[a]	60000[a]	4.7[b]	13000[a]	nd	nd	1100[a]
N241D+W312F	3000[a]	0.37[a]	8200[a]	58000[c]	11[c]	5200[a]	nd	nd	1800[a]

nd: not determined due to high K_M. The standard deviation of values marked by [a] is ±0 to 10% while [b] indicates a deviation of ±10 to 20% and [c] indicates ±20 to 30%.

Table 2 Enzymes engineered for activity with acidic P_1 residues

Enzyme	FA-Leu-Ala-OH			FA-Glu-Ala-OH			FA-Asp-Ala-OH		
	k_{cat} (min^{-1})	K_M (mM)	k_{cat}/K_M (min^{-1} mM^{-1})	k_{cat} (min^{-1})	K_M (mM)	k_{cat}/K_M (min^{-1} mM^{-1})	k_{cat} (min^{-1})	K_M (mM)	k_{cat}/K_M (min^{-1} mM^{-1})
Wild type	3800[a]	0.11[b]	35000[b]	nd	nd	41[a]	300[a]	9.8[a]	31[a]
L178A	2100[a]	0.33[a]	6400[a]	nd	nd	93[a]	–	–	–
L178K	nd	nd	92[a]	nd	nd	62[a]	–	–	–
W312S	3600[a]	0.51[a]	7000[a]	nd	nd	84[a]	–	–	–
W312N	5000[a]	0.45[a]	11000[a]	nd	nd	85[a]	–	–	–
W312D	3400[a]	0.43[a]	8000[a]	nd	nd	120[a]	–	–	–
W312K	3000[a]	1.7[b]	1800[a]	850[a]	2.9[b]	290[a]	–	–	–
L178D+W312D	70[a]	1.4[a]	51[a]	6.3[a]	0.068[a]	93[a]	–	–	–
N241K	3400[a]	0.64[a]	5300[a]	nd	nd	160[a]	nd	nd	41[a]
N241K+W312N	5000[a]	0.84[a]	6000[a]	11000[a]	4.5[c]	2500[a]	130[a]	0.84[a]	160[a]
L245R	2900[a]	0.54[a]	5200[a]	840[a]	1.5[b]	570[a]	830[a]	0.30[a]	2800[a]
L245R+W312S	5200[a]	0.68[a]	7700[a]	1600[a]	0.32[a]	5100[a]	180[a]	0.035[a]	5300[a]
L245R+W312N	5200[a]	0.72[a]	7200[a]	2000[a]	0.44[b]	4500[a]	360[a]	0.11[a]	3300[a]

nd: not determined due to high K_M. The standard deviation of values marked by [a] is ±0 to 10% while [b] indicates a deviation of ±10 to 20% and [c] indicates ±20 to 30%.

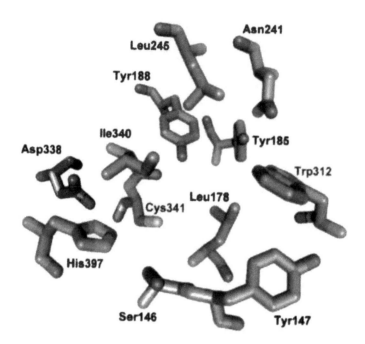

Figure 1 Structure of the S_1 binding pocket and catalytic triad (Ser146-His397-Asp338) of CPD-Y seen from the direction of the catalytic essential Ser146

that all enzymes with low activity with basic P_1 residues possessed a Trp at position 312 while those with high activity toward such substrates possessed smaller residues like Asn, Ile, and Gly. Furthermore, most of these enzymes possessed substitutions of Tyr185, e.g. to Glu as in CPD-S1 (Figure 2).

Thus, residues 185, 188, and 312 were selected as targets for degenerated oligonucleotide directed mutagenesis followed by activity and specificity screens as outlined for position 178. As expected, no position 188 mutants with pronounced changes in substrate preference were found. Likewise, no position 185 mutants were found, while mutants were selected that had Trp312 substituted for either Asp or Glu. Thus, position 312 was selected for a thorough investigation and mutants replacing Trp312 for Phe, Leu, Ala, Ser, Asn, Asp, Glu, or Lys were produced while for position 185 only the Tyr to Glu substitution was produced to investigate the importance of this residue for the substrate preference of CPD-S1. Following purification by affinity chromatography these enzymes were kinetically characterised using FA-Xaa-Ala-OH substrates with Phe, Leu, Val, Ala, Ser, Glu, Arg, or Lys in the P_1 position (Olesen and Breddam, 1995).

With Y185E only small alterations in activity with substrates containing charged P_1 residues were observed, i.e. a 2-fold and 6-fold increase in activity with Arg and Lys in P_1, respectively, while k_{cat}/K_M with FA-Glu-Ala-OH decreased 4-fold. Much larger effects on activity were observed with the Trp312 mutants. All of these enzymes exhibited slightly increased activity with Glu in P_1 from 2-fold with W312S and W312N to 7-fold with W312K (Table 2). The largest effects of the Trp312 substitutions were observed with basic P_1 residues. With such substrates decreased

	enzyme	147	178	185	188	241	245	312	340	341	species / database entry / reference
Fungi	CPD-Y	Y	L	Y	Y	N	L	W	I	C	Bakers yeast (*Saccharomyces cerevisiae*) / SW:CBPY YEAST / Valls *et al.*, 1987; Svendsen *et al.*, 1982
	YBY9	Y	I	A	Y	D	L	G	I	C	Bakers yeast (*Saccharomyces cerevisiae*) / SW:YBY9_YEAST / Becam *et al.*, 1994.
	KEX1	Y	W	S	Y	E	N	I	I	C	Bakers yeast (*Saccharomyces cerevisiae*) / SW:KEX1_YEAST / Dmochowska *et al.*, 1987.
	CPD-Pp*	Y	L	Y	Y	N	M	W	I	C	Yeast (*Pichia pastoris*) / SW:CBPY_PICPA / Ohi *et al.*, 1996
	CPY1	Y	L	Y	Y	N	M	W	I	C	Yeast (*Candida albicans*) / SW:CBPCANAL / Mukhtar *et al.*, 1992.
	CPD-S1	Y	W	E	Y	Y	E	N	I	C	*Penicillium janthinellum* (fungus) / SW:CPS1_PENJA / Svendsen *et al.*, 1993.
	CPD-Z	Y	F	Y	Y	G	E	G	I	C	*Absidia zychae* (zygomycete) / EMBL:AZSCPZ / Lee *et al.*, 1995.
Plants	CPD-M3	Y	L	Y	Y	N	N	W	I	C	Barley (*Hordeum vulgare*) / SW:CBP3_HORVU / Sørensen *et al.*, 1989.
	CPD-W3	Y	L	Y	Y	N	N	W	I	C	Wheat (*Triticum aestivum*) / SW:CBP3_WHEAT / Baulcombe *et al.*, 1987.
	CPD-R3	Y	L	Y	Y	N	S	W	I	C	Rice (*Oryza sativa*) / SW:CBP3_ORYSA / Washio & Ishikawa, 1992.
	CPD-R3-1	Y	L	Y	Y	N	S	W	I	C	Rice (*Oryza sativa*) / SW:CBPX_ORYSA / Washio & Ishikawa, 1993.
	VRU49741*	Y	L	Y	Y	N	N	W	I	C	Mung bean (*Vigna radiata*) / EMBL:VR49382+VR49741 / Tian *et al.*, 1996; Lee *et al.*, 1996.
	CPD-Pea*	Y	L	Y	Y	N	S	W	I	C	Pea (*Pisum sativum*) / EMBL:PSCARBXPD / Jones *et al.*, 1996.
	CPD-X	Y	L	Y	Y	N	S	W	I	C	Mouse-ear cress (*Arabidopsis thaliana*) / SW:CBPX_ARATH / Bradley, 1992.
Insects	CPD-Dm*	Y	L	L	Y	P	G	?	?	?	Fruit fly (*Drosophila melanogaster*) / EMBL:DMAA40919 / Kopczynski *et al.*, 1997.
	CPD-Ae	Y	Y	L	Y	D	G	T	I	C	Yellow fever mosquito (*Aedes aegypti*) / PIR2:A41612 / Cho *et al.*, 1991.

Figure 2 Partial alignment of amino acids comprising the S_1 binding pocket of serine carboxypeptidases, that based on their primary sequence are predicted to be single chained and share the CPD-Y like structure. Numbers refer to residue positions of mature CPD-Y. In enzymes having Trp at position 312 interactions between a P_1 side chain and residues 241 and 245 are probably hindered. *New entries relative to alignment of Olesen and Breddam, 1995. Complete alignment, now containing 42 sequences including two chained enzymes, is available on request or at ftp://ftp.crc.dk/pub/carboxypeptidase/Alignmnt.htm

activity was observed only with W312K while all other mutations increased activity from 2.5-fold to 1150-fold. The largest activity with Arg in P_1 was observed with W312D (increased 60-fold) while with Lys in P_1 the highest activity was observed with W312E (increased 1150-fold) (Table 1). It was demonstrated that the beneficial effects of these mutations on the activity with basic P_1 residues were entirely on k_{cat}. Three double mutants, L178S+W312N, L178D+W312N, and L178D+W312D, were also produced to investigate the interplay between these residues. The enzyme L178D+W312D exhibited a 700-fold decrease in activity with Leu in P_1 (similar to L178D) and a 550-fold increase in activity with Lys in P_1 (similar to W312D). Thus, the P_1 substrate preference of this enzyme for Leu versus Lys had changed 3.8×10^5-fold relative to wild type CPD-Y (Table 1).

In phase 3 the binding of charged P_1 residues to the S_1 pocket of serine carboxypeptidases was reevaluated. Although the activity of the W312X mutants with basic P_1 residues had increased significantly relative to wild type CPD-Y the activity was still not comparable to that of enzymes like CPD-S1, especially with Arg in P_1. It was speculated that in an enzyme with a small residue at position 312, interactions between P_1 and residues 241 and 245, residing in S_3 in an α-helix above S_1 in CPD-Y, might be facilitated (Olesen and Breddam, 1997). In enzymes with a large residue at position 312, like CPD-Y, these residues would be part of S_3 and not of S_1, while in enzymes with a small residue at position 312, like CPD-S1, these residues would be part of S_1 as well as S_3 (see Figure 1) To investigate this, 35 enzymes were produced that incorporated a charge at either of positions 241 and 245, optionally in combination with a reduction in size of residue 312. The purified enzymes were kinetically characterised using FA-Xaa-Ala-OH substrates with either Leu, Arg, and Lys or Leu, Glu, and Asp in the P_1 position (Olesen and Breddam, 1997).

All of these mutant enzymes exhibited increased activity with substrates containing P_1 residues of the opposite charge. With basic P_1 residues it was possible to increase activity with Lys as well as Arg in P_1 up to about 900-fold. Thus, with Lys in P_1 the best effects were about equal to those obtained with the Trp312 mutants while the best effects with Arg in P_1 were about 15 times higher (Table 1). The k_{cat}/K_M value for the hydrolysis of FA-Arg-Ala-OH with N241D+W312L was 13000 $min^{-1}mM^{-1}$ which is comparable to the activity of CPD-S1. With acidic P_1 residues these enzymes exhibited up to 170-fold increased activity, both with Glu and Asp in P_1. This value is about 25 times higher than the highest obtained with Leu178 and Trp312 mutants (Table 2).

The two enzymes with the highest activity with basic or acidic P_1 residues, N241D+W312L and L245R+W312S respectively, were chosen for further kinetic characterisation with respect to dependency of ionic strength, pH, and P_1 substrate preference. It was found that the activity of both enzymes with charged substrates was essentially independent of ionic strength. Thus, the high activity with such substrates is probably not dependent on the formation of salt bridges between the charged residues involved.

The pH profile of k_{cat} for the hydrolysis of FA-Lys-Leu-OH with N241D+W312L deviated from that observed with wild type CPD-Y. This parameter was found to fit the dissociation of a single residue with a pK_a value of 5.7 with low k_{cat} at low pH. Thus, as the dissociating group must be assigned to Asp241 the pK_a of this residue is perturbed about 1.8 pH units relative to the free form of aspartic acid. As a consequence of this altered pH dependency, the P_1 substrate preference of N241D+W312L for hydrophobic versus basic residues varies with pH. At pH 6 to 7 the enzyme was found to display a 2-fold preference for Lys relative to Ala while at pH 4.0 this preference was reversed to a 10-fold preference for Ala.

The P_1 substrate preference of both enzymes were further characterised using FA-Xaa-Ala-OH substrates with Phe, Leu, Val, Ala, Ser, Glu, Asp, Arg, or Lys in the P_1 position. It was observed that the activity of L245R+W312S with FA-Arg-Ala-OH had decreased 14-fold while the activity of N241D+W312L with acidic P_1 residues was only slightly affected. Thus, as the activity of the two enzymes with FA-Arg-Ala-OH and FA-Asp-Ala-OH differs 13000-fold and 190-fold, respectively, their P_1 substrate preference for Arg versus Asp varies 2.5×10^6-fold.

CONCLUDING REMARKS

Mutational Effects

It is noteworthy that it has been possible to increase the activity of CPD-Y with basic P_1 residues by mutational substitutions at several positions in the S_1 binding pocket, i.e. positions 178, 185, 241, 245, and 312. The reason for this may be that the S_1 pocket of CPD-Y has been designed for low activity with basic P_1 residues. However, the over expression of mutant forms of CPD-Y, like N241D+W312L, in both *vpl1* and *VPL1* strains of yeast does not seem to affect the growth of the cells. Alternatively, activity with basic P_1 residues may have been sacrificed in the evolutionary design for optimal activity with hydrophobic P_1 residues.

Attempts to co-crystallise CPD-Y with peptide derivatives to evaluate the effects of these mutations have not been successful but we assume that the mutations at different positions within the S_1 pocket affect activity with basic P_1 residues in different ways. A reduction in size of residue 312 presumably assist catalysis of such substrates by relieving steric constraints against large P_1 side chains like Trp, Phe, Lys, and Arg and perhaps also by increasing accessibility of the distal charge of the P_1 side chain to solvent water. The beneficial effects of mutations at position 178 are likely mediated in an indirect manner by inducing a more favourable steric conformation of the neighbouring Trp312. Thus, the beneficial effects of the above mutations is to remove the inherent discrimination of the S_1 pocket against basic P_1 residues. However, the additional introduction of negatively charged residues at position 241 or 245 (or 312 in the case of Lys in P_1) constitutes the positive selection for basic P_1 residues, presumably by increasing the hydrophilic character in the vicinity of the charge of the P_1 side chain and long range charge-charge interactions. The optimal composition of the S_1 pocket of CPD-Y for high activity with basic P_1 residues thus closely resembles the composition in enzymes like CPD-S1.

In the case of acidic P_1 residues the only successful design to significantly increase activity was the introduction of basic residues at either of positions 241 or 245 in combination with the substitution of Trp312 for a smaller polar residue. The beneficial effect of this design is probably also achieved through an increase in the hydrophilic character of the S_1 pocket in combination with long range charge-charge interactions that requires that position 312 is not a large Trp. Importantly, no naturally occurring serine carboxypeptidase has been found to display such an activity or to have a primary sequence indicative of such activity.

Applicability of the Enzymes

The applicability of a serine carboxypeptidase for a particular enzymatic application, e.g. C-terminal sequencing or C-terminal modification, is governed by its activity and/or substrate preference. With respect to C-terminal sequencing it is desirable that individual peptide bonds of the substrate are not hydrolysed at very different rates. Thus, it is desirable that the applied enzyme exhibits low substrate preference which can be achieved by increasing poor activities and/or by decreasing good activities. During this investigation of the S_1 pocket of CPD-Y examples of both types of substrate preference relaxation have been observed. However, it has not been possible to develop enzymes completely lacking substrate preference. Even so, enzymes like L178S, which has a P_1 substrate preference about 100-fold lower than CPD-Y, have been shown to be superior relative to wild type CPD-Y for the C-terminal sequencing of peptides containing basic residues like the S6 phosphate receptor peptide (Olesen et al., 1994).

Perhaps the most important application of serine carboxypeptidases is in C-terminal modification of peptides. In this stereo specific reaction peptide precursors can be C-terminally amidated (Breddam et al., 1981a), thus converting them into stable bioactive messengers (Breddam et al., 1981b), or C-terminally labelled with affinity or fluorescent compounds, such as lysine amide derivatized at the ε-amino group with biotin or anthraniloyl, which may facilitate purification or tracking of labelled peptide fragments (Stennicke et al., in press). Obviously, if the substrate preference of the applied enzyme match the sequence of the precursor peptide the catalytic turnover can be increased and problems with secondary degradation and side reactions decreased. The applicability of mutant forms of CPD-Y with increased activity with basic or acidic P_1 residues is currently being investigated.

REFERENCES

Baulcombe, D.C., Barker, R.F. and Jarvis, M.G. (1987) A gibberellin responsive wheat gene has homology to yeast carboxypeptidase Y. *J. Biol. Chem.*, **262**, 13726–13735.

Becam, A.-M., Cullin, C., Grzybowska, E., Lacroute, F., Nasr, F., Ozier-Kalogeropoulos, O., Palucha, A., Slonimski, P.P., Zagulski, M. and Herbert, C.J. (1994) The sequence of 29.7 kb from the right arm of chromosome II reveals 13 complete open reading frames, of which ten correspond to new genes. *Yeast*, **10**, 1–11.

Bradley, D. (1992) EMBL:ATCPYLP.

Breddam, K., Widmer, F. and Johansen, J.T. (1981a) Carboxypeptidase Y catalyzed C-terminal modification of peptides. *Carlsberg Res. Commun.*, **46**, 121–128.

Breddam, K., Widmer, F. and Johansen, J.T. (1981b) Carboxypeptidase Y catalyzed C-terminal modification in the B-chain of porcine insulin. *Carlsberg Res. Commun.*, **46**, 361–372.

Breddam, K. and Johansen, J.T. (1984) Semisynthesis of human insulin utilizing chemically modified carboxypeptidase Y. *Carlsberg Res. Commun.*, **49**, 463–472.

Breddam, K. and Ottesen, M. (1987) Determination of C-terminal sequences by digestion with serine carboxypeptidases: The influence of enzyme specificity. *Carlsberg Res. Commun.*, **52**, 55–63.

Breddam, K. (1988) Carboxypeptidase S-1 from Penicillium janthinellum: enzymatic properties in hydrolysis and aminolysis reactions. *Carlsberg Res. Commun.*, **53**, 309–320.

Cho, W.L., Deitsch, K.W. and Raikhel, A.S. (1991) An extraovarian protein accumulated in mosquito oocytes is a carboxypeptidase activated in embryos. *Proc. Natl. Acad. Sci. USA*, **88**, 10821–10824.

Dmochowska, A., Dignard, D., Henning, D., Thomas, D.Y. and Bussey, H. (1987) Yeast KEX1 gene encodes a putative protease with a carboxypeptidase B-like function involved in killer toxin and alpha-factor precursor processing. *Cell*, **50**, 573–584.

Endrizzi, J.A., Breddam, K. and Remington, S.J. (1994) 2.8-Å structure of yeast serine carboxypeptidase. *Biochemistry*, **33**, 11106–11120.

Jones, E.W. (1977) Proteinase mutants of Saccharomyces cerevisiae. *Genetics*, **85**, 23–33.

Jones, C.G., Lycett, G.W. and Tucker, G.A. (1996) Protease inhibitor studies and cloning of a serine carboxypeptidase cDNA from germinating seeds of pea (Pisum sativum L.) *Eur. J. Biochem.*, **235**, 574–578.

Kopczynski, C., Serrano, T., Rubin, G. and Goodman, C. (1997) EMBL:DMAA40920

Latchinian-Sadek, L. and Thomas, D.Y. (1994) Secretion, purification and characterization of a soluble form of the yeast KEX1-encoded protein from insect cell cultures. *Eur. J. Biochem.*, **219**, 647–652.

Lee, B., Takeuchi, M. and Kobayashi, Y. (1995) Molecular cloning and sequence analysis of the scpZ gene encoding the serine carboxypeptidase of Absidia zychae. *Curr. Genet.*, **27**, 159–165.

Lee, K., Tan-Wilson, A.L. and Wilson, K.A. (1996) EMBL:VRU49382.

Lewis, W.H and Harris, H. (1967) Human red cell peptidases. *Nature*, **215**, 351–359.

Liao, D.I. and Remington, S.J. (1990) Structure of wheat serine carboxypeptidase II at 3.5-Å resolution. A new class of serine proteinase. *J. Biol. Chem.*, **265**, 6528–6531.

Liao, D.I., Breddam, K., Sweet, R.M., Bullock, T. and Remington, S.J. (1992) Refined atomic model of wheat serine carboxypeptidase II at 2.2-Å resolution. *Biochemistry*, **31**, 9796–9812

Mukhtar, M., Logan, D.A. and Kaeufer, N.F. (1992) The carboxypeptidase Y-encoding gene from Candida albicans and its transcription during yeast-to-hyphae conversion. *Gene*, **121**, 173–177.

Ohi, H., Ohtani, W., Okazaki, N., Furuhata, N. and Ohmura, T. (1996) Cloning and characterization of the Pichia pastoris PRC1 gene encoding carboxypeptidase Y. *Yeast*, **12** 31–40.

Olesen, K. and Kielland-Brandt, M.C. (1993) Altering substrate preference of carboxypeptidase Y by a novel strategy of mutagenesis eliminating wild type background. *Protein Eng.*, **6**, 409–415.

Olesen, K., Mortensen, U.H., Aasmul-Olsen, S., Kielland-Brandt, M.C., Remington, S.J. and Breddam, K. (1994) The activity of carboxypeptidase Y toward substrates with basic P_1 amino acid residues is drastically increased by mutational replacement of leucine 178. *Biochemistry*, **33**, 11121–11126.

Olesen, K. and Breddam, K. (1995) Increase of the P_1 Lys/Leu substrate preference of carboxypeptidase Y by rational design based on known primary and tertiary structures of serine carboxypeptidases *Biochemistry*, **34**, 15689–15699.

Olesen, K. and Breddam, K. (1997) Substrates with charged P_1 residues are efficiently hydrolyzed by serine carboxypeptidases when S_3-P_1 interactions are facilitated. *Biochemistry*, **36**, 12235–12241.

Schechter, I. and Berger, A. (1967) On the size of the active site in proteases. I. Papain. *Biochem. Biophys. Res. Commun.*, **27**, 157–162.

Stennicke, H.R., Olesen, K., Sørensen, S.B. and Breddam, K. (1997) C-terminal incorporation of fluorogenic and affinity labels using wild type and mutagenized carboxypeptidase Y. *Anal. Biochem.*, **248**, 141–148.

Sørensen, S.B., Svendsen, I. and Breddam, K. (1987) Primary structure of carboxypeptidase II from malted barley. *Carlsberg Res. Commun.*, **52**, 285–295.

Sørensen, S.B., Svendsen, I. and Breddam, K. (1989) Primary structure of carboxypeptidase III from malted barley. *Carlsberg Res. Commun.*, **54**, 193–202.

Svendsen, I., Martin, M., Viswanatha, T. and Johansen, J.T. (1982) Amino acid sequence of carboxypeptidase Y. II. Peptides from enzymatic cleavages. *Carlsberg Res. Commun.*, **47**, 15–27.

Svendsen, I., Hofmann, T., Endrizzi, J., Remington, S.J. and Breddam, K. (1993) The primary structure of carboxypeptidase S$_1$ from Penicillium janthinellum. *FEBS Lett.*, **333**, 39–43.

Tian, Z., Wu, J.J., Lee, K., Tan-Wilson, A.L. and Wilson, K.A. (1996) EMBL:VRU49741.

Valls, L.A., Hunter, C.P., Rothman, J.H. and Stevens, T.H. (1987) Protein sorting in yeast: the localization determinant of yeast vacuolar carboxypeptidase Y resides in the propeptide. *Cell*, **48**, 887–897.

Washio, K. and Ishikawa, K. (1992) Structure and expression during the germination of rice seeds of the gene for a carboxypeptidase. *Plant Mol. Biol.*, **19**, 631–640.

Washio, K. and Ishikawa, K. (1993) EMBL:OSCBP31.

10. ENGINEERING β-GLYCOSIDE HYDROLASES

R.A.J. WARREN

Department of Microbiology and Immunology, and Protein Engineering Network of Centres of Excellence, University of British Columbia, Vancouver, B.C., Canada V6T 1Z3

INTRODUCTION

β-glycoside hydrolases include β-glycosidases, which hydrolyse β-glycosides and, in some instances, β-linked oligosaccharides, and β-1,4-glycanases, which hydrolyse β-1,4-linked polysaccharides such as chitin, cellulose, xylan and mannan. Most β-1,4-glycanases exhibit some measure of activity against oligosaccharides. β-1,4-glycanases are used in the detergent, textile food processing and paper industries.

The enzymes fall into two large groups: retaining enzymes use a double-displacement mechanism involving a covalent enzyme-substrate intermediate (White *et al.*, 1996) to hydrolyse glycosidic bonds with retention of configuration at the anomeric centre; inverting enzymes use a concerted mechanism to hydrolyse glycosidic bonds with inversion of configuration at the anomeric centre. Retaining but not inverting enzymes can transglycosylate (Figure 1). The enzymes exhibit varying degrees of specificity for the glycone moiety of the bond being hydrolysed; they are classified as β-glucosidases, β-xylosidases, chitinases, cellulases, mannanases, and so on, but the specificity is rarely, if ever, absolute other than requiring a β-1 glycosidic bond. Nonetheless, the activity is usually greatest against a particular glycone. In general, the specificity for the aglycone moiety of the substrate is looser, so aryl β-glycosides and aryl β-oligosaccharides can be used as substrates for the quantification of enzyme activity.

The β-1,4-glycanases exhibit either an endo- or an exoglycanolytic action on their polysaccharide substrates. This distinction is not absolute and it is probably more accurate to say that an enzyme has predominantly endo- or exoglycanolytic activity.

The activities of β-1,4-glycanases that hydrolyse chitin, cellulose, mannan and other insoluble polycaccharides are dictated by the insolubility and aggregated nature of their substrates. Although quite efficient as glycoside hydrolases, the activities against the insoluble substrates are determined by the efficiency with which the enzymes can sequester single molecules of substrate into their active sites.

MODULES AND FAMILIES

Domains are independently folding, spatially distinct structural units in proteins. The sequence need not be contiguous. Modules are a subset of domains that are contiguous in sequence and that are used repeatedly as domains in functionally diverse proteins (Bork *et al.*, 1996). Many of the β-glycoside hydrolases, especially the β-1,4-glycanases, are modular proteins, comprising two or more discrete domains that retain their functions when separated by proteolysis or by manipulation of the DNA sequences encoding them (Gilkes *et al.*, 1991; Tomme *et al.*, 1995b). As with

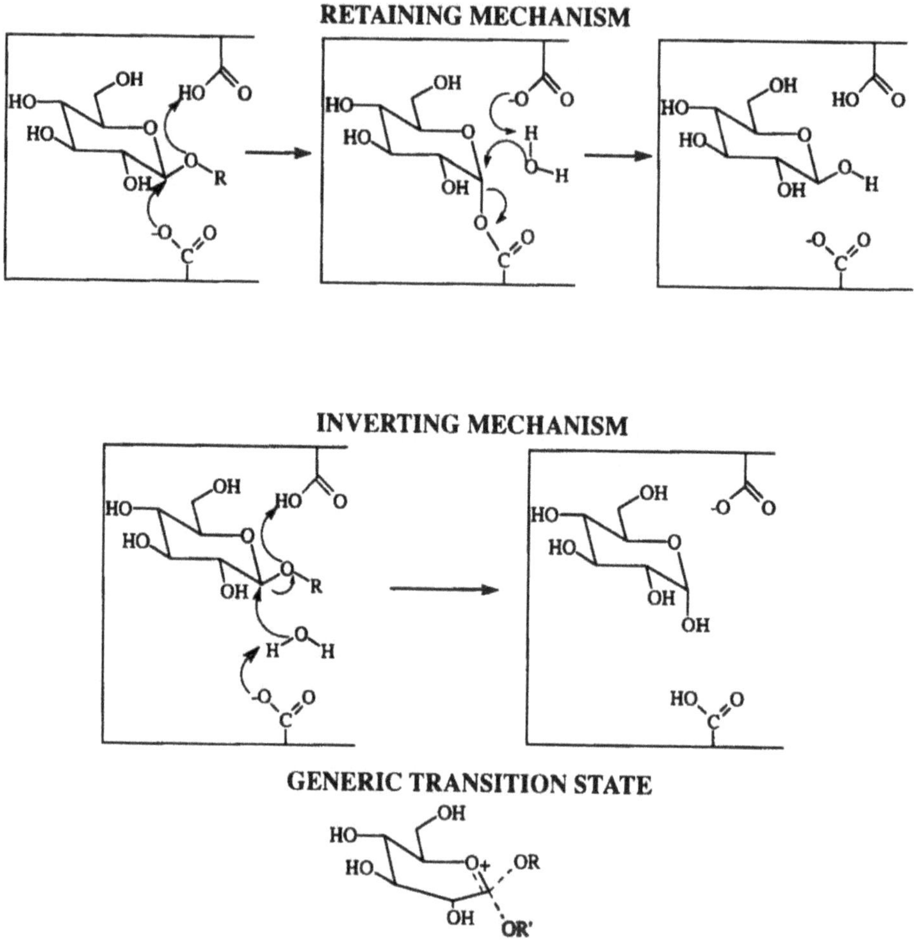

RETAINING MECHANISM

INVERTING MECHANISM

GENERIC TRANSITION STATE

Figure 1 Mechanisms of β-glycosyl hydrolases. Note that the cartoons are not accurate representations of the spatial relationships of the catalytic carboxyl groups.

other modular proteins (Bork *et al.*, 1996), the modules appear to have been spread amongst β-glycoside hydrolases by genetic shuffling.

All of the enzymes have at least a catalytic domain (CD) or module. The commonest ancillary module is a substrate-binding domain. Where present, the substrate-binding domain is often a cellulose-binding domain (CBD). CBDs are components of hydrolases attacking the polysaccharides in plant cell walls, such as cellulases, mannanases and xylanases. Some chitinases have chitin-binding domains that are related to CBDs; many CBDs can bind to chitin. Therefore, CBD is used here to designate substrate-binding domain in general.

Glycoside hydrolases are classified into families of related amino acid sequences, using the sequences of the CDs only (Henrissat, 1991; Henrissat and Bairoch, 1993). Where analysed, all of the CDs in a family have similar three-dimensional structures and exhibit the same sterospecificity of hydrolysis (Tomme *et al.*, 1995b; Warren, 1996). Similarities in the amino acid sequences of enzymes in families 1, 5 and 10 prompted the prediction that the enzymes in these families are structurally related (Liebl *et al.*, 1994). Structure determinations subsequently confirmed the prediction: at least nine families of retaining β-glycoside hydrolases, including families 1, 5 and 10, form a superfamily of 4/7 $(\beta/\alpha)_8$ barrel structures (Jenkins *et al.*, 1995; Dominguez *et al.*, 1995; Ducros *et al.*, 1995; Henrissat *et al.*, 1995). All of the enzymes in the superfamily are retaining enzymes in which the nucleophilic and acid/base catalytic glutamate residues are at the carboxyl termini of β-strands 4 and 7.

Although the enzymes in a family have similar structures and the same stereospecificity, hence the same mechanism of glycoside hydrolysis, they may have different specificities and modes of action. Family 5 of the retaining enzymes contains cellulases, mannanases, xylanases and β-1,3-glucanases; family 2 of retaining enzymes contains β-galactosidases and β-glucosidases; family 6 of inverting enzymes contains endoglucanases and enzymes that act predominantly as exoglucanases (Tomme *et al.*, 1995b; Henrissat, 1991; Henrissat and Bairoch, 1993). The substrate specificities and modes of action of β-glycoside hydrolases are determined by the fine structures of their active sites rather than by their global folds (Davies and Henrissat, 1995).

CBDs are also classified into families of related amino acid sequences (Tomme *et al.*, 1995c).

Some β-glycoside hydrolases contain modules in addition to a CBD and a CD. The additional modules can include a second CD and/or a second CBD, fibronectin type III modules, and thermal-stabilizing domains (Tomme *et al.*, 1995b; Warren, 1996; Fontes *et al.*, 1995; Clarke *et al.*, 1996). The modules may be connected by linkers of different lengths (Gilkes *et al.*, 1991; Tomme *et al.*, 1995b).

ENGINEERING β-GLYCOSIDE HYDROLASES

β-glycoside hydrolases can be engineered either by altering the CD or by moving modules. Potential changes to a CD include changing the thermal stability, changing the pH optimum, increasing the catalytic activity, altering the substrate specificity, changing the stereospecificity, and, for a retaining enzyme, increasing or decreasing the transglycosylating activity. Modules can be interchanged to give new combinations of CDs and CBDs, and CBDs can be fused to virtually any protein to serve as affinity tags. Many of these manipulations prove to be of little more than academic interest, but some do have applications.

ENGINEERING CATALYTIC DOMAINS

Changing Thermal Stability

The thermal stability of an enzyme can be altered in two ways: its thermal denaturation temperature or T_m can be raised; and its half-life at an inactivating temperature can be lengthened. This may be achieved by either random or site-directed mutation. Sites can be selected for directed mutation on the basis of the three-dimensional structure of an enzyme, on the basis of comparison of its amino acid sequence with that of a more thermostable enzyme from the same family, or on both structure and sequence relatedness, depending on the information available. It should be borne in mind, however, that site-directed mutations based on the three-dimensional structure may produce unexpected, and perhaps unwanted, effects, and that single mutations rarely increased T_m significantly (Cowan, 1996). Site-directed mutation based on sequence is not straightforward. A mesophilic and a thermophilic enzyme in the same family may have amino acid sequences that are 35–85% similar, superimposable three-dimensional structures, and identical mechanisms (Vieille and Zeikus, 1996). The difference in thermostability between the enzymes may be the consequence of the thermophilic enzyme containing only a few additional H-bonds, salt bridges or hydrophobic interactions, but it is difficult to predict which are crucial by comparing the sequences (Vieille and Zeikus, 1996). Each thermophilic enzyme is stabilized by a unique combination of bonds and interactions (Vieille and Zeikus, 1996). There is no strong evidence that any particular interaction is more important than others in determining thermostability, or that the structures of very stable enzymes are systematically different from those of less stable proteins (Daniel et al., 1996). Nonetheless, the T_ms of at least two β-glycoside hydrolases have been raised by random or site-directed mutations, although the increases were not striking (Arase et al., 1993; Wakarchuk et al., 1994).

Chimaeras of a mesophilic and a thermophilic enzyme from the same family can have thermal stabilities between those of the parent enzymes (Singh et al., 1995; Singh and Hayashi, 1995; Welfle et al., 1996). Such chimaeras, however, are no more useful than the parent enzymes, since they have similar catalytic activities. They are useful for assessing global determinants of stability and folding (Hahn et al., 1994; Hahn et al., 1995; Heinemann et al., 1996).

Overall, random mutation seems to be the best approach to raising the thermal stability of an enzyme. Error-prone PCR, sexual PCR, and combinations of these are currently the most powerful methods for generating the mutants (Stemmer, 1994a; Stemmer, 1994b). However, the availability of highly thermal tolerant enzymes from extremophiles makes it unnecessary to raise the thermal stability of enzymes for particular applications. It is likely that for every enzyme currently used commercially, a more stable version is produced by an extremophile (Adams et al., 1995).

β-glycoside hydrolases are produced by thermophiles and hyperthermophiles (Bauer et al., 1996). The enzymes retain their characteristics when produced in a mesophile, such as *Escherichia coli*. The great majority of such enzymes characterized to date can be assigned to known families of β-glycoside hydrolases. They include β-glucosidases (Liebl et al., 1994; Love et al., 1988; Fischer et al., 1996; Voorhorst et al., 1995; Morana et al., 1995), cellulases (Ruttersmith and Daniel, 1991; Hreggvidsson et al., 1996; Liebl et al., 1996), xylanases (Winterhalter et al., 1995), a mannosidase and a mannanase (Duffaud et al., 1997). Initially, such enzymes were characterized from cultivable extremophiles, but these are a very small fraction of the organisms in such environments (Hugenholtz and Pace, 1996). A more useful approach than attempting to cultivate further organisms from such environments is to isolate the genes encoding potentially useful enzymes directly from those environments, using PCR-based methods (Hugenholtz and Pace, 1996; Bergquist et al., 1996; Brennan, 1996).

Two approaches can be taken. Genes can be amplified from DNA isolated from the environment using primers based on highly conserved sequences in a family of enzymes. This will yield further members of that family. Such enzymes may have different specificities and modes of action to those of the previous enzymes in the family. An alternative approach is to amplify the DNA at random, clone the fragments, then screen the clones for a particular activity. This may yield enzymes from an as yet unidentified family.

Given the rate at which thermostable β-glycoside hydrolases are being reported, raising the T_ms of mesophilic enzymes is of little practical value. It has been pointed out that increasing the half-lives of themophilic enzymes rather than raising the T_ms of mesophilic enzymes will be a more useful exercise (Vieille and Zeikus, 1996).

Changing the Optimum pH

The great majority of β-glycoside hydrolases have two catalytic carboxyl groups in their active sites. In retaining enzymes, one functions as a nucleophile, the other as an acid/base catalyst. In inverting enzymes, one functions as a base catalyst, the other as an acid catalyst. The two carboxyls have different pK_as, determined by their environments within the enzyme active-site (McCarter and Withers, 1994). The pK_a of the acid/base catalyst may change by as much as 3.5 pH units during the reaction (McIntosh et al., 1996). There is no nucleophile in a retaining chitinase and a retaining chitobiase because substrate-assisted catalysis is used in the first step of the reaction (Terwisscha van Scheltinga et al., 1995; Tews et al., 1996). If the pH optimum of an enzyme is to be changed, the ionization states of the catalytic carboxyls must be the same at the new optimum as they are at the original optimum. This will require significant changes in the environments of the carboxyls (Krengel and Dijkstra, 1996). Comparison of the sequences of related enzymes with significantly different pH optima does not indicate which amino acids are responsible for the difference. Although chimaeric enzymes can offer clues (Nakamura et al., 1991), it is probably easier to isolate enzymes from organisms that grow at extremes of pH than it is to change a pH optimum significantly by mutation (Ingledew, 1990; Kroll, 1990). If mutation is attempted, it should be random rather than site-directed.

Increasing the Catalytic Activity

The catalytic carboxyls in β-glycoside hydrolases are identified by a combination of methods, including biochemistry, X-ray crystallography and site-directed mutation (McCarter and Withers, 1994; Svensson and Søgaard, 1993; Withers and Aebersold, 1995; Bray et al., 1996a). Mutation of the catalytic carboxyl amino acids should decrease activity at least several orders of magnitude (Tomme et al., 1995b; Svensson and Søgaard, 1993; Withers and Aebersold, 1995; Bray et al., 1996a; Ståhlberg et al., 1996; MacKenzie et al., 1997). In general, the catalytic carboxyls are about 5 Å apart in retaining and about 10 Å apart in inverting enzymes (McCarter and Withers, 1994), but this is not invariant. The catalytic carboxyls in endoglucanase CelC from Clostridium thermocellum, a retaining enzyme, are 10 Å apart. Hydrolysis by the enzyme proceeds via an induced fit to the substrate: the loop carrying the acid/base catalyst reorients to move the carboxyl to within 5 Å of the nucleophile (Dominguez et al., 1996). In other enzymes for which induced fit has not been reported, and in which the catalytic carboxyls are about 5 Å apart, changing the distance between the carboxyls by site-directed mutation and/or chemical modification reduces the catalytic efficiency severely (Withers et al., 1992; Yuan et al.,

1994; Lawson *et al.,* 1996; MacLeod *et al.,* 1996). In endoglucanase EGV from *Humicola insolvens,* there is a large conformational change on binding substrate; although the distance between the catalytic carboxyls does not change, a loop closes over the acid catalyst, thereby increasing the hydrophobicity of its environment (Davies *et al.,* 1993; Davies *et al.,* 1995). In retaining endoglucanase I (EGI) from *Fusarium oxysporum* complexed with a non-hydrolyzable thio-oligosaccharide substrate, the sugar ring in the –1 site is distorted towards a boat conformation to give a quasi-axial orientation for the glycosidic bond and leaving aglycone (Sulzenbacher *et al.,* 1996). A similar distortion is proposed for retaining endocellulase E1 from *Acidothermus cellulolyticus* (Sakon *et al.,* 1996). The substrate is also bent at the scissile bond in the inverting endoglucanase CelA from *C. thermocellum,* but the conformation of the glycosidic residue is yet to be resolved (Alzari *et al.,* 1995). Finally, the catalytic sites of β-1,4-glycanases comprise more than two sugar-binding sites (Dominguez *et al.,* 1995; Ducros *et al.,* 1995; Sulzenbacher *et al.,* 1996; Sakon *et al.,* 1996; Alzari *et al.,* 1995; Rouvinen *et al.,* 1990; Juy *et al.,* 1992; Spezio *et al.,* 1993; Varghese *et al.,* 1994; Divne *et al.,* 1994; Derewenda *et al.,* 1994; White *et al.,* 1994; Harris *et al.,* 1994). The amino acids conserved within families and superfamilies are generally within the sugar-binding sites. The eight conserved residues in the enzymes of family 5, for example, are involved in recognition of the sugar in the –1 site (Sakon *et al.,* 1996).

All of this suggests that increasing the catalytic efficiency of a β-glycoside hydrolase by mutation will be difficult. Can it be done by site-directed mutation? The amino acids to be mutated can be chosen on the basis of the three-dimensional structure of the enzyme, or they can be conserved amino acids in the enzymes of a particular family. Mutation of any of the conserved residues in an endoglucanase from family 5 decreases catalytic activity (Bortoli-German *et al.,* 1995), whereas mutation of conserved residues in two xylanases from family 10 increases activity 25–50 per cent in some instances, reduces activity in others, or has little effect on activity (Moreau *et al.,* 1994b; Dupont *et al.,* 1996; Charnock *et al.,* 1997). Even with methods such as overlap extension PCR for the simultaneous introduction of multiple site-directed mutations (Ge and Rudolph, 1997), the outcome of constructing strains with multiple mutations is unpredictable.

Random mutation of an inactive or sluggish mutant of an enzyme is an alternative approach to isolating potentially enhancing mutations, as demonstrated elegantly with triose phosphate isomerase. Six random, second-site suppressor mutations in a sluggish mutant of this enzyme increase the activity of the mutant between 1.3 and 19-fold. The suppressor mutations are scattered throughout the amino acid sequence of the enzyme but each is close to the active site. Combinations of pairs of the mutations are additive or subtractive, but not synergistic. Although none of the mutations increases the activity of the wild-type enzyme because the reaction is diffusion controlled, the mutations do show that at least in this instance mutations can improve an enzyme incrementally (Hermes *et al.,* 1989; Hermes *et al.,* 1990; Blacklow *et al.,* 1991). Such experiments suggest that the catalytic efficiency of an enzyme might be enhanced significantly by random mutation of the wild-type enzyme by sexual and/or error-prone PCR. Although some of the mutations in such multiply mutated proteins may confer undesirable characteristics on the enzyme, they can be eliminated by back-crossing to the wild-type (Stemmer, 1994a; Stemmer, 1994b). Screening for enhanced activity may be the limiting factor here. One approach is to screen for enhanced growth, but this will work best if a crippled mutant is the starting point and if a suitable substrate is available (Venekei *et al.*, 1996).

Changing the Specificity

Determination of the three-dimensional structure of an enzyme complexed with a substrate or an inhibitor will reveal the hydrogen-bonding and other interactions between enzyme and substrate (White *et al.*, 1996; Dominguez *et al.*, 1996; Davies *et al.*, 1995; Sulzenbacher *et al.*, 1996; Sakon *et al.*, 1996; Alzari *et al.*, 1995; Rouvinen *et al.*, 1990; Juy *et al.*, 1992; Spezio *et al.*, 1993; Varghese *et al.*, 1994; Divne *et al.*, 1994; Derewenda *et al.*, 1994; White *et al.*, 1994; Harris *et al.*, 1994). In theory, this should allow rational alteration of the specificity by site-directed mutation. In practice, random mutation is more likely to succeed, providing a suitable screen or selection is available. In other words, knowledge of the structure may be unnecessary to effect such changes in an enzyme. It should be borne in mind that the specificity of glycoside hydrolases for the glycone is often not absolute, and for the aglycone may be quite loose. Given the enormous number of such enzymes identified to date, and the range of specificities they have collectively, changing the specificity of an enzyme may be unnecessary. The mutated enzyme may be inferior to enzymes with natural specificity for the substrate.

Mutants of LacZ, the β-galactosidase from *Escherichia coli* can be selected for growth on lactobionate; some of them are heat-labile. In one such mutant, the only mutation is G794D. The mutation increases k_2, the rate constant for glycoside bond cleavage in lactose some 25-fold, but decreases k_3, the rate constant for hydrolysis of the enzyme-substrate covalent intermediate, some 4-fold. Since k_2 is rate-limiting, the k_{cat} for lactose hydrolysis is increased some 6-fold. The K_m for lactose is increased (Martinez-Bilbao *et al.*, 1991). In this instance, the specificity for the aglycone is changed; it illustrates the difficulty of predicting the effects of such mutations.

Changing the Stereospecificity

The stereospecificity of an enzyme might be changed in either of two senses: changing the anomeric configuration of the glycosidic bond hydrolysed; or changing the anomeric configuration of the glycone produced. There are numerous α-glycoside and β-glycoside hydrolases, so attempting to change the specificity in this sense is largely of academic interest. Changing an inverting to a retaining enzyme, or vice versa, may have practical applications. Retaining enzymes transglycosylate, and there is great interest in using the enzymes for glycoside synthesis rather than hydrolysis. Changing the stereospecificity of an inverting enzyme with the desired substrate specificity may allow its use as a synthase. Conversely, a retaining enzyme used for hydrolysis may form unwanted products by transglycosylation, and changing it to an inverting enzyme may eliminate their formation.

Hydrolysis of glycosyl fluorides of the "wrong" anomeric configuration is characteristic of inverting but not of retaining glycoside hydrolases. Abg, a retaining β-glucosidase from *Agrobacterium*, does not hydrolyse α-glucosyl fluoride; the nucleophile mutant, E358A, of Abg hydrolyses it rapidly. The reaction occurs at the active site of the enzyme because it is inhibited by 1-deoxynojirimycin (Wang *et al.*, 1994).

The k_{cat} for the enzyme on 2,4-dinitrophenyl-β-D-glucoside is decreased 10^7-fold by the mutation; the k_{cat} of the mutant on this substrate is increased 10^5-fold in the presence of azide or formate, and the product is α-glucosyl azide or formate. Larger nucleophiles do not increase k_{cat}. Azide and formate do not affect wild-type Abg. In the mutant, these nucleophiles can occupy the space filled by the nucleophilic glutamate in wild-type Abg, and, by substitution for it, allow the mutant to act as an inverting enzyme (Wang *et al.*, 1994). The nucleophile mutant, E233A, of Cex, a retaining xylanase from *Cellulomonas fimi*, also acts as an inverting enzyme in the presence of azide and formate, producing α-cellobiosyl azide and formate with 2,4-dinitrophenyl-β-D cellobioside as substrate (MacLeod *et al.*, 1996).

T4L, the bacteriophage T4 lysozyme, is an inverting enzyme, releasing the α-anomer from a bacterial cell-wall fragment. The mutant T26H releases the β-anomer (Kuroki *et al.*, 1993). In T4L, a water molecule is H-bonded between D20 and T26, well-positioned to attack the substrate from the α-side. In the T26H mutant, the water molecule is absent; its position is occupied by the N^{ϵ} of the histidine, which is positioned to act as a nucleophile in the retaining mechanism exhibited by the mutant (Kuroki *et al.*, 1993; Kuroki *et al.*, 1995).

This suggests a general approach for converting retaining to inverting glycosidases, and vice versa (Kuroki *et al.*, 1995). In a retaining enzyme, the nucleophile should be replaced with a smaller amino acid such as serine, alanine or glycine, so that water can attack from the α-side. In an inverting enzyme, the replacement should block attack of water from the α-side while providing an appropriate group, such as an imidazole or carboxylate, as an alternative nucleophile, as seen in the T26H mutant of T4L (Kuroki *et al.*, 1993; Kuroki *et al.*, 1995).

Nucleophiles such as azide and formate can also substitute for the acid/base catalyst in the deglycosylation of the covalent enzyme-substrate complex formed during glycoside hydrolysis by retaining enzymes (MacLeod *et al.*, 1994; Wang *et al.*, 1995). The product, however, is the β-glycosyl azide or formate.

Changing the Action

Cellobiohydrolases remove cellobiose units processively from either the reducing or the non-reducing ends of cellulose molecules (Bray *et al.*, 1996; Vranská and Biely, 1992; Barr *et al.*, 1996; Gilkes *et al.*, 1997). Endoglucanases cleave cellulose molecules at random, releasing fragments of varying length (Tomme *et al.*, 1995b; Warren, 1996). Cellobiohydrolases may have endoglucanase activity (Ståhlberg *et al.*, 1993; Shen *et al.*, 1995), but the predominant action is exoglycanolytic (Irwin *et al.*, 1993). Amongst enzymes for which the three-dimensional structures are known, the active-sites of endoglucanases are clefts (Davies *et al.*, 1993; Davies *et al.*, 1995; Sulzenbacher *et al.*, 1996; Sakon *et al.*, 1996; Alzari *et al.*, 1995; Juy *et al.*, 1992; Spezio *et al.*, 1993), those of cellobiohydrolases are tunnels (Rouvinen *et al.*, 1990; Divne *et al.*, 1994). It is proposed that cellulose molecules can enter the tunnels only by being threaded in from one end (Rouvinen *et al.*, 1990; Divne *et al.*, 1994). This implies some flexibility in the tunnel if the enzyme is to be processive, and the enzyme may exhibit low endoglucanase activity if cellulose molecules can enter the active site through the roof of the tunnel. Cellobiohydrolase CbhA from *C. fimi* has detectable endoglucanolytic activity (Shen *et al.*, 1995); deletion of a loop forming part of the roof of the tunnel enhances the endoglucanolytic activity but the overall activity of the enzyme is greatly reduced (Meinke *et al.*, 1995). The converse experiment has not been reported.

Cellobiohydrolases CbhI and CbhII from *Humicola insolens* hydrolyse a bifunctionalized tetrasaccharide carrying non-carbohydrate constituents at each end (Armand *et al.*, 1997). This suggests that initially the enzymes are not recognizing the ends of the model substrate but acting endoglucanolytically and that the loops forming the tunnel are flexible. Endoglucanase CenC from *C. fimi* behaves as a processive endoglucanase, acting as a cellobiohydrolase after making an initial cut within a cellulose molecule (Tomme *et al.*, 1996b).

Loops may be inserted into or deleted from proteins without affecting stability significantly; they may allow structural modifications that are inaccessible using substitutions alone (Shortle and Sondek, 1995). Although the most striking differences between endoglucanases and cellobiohydrolases are extra loops covering the active-sites of the latter, interconverting the two types of enzyme is of little practical significance. It could, however, help to explain their different actions.

Changing the Transglycosylating Activities of Retaining Enzymes

A retaining glycoside hydrolase may yield products from an oligosaccharide that are not expected from hydrolysis of the substrate. Such products are a consequence of transglycosylation. If they are undesirable, it may be possible to reduce the transglycosylating activity of the enzyme. Xylanase XlnA from *Streptomyces lividans* produces significant quantities of xylohexa-, -hepta- and -octaose from high concentrations of xylopentaose. The mutant N173D produces only small amounts of hepta- and octaose from this substrate (Moreau *et al.*, 1994a).

However, increasing the transglycosylating activities of retaining enzymes is of much greater utility than decreasing them. The chemical synthesis of glycosides is difficult and time-consuming. The need for defined oligosaccharides has turned attention towards enzymatic synthesis. Two types of enzyme can be used: sugar transferases; and retaining glycoside hydrolases. Both have disadvantages: transferases use expensive sugar nucleotides; hydrolases degrade the desired product.

An approach to using retaining glycoside hydrolases for synthesis is to adjust the reaction conditions to enhance transglycosylation (Rastall *et al.*, 1992; Baker *et al.*, 1994; Thiem, 1995; Monsan and Paul, 1995), and to immobilize the enzyme (Ravet *et al.*, 1993; Ljunger *et al.*, 1994). Although significant yields of the desired product can be obtained, product is still lost by hydrolysis. Transglycosylation may be enhanced by mutations in the active-site that increase the hydrophobicity around the attacking water molecule in bond hydrolysis (Kuriki *et al.*, 1996).

ENGINEERING CELLULOSE-BINDING DOMAINS

Properties of CBDs

CBDs enhance the activities of CDs on insoluble substrates (Gilkes *et al.*, 1988; Hall *et al.*, 1995; Black *et al.*, 1996; Maglione *et al.*, 1992; Karita *et al.*, 1996), probably by increasing the effective concentration of a CD on the substrate. The CBD can also influence the activity of a CD on different forms of cellulose (Coutinho *et al.*, 1993).

All CBDs characterized to date are anti-parallel β-strand polypeptides. The most understood families are I, II, III and IV. Family I contains only CBDs from fungal enzymes; they are ~36 amino acids long, with 3 β-strands (Kraulis *et al.*, 1989). Except for CBDs from a slime mold and a crab in family II, all other CBDs are from bacterial enzymes or structural proteins. Family II CBDs are ~110 amino acids long with 9 β-strands (Xu *et al.*, 1995). Family III CBDs are ~150 amino acids long with 12 β-strands (Tormo *et al.*, 1996). Family IV CBDs are ~150 amino acids long with 10 β-strands (Johnson *et al.*, 1996a).

The CBDs in families I, II and III bind to amorphous and to crystalline cellulose. The binding involves conserved aromatic amino acids on one side of the folded β-sheet structures (Kraulis *et al.*, 1989; Xu *et al.*, 1995; Tormo *et al.*, 1996; Reinikainen *et al.*, 1992; Reinikainen *et al.*, 1995; Linder *et al.*, 1995a; Linder *et al.*, 1995b; Poole *et al.*, 1993; Din *et al.*, 1994; Bray *et al.*, 1996b; Goldstein and Doi, 1994). Crystalline cellulose presents an array of overlapping binding sites to the binding faces of these CBDs (Gilkes *et al.*, 1992). A CBD from family IV binds only to amorphous cellulose (Coutinho *et al.*, 1992). Unlike the CBDs in families I, II and III, it does not have a binding face but a 5-stranded binding cleft on one side with a central strip of hydrophobic residues flanked on both sides by polar H-bonding residues (Johnson *et al.*, 1996a). The cleft binds single cellulase molecules whereas the binding faces of the CBDs in families I, II and III bind to the parallel, close packed molecules at the surface of crystalline cellulose. The family IV CBD but not those from families I, II and III also binds soluble cellulose, such as hydroxyethylcellulose, and soluble oligosaccharides (Tomme *et al.*, 1996a; Johnson *et al.*, 1996b).

Despite their differences in size, the affinities of CBDs from families I, II and III for crystalline cellulose are very similar, with K_as in the 10^6 M^{-1} range (Gilkes *et al.*, 1992; Lindner *et al.*, 1996; Goldstein *et al.*, 1993; Tomme *et al.*, 1995a). Fused family I CBDs bind cooperatively to cellulose (Lindner *et al.*, 1996). The binding for a family II CBD is entropically driven (Creagh *et al.*, 1996). The K_as for the binding of the family IV CBD to cellopentaose, to hydroxyethylcellulose and to amorphous cellulose are ~$10^4 M^{-1}$ (Tomme *et al.*, 1996a; Johnson *et al.*, 1996b); the binding is enthalpically driven (Tomme *et al.*, 1996a). CBDs in families II and III bind irreversibly to cellulose (Gilkes *et al.*, 1992; Ong *et al.*, 1991), some of those of family I bind reversibly (Linder and Teeri, 1996).

Family II CBDs adsorb to cellulose in 10^{-2} M to >1 M salt solutions, from pH 3–8 and from 4°C–70°C (Ong *et al.*, 1991; Kilburn *et al.*, 1993). The conditions required for desorption vary with the CBD and with the polypeptide to which a CBD is fused. CBDs may be desorbed with water, with 6 M guanidinium hydrochloride or with pure ethylene glycol (Ong *et al.*, 1991; Shen *et al.*, 1991; Black *et al.*, 1996).

The properties of CBDs make them useful as affinity tags for protein purification and enzyme immobilization. Furthermore, cellulose is inexpensive, is available in a variety of forms, and does not have to be modified for use as an affinity matrix.

CBDs in Protein Purification

The native enzymes, the isolated CBDs and the CBDs fused to heterologous proteins all bind to cellulose following production in a variety of host cells (Greenwood *et al.*, 1989; Ong *et al.*, 1989; MacLeod *et al.*, 1992; Assouline *et al.*, 1993). The proteins

Table 1 Cellulose-binding domain fusion proteins

CBD[1]	Fusion[2]	Partner
CenA	CBD$_{CenA}$-PhoA	alkaline phosphatase
CenA	CBD$_{CenA}$-IL2	human interleukin 2
CenA	CBD$_{CenA}$-IL3	human interleukin 3
CenA	CBD$_{CenA}$-Benzonase™	endonuclease
CenA	CBD$_{CenA}$-RGD	adhesion peptide
CenA	XynA-CBD$_{CenA}$-CD$_{CenA}$	xylanase
Cex	Abg-CBD$_{Cex}$	β-glucosidase
Cex	Cbg-CBD$_{Cex}$	β-glucosidase (thermostable)
Cex	Glu oxidase-CBD$_{Cex}$	glucose oxidase (chemically linked)
Cex	Prot A-CBD$_{Cex}$	protein A
Cex	Avid-CBD$_{Cex}$	streptavidin
Cex	factor X-CBD$_{Cex}$	factor X
Cex	E. coli cells-CBD$_{Cex}$	Lpp-OmpA
Cex	XylA-CBD$_{Cex}$	xylanase
CenB	SF-CBD$_{CenB}$	stem cell factor
CenC	CBD$_{CenC}$ (N1)-CD$_{CenA}$	CenA catalytic domain
CenC	CBD$_{CenC}$-PhoA	alkaline phosphatase

[1]CenA is endoglucanase CenA, Cex is xylanase B, CenB is endoglucanase CenB and CenC is endoglucanase CenC, all from C. fimi.

[2]The order indicates the position of the CBD: in CBDCenA-PhoA it is at the N-terminus of the fusion protein; in Abg-CBDCex it is at the C-terminus.

can be purified virtually to homogeneity from crude extracts in a single step (Ong et al., 1991; Greenwood et al., 1994; Ramirez et al., 1993). CBDs occur naturally in extracellular proteins; when fused to other extracellular proteins, the fusion proteins are exportable (Assouline et al., 1993; Greenwood et al., 1994; Ramirez et al., 1993). In E. coli the proteins are exported to the periplasm. If produced in the periplasm at high levels, the protein may leak into the medium without cell lysis, simplifying its recovery (Greenwood et al., 1994; Ramirez et al., 1993; Ong et al., 1993; Hasenwinkle et al., 1997).

The conditions required to desorb a CBD fusion protein from cellulose cannot be predicted; they must be determined empirically, and may require unfolding of the protein. Nonetheless, a number of CBD fusion proteins can be recovered in an active state (Table 1).

A CBD can be fused to the N- or C-terminus of a target polypeptide (Table 1). If the target polypeptide must be obtained in its native state, the CBD is fused to the N-terminus by a factor Xa cleavage sequence. The target polypeptide can be digested with factor Xa in solution and the CBD removed with cellulose. Alternatively, the target polypeptide can be released with factor Xa while the fusion protein is adsorbed to cellulose (Assouline et al., 1993; Assouline et al., 1995), thereby obviating desorption of the fusion protein.

Enzyme Immobilization with CBDs

Immobilization of an enzyme on cellulose with a CBD has the advantage that purification and immobilization can be accomplished in a single step. The procedure is applicable to enzymes over the ranges of temperature and pH at which the CBD remains bound. The immobilization efficiency is comparable to those of other methods for immobilization (Ong *et al.*, 1991; Ong *et al.*, 1989). If the enzyme is stable, it can be used continuously for at least 2–3 weeks (Ong *et al.*, 1991; Ong *et al.*, 1989). A factor Xa-CBD fusion is active against susceptible proteins either in solution or when it is immobilized on cellulose (Assouline *et al.*, 1993; Assouline *et al.*, 1995).

A protein A-CBD fusion and a streptavidin-CBD fusion can be used to immobilize IgG and biotinylated proteins, respectively, on cellulose (Ramirez *et al.*, 1993; Le *et al.*, 1994). Fusion of a CBD to OmpA allows *E. coli* to bind to cellulose (Francisco *et al.*, 1993). Such fusions have potential applications in both enzyme immobilization and diagnostics.

CBDs and Mammalian Cell Culture

CBDs can be used as affinity tags for the purification of growth factors (Ong *et al.*, 1995; Doheny *et al.*, 1997). A CBD fused to a peptide containing the sequence RGD promotes the attachment of Vero cells to cellulose acetate and to cellulose-based microcarriers. The fusion protein is more efficient than collagen in promoting attachment to the microcarriers (Wierzba *et al.*, 1995). Although irreversibly bound to cellulose, CBDs are mobile on the surface of the cellulose (Jervis *et al.*, 1997). A stem cell factor-CBD fusion protein when adsorbed to cellulose has significantly greater biological activity than the soluble factor. The strong binding to and mobility of the fusion protein on the cellulose surface allows effective, continuous, localized stimulation of the target cells (Doheny *et al.*, 1997).

CONCLUSIONS

β-Glycoside hydrolases offer two major possibilities for protein engineering. The first is to develop them as agents for the synthesis rather than the hydrolysis of glycosides. The second is to use their CBDs as affinity tags for a variety of applications; this goes beyond simply constructing appropriate gene fusions; it may be necessary to modify the binding or desorption characteristics of a CBD for a particular application. Random mutation by sexual PCR (28,29) is applicable in both cases. It may also be possible to minimize the sizes of CBDs, as was done for an IgG-binding domain from protein A (Braisted and Wells, 1996), for other applications.

ACKNOWLEDGMENTS

I thank Carrie Hirsch and Laura Minato for typing the manuscript.

REFERENCES

Adams, M.W.W., Perler, F.B. and Kelly, R.M. (1995) Extremozymes: expanding the limits of biocatalysis. *Bio/Technology*, 13, 662–668.

Alzari, P., Souchon, H. and Dominguez, R. (1995) The crystal structure of endoglucanase CelA, a family 8 glycosyl hydrolase from *Clostridium thermocellum. Structure*, 4, 265–275.

Arase, A., Yomo, T., Urabe, I., Hata, Y., Katsube, Y. and Okada, H. (1993) Stabilization of xylanase by random mutagenesis. *FEBS Lett.*, 316, 123–127.

Armand, S., Drouillard, S., Schülein, M., Henrissat, B. and Driguez, H. (1997) A bifunctionalized fluorogenic tetrasaccharide as a substrate to study cellulases. *J. Biol. Chem.*, 272, 2709–2713.

Assouline, Z., Graham, R., Miller, R.C., Jr., Warren, R.A.J. and Kilburn, D.G. (1995) Processing of fusion proteins with immobilized factor Xa. *Biotechnol. Prog.*, 11, 45–49.

Assouline, Z., Shen, H., Kilburn, D.G. and Warren, R.A.J. (1993) Production and properties of a factor X-cellulose-binding domain fusion protein. *Prot. Engng.*, 6, 787–792.

Baker, A., Turner, N.J. and Webberley, M.C. (1994) An improved strategy for the stereoselective synthesis of glycosides using glycosidases as catalysts. *Tetrahedron: Asymmetry*, 5, 2517–2522.

Barr, B.K., Hsieh, Y.-L., Ganem, B. and Wilson, D.B. (1996) Identification of two functionally different classes of exocellulases. *Biochemistry*, 35, 586–592.

Bauer, M.W., Halio, S.B. and Kelly, R.M. (1996) Proteases and glycosyl hydrolases from hyperthermophilic microorganisms. *Adv. Prot. Chem.*, 48, 271–310.

Bergquist, P.L., Gibbs, M.D., Saul, D.J., Reeves, R.A., Morris, D.S. and Te'O, V.S.J. (1996) Families and functions of novel thermophilic xylanases in the facilitated bleaching of pulp. In *Enzymes for Pulp and Paper Processing*, edited by T.W. Jeffries and L. Viikari, pp. 85–100, American Chemical Society. Symposium Series 655.

Black, G.W., Rixon, J.E., Clarke, J.H., Hazlewood, G.P., Theodorou, M.K., Morris, P. and Gilbert, H.J. (1996) Evidence that linker sequences and cellulose-binding domains enhance the activity of hemicellulases against complex substrates. *Biochem. J.*, 319, 515–520.

Blacklow, S.C., Liu, K.D. and Knowles, J.R. (1991) Stepwise improvements in catalytic effectiveness: independence and interdependence in combinations of point mutations of a sluggish triosephosphate isomerase. *Biochemistry*, 30, 8470–8476.

Bork, P., Downing, A.K., Kieffer, B. and Campbell, I.D. (1996) Structure and distribution of modules in extracellular proteins. *Quart. Rev. Biophys.*, 29, 119–167.

Bortoli-German, I., Haiech, J., Chippaux, M. and Barras, F. (1995) Informational suppression to investigate structural, functional and evolutionary aspects of the *Erwinia chrysanthemi* cellulase EG2. *J. Mol. Biol.*, 246, 82–94.

Braisted, A.C. and Wells, J.A. (1996) Minimizing a binding domain from protein A. *Proc. Nat. Acad. Sci. USA*, 93, 5688–5692.

Bray, M.R., Creagh, A.L., Damude, H.G., Gilkes, N.R., Haynes, C.A., Jervis, E., Kilburn, D.G., MacLeod, A.M., Meinke, A., Miller, R.C., Jr, Rose, D.R., Shen, H., Tomme, P., Tull, D., White, A., Withers, S.G. and Warren, R.A.J. (1996a) β-1,4-Glycanases from *Cellulomonas fimi*: families, mechanisms, and kinetics. In *Enzymes for Pulp and Paper Processing*, edited by T.W. Jeffries and L. Viikari, pp. 65–84, American Chemical Society. Symposium Series 655.

Bray, M.R., Johnson, P.E., Gilkes, N.R., McIntosh, L.P., Kilburn, D.G. and Warren, R.A.J. (1996b) Probing the role of tryptophan residues in a cellulose-binding domain by chemical modification. *Protein Sci.*, 5, 2311–2318.

Brennan, M.B. (1996) Enzyme discovery heats up: enzymes that could maximize industrial processing reactions are reaped from exotic microorganisms. *Chem. Eng. News*, 74, 31–33.

Charnock, S.J., Lakey, J.H., Virden, R., Hughes, N., Sinnott, M.L., Hazlewood, G.P., Pickersgill, R.P. and Gilbert, H.J. (1997) Key residues in subsite F play a critical role in the activity of *Pseudomonas fluorescens* subsp. *cellulosa* xylanase A against xylooligosaccharides but not against highly polymeric substrates such as xylan. *J. Biol. Chem.*, 272, 2942–2951.

Clarke, J.H., Davidson, K., Gilbert, H.J., Fontes, C.M.G.A. and Hazlewood, G.P. (1996) A modular xylanase from mesophilic *Cellulomonas fimi* contains the same cellulose-binding and thermostabilizing domains as xylanases from thermophilic bacteria. *FEMS Microbiol. Lett.*, 139, 27–35.

Coutinho, J.B., Gilkes, N.R., Kilburn, D.G. Warren, R.A.J. and Miller, R.C., Jr. (1993) The nature of the cellulose-binding domain affects the activities of a bacterial endoglucanase on different forms of cellulose. *FEMS Microbiol. Lett.*, 113, 211–218.

Coutinho, J.B., Gilkes, N.R., Warren, R.A.J., Kilburn, D.G. and Miller, R.C., Jr. (1992) The binding of *Cellulomonas fimi* endoglucanase C (CenC) to cellulose and sephadex is mediated by the N-terminal repeats. *Mol. Microbiol.*, 6, 1243–1252.

Cowan, D. (1996) Industrial enzyme technology: meeting report. *Trends Biotechnol.*, 14, 177–178.

Creagh, L., Ong, E., Jervis, E., Kilburn, D.G. and Haynes, C.A. (1996) Binding of the cellulose-binding domain of exoglucanase Cex from *Cellulomonas fimi* to insoluble microcrystalline cellulose is entropically driven. *Proc. Nat. Acad. Sci. USA*, 93, 12229–12234.

Daniel, R.M., Dines, M. and Petach, H.H. (1996) The denaturation and degradation of stable enzymes at high temperatures. *Biochem. J.*, 317, 1–11.

Davies, G. and Henrissat, B. (1995) Structure and mechanism of glycosyl hydrolases. *Structure*, 3, 853–859.

Davies, G.J., Dodson, G.G., Hubbard, R.E., Tolley, S.P., Dauter, Z., Wilson, K.S., Hjort, C., Mikkelsen, J.M. Rasmussen, G. and Schülein, M. (1993) Structure and function of endoglucanase V. *Nature*, 365, 362–364.

Davies, G.J., Tolley, S.P., Henrissat, B., Hjort, C. and Schülein, M. (1995) Structures of oligosaccharide-bound forms of endoglucanase V from *Humicola insolens* at 1.9 Å resolution. *Biochemistry*, 34, 16210–16220.

Derewenda, U., Swenson, L., Green, R., Wei, Y., Morosoli, R., Shareck, F., Kluepfel, D. and Zerewenda, Z.S. (1994) Crystal structure, at 2.6 Å resolution, of the *Streptomyces lividans* xylanase A, a member of the F family of β-1,4-D-glucanases. *J. Biol. Chem.*, 269, 20811–20814.

Din, N., Forsythe, I.J., Burtnick, L.J., Gilkes, N.R., Miller, R.C., Jr., Warren, R.A.J. and Kilburn, D.G. (1994) The cellulose-binding domain of endoglucanase A (CenA) from *Cellulomonas fimi*: evidence for the involvement of tryptophan residues in binding. *Mol. Microbiol.*, 11, 747–755.

Divne, C., Ståhlberg, J., Reinikainen, T., Ruohonen, L., Pettersson, G., Knowles, J.K.C., Teeri, T.T. and Jones, A. (1994) The three-dimensional crystal structure of the catalytic core of cellobiohydrolase I from *Trichoderma reesei*. *Science*, 265, 524–528.

Doheny, J.G., Jervis, E.J., Guarna, M.M., Humphries, R.K., Warren, R.A.J. and Kilburn, D.G. Dynamic cytokine localization onto cellulose by genetic fusion to a cellulose-binding domain. *Submitted to Bio/Technology*.

Dominguez, R., Souchon, H., Lascombe, M.-B. and Alzari, P.M. (1996) The crystal structure of a family 5 endoglucanase mutant in complexed and uncomplexed forms reveals an induced fit activation mechanism. *J. Mol. Biol.*, 257, 1042–1051.

Dominguez, R., Souchon, H., Spinelli, S., Dauter, Z., Wilson, K.S., Chauvaux, S., Béguin, P. and Alzari, P.M. (1995) A common protein fold and similar active site in two distinct families of β-glycanases. *Nature Struct. Biol.*, 2, 569–576.

Ducros, V., Czjzek, M., Belaich, A., Gaudin, C., Fierobe, H.-P., Belaich, J.-P., Davies G.J. and Haser, R.J. (1995) Crystal structure of the catalytic domain of a bacterial cellulase belonging to family 5. *Structure*, 3, 939–949.

Duffaud, G.D., McCutchen, C.M., Leduc, P., Parker, K.N. and Kelly, R.M. (1997) Purification and characterization of extremely thermostable β-mannanase, β-mannosidase, and α-galactosidase from the hyperthermophilic eubacterium *Thermotoga neopolitana* 5068. *Appl. Environ. Microbiol.*, 63, 169–177.

Dupont, C., Roberge, M., Morsoli, R., Shareck, F., Moreau, A. and Kluepfel, D. (1996) Structural and functional modifications of a xylanase from *Streptomyces lividans* belonging to glycanase family 10. In *Enzymes for Pulp and Paper Processing*, edited by T.W. Jeffries and L. Viikari, pp. 116–128, American Chemical Society. Symposium Series 655.

Fischer, L., Bromann, R., Kengen, S.W.M., de Vos, W.M. and Wagner, F. (1996) Catalytic potency of β-glucosidase from the extremophile *Pyrococcus furiosus* in glucoconjugate synthesis. *Bio/Technology*, 14, 88–91.

Fontes, C.M.G.A., Hazlewood, G.P., Morag, E., Hall, J., Hirst, B.H. and Gilbert, H.J. (1995) Evidence for a general role for non-catalytic thermostabilizing domains in xylanases from thermophilic bacteria. *Biochem. J.*, 307, 151–158.

Francisco, J.A., Stathopoulos, C., Warren, R.A.J., Kilburn, D.G. and Georgiou, G. (1993) Specific adhesion and hydrolysis of cellulose by intact *Escherichia coli* cells expressing surface-anchored cellulase or cellulose-binding domains. *Bio/Technology*, 11, 491–495.

Ge, L. and Rudolph, P.H. (1997) Simultaneous introduction of multiple mutations using overlap extension PCR. *Biotechniques*, 22, 28–30.

Gilkes, N.R., Henrissat, B., Kilburn, D.G., Miller, R.C. Jr. and Warren, R.A.J. (1991) Domains in microbial β-1,4-glycanases: sequence conservation, function and enzyme families. *Microbiol. Rev.*, 55, 305–315.

Gilkes, N.R., Jervis, E., Henrissat, B., Tekant, B., Miller, R.C., Jr., Warren, R.A.J. and Kilburn, D.G. (1992) The adsorption of a bacterial cellulase and its two isolated domains to crystalline cellulose. *J. Biol. Chem.*, 267, 6743–6749.

Gilkes, N.R., Kwan, E., Kilburn, D.G., Miller, R.C. Jr. and Warren, R.A.J. (1997) Attack of carboxymethylcellulose at opposite ends by two cellobiohydrolases from *Cellulomonas fimi. J. Biotechnol.*, In press.

Gilkes, N.R., Warren, R.A.J., Miller, R.C., Jr. and Kilburn, D.G. (1988) Precise excision of the cellulose-binding domains from two *Cellulomonas fimi* cellulases by a homologous protease and the effect on catalysis. *J. Biol. Chem.*, 263, 10401–10407.

Goldstein, M.A. and Doi, R.H. (1994) Mutation analysis of the cellulose-binding domain of the *Clostridium cellulovorans* cellulose-binding protein A. *J. Bacteriol.*, 176, 7328–7334.

Goldstein, M.A., Takagi, M., Hashida, S., Shoseyov, O., Doi, R.H. and Segel, I.H. (1993) Characterization of the cellulose-binding domain of the *Clostridium cellulovorans* cellulose-binding protein A. *J. Bacteriol.*, 175, 5762–5768.

Greenwood, J.M., Gilkes, N.R., Kilburn, D.G., Miller, R.C., Jr. and Warren, R.A.J. (1989) Fusion to an endoglucanase allows alkaline phosphatase to bind to cellulose. *FEBS Lett.*, 244, 127–131.

Greenwood, J.M., Gilkes, N.R., Miller, R.C., Jr., Kilburn, D.G. and Warren, R.A.J. (1994) Purification and processing of cellulose-binding domain-alkaline phosphatase fusion proteins. *Biotechnol. Bioeng.*, 44, 1295–1305.

Hahn, M., Keitel, T. and Heinemann, U. (1995) Crystal and molecular structure at 0.16-nm resolution of the hybrid *Bacillus* endo-1,3-1,4-β-D-glucan 4-glucanohydrolase H(A16-M). *Eur. J. Biochem.*, 232, 849–858.

Hahn, M., Piotukh, K., Borriss, R. and Heinemann, U. (1994) Native-like *in vivo* folding of a circularly permuted jellyroll protein shown by crystal structure analysis. *Proc. Nat. Acad. Sci. USA*, 91, 10417–10421.

Hall, J., Black, G.W., Ferreira, L.M.A., Millward-Sadler, S.J., Ali, B.R.S., Hazlewood, G.P. and Gilbert, H.J. (1995) The non-catalytic cellulose-binding domain of a novel cellulase from *Pseudomonas fluorescens* subsp. *cellulosa* is important for the efficient hydrolysis of Avicel. *Biochem. J.*, 309, 749–756.

Harris, G.W., Jenkins, J.A., Connerton, I., Cummings, N., Leggio, L.L., Scott, M., Hazlewood, G.P., Laurie, J.I., Gilbert, H.J. and Pickersgill, R.W. (1994) Structure of the catalytic core of the family F xylanase from *Pseudomonas fluorescens* and identification of the xylopentaose-binding sites. *Structure*, 2, 1107–1116.

Hasenwinkle, D., Jervis, E., Kops, O., Liu, C., Lesnicki, G., Haynes, C.A. and Kilburn, D.G. (1997) Very high level production and export in *Escherichia coli* of a cellulose-binding domain for use in a generic secretion-affinity fusion system. Accepted by *Biotechnol. Bioeng.*

Heinemann, U., Aÿ, J., Gaiser, O., Müller, J.J. and Ponnuswamy, M.N. (1996) Enzymology and folding of natural and engineered bacterial β-glucanases studied by X-ray crystallography. *Biol. Chem.*, 377, 447–454.

Henrissat, B. (1991) A classification of glycosyl hydrolases based on amino acid sequence similarities. *Biochem. J.*, 280, 309–316.

Henrissat, B. and Bairoch, A. (1993) New families in the classification of glycosyl hydrolases based on amino acid sequence similarities. *Biochem. J.*, 293, 781–788.

Henrissat, B., Callebaut, I., Fabrega, S., Lehn, P., Mornon, J.-P. and Davies, G. (1995) Conserved catalytic machinery and the prediction of a common fold for several families of glycosyl hydrolases. *Proc. Nat. Acad. Sci. USA*, 92, 7090–7094.

Hermes, J.D., Blacklow, S.C. and Knowles, J.R. (1990) Searching sequence space by definably random mutagenesis: improving the catalytic potency of an enzyme. *Proc. Nat. Acad. Sci. USA*, 87, 696–700.

Hermes, J.D., Parekh, S.M., Blacklow, S.C., Köster, H. and Knowles, J.R. (1989) A reliable method for random mutagenesis: the generation of mutant libraries using spiked oligodeoxyribonucleotide primers. *Gene*, 84, 143–151.

Hreggvidsson, G.O., Kaiste, E., Holst, O., Eggertsson, G., Palsodottir, A. and Kristjansson, J.K. (1996) An extremely thermostable cellulase from the thermophilic eubacterium *Rhodothermus marinus*. *Appl. Environ. Microbiol.*, 62, 3047–3049.

Hugenholtz, P. and Pace, N.R. (1996) Identifying microbial diversity in the natural environment: a molecular phylogenetic approach. *Trends Biotechnol.*, 14, 190–197.

Ingledew, W.J. (1990) Acidophiles. In *Microbiology of Extreme Environments*, edited by in C. Edwards, pp. 33–54. London: McGraw-Hill.

Irwin, D.C., Spezio, M., Walker, L.P. and Wilson, D.B. (1993) Activity studies of eight purified cellulases: specificity, synergism, and binding domain effects. *Biotechnol. Bioeng.*, 42, 1002–1013.

Jenkins, J., Leggio, L.L., Harris, G. and Pickersgill, R. (1995) β-glucosidase, β-galactosidase, family A cellulases, family F xylanases, and two barley glycanases form a superfamily of enzymes with 8-fold β/α architecture and with two conserved glutamates near the carboxy-terminal ends of β-strands four and seven. *FEBS Lett.*, 362, 281–285.

Jervis, E., Haynes, C. and Kilburn, D.G. Surface diffusion of cellulases and their isolated binding domains on cellulose. *Submitted to J. Biol. Chem.*

Johnson, P.E., Joshi, M.D., Tomme, P., Kilburn, D.G. and McIntosh, L.P. (1996a) Structure of the N-terminal cellulose-binding domain of *Cellulomonas fimi* CenC determined by nuclear magnetic resonance spectroscopy. *Biochemistry*, 35, 14381–14394.

Johnson, P.E., Tomme, P., Joshi, M.D. and McIntosh, L.P. (1996b) Interaction of soluble cellooligosaccharides with the N-terminal cellulose-binding domain of *Cellulomonas fimi* CenC. 2. NMR and ultraviolet absorption spectroscopy. *Biochemistry*, 35, 13895–13906.

Juy, M., Amit, A.G., Alzari, P.M., Poljak, R.J., Claeyssens, M., Béguin, P. and Aubert, J.-P. (1992) Three-dimensional structure of a thermostable bacterial cellulase. *Nature*, 357, 89–91.

Karita, S., Sakka, K. and Ohmiya, K. (1996) Cellulose-binding domains confer an enhanced activity against insoluble cellulose to *Ruminococcus albus* endoglucanase IV. *J. Ferment. Bioeng.*, 81, 553–556.

Kilburn, D.G., Assouline, Z., Din, N., Gilkes, N.R., Ong, E., Tomme, P. and Warren, R.A.J. (1993) Cellulose-binding domains: properties and applications. Proceedings 2nd Tricel Symposium: *Trichoderma reesei* cellulases and other hydrolases, edited by P. Suominen and T. Reinikainen. *Foundation for Biotechnical and Industrial Fermentation Research*, 8, 281–290.

Kraulis, P.J., Clore, G.M., Nilges, M., Jones, T.A., Petterson, G., Knowles, J. and Gronenborn, A.M. (1989) Determination of the three-dimensional solution structure of the C-terminal domain of cellobiohydrolase I from *Trichoderma reesei*. A study using nuclear magnetic resonance and hybrid distance geometry-dynamical simulated annealing. *Biochemistry*, 28, 7241–7257.

Krengel, U. amd Dijkstra, B.W. (1996) Three-dimensional structure of endo-1,4-β-xylanase I from *Aspergillus niger*: molecular basis for its low pH optimum. *J. Mol. Biol.*, 263, 70–78.

Kroll, R.G. (1990) Alkalophiles. *Microbiology of Extreme Environments*, edited by C. Edwards, pp. 55–92. London: McGraw-Hill.

Kuriki, T., Kaneko, H., Yanase, M., Takata, H., Shimada, J., Handa, S., Takada, T., Umeyama, H. and Okada, S. (1996) Controlling substrate preference and transglycosylation activity of neopullulanase by manipulating steric constraint and hydrophobicity in active centre. *J. Biol. Chem.*, 271, 17321–17329.

Kuroki, R., Weaver, L.H. and Matthews, B.W. (1993) A covalent enzyme-substrate intermediate with disaccharide distortion in a mutant T_4 lysozyme. *Science*, 262, 2030–2033.

Kuroki, R., Weaver, L.H. and Matthews, B.W. (1995) Structure-based design of a lysozyme with altered catalytic activity. *Nature Struct. Biol.*, 2, 1007–1011.

Lawson, S.L., Wakarchuk, W.W. and Withers, S.G. (1996) Effects of both shortening and lengthening the active site nucleophile of *Bacillus circulans* xylanase on catalytic activity. *Biochemistry*, 35, 10110–10118.

Le, K.D., Gilkes, N.R., Kilburn, D.G., Miller, R.C., Jr., Saddler, J.N. and Warren, R.A.J. (1994) A streptavidin-cellulose-binding domain fusion protein that binds biotinylated proteins to cellulose. *Enzyme Microb. Technol.*, 16, 496–500.

Liebl, W., Gabelsberger, J. and Schleifer, K.-H. (1994) Comparative amino acid sequence analysis of *Thermotoga* maritima β-glucosidase (BglA) deduced from the nucleotide sequence of the gene indicates distant relationship between β-glucosidases of the BGA family and other families of β-1,4-glycosyl hydrolases. *Mol. Gen. Genet.*, 242, 111–115.

Liebl, W., Ruile, P., Bronnenmeier, K., Riedel, K., Lottspeich, F. and Greif, I. (1996) Analysis of a *Thermotoga maritima* DNA fragment encoding two similar thermostable cellulases, CelA and CelB, and characterization of the recombinant enzymes. *Microbiol.*, 142, 2533–2542.

Linder, M. and Teeri, T. (1996) The cellulose-binding domain of the major cellobiohydrolase of *Trichoderma reesei* exhibits true reversibility and a high exchange rate on crystalline cellulose. *Proc. Nat. Acad. Sci. USA*, 93, 12251–12255.

Linder, M., Lindeberg, G., Reinikainen, T., Teeri, T.T. and Pettersson, G. (1995a) The difference in affinity between two fungal cellulase-binding domains is dominated by a single amino acid substitution. *FEBS Lett.*, **372**, 96–98.

Linder, M., Mattinen, M.-L., Kontteli, M., Lindeberg, G., Ståhlberg, J., Drakenberg, T., Reinikainen, T., Pettersson, G. and Annila, A. (1995b) Identification of functionally important amino acids in the cellulose-binding domain of *Trichoderma reesei* cellobiohydrolase I. *Prot. Sci.*, **4**, 1056–1064.

Lindner, M., Salovuori, I., Ruohonen, L. and Teeri, T.T. (1996) Characterization of a double cellulose-binding domain: synergistic high affinity binding to crystalline cellulose. *J. Biol. Chem.*, **271**, 21268–21272.

Ljunger, G., Adlercrentz, P. and Mattiasson, B. (1994) Enzymatic synthesis of octyl-β-glucoside in octanol at controlled water activity. *Enzyme Microb. Technol.*, **16**, 751–755.

Love, D.R., Fisher, R. and Bergquist, P.L. (1988) Sequence structure and expression of a cloned β-glucosidase gene from an extreme thermophile. *Mol. Gen. Genet.*, **213**, 84–92.

MacKenzie, L.F., Brook, G.S., Cutfield, J.R., Sullivan, P.A. and Withers, S.G. (1997) Identification of Glu-330 as the catalytic nucleophile of *Candida albicans* exo-β-(1,3)-glucanase. *J. Biol. Chem.*, **272**, 3161–3167.

MacLeod, A.M., Gilkes, N.R., Escote-Carlson, L., Warren, R.A.J., Kilburn, D.G. and Miller, R.C., Jr. (1992) *Streptomyces lividans* glycosylates an exoglucanase (Cex) from *Cellulomonas fimi*. Gene **121**, 143–147.

MacLeod, A.M., Lindhorst, T., Withers, S.G. and Warren, R.A.J. (1994) The acid/base catalyst in the exoglucanase/xylanase (Cex) from *Cellulomonas fimi*: evidence from detailed kinetic studies of mutants. *Biochemistry*, **33**, 6371–6376.

Macleod, A.M., Tull, D., Rupitz, K., Warren, R.A.J. and Withers, S.G. (1996) Mechanistic consequences of mutation of active site carboxylates in a retaining β-1,4-glycanase from *Cellulomonas fimi*. *Biochemistry*, **35**, 13165–13172.

Maglione, G., Matsushita, O., Russell, J.B. and Wilson, D.B. (1992) Properties of a genetically reconstructed *Prevotella ruminicola* endoglucanase. *Appl. Environ. Microbiol.*, **58**, 3593–3597.

Martinez-Bilbao, M., Holdsworth, R.E., Edwards, L.A. and Huber, R.E. (1991) A highly reactive β-galactosidase (*Escherichia coli*) resulting from a substitution of an aspartic acid for Glu794. *J. Biol. Chem.*, **266**, 4979–4986.

McCarter, J.D. and Withers, S.G. (1994) Mechanisms of enzymatic glycoside hydrolysis. *Curr. Opin. Struct. Biol.*, **4**, 885–892.

McIntosh, L.P., Hand, G., Johnson, P.E., Joshi, M.D., Körner, M., Plesniak, L.A., Ziser, L., Wakarchuk, W.W. and Withers, S.G. (1996) The pK_a of the general acid/base carboxyl group of a glycosidase cycles during catalysis: a ^{13}C-NMR study of *Bacillus circulans* xylanase. *Biochemistry*, **35**, 9958–9966.

Meinke, A., Damude, H., Tomme, P., Kwan, E., Kilburn, D.G., Miller, R.C., Jr., Warren, R.A.J. and Gilkes, N.R. (1995) Enhancement of the endo-β-1,4-glucanase activity of an exocellobiohydrolase by deletion of a surface loop. *J. Biol. Chem.*, **270**, 4384–4386.

Monsan, P. and Paul, F. (1995) Enzymatic synthesis of oligosaccharides. *FEMS Microbiol. Rev.*, **16**, 187–192.

Morana, A., Moracci, M., Ottombrino, A., Ciaramella, M., Rossi, M. and De Rosa, M. (1995) Industrial-scale production and rapid purification of an archael β-glucosidase expressed in *Saccharomyces cerevisiae*. *Biotechnol. Appl. Biochem.*, **22**, 261–268.

Moreau, A., Shareck, F., Kluepfel, D. and Morosoli, R. (1994a) Alteration of the cleavage mode and of the transglycosylation reactions of the xylanase A of *Streptomyces lividans* 1326 by site-directed mutagenesis of the Asn173 residue. *Eur. J. Biochem.*, **219**, 261–266.

Moreau, A., Shareck, F., Kluepfel, D. and Morosoli, R. (1994b) Increase in catalytic activity and thermostability of the xylanase A of *Streptomyces lividans* 1326 by site-specific mutagenesis. *Enzyme Microb. Technol.*, **16**, 420–424.

Nakamura, A., Fukumori, F., Horinouchi, S., Masaki, H., Kudo, T., Uozumi, T., Horikoshi, K. and Beppu, T. (1991) Construction and characterization of the chimeric enzymes between the *Bacillus subtilis* cellulase and an alkalophilic *Bacillus* cellulase. *J. Biol. Chem.*, **266**, 1579–1583.

Ong, E., Alimonti, J.B., Greenwood, J.M., Miller, R.C., Jr., Warren, R.A.J. and Kilburn, D.G. (1995) Purification of human interleukin-2 using the cellulose-binding domain of a procaryotic cellulase. *Bioseparation*, **5**, 95–104.

Ong, E., Gilkes, N.R., Miller, R.C., Jr., Warren, R.A.J. and Kilburn, D.G. (1991) Enzyme immobilization using a cellulose-binding domain: properties of a β-glucosidase fusion protein. *Enzyme Microb. Technol.*, **13**, 59–65.

Ong, E., Gilkes, N.R., Miller, R.C., Jr., Warren, R.A.J. and Kilburn, D.G. (1993) The cellulose-binding domain (CBD$_{Cex}$) of an exoglucanase from *Cellulomonas fimi*: production in *Escherichia coli* and characterization of the polypeptide. *Biotechnol. Bioeng.*, **42**, 401–409.

Ong, E., Gilkes, N.R., Warren, R.A.J., Miller, R.C., Jr., and Kilburn, D.G. (1989) Enzyme immobilization with the cellulose-binding domain of a *Cellulomonas fimi* exoglucanase. *Bio/Technology*, **7**, 604–607.

Poole, D.M., Hazlewood, G.P., Huskisson, N.S., Virden, R. and Gilbert, H.J. (1993) The role of conserved tryptophan residues in the interaction of a bacterial cellulose-binding domain with its ligand. *FEMS Microbiol. Lett.*, **106**, 77–84.

Ramirez, C., Fung, J., Miller, R.C. Jr., Warren, R.A.J. and Kilburn, D.G. (1993) A bifunctional affinity linker to couple antibodies to cellulose. *Bio/Technology*, **11**, 1570–1573.

Rastall, R.A., Pikett, S.F., Adlard, M.W. and Bucke, C. (1992) Synthesis of oligosaccharides by reversal of a fungal β-glucanase. *Biotechnol. Lett.*, **14**, 373–378.

Ravet, C., Thomas, D. and Legoy, M.D. (1993) Gluco-oligosaccharide synthesis by free and immobilized β-glucosidase. *Biotechnol. Bioeng.*, **42**, 303–308.

Reinikainen, T., Ruohonen, L., Nevanen, T., Laaksonen, L., Kraulis, P., Jones, T.A., Knowles, J.K.C. and Teeri, T.T. (1992) Investigation of the function of mutated cellulose-binding domains of *Trichoderma reesei* cellobiohydrolase I. *Proteins Struct. Func. Genet.*, **14**, 475–482.

Reinikainen, T., Teleman, O. and Teeri, T.T. (1995) Effects of pH and high ionic strength on the adsorption and activity of native and mutated cellobiohydrolase I from *Trichoderma reesei*. *Proteins Struct. Func. Genet.*, **22**, 392–403.

Rouvinen, J., Bergfors, T., Teeri, T., Knowles, J.K.C. and Jones, T.A. (1990) Three-dimensional structure of cellobiohydrolase II from *Trichoderma reesei*. *Science*, **249**, 380–386.

Ruttersmith, L.D. and Daniel, R.M. (1991) Thermostable cellobiohydrolase from the thermophilic eubacterium *Thermotoga* sp. strain FjSS3-B.1. *Biochem. J.*, **277**, 887–890.

Sakon, J., Adney, W.S., Himmel, M.E., Thomas, S.R. and Karplus, P.A. (1996) Crystal structure of thermostable family V endocellulase E1 from *Acidothermus cellulolyticus* in complex with cellotetraose. *Biochemistry*, **35**, 10648–10660.

Shen, H., Gilkes, N.R., Kilburn, D.G., Miller, R.C., Jr. and Warren, R.A.J. (1995) Cellobiohydrolase B, a second exocellobiohydrolase from the cellulolytic bacterium *Cellulomonas fimi*. *Biochem. J.*, **311**, 67–74.

Shen, H., Schmuck, M., Pilz, I., Gilkes, N.R., Kilburn, D.G., Miller, R.C., Jr. and Warren, R.A.J. (1991) Deletion of the linker connecting the catalytic and cellulose-binding domains of endoglucanase A (CenA) of *Cellulomonas fimi* alters its conformation and catalytic activity. *J. Biol. Chem.*, **266**, 11335–11340.

Shortle, D. and Sondek, J. (1995) The emerging role of insertions and deletions in protein engineering. *Curr. Opin. Biotechnol.*, **6**, 387–393.

Singh, A. and Hayashi, K. (1995) Construction of chimeric β-glucosidases with improved enzymatic properties. *J. Biol. Chem.*, **270**, 21928–21933.

Singh, A., Hayashi, K., Hoa, T.T., Kashiwagi, Y. and Tokuyasu, K. (1995) Construction and characterization of a chimeric β-glucosidase. *Biochem. J.*, **305**, 715–719.

Spezio, M., Wilson, D.B. and Karplus, P.A. (1993) Crystal structure of the catalytic domain of a thermophilic endocellulase. *Biochemistry*, **32**, 9906–9916.

Ståhlberg, J., Divne, C., Koivula, A., Piens, K., Claeyssens, M., Teeri, T.T. and Jones, T.A. (1996) Activity studies and crystal structures of catalytically deficient mutants of cellobiohydrolase I from *Trichoderma reesei*. *J. Mol. Biol.*, **264**, 337–349.

Ståhlberg, J., Johansson, G. and Pettersson, G. (1993) *Trichoderma reesei* has no true exocellulase: all intact and truncated cellulases produce new reducing end groups on cellulose. *Biochim. Biophys. Acta*, **1157**, 107–113.

Stemmer, W.P.C. (1994a) DNA shuffling by random fragmentation and reassembly: *in vitro* recombination for molecular evolution. *Proc. Nat. Acad. Sci. USA*, **91**, 10747–10751.

Stemmer, W.P.C. (1994b) Rapid evolution of a protein *in vitro* by DNA shuffling. *Nature*, **370**, 389–391.

Sulzenbacher, G., Driguez, H., Henrissat, B., Schülein, M. and Davies, G.J. (1996) Structure of the *Fusarium oxysporum* endoglucanase I with a non-hydrolyzable substrate analogue: substrate distortion gives rise to the preferred axial orientation for the leaving group. *Biochemistry*, **35**, 15280–15287.

Svensson, B. and Søgaard, M. (1993) Mutational analysis of glycosylase function. *J. Biotechnol.*, **29**, 1–37.

Terwisscha van Scheltinga, A.C., Armand, S., Kalk K.H., Isogai, A., Henrissat, B. and Dijkstra, B. (1995) Stereochemistry of chitin hydrolysis by a plant chitinase/lysozyme and X-ray structure of a complex with allosamidin: evidence for substrate-assisted catalysis. *Biochemistry*, **34**, 15619–15623.

Tews, I., Perrakis, A., Oppenheim, A., Dauter, Z., Wilson, K.S. and Vorgias, C.E. (1996) Bacterial chitobiase structure provides insight into catalytic mechanism and the basis of Tay-Sachs disease. *Nature Struct. Biol.*, **3**, 638–648.

Thiem, J. (1995) Applications of enzymes in synthetic carbohydrate chemistry. *FEMS Microbiol. Rev.*, **16**, 193–211.

Tomme, P., Creagh, A.L., Kilburn, D.G. and Haynes, C.A. (1996a) Interaction of polysaccharides with the N-terminal cellulose-binding domain of *Cellulomonas fimi* CenC. 1. Binding specificity and calorimetric analysis. *Biochemistry*, **35**, 13885–13894.

Tomme, P., Driver, D.P., Amandoron, E., Miller, R.C., Jr., Warren, R.A.J. and Kilburn, D.G. (1995a) Domain swapping allows the comparison of a fungal (type I) and a bacterial (type II) cellulose-binding domain. *J. Bacteriol.*, **177**, 4356–4363.

Tomme, P., Kwan, E., Gilkes, N.R., Kilburn, D.G. and Warren, R.A.J. (1996b) Characterization of CenC, an enzyme from *Cellulomonas fimi* with both endo- and exoglucanase activities. *J. Bacteriol.*, **178**, 4216–4223.

Tomme, P., Warren, R.A.J. and Gilkes, N.R. (1995b) Cellulose hydrolysis by bacteria and fungi. *Adv. Microb. Physiol.*, **37**, 1–81.

Tomme, P., Warren, R.A.J., Miller, R.C. Jr., Kilburn, D.G. and Gilkes, N.R. (1995c) Cellulose-binding domains: classification and properties. In *Enzymatic Degradation of Insoluble Carbohydrates*, edited by J.N. Saddler and M.H. Penner, pp. 142–163, American Chemical Society. Symposium Series 618.

Tormo, J., Lamed, R., Chirino, A.J., Morag, E., Bayer, E.A., Shoham, Y. and Steitz, T.A. (1996) Crystal structure of a bacterial family III cellulose-binding domain: a general mechanism for attachment to cellulose. *EMBO Jour.*, **15**, 5739–5751.

Varghese, J.N., Garrett, T.P.J., Coleman, P.M., Chen, L., Høj, P.B. and Fincher, G.B. (1994) Three-dimensional structures of two plant β-glucan endohydrolases with distinct substrate specificities. *Proc. Nat. Acad. Sci. USA*, **91**, 2785–2789.

Venekei, I., Hedstrom, L. and Rutter, W.J. (1996) A rapid and effective procedure for screening protease mutants. *Protein Engng.*, **9**, 85–94.

Vieille, C. and Zeikus, J.G. (1996) Thermozymes: identifying molecular determinants of protein structural and functional stability. *Trends Biotechnol.*, **14**, 183–190.

Voorhorst, W.G.B., Eggen, R.I.L., Luesink, E.J. and de Vos, W.M. (1995) Characterization of the *celB* gene coding for β-glucosidase from the hyperthermophilic Archaeon *Pyrococcus furiosus* and its expression and site-directed mutation in *Escherichia coli. J. Bacteriol.*, **177**, 7105–7111.

Vranská, M. and Biely, P. (1992) The cellobiohydrolase I from *Trichoderma reesei* QM 9414: action on cellooligosaccharides. *Carb. Res.*, **227**, 19–27.

Wakarchuk, W.W., Sung, W.L., Campbell, R.L., Cunningham, A., Watson, D.C. and Yaguchi, M. (1994) Thermostabilization of the *Bacillus circulans* xylanase by the introduction of disulfide bonds. *Prot. Engng.*, **7**, 1379–1386.

Wang, Q., Graham, R.W., Trimbur, D., Warren, R.A.J. and Withers, S.G. (1994) Changing enzymatic reaction mechanisms by mutagenesis: conversion of a retaining glucosidase to an inverting enzyme. *J. Am. Chem. Soc.*, **116**, 11594–11595.

Wang, Q., Trimbur, D., Graham, R., Warren, R.A.J. and Withers, S.G. (1995) Identification of acid/base catalyst in *Agrobacterium faecalis* β-glucosidase by kinetic analysis. *Biochemistry*, **34**, 14554–14562.

Warren, R.A.J. (1996) Microbial hydrolysis of polysaccharides. *Annu. Rev. Microbiol.*, **50**, 183–212.

Welfle, K., Misselwitz, R., Politz, O., Borriss, R. and Welfle, H. (1996) Individual amino acids in the N-terminal loop region determine the thermostability and unfolding characteristics of bacterial glucanases. *Protein Sci.*, **5**, 2255–2265.

White, A., Tull, D., Johns, K., Withers, S.G. and Rose, D.R. (1996) Crystallographic observation of a covalent catalytic intermediate in a β-glycosidase. *Nature Struct. Biol.*, **3**, 149–154.

White, A., Withers, S.G., Gilkes, N.R. and Rose, D.R. (1994) Crystal structure of the catalytic domain of the β-1,4-glycanase Cex from *Cellulomonas fimi. Biochemistry*, **33**, 12546–12552.

Wierzba, A., Reichl, V., Turner, R.F.B., Warren, R.A.J. and Kilburn, D.G. (1995) Production and properties of a bi-functional fusion protein that mediates attachment of Vero cells to cellulosic matrices. *Biotechnol. Bioeng.*, **47**, 147–154.

Winterhalter, C., Heinrich, P., Candussio, A., Wich, G. and Liebl, W. (1995) Identification of a novel cellulose-binding domain within the multidomain 120 kDa xylanase XynA of the hyperthermophilic bacterium *Thermotoga maritima. Mol. Microbiol.*, **15**, 431–444.

Withers, S.G. and Aebersold, R. (1995) Approaches to labelling and identification of active site residues in glycosidases. *Protein Sci.*, **4**, 361–372.

Withers, S.G., Rupitz, K., Trimbur, D. and Warren, R.A.J. (1992) Mechanistic consequences of mutation of the active site nucleophile Glu358 in *Agrobacterium* β-glucosidase. *Biochemistry*, **31**, 9979–9985.

Xu, G.-Y., Ong, E., Gilkes, N.R., Kilburn, D.G., Muhindaram, D.R., Harris-Brandts, M., Carver, J.P., Kay, L.E. and Harvey, T.S. (1995) Solution structure of a cellulose-binding domain from *Cellulomonas fimi* by nuclear magnetic resonance spectroscopy. *Biochemistry*, **34**, 6993–7009.

Yuan, J., Martinez-Bilbao, M. and Huber, R.E. (1994) Substitutions for Glu537 of β-galactosidase from *Escherichia coli* cause large decreases in catalytic activity. *Biochem. J.*, **299**, 527–531.

11. ENGINEERING SPECIFICITY AND STABILITY IN GLUCOAMYLASE FROM *ASPERGILLUS NIGER*

TORBEN P. FRANDSEN, HENRI-PIERRE FIEROBE[a] and BIRTE SVENSSON*

Carlsberg Laboratory, Department of Chemistry, Gamle Carlsberg Vej 10,
DK-2500 Copenhagen Valby, Denmark

INTRODUCTION

Glucoamylases (GAs)[1] (1,4-α-D-glucan glucohydrolase, EC 3.2.1.3) constitute glycoside hydrolase family 15 (Henrissat, 1991; Henrissat and Bairoch, 1993) and catalyze the release of β-D-glucose from nonreducing ends of starch and related oligo- and polysaccharides. Fungal GAs are applied in the commercial production of high glucose and fructose syrups (Saha and Zeikus, 1989), and the widely used *Aspergillus niger* GA is an obvious target for improvement of industrial properties through protein engineering. GA, at the same active site, cleaves the α-1,4 linked with about 500 times higher specificity than α-1,6 linked substrates (Hiromi *et al.*, 1966; Sierks and Svensson, 1994; Frandsen *et al.*, 1995; Fierobe *et al.*, 1996). One major goal is to raise the current yield of approximately 96% glucose in industrial saccharification by eliminating the GA catalyzed condensation reactions that lead to accummulation of α-1,6 linked oligosaccharides (Nikolov *et al.*, 1989). Other goals include elevated thermostability or high activity in the neutral pH-range, which would advance application of immobilized GA and enable a single-step liquefaction, saccharification, and isomerization process (Bhosale *et al.*, 1996).

Protein engineering of GA from *A. niger* (identical to *A. awamori* GA) is based on structure/function relationship investigations using site-directed mutagenesis (Sierks *et al.*, 1989, 1990; Bakir *et al.*, 1993; Sierks and Svensson, 1993, 1994, 1996; Chen *et al.*, 1994, 1995, 1996; Frandsen *et al.*, 1994, 1995; Natajaran *et al.*, 1996; Stoffer *et al.*, 1997) coupled with analysis of substrate binding kinetics (Hiromi *et al.*, 1983; Olsen *et al.*, 1992, 1993; Christensen *et al.*, 1996), inhibitor binding thermodynamics (Sigurskjold *et al.*, 1994; Berland *et al.*, 1995), molecular recognition of α-1,4 and α-1,6 linked substrate analogs (Sierks and Svensson, 1992; Sierks *et al.*, 1992; Svensson *et al.*, 1994; Frandsen *et al.*, 1996; Lemieux *et al.*, 1996), and high-resolution structural models of free and inhibitor complexed GA (Aleshin *et al.*, 1992, 1994ab, 1996; Harris *et al.*, 1993; Stoffer *et al.*, 1995). Thanks to efficient heterologous expression systems for GA production (Innis *et al.*, 1985; Christensen *et al.*, 1988; Fierobe *et al.*, 1997), a series of well characterized mutants provide an excellent basis for protein engineering of GA and related enzymes.

GA from *A. niger* has a catalytic (aa 1-440) and a starch binding domain (aa 509-616) which are linked by a long and highly O-glycosylated region (Svensson *et al.*, 1983, 1986). The catalytic domains of GA from *A. niger* (B.B. Stoffer, A.E. Aleshin, M. Gajhede, B. Svensson and R.B. Honzatko, unpublished) and of the 94% identical GA from *A. awamori* var. *X100* GA (Aleshin *et al.*, 1992) adopt an (α/α)$_6$-fold in which six well-conserved α→α loop segments, that form the active site, connect the

---

* To whom correspondence should be addressed.

[a] Present address: BIP-CNRS, 31 Chemin Joseph Aiguier, F-13402 Marseille, France

a)

b)

c)

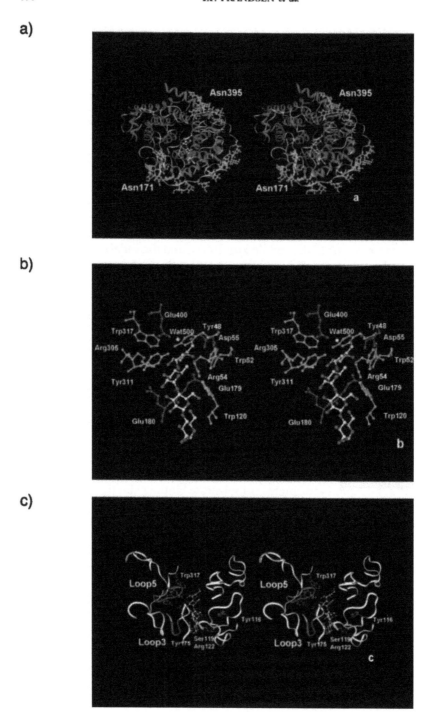

Figure 1 Stereoview of glucoamylase from *Aspergillus awamori* var. *X100* with the pseudotetrasaccharide inhibitor acarbose bound in the active site (Aleshin *et al.*, 1994), a. $(\alpha/\alpha)_6$-barrel fold with *O*- and *N*-linked glycosylation in blue; b. active site region as seen in the glucoamylase-acarbose crystal structure; c. loop replacement and mutations at a distance from the site of catalysis, in loop region 5 are displayed Arg305 in blue, Tyr306 in green, and Asp309 in pink

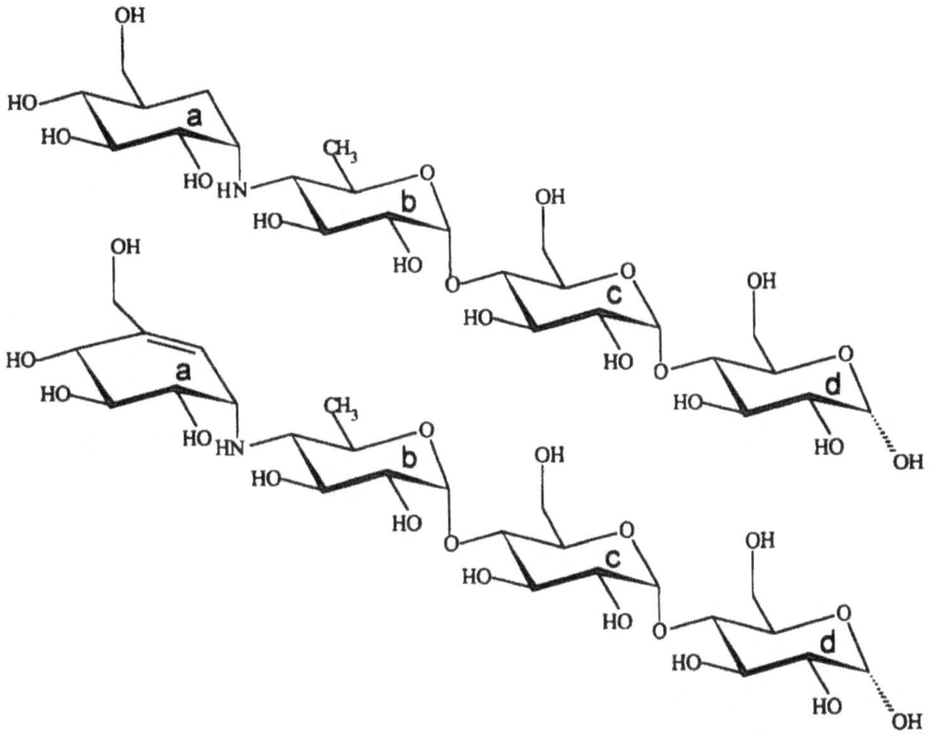

Figure 2 Structures of glucoamylase inhibitors, D-*gluco*-dihydroacarbose (top) and acarbose

outer and inner α_6-barrels (Figure 1a). The starch binding domain is an eight-stranded antiparallel open-sided β-barrel (Sorimachi *et al.*, 1996). The three-dimensional structure of the proteolytically produced catalytic domain (aa 1-471) includes the start of the *O*-glycosylated linker (Figure 1a). The bound pseudomaltotetrasaccharide inhibitors, acarbose and D-*gluco*-dihydroacarbose (Figure 2), highlight protein-carbohydrate interactions (Figure 1b,c; Aleshin *et al.*, 1994a, 1996; Stoffer *et al.*, 1995) at four glucosyl residue binding subsites, −1, +1, +2, and +3.

MUTATIONAL ANALYSIS OF RESIDUES AT THE SITE OF CATALYSIS

Crystal structures of GA in complex with acarbose-type inhibitors (Figure 2) and 1-deoxynojirimycin (Harris *et al.*, 1993; Aleshin *et al.*, 1994a, 1996; Stoffer *et al.*, 1995) are compatible with Glu179 acting as the general acid catalyst which protonates the glucosidic oxygen of the bond to be cleaved, and Glu400 as catalytic base that activates water (Wat 500, Figure 1b) for nucleophilic attack at C1 at subsite −1. The catalytic groups are approx. 10 Å apart, which is a typical distance observed in inverting carbohydrases making room for the water molecule (Figure 3; for reviews see McCarter and Withers, 1994; Davies and Henrissat, 1995). The OE1 of Glu400 is within hydrogen bond distance of OH of Tyr48. Moreover, since the phenol ring of Tyr48 is very close to C1 and N of 1-deoxynojirimycin (O5 in the substrate),

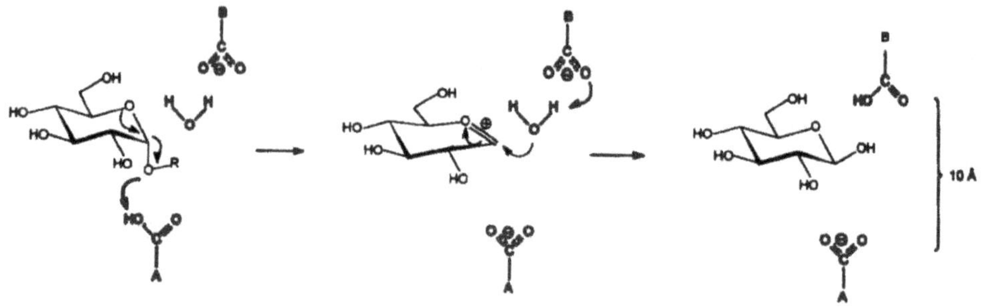

Figure 3 Schematic representation of the catalytic mechanism proceeding, via an oxocarbonium ion intermediate, with inversion of the anomeric configuration

nonbonding electrons at OH of Tyr48 may stabilize a substrate oxycarbonium intermediate. Similarly, a tyrosine in crystal structures of viral influenza B neuraminidase (Burmeister *et al.*, 1993) and bacterial sialidase (Crennell *et al.*, 1993) is suggested to stabilize an oxycarbonium intermediate, and Tyr298 in the retaining β-glucosidase from *Agrobacterium sp.* assists in catalysis by orienting the nucleophile Glu358 (Gebler *et al.*, 1995).

The catalytic role of Glu179 is confirmed by the very low k_{cat} values of about 0.005% and 0.08% (relative to wild-type) in maltoheptaose hydrolysis (Table 1) by the GA mutants Glu179→Asp (Bakir *et al.*, 1993) and the isosteric Glu179→Gln, respectively (Sierks *et al.*, 1990). The Glu400→Gln (Frandsen *et al.*, 1994) and Glu400→Cys GAs (H-P. Fierobe, A.J. Clarke, and B. Svensson, unpublished) show 1.8% and 0.2% of wild-type k_{cat}, and 12 and 3 fold increased K_m (Table 1). The isosteric Gln400 apparently exerts some activation of Wat 500, while Cys400 is a very poor base catalyst. Furthermore, pH-activity profiles of Glu400→Gln and Tyr48→Trp GAs (Figure 1b) reflect that a catalytic group titrating with low pK_a is lost in both of these mutants and support the proposed functionally important interaction between Tyr48 and Glu400 (Frandsen *et al.*, 1994). In conclusion, while a retained proton donor is vital to the GA activity, considerable flexibility is allowed for the catalytic base.

Table 1 Enzymatic properties of mutants at the site of catalysis in *Aspergillus* glucoamylase[a]

	Maltose			Maltoheptaose		
	k_{cat} (s^{-1})	K_m (mM)	k_{cat}/K_m $(s^{-1}mM^{-1})$	k_{cat} (s^{-1})	K_m (mM)	k_{cat}/K_m $(s^{-1}mM^{-1})$
Wild-type[b]	10.7	1.21	8.84	59.7	0.12	498
Tyr48→Trp[b]	0.120	3.92	0.031	0.762	0.168	4.54
Glu179→Asp[c]	n.d.			0.0029	0.09	0.03
Glu179→Gln[d]	n.d.			0.047	0.15	0.32
Glu400→Cys[e]	0.015	2.70	0.0055	0.116	0.45	0.26
Glu400→Gln[b]	0.299	14.8	0.020	1.05	0.380	2.76

n.d.: no detectable activity, the mutated side-chains are highlighted in Figure 1b
[a]Determined at 45°C in 0.05 M sodium acetate, pH 4.5; [b]Frandsen *et al.*, 1994; [c]Bakir *et al.*, 1993 (35°C); [d]Sierks *et al.*, 1990 (50°C); [e]H-P. Fierobe, A.J. Clarke, and B. Svensson, unpublished

MUTATIONAL ANALYSIS OF SUBSTRATE BINDING RESIDUES

GA binds substrate in a two-step mechanism (Hiromi *et al.*, 1983), in which an initial fast association to ES followed by a slower conformational change to the activated complex, ES*, leads to hydrolysis and release of products (Olsen *et al.*, 1992).

$$E + S \underset{k_{-1}}{\overset{k_1}{\rightleftharpoons}} ES \underset{k_{-2}}{\overset{k_2}{\rightleftharpoons}} ES^* \overset{k_3}{\rightleftharpoons} E + P$$

k_1, k_2, k_{-2}, and k_3 were determined for wild-type, Trp120→Phe, Glu180→Gln, and Trp317→Phe GAs (Figure 1bc) by stopped flow kinetic analysis (Olsen *et al.*, 1992, 1993; Christensen, 1995; Christensen *et al.*, 1996). The mutation of Trp317 primarily affects the first association to ES (Christensen, 1995), while both Glu180 and Trp120 are important for the transition from ES to ES* (Olsen *et al.*, 1993; Christensen *et al.*, 1996). Thus, the hydrogen bond from Glu180 to OH-2 in maltose and the hydrophobic stacking between Trp120 and longer maltooligosaccharides stabilize the Michaelis complex, ES*.

The binding of 1-deoxynojirimycin and acarbose to GA and several mutants is shown by isothermal titration calorimetry to be both enthalpically and entropically favorable (Sigurskjold *et al.*, 1994; Berland *et al.*, 1995). Rings a and b of the pseudomaltotetraose inhibitor are sequestered from solvent at subsites −1 and +1 (Figure 1b,c; Aleshin *et al.*, 1994a, 1996; Stoffer *et al.*, 1995), whereas rings c and d, at the reducing end, are solvent exposed. These glucosyl residues can occupy subsites +2 and +3 in two ligand binding modes that involve hydrogen bonds to different areas of the funnel-shaped GA active site (Figure 4; Stoffer *et al.*, 1995; Aleshin *et al.*, 1996). Although the stronger protein-carbohydrate interaction clearly occurs at subsites −1 and +1, the more distant subsites can influence events at the inner subsites (Svensson *et al.*, 1995).

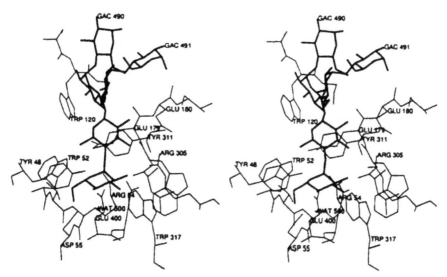

Figure 4   Stereoview of the active site of glucoamylase from *Aspergillus awamori* var. *X100* with bound D-*gluco*-dihydroacarbose (Stoffer *et al.*, 1995; copyright permission by Elsevier)

Table 2 Enzymatic properties of mutants at the charged substrate binding residues at subsites −1 and +1 in *Aspergillus* glucoamylase[a]

	Maltose			Isomaltose			Maltoheptaose		
	k_{cat} (s^{-1})	K_m (mM)	k_{cat}/K_m $(s^{-1}mM^{-1})$	k_{cat} (s^{-1})	K_m (mM)	k_{cat}/K_m $(s^{-1}mM^{-1})$	k_{cat} (s^{-1})	K_m (mM)	k_{cat}/K_m $(s^{-1}mM^{-1})$
Wild-type[b]	10.7	1.21	8.84	0.41	19.8	0.021	59.7	0.12	498
Arg54→Leu[b]	0.0094	3.03	0.0031	n.d.			0.043	0.21	0.200
Arg54→Lys[b]	0.0079	0.59	0.013	n.d.			0.032	0.031	1.03
Asp55→Gly[c]	0.040	1.50	0.030	0.0017	38.5	4.41×10^{-5}	0.23	0.19	1.23
Asp55→Val[d]	0.0020	0.90	0.0022	n.d.			0.011	0.082	0.13
Glu180→Asp[e]	n.d			n.d.			21.5	0.461	46.3
Glu180→Gln[f]	1.53	41.4	0.0369	0.184	95.0	0.00193	30.9	9.39	3.29
Arg305→Lys[g]		>400	0.0024[g]	4.4×10^{-4}	21.0	2.1×10^{-5}			0.0066[g]

[a]Determined at 45°C in 0.05 M sodium acetate, pH 4.5; [b]Frandsen *et al.*, 1995; [c]Sierks & Svensson, 1993; [d]T.P. Frandsen, J. Lehmbeck, B.B. Stoffer, and B. Svensson, unpublished; [e]Bakir *et al.*, 1993 (35°C); [f]Sierks *et al.*, 1990 (50°C); [g]Determined at low, unsaturating substrate concentration as the second-order rate constant (k_{cat}/K_m). Substrate saturation of this mutant was not obtained, i.e., $K_m \gg$ [S].

The mutated side-chains are highlighted in Figure 1b.

Molecular recognition of deoxygenated substrate analogs by GA has identified OH-4', -6', and -3 in maltose (Bock and Pedersen, 1987; Sierks and Svensson, 1992; Sierks *et al.*, 1992; Svensson *et al.*, 1994) and OH-4', -6', and -4 in isomaltose (Frandsen *et al.*, 1996; Lemieux *et al.*, 1996) as so-called key polar groups, which are needed for efficient catalysis. Mutation of Asp309 and Glu180, located at subsites −1 and +1 (Figure 1b), enables estimation of the contribution to GA activity of their sugar-OH bonding and elucidates protein structural features implicated in α-1,4 versus α-1,6 bond-type specificity (Sierks *et al.*, 1990; Sierks and Svensson, 1992; Frandsen *et al.*, 1995, 1996).

Mutational Analysis of Residues Engaged in Protein-Carbohydrate Interactions at Subsites −1 and +1

Insight in the significance of individual protein and substrate groups for activity is crucial for protein engineering. In GA-inhibitor complexes NE and NH2 of Arg54 bind with OH-4' in ring a; OD1 and OD2 of Asp55 with OH-4' and -6'; and Arg305 NH1 and Glu180 OE2 with OH-3 and 2 in ring b, respectively (Figures 1b and 5; Aleshin *et al.*, 1994a, 1996; Stoffer *et al.*, 1995). Site-directed mutagenesis confirms that Arg54 and Asp55 are critical for activity (Table 2). Compared to wild-type, loss in transition-state stabilisation, $\Delta\Delta G^{\ddagger}$, of 16–21 kJ mol^{-1} (Figure 5) in maltooligosaccharide hydrolysis by Arg54→Leu/Lys GAs stemmed mainly from $1.2 - 1.8 \times 10^3$ fold reduction in k_{cat} (Table 2; Frandsen *et al.*, 1995). This magnitude of $\Delta\Delta G^{\ddagger}$ is compatible with elimination of a charged hydrogen bond in the transition-state complex (Fersht *et al.*, 1985). Mutation of Asp55 (Table 2) similarly resulted in $\Delta\Delta G^{\ddagger}$ of 15–22 kJ mol^{-1} (Figure 5), due to 5.5×10^3 and 2.5×10^5 fold decreases, respectively, of k_{cat} in Asp55→Gly (Sierks and Svensson, 1993) and Asp55→Val GAs

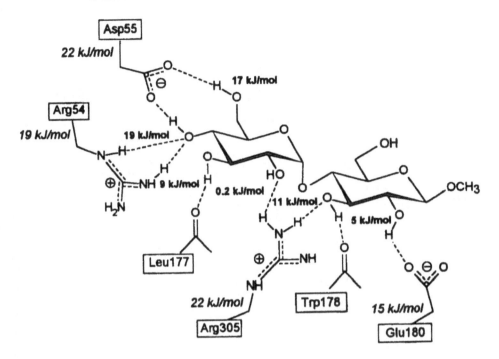

Figure 5 Schematic representation of protein-carbohydrate interactions showing the energetics of glucoamylase-maltose complexation as determined by mutagenesis ($\Delta\Delta G^{\ddagger}$ in italics) and from substrate analog ($\Delta\Delta G^{\ddagger}$ in bold) studies

(T.P. Frandsen, J. Lehmbeck, B.B. Stoffer and B. Svensson, unpublished). Even though Asp55 and Glu179 are 6–7 Å apart, Asp55→Gly GA that lacks a negatively charged side-chain, has lowered the pK_a assigned to Glu179 by approx. one unit. The electrostatic field at the active site of GA has not been thoroughly analyzed, but it is known from other glycosyl hydrolases that such forces can delicately control the pK_a of catalytic groups (Baptista *et al.*, 1995).

The thermodynamics of acarbose binding to Asp55→Val GA suggests that Asp55 has another important role in the mechanism. This purely enthalpy-driven association has a K_a of 10^5 M^{-1} (C.R. Berland, T.P. Frandsen, B.W. Sigurskjold and B. Svensson, unpublished), which is 10^9 times weaker than of the both enthalpy- and entropy-favored wild-type binding (Sigurskjold *et al.*, 1994). Asp55 adopts two conformations in the crystal structure of free GA at both pH 4 and 6, leading to different hydrogen bonding interactions with water (Aleshin *et al.*, 1994b). These authors suggest Asp55 participates in a mechanism in which bound water is expelled from the active site through a channel that leads to the surface of the protein, upon substrate or ligand binding (Aleshin *et al.*, 1994b). This is consistent with water transportation to bulk, that contributes a gain in entropy in wild-type GA, being lost in the Asp55→Val mutant.

Arg305 (Figure 1b,c) binds to ring a and b at OH-2′ and -3 (Aleshin *et al.*, 1994, 1996; Stoffer *et al.*, 1995) and plays a key role in discrimination of α-1,4 and α-1,6-linked substrates (Frandsen *et al.*, 1995). While Arg305→Lys GA lost 22–24 and 18 kJ mol^{-1} in transition-state stabilization of maltooligosaccharides and isomaltose, respectively, this mutation has very different effects on the kinetic parameters for hydrolysis of these substrates. Most remarkably, K_m is the same as of wild-type GA for isomaltose, but increases at least 400 fold for maltose (Table 2; Frandsen *et al.*, 1995). A critical interaction for α-1,4 linked ligands is between Arg305 and OH-3 at subsite +1, as found in the structure of GA-D-*gluco*-dihydroacarbose (Aleshin *et al.*, 1996). Deoxygenation of maltose OH-3 thus causes a loss in transition-state stabilization of 11 kJ mol^{-1} (Sierks *et al.*, 1992). A greater $\Delta\Delta G^{\ddagger}$ of 22 kJ mol^{-1} for maltose hydrolysis by the mutant Arg305→Lys GA (Figure 5) suggests, however, that loss of additional hydrogen bonding in the mutant reduces the complementarity of GA and ligand as reflected also by the lack of entropy-gain in acarbose binding (Berland *et al.*, 1995).

Glu180, adjacent to the catalytic Glu179 in the third conserved $\alpha\rightarrow\alpha$ loop, binds to OH-2 of ring b (Figures 1b and 5), and is important in substrate specificity (Sierks *et al.*, 1990). Glu180→Gln GA gives 34–75 and 3 fold increases in K_m for malto-oligosaccharides and isomaltose, respectively, and 2–7 fold decreases in k_{cat} (Table 1). In Glu180→Gln and Glu180→Asp GAs pK_a values, assigned to the catalytic Glu400 and Glu179, are 2.2 and 4.9 and 2.4 and 6.8, respectively, compared to 2.7 and 5.9 in the wild-type GA-substrate complex (Sierks *et al.*, 1990; Bakir *et al.*, 1993). In fact, Glu180→Gln has the higher activity at around pH 3 as compared to pH 4.5 for wild-type GA (Sierks *et al.*, 1990). Although this shows the potential to shift the pH optimum, extensive introduction of negative charges in the active site of GA failed to displace the profile towards neutral pH optimum (Bakir *et al.*, 1993), a property which would facilitate a single step industrial saccharification and isomerization (Bhosale *et al.*, 1996). Analysis of Glu180→Gln GA using a series of deoxygenated maltose and isomaltose analogs demonstrated transition-state stabilizing hydrogen bonds between Glu180 and OH-2 in maltose (Figure 5; Sierks and Svensson, 1992) and OH-4 and OH-3 in isomaltose (Frandsen *et al.*, 1996). The action of wild-type and Glu180→Gln GAs on conformationally biased isomaltose analogs showed that Glu180 via subsite +1 interaction induces a conformational change of the bound substrate that optimizes transition-state stabilization by charged hydrogen bonds to substrate OH-4' and -6' in subsite −1 (Frandsen *et al.*, 1996). Key polar group binding in one sugar ring thus has impact on utilization of key polar groups in another ring. This example demonstrates the complexity encountered in rational protein engineering and argues in favor of thorough analyses of the energetics of enzyme-ligand complexes.

Mutational Analysis of Residues at a Distance from the Site of Catalysis

Protein engineering of GA residues that either interact indirectly with substrate, i.e. via other residues, or belong to subsites at a distance from the catalytic site represents an excellent route to active GA variants with modified substrate specificity. Mutations at Tyr116, Ser119, Arg122, Tyr175, Gly183, Ser184, Tyr306, Asp309, and Trp317 (Figure 1c) are examples on modulation of i) α-1,4 versus α-1,6 bond-type specificity, ii) substrate size preference, iii) tendency to form condensation products, and iv) thermodynamics of ligand binding.

Asp309 and Trp317 provide electrostatic and hydrophobic stabilization of Arg305 (Figure 1b), that is crucial for recognition of α-1,4 linked substrates (Frandsen *et al.*, 1995). While Arg305 mutants have very low activity, also for isomaltose, the Asp309→Glu and Trp317→Phe GAs show for α-1,4 linked substrates essentially unaffected k_{cat} and increased K_m, but minor decreases in both affinity and activity for isomaltose (Tables 2 and 3; Frandsen *et al.*, 1995). The mutations affect protein-carbohydrate interaction at subsite +1 through Arg305, as demonstrated by $\Delta\Delta G^{\ddagger}$ values of 7 kJ mol^{-1} to OH-4 in hydrolysis of an isomaltoside analog catalyzed by Asp309→Glu compared to wild-type GA (Frandsen *et al.*, 1996). Pre-steady-state kinetic analysis of Trp317→Phe GA indicated a 10-fold increased K_1 and hence a weakened formation of the first association complex, ES, with α-1,4 linked substrates (Christensen, 1995).

The mutants Ser119→Tyr GA (Sierks and Svensson, 1994), at subsite +3 (Figure 1c; Stoffer *et al.*, 1995; Aleshin *et al.*, 1996), and Gly183→Lys and Ser184→His at a more distant α→α loop 3 location have 2.3–3.5 fold increased selectivity for α-1,4 over α-1,6 linked disaccharides as defined by $(k_{cat}/K_m)_{G2}/(k_{cat}/K_m)_{iG2}$ (Table 3). Because reduced formation of α-1,6 linked condensation products would importantly advance industrial saccharification, Ser119→Tyr GA was tested, but surprisingly found to enhance the formation of multiply branched oligosaccharides (Svensson *et al.*, 1995). Ser119→Tyr is the only mutant found so far that retains high affinity for acarbose with a favorable entropy contribution which is superior to that of wild-type GA (Berland *et al.*, 1995). Such improved complementarity in the mutant-oligosaccharide complex probably facilitates oligosaccharide accommodation at subsites +1 through +2, and more remote ones and thereby elevates condensation with glucose bound at subsite −1 (Svensson *et al.*, 1995).

For wild-type GA the specificity constant, k_{cat}/K_m, increases with the length of the oligosaccharide substrate (Sierks *et al.*, 1989; Fierobe *et al.*, 1996). Among ca. 50 GA mutants, a few selectively suppress hydrolysis of shorter substrates. Most conspicuously k_{cat} of Tyr116→Ala (Sierks and Svensson, 1993) and Arg122→Tyr GAs (Natarajan and Sierks, 1996) decreased 17 and 38 fold towards maltose as opposed to 2.5 and 10.5 fold with maltoheptaose (Table 3). Tyr306→Phe GA, in contrast, retains wild-type level activity towards maltose, but displays close to 2-fold increase in activity for maltoheptaose. Preference for short substrates resulted from replacement of four residues in the α→α loop 5 (see below; Fierobe *et al.*, 1996). This mutant, L5 GA, has k_{cat} of 36% of wild-type for maltose, but only 5–8% for maltotriose through maltoheptaose (Table 3; Fierobe *et al.*, 1996) and is so far the only GA mutant with higher k_{cat} for maltose than maltoheptaose.

In an attempt to confer *A. niger* GA the 1.8 fold higher k_{cat} reported for *Rhizopus oryzae* GA (Ohnishi *et al.*, 1990), the remarkable sequence variation in Trp170-Tyr175 of α→α loop 3 has been mimicked by singly mutating the *A. niger* to the corresponding *Rh. oryzae* residues (Stoffer *et al.*, 1997). Tyr175, at subsite +3 is closest to the catalytic site and has CE2 3.46 Å from C6 of ring d in GA-D-*gluco*-dihydroacarbose (Stoffer *et al.*, 1995; Aleshin *et al.*, 1996). Tyr175→Phe GA displays a modest 1.3 fold increase in k_{cat} towards maltoheptaose (Table 3; Stoffer *et al.*, 1997), and an unusual 10 fold higher affinity for acarbose, (Berland *et al.*, 1995), while 1-deoxynojirimycin bound at subsite −1 (Harris *et al.*, 1993), has 1.6 fold lower affinity. The more distant Trp170→Phe mutation displays 1.7 fold increased activity towards maltotetraose (Stoffer *et al.*, 1997). Thus, minor structural changes at a distance from the catalytic site can improve the GA activity.

Table 3 Enzymatic properties of mutants of selected residues at a distance from the site of catalysis in *Aspergillus* glucoamylase[a]

	Maltose			Isomaltose			Maltoheptaose			Characteristics[b]
	k_{cat} (s^{-1})	K_m (mM)	k_{cat}/K_m $(s^{-1}mM^{-1})$	k_{cat} (s^{-1})	K_m (mM)	k_{cat}/K_m $(s^{-1}mM^{-1})$	k_{cat} (s^{-1})	K_m (mM)	k_{cat}/K_m $(s^{-1}mM^{-1})$	
Wild-type[c]	10.7	1.21	8.84	0.41	19.8	0.021	59.7	0.12	498	–
Tyr116→Ala[d]	0.62	0.57	1.09	0.11	23.2	0.0047	23.6	0.29	81.4	↑G_7/G_2
Ser119→Tyr[e]	10.1	1.10	9.20	0.36	42.0	0.0086	57.0	0.16	360	↑α-1,4/α-1,6
Arg122→Tyr[f]	0.28	0.77	0.36	0.025	13.2	0.0019	5.7	0.83	6.9	↑G_7/G_2
Tyr175→Phe[g]	10.6	0.97	10.9	0.53	28.5	0.019	71.0	0.11	650	↑G_4
Gly183→Lys[g]	10.4	1.1	9.6	0.53	39.0	0.014	79.0	0.14	514	↑α-1,4/α-1,6; ↑G_7
Ser184→His[g]	9.8	0.9	10.9	0.29	27.0	0.011	66.0	0.14	567	↑α-1,4/α-1,6
Tyr306→Phe[h]	8.2	3.68	2.23	0.24	47.2	5×10^{-3}	99.0	0.67	148	↑G_7
L5[i]	4.1	6.4	0.64	0.11	67.4	0.0016	2.7	0.13	21.7	↓G_7
Asp309→Glu[j]	9.1	52.4	0.174	0.09	61.9	0.0015	67.1	9.44	7.11	↓α-1,4/α-1,6
Trp317→Phe[j]	6.0	14.0	0.43	0.17	153.0	0.0011	51.8	3.30	15.70	↓α-1,4/α-1,6

[a] Determined at 45°C in 0.05 M sodium acetate, pH 4.5; [b] typical alteration of substrate specificity by mutation; [c] Frandsen *et al.*, 1994; [d] Sierks & Svensson, 1996; [e] Sierks & Svensson, 1994; [f] Natarajan & Sierks, 1996; [g] Stoffer *et al.*, 1997, this mutant shows highest activity towards maltotetraose; [h] Sierks & Svensson, 1993; [i] Fierobe *et al.*, 1996; [j] Frandsen *et al.*, 1995.

The mutated side-chains are highlighted in Figure 1c

	Loop 3	Loop 5
	181	307
Aspergillus niger	DLWEEVNGSSFFT-/-AVGRYP	EDTYYNG
Hormoconis resinae	DLWEETYASSFFT-/-AVGRYA	EDVYMGG
Aspergillus oryzae	DLWEEVQGTSFYT-/-AVGRYP	EDSYYNG
Aspergillus terreus	DLWEEVNGSSFFT-/-AVGRYP	EDSYYNG
Humicola grisea	DLWEEVPGSSFFT-/-AVGRYS	EDVYYNG
Saccharomycop. fibuligera	DLWEENQGRHFFT-/-AIGRYP	EDVY-NG
Rhizopus oryzae	DLWEEVNGVHFYT-/-SIGRYP	EDTY-NG
Saccharomyces cerevisiae	DLWEEVNGMHFFT-/-ALGRYP	EDVY-DG

Figure 6   Sequence alignment of fungal GAs in $\alpha \rightarrow \alpha$ loops 3 and 5, modified from Fierobe *et al.*, (1996)

HOMOLOGUE LOOP REPLACEMENT MODULATES THE SUBSTRATE BOND-TYPE SPECIFICITY

Mutational analysis, modelling, and molecular recognition of substrate analogs clearly indicate that certain residues in the conserved $\alpha \rightarrow \alpha$ loops 3 and 5 are important for bond-type specificity (Sierks *et al.*, 1990; Svensson *et al.*, 1994, 1995; Frandsen *et al.*, 1995, 1996). Because mutation at substrate binding residues dramatically reduces the activity, a strategy based on sequence comparison was chosen to design GA with engineered bond-type specificity. Fungal GAs generally have 400–1000 fold lower specificity toward the α-1,6 compared to α-1,4 linked substrates (Meagher *et al.*, 1989; Sierks *et al.*, 1989; Fagerström and Kalkkinen, 1995; Fierobe *et al.*, 1997). However, for GA P from *Hormoconis resinae* these specificities differ only 50 times (Fagerström, 1991). GA P shows remarkable sequence variation in the otherwise conserved $\alpha \rightarrow \alpha$ loops 2 (not shown), 3, and 5 (Figure 6), and putative structural mimics have been created by replacing three and four residues, respectively, in *A. niger* GA loops 3 and 5 with the GA P counterparts (Figure 1c; Figure 6; Fierobe *et al.*, 1996). In addition a tetrapeptide in loop 3 has been changed to a functional motif from α-amylases which are strictly α-1,4 specific (C. Dupont, T.P. Frandsen, J. Lehmbeck and B. Svensson, unpublished).

The replacements in either $\alpha \rightarrow \alpha$ region 3 or 5 affects hydrolysis of α-1,4 more than α-1,6-linkages resulting in 3–11 fold reduced relative specificity, $(k_{cat}/K_m)_{G2}/(k_{cat}/K_m)_{iG2}$ (Table 4; Fierobe *et al.*, 1996; C. Dupont, T.P. Frandsen, J. Lehmbeck and B. Svensson, unpublished). For Val181→Ala/Asn182→Ala/Gly183→Lys/Ser184→His (L3A) GA the effect originates exclusively from reduced k_{cat}/K_m for maltose hydrolysis. In the case of Val181→Thr/Asn182→Met/Gly183→Ala GA (L3) (Table 3; Fierobe *et al.*, 1996), hydrolysis of α-1,6 linkages is most affected and 2–3 fold increase in relative specificity is obtained. Finally Pro307→Ala/Thr310→Val/Tyr312→Met/Asn313→Gly GA (L5) has drastically reduced activity towards both substrates and retains the relative specificity. Very surprisingly, when both loops 3 and 5 are replaced, in L3L5 GA, wild-type activity is restored (Table 4) and a relative specificity only 3 times higher than of GA P is obtained for maltotriose over isomaltotriose (Table 4). Apparently these two active site segments cooperate. In the

Table 4 Engineering substrate bond-type specificity by homologue binding loop replacement in *Aspergillus* glucoamylase[a]

	Maltose			Isomaltose			Relative specificity
	k_{cat} (s^{-1})	K_m (mM)	k_{cat}/K_m (s^{-1}mM^{-1})	k_{cat} (s^{-1})	K_m (mM)	k_{cat}/K_m (s^{-1}mM^{-1})	$(k_{cat}/K_m)_{G2}/(k_{cat}/K_m)_{iG2}$
Wild-type[b]	10.7	1.21	8.84	0.41	19.8	0.021	421 (27)[e]
L3[c]	5.65	3.58	1.6	0.23	22.9	0.010	160
L3A[d]	1.68	1.83	0.92	0.63	24.8	0.025	37
L5[c]	4.10	6.40	0.64	0.11	67.4	0.0016	400
L3L5[c]	5.69	1.61	3.5	0.59	27.6	0.027	130

	Maltotriose			Isomaltotriose			Relative specificity
	k_{cat} (s^{-1})	K_m (mM)	k_{cat}/K_m (s^{-1}mM^{-1})	k_{cat} (s^{-1})	K_m (mM)	k_{cat}/K_m (s^{-1}mM^{-1})	$(k_{cat}/K_m)_{G2}/(k_{cat}/K_m)_{iG2}$
Wild-type b	30.9	0.25	124	1.29	11.5	0.112	1107 (109)[e]
L3[c]	21.4	0.44	48	0.35	20.9	0.017	2823
L5[c]	1.7	0.33	5.2	0.48	108	0.0045	1156
L3L5[c]	20.6	0.26	78	1.16	5.4	0.229	341

[a] Determined at 45°C in 0.05 M sodium acetate, pH 4.5; [b] Frandsen *et al.*, 1994; [c] Fierobe *et al.*, 1996; [d] C. Dupont, T.P., Frandsen, J. Lehmbeck and B. Svensson, unpublished; [e] *H. resinae* GA P (Fagerström, 1991)

The loops 3 and 5 are highlighted in Figure 1c.

structure Arg305 NH2 (loop 5) and Glu180 OE2 (loop 3) are indeed hydrogen-bonded. The concept of homologue loop engineering coupled with site-directed mutagenesis in GA deserves further attention.

PROTEIN ENGINEERING TO ALTER THE STABILITY OF GLUCOAMYLASE

Four different engineering strategies are reported to enhance the stability of GA; i) replacement of glycines in α-helices (Chen *et al.*, 1996); ii) elimination of fragile Asp-X bonds (Chen *et al.*, 1995); iii) substitution of asparagine in Asn-Gly sequences (Chen *et al.*, 1994); and iv) introduction of disulfide bonds (Fierobe *et al.*, 1996).

The high conformational flexibility of glycine makes it an α-helix breaker that can lead to destabilization. GA contains five glycines in helices; Gly57 (helix 2), Gly137 and Gly139 (helix 4), Gly251 (helix 8), and Glu383 (helix 12). Only Gly137→Ala and Gly139→Ala GAs, however, show improved stability, amounting to a modest increase in the free energy of inactivation of 2 kJ mol^{-1} (Table 5; Chen *et al.*, 1996). This effect is not additive as seen for Gly137→Ala/Gly139→Ala GA; moreover, other Gly→Ala mutants have lower stability than wild-type (Table 5).

The three Asp126-Gly127, Asp257-Pro258, and Asp293-Gly294 bonds in *A. awamori* GA are cleaved under mild acidic conditions and heating (Chen *et al.*, 1995). While the Asp257→Glu and Asp293→Glu/Gln mutants display close to wild-type activity

Table 5 Thermostability (T_m) and free energy ($\Delta G^{\ddagger}$) of thermoinactivation of selected glucoamylase mutants

Enzyme	T_m (°C)	$\Delta G^{\ddagger}$ (kJ/mol)	Reference
Wild-type[a]	69.0	105	Fierobe et al., 1996
Wild-type[b]	71.0	–	Frandsen et al., 1995
Ala39→Val[a]	–	97	Flory et al., 1994
Gly57→Ala[a]	–	105	Chen et al., 1996
Asp112→Tyr[b]	66.0	–	Frandsen, 1993
Asp126→Glu[a]	–	101	Chen et al., 1995
Gly127→Ala[a]	–	102	Chen et al., 1995
Pro128→Ser[a]	–	105	Flory et al., 1994
Gly137→Ala[a]	–	107	Chen et al., 1996
Gly139→Ala[a]	–	107	Chen et al., 1996
Gly137→Ala/Gly139→Ala[a]	–	107	Chen et al., 1996
Asn171→Ser[b]	71.0	–	Stoffer et al., 1997
Asn182→Asp[a]	–	106	Chen et al., 1994
Ala246→Cys[a]	73.0	–	Fierobe et al., 1996
Gly251→Ala[a]	–	105	Chen et al., 1996
Asp257→Glu[a]	–	107	Chen et al., 1995
Asp293→Glu[a]	–	106	Chen et al., 1995
Asp293→Gln[a]	–	106	Chen et al., 1995
Ala302→Val[a]	–	94	Flory et al., 1994
Asp309→Glu[b]	58.0	–	Frandsen et al., 1995
Trp317→Phe[b]	64.0	–	Frandsen et al., 1995
Cys320→Ala[a]	70.5	–	Fierobe et al., 1996
Gly383→Ala[a]	–	103	Chen et al., 1996
Glu389→Asp[b]	71.0	–	Frandsen, 1993
Glu389→Asn[b]	67.0	–	Frandsen, 1993
His391→Trp[b]	68.0	–	Frandsen, 1993
Ala392→Asp[b]	61.0	–	Frandsen, 1993
Asn395→Gln[b]	–	95	Chen et al., 1994
Gly396→Ser[a]	–	93	Flory et al., 1994
Gly407→Asp[a]	–	98	Flory et al., 1994
Leu410→Phe[a]	–	98	Flory et al., 1994
Ala442→Thr[a]	–	104	Flory et al., 1994

Recombinant GA variants produced in [a] *Saccharomyces cerevisiae* (Innis et al., 1985; Sierks et al., 1989) or [b] *Aspergillus niger* (Christensen et al., 1988; Frandsen et al., 1994).

T_m is defined as the temperature at which 50% activity is retained after 5 min incubation

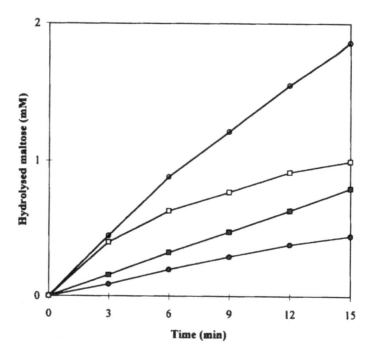

Figure 7 Thermoperformance of Ala246→Cys (●,○) and wild-type (■,) GA at 45°C (●,■) and 66°C (○,■), modified from Fierobe *et al.*, (1996)

and a small increase in free energy ($\Delta G^{\ddagger}$) of thermoinactivation by 1–2 kJ mol^{-1}, Asp126→Glu and Gly127→Ala GAs both have reduced activity and thermostability (Table 5).

Another source of protein destabilization is deamidation of acidic side chains which occurs especially at Asn-Gly sequences. Two such sequences are present in GA and while Asn182→Asp GA has a slight 1 kJ mol^{-1} higher free energy of inactivation than wild-type, Asn395→Gln GA, in which an N-linked sugar unit was eliminated, is less thermostable, less resistant to subtilisin, and is secreted from *S. cerevisiae* in reduced amounts (Chen *et al.*, 1994). In contrast, deletion of the N-linked carbohydrate in Asn171→Ser GA has no adverse effect on enzyme activity, thermostability, or protease-sensitivity (Stoffer *et al.*, 1997).

Extra disulfide bonds have increased the thermostability of a number of proteins. An impressive raise in T_m of 23.4°C thus occurs for T_4 lysozyme by introduction of three disulfide bridges (Matsumura *et al.*, 1989). While this approach has been developed for disulfide-free proteins and domains below 25 kDa, GA is 82 kDa and has two domains and four disulfide bridges. The free thiol group at Cys320 is located at the N-terminus of helix 10 of the catalytic domain and close to Ala246, a variable residue at the N-terminus of helix 8. Ala246→Cys GA spontaneously forms the disulfide bridge in 80% yield and the mutation is accompanied by an increase in T_m of 4°C and by improved enzymatic performance at elevated temperatures (Fierobe *et al.*, 1996). Thus at 66°C the initial rate of maltose hydrolysis by wild-type and Ala246→Cys GAs is the same, but whereas the wild-type activity declines rapidly, Ala246→Cys remains fully active for a longer period (Figure 7; Fierobe *et al.*, 1996).

This mutant therefore has an interesting industrial potential. At the standard assay temperature (45°C) Ala246→Cys has only 50% activity of wild-type GA, the disulfide bond possibly impaired the dynamics of the active site.

Finally, GA with low thermostability is of interest in food and beverage industries. Thermosensitive mutants of GA may thus be used in production of light beer, to secure their complete inactivation during pasteurization. Random mutagenesis coupled with screening identified seven thermosensitive mutants. Of these, Ala302→Val and Gly396→Ser, which involves alteration of hydrophobic subdomains, leads to dramatic decreases in free energy of thermoinactivation of 11–12 kJ mol^{-1} (Table 5; Flory *et al.*, 1994). In addition, two replacements at helix 12 (Frandsen, 1993), His391→Trp and Glu389→Gln GAs, have reduced T_m of 3–4°C; while the charged-conserved Glu389→Asp GA retains the wild-type thermostability (Table 5). The large ΔT_m of 10°C for Ala392→Asp (Table 5) is perhaps elicited by the negative charge introduced in a small antiparallel β-sheet formed by Ala392, Ala393, Gly396, and Ser397 (Frandsen, 1993).

CONCLUSION AND PROSPECTS

Major improvement of industrial properties using protein engineering has not yet been reported in the small glycoside hydrolase family 15 that comprises only glucoamylases. However, available GA mutants have provided a large knowledge base on the roles of individual residues and their interplay in the extended substrate binding site including information on the energetics and structure/function relations in the mechanism of action. In particular, deeper understanding of complementarity among reactants and enzyme that takes into account the role of water is anticipated to be useful in design of glucoamylase with suppressed condensation activity, which in conjunction with elevated thermostability may advance industrial saccharification, including the use of immobilized glucoamylase.

ACKNOWLEDGEMENTS

We are grateful to Dr. Anthony J. Clarke for carefully reading the manuscript and to Dr. Niels K. Thomsen for help in the preparation of figure 1. Henri-Pierre Fierobe received a Human Capital and Mobility fellowship (ERBCHBICT941224, E.U.) in 1995–1996. This work was supported in part by the Danish Research Council; Committee on Biotechnology, grant no 9502114.

REFERENCES

Aleshin, A., Golubev, A., Firsov, L.M. and Honzatko, R.B. (1992) Crystal structure of glucoamylase from *Aspergillus awamori* var. *X100* to 2.2-Å resolution. *J. Biol. Chem.*, **267**, 19291–19298.

Aleshin, A.E., Firsov, L.M. and Honzatko, R.B. (1994a) Refined structure for the complex of acarbose with glucoamylase from *Aspergillus awamori* var. *X100* to 2.4 Å resolution. *J. Biol. Chem.*, **269**, 15631–15639.

Aleshin, A.E., Hoffman, C., Firsov, L.M. and Honzatko, R.B. (1994b) Refined crystal structures of glucoamylase from *Aspergillus awamori* var. *X100*. *J. Mol. Biol.*, **238**, 575–591.

Aleshin, A.E., Stoffer, B.B., Firsov, L.M., Svensson, B. and Honzatko, R.B. (1996) Crystallographic complexes of glucoamylase with maltooligosaccharide analogs: relationship of stereochemical distortions at the nonreducing end to the catalytic mechanism. *Biochemistry*, **35**, 8319–8328.

Bakir, U., Coutinho, P.M., Sullivan, P.A., Ford, C. and Reilly, P.J. (1993) Cassette mutagenesis of *Aspergillus awamori* glucoamylase near its general acid residue to probe its catalytic and pH properties. *Prot. Engng.*, **6**, 939–946.

Baptista, A., Brautaset, T., Drabløs, F., Martel, P., Valla, S. and Pedersen, S.B. (1995) Electrostatic studies of carbohydrate active enzymes. In *Carbohydrate Bioengineering*, edited by S.B. Petersen, B. Svensson and S. Pedersen, pp. 181–204, Progress in Biotechnology 10. Amsterdam, Elsevier.

Berland, C.R., Sigurskjold, B.W., Stoffer, B., Frandsen, T.P. and Svensson, B. (1995) Thermodynamics of inhibitor binding to mutant forms of glucoamylase from *Aspergillus niger* determined by isothermal titration calorimetry. *Biochemistry*, **34**, 10153–10161.

Bhosale, S.H., Rao, M.B. and Deshipande, V.V. (1996) Molecular and industrial aspects of glucose isomerase. *Microbiological reviews*, **60**, 280–300.

Bock, K. and Pedersen, H. (1987) The substrate specificity of the enzyme amyloglucosidase (AMG). Part I. Deoxy derivatives. *Acta. Chem. Scand. Ser. B*, **41**, 617–628.

Burmeister, W.P., Henrissat, B., Bosso, C., Cusack, S., and Ruigrok, R.W.H. (1993) Influenza B virus neuraminidase can synthesize its own inhibitor. *Structure*, **1**, 19–26.

Chen, H.-M., Ford, C. and Reilly, P.J. (1994) Substitution of asparagine residues in *Aspergillus awamori* glucoamylase by site-directed mutagenesis to eliminate N-glycosylation and inactivation by deamidation. *Biochem. J.*, **301**, 275–281.

Chen, H.-M., Ford, C. and Reilly, P.J. (1995) Identification and elimination by site-directed mutagenesis of thermolabile aspartyl bonds in *Aspergillus awamori* glucoamylase. *Prot. Engng.*, **8**, 575–582.

Chen, H.-M., Li, Y., Panda, T., Buehler, F.U., Ford, C. and Reilly, P.J. (1996) Effect of replacing helical glycine residues with alanines on reversible and irreversible stability and production of *Aspergillus awamori* glucoamylase. *Prot. Engng.*, **9**, 499–505.

Christensen, T. (1995) Presteady-state and steady-state kinetic characterization of Trp52→Phe, Trp317→Phe, and Trp417→Phe glucoamylase mutants from *Aspergillus niger*. M. Sc. Thesis, University of Copenhagen.

Christensen, T., Wöeldike, H., Boel, E., Mortensen, S. B., Hjortshoej, K., Thim, L. and Hansen, M.T. (1988) High level expression of recombinant genes in *Aspergillus oryzae*. *Bio/Technology*, **6**, 1419–1422.

Christensen, U., Olsen, K., Stoffer, B.B. and Svensson, B. (1996) Substrate binding mechanism of Glu180→Gln, Asp176→Asn, and wild-type glucoamylases from *Aspergillus niger*. *Biochemistry*, **35**, 15009–15018.

Crennell, S.J., Garman, E.F., Laver, W.G., Vimr, E.R. and Taylor, G.L. (1993) Crystal structure of a bacterial sialidase (from *Salmonella typhimurium* LT2) shows the same fold as an influenza virus neuraminidase. *Proc. Natl. Acad. Sci.*, **90**, 9852–9856.

Davies, G. and Henrissat, B. (1995) Structure and mechanisms of glycosyl hydrolases. *Structure*, **3**, 853–859.

Fagerström, R. (1991) Subsite mapping of *Hormoconis resinae* glucoamylases and their inhibition by gluconolactone. *J. Gen. Microbiol.*, **137**, 1001–1008.

Fagerström, R. and Kalkkinen, N. (1995) Characterization, subsite mapping and partial amino acid sequence of glucoamylase from the filamentous fungus *Trichoderma reesei*. *Biotechnol. Appl. Biochem.*, **21**, 223–231.

Fersht, A.R., Shi, J.-P., Knill-Jones, J., Lowe, D.M., Wilkinson, A.J., Blow, D.M., Brick, P., Carter, P., Waye, M.M.Y. and Winter, G. (1985) Hydrogen bonding and biological specificity analyzed by protein engineering. *Nature*, **314**, 235–238.

Flory, N., Gorman, M., Coutinho, P.M., Ford, C. and Reilly, P.J. (1994) Thermosensitive mutants of *Aspergillus awamori* glucoamylase by random mutagenesis: inactivation kinetics and structural interpretation. *Prot. Engng.*, **7**, 1005–1012.

Fierobe, H.-P., Stoffer, B.B., Frandsen, T.P. and Svensson, B. (1996) Mutational modulation of substrate bond-type specificity and thermostability of glucoamylase from *Aspergillus awamori* by replacement with short homologue active site sequences and thiol/disulfide engineering. *Biochemistry*, **35**, 8698–8704.

Fierobe, H.-P., Mirgorodskaya, E., Frandsen, T.P., Roepstorff, P. and Svensson, B. (1997) Overexpression and characterization of *Aspergillus awamori* wild-type and mutant glucoamylase secreted by the methylotrophic yeast *Pichia pastoris*: comparison with wild-type recombinant glucoamylase produced using *Saccharomyces cerevisiae* and *Aspergillus niger* as hosts. *Protein Expression Purif.*, **9**, 159–170.

Frandsen, T.P. (1993) Structure/function studies of glucoamylase from *A. niger*. M. Sc. Thesis, University of Copenhagen.

Frandsen, T.P., Dupont, C., Lehmbeck, J., Stoffer, B., Sierks, M.R., Honzatko, R.B. and Svensson, B. (1994) Site-directed mutagenesis of the catalytic base glutamic acid 400 in glucoamylase from *Aspergillus niger* and of tyrosine 48 and glutamine 401, both hydrogen-bonded to the γ-carboxylate group of glutamic acid 400. *Biochemistry*, **33**, 13808–13816.

Frandsen, T.P., Christensen, T., Stoffer, B., Lehmbeck, J., Dupont, C., Honzatko, R.B. and Svensson, B. (1995) Mutational analysis of the roles in catalysis and substrate recognition of arginines 54 and 305, aspartic acid 309, and tryptophan 317 located at subsites 1 and 2 in glucoamylase from *Aspergillus niger*. *Biochemistry*, **34**, 10162–10169.

Frandsen, T.P., Stoffer, B.B., Palcic, M.M., Hof, S. and Svensson, B. (1996) Structure and energetics of the glucoamylase-isomaltose transition-state complex probed by using modeling and deoxygenated substrates coupled with site-directed mutagenesis. *J. Mol. Biol.*, **263**, 79–89.

Gebler, J.C., Trimbur, D.E., Warren, A.J., Aebersold, R., Namchuk, M. and Withers, S.G. (1995) Substrate-induced inactivation of a crippled β-glucosidase mutant: identification of the labeled amino acid and mutagenic analysis of its role. *Biochemistry*, **34**, 14547–14553.

Harris, E.M.S., Aleshin, A.E., Firsov, L.M. and Honzatko, R.B. (1993) Refined structure for the complex of 1-deoxynojirimycin with glucoamylase from *Aspergillus awamori* var. *X100* to 2.4-Å resolution. *Biochemistry*, **32**, 1618–1626.

Henrissat, B. (1991) A classification of glycosyl hydrolases based on amino acid sequence similarity. *Biochem. J,*. **280**, 309–316.

Henrissat, B. and Bairoch, A. (1993) New families in the classification of glycosyl hydrolases based on amino acid similarities. *Biochem J,*. **293**, 781–788.

Hiromi, K., Hamauzu, Z.-I., Takahashi, K. and Ono, S. (1966) Kinetic studies on gluc-amylase. II. Competition between two types of substrate having α-1,4 and α-1,6 glucosidic linkage. *J. Biochem.*, **59**, 411–418.

Hiromi, K., Ohnishi, M. and Tanaka, A. (1983) Subsite structure and ligand binding mechanism of glucoamylase. *Mol. Cell. Biol.*, **51**, 79–95

Innis, M.A., Holland, M.H., McCabe, P.C., Cole, G.E., Wittman, V.P., Tal, R., Watt, K.W.K., Gelfand, D.H., Holland, J.P. and Meade, J.H. (1985) Expression, glycosylation, and secretion of an *Aspergillus* glucoamylase by *Saccharomyces cerevisiae*. *Science*, **228**, 21–26.

Lemieux, R.U., Spohr, U., Bach, M., Cameron, D.R., Frandsen, T.P., Stoffer, B.B., Svensson, B. and Palcic, M.M. (1996) Chemical mapping of the active site of glucoamylase of *Aspergillus niger*. *Can. J. Chem.*, **74**, 319–335.

Matsumura, M., Signor, G. and Matthews, B.W. (1989) Substantial increase of protein stability by multiple disulphide bonds. *Nature*, **342**, 291–293.

McCarter, J.D. and Withers, S. (1994) Mechanisms of enzymatic glucoside hydrolysis. *Curr. Opin. Struc. Biol.*, **4**, 885–892.

Natarajan, S. and Sierks, M.R. (1996) Functional and structural roles of the highly conserved Trp120 loop region of glucoamylase from *Aspergillus awamori*. *Biochemistry*, **35**, 3050–3058.

Nikolov, Z.L., Meagher, M.M. and Reilly, P.J. (1989) Kinetics, equilibria, and modeling of the formation of oligosaccharides from D-glucose with *Aspergillus niger* glucoamylases I and II. *Biotech. Bioeng.*, **34**, 694–704.

Ohnishi, M., Matsumoto, T., Yamanaka, T. and Hiromi, K. (1990) Binding of isomaltose and maltose to the glucoamylase from *Asperillus niger*, as studied by fluorescence spectrophotometry and steady-state kinetics. *Carbohy. Res.*, **204**, 187–196.

Olsen, K., Christensen, U., Sierks, M.R. and Svensson, B. (1993) Reaction mechanism of Trp120→Phe and wild-type glucoamylases from *Aspergillus niger*. Interactions with acarbose and maltooligodextrins. *Biochemistry*, **32**, 9686–9693.

Olsen, K., Svensson, B. and Christensen, U. (1992) Stopped-flow fluorescence and steady-state kinetic studies of ligand-binding reactions of glucoamylase from *Aspergillus niger*. *Eur. J. Biochem.*, **209**, 777–784.

Saha, B.C. and Zeikus, J.G. (1989) Microbial glucoamylases: Biochemical and biotechnological features. *Starch/Stärke*, **41**, 57–64.

Sierks, M.R. and Svensson, B. (1992) Kinetic identification of a hydrogen bonding pair in the glucoamylase/maltose transition state complex. *Prot. Engng.*, **5**, 185–188.

Sierks, M.R. and Svensson, B. (1993) Functional roles of the invariant aspartic acid 55, tyrosine 306, and aspartic acid 309 in glucoamylase from *Aspergillus awamori* studied by mutagenesis. *Biochemistry*, **32**, 1113–1117.

Sierks, M.R. and Svensson, B. (1994) Protein engineering of the relative specificity of glucoamylase from *Aspergillus awamori* based on sequence similarities between starch-degrading enzymes. *Prot. Engng.*, **7**, 1479–1484.

Sierks, M.R. and Svensson, B. (1996) Catalytic mechanism of glucoamylase probed by mutagenesis in conjunction with hydrolysis of α-D-glucopyranosyl fluoride and maltooligosaccharides. *Biochemistry*, **35**, 1865–1871.

Sierks, M.R., Bock, K., Refn, S. and Svensson, B. (1992) Active site similarities of glucose dehydrogenase, glucose oxidase and glucoamylase probed by deoxygenated substrates. *Biochemistry*, **31**, 8972–8977.

Sierks, M.R., Ford, C., Reilly, P.J., and Svensson, B. (1989) Site-directed mutagenesis at the active site Trp 120 of *Aspergillus awamori* glucoamylase. *Prot. Engng.*, **2**, 621–625.

Sierks, M.R., Ford, C., Reilly, P.J. and Svensson, B. (1990) Catalytic mechanism of fungal glucoamylase as defined by mutagenesis of Asp176, Glu179, and Glu180 in the G1 enzyme from *Aspergillus awamori*. *Prot. Engng.*, **3**, 193–198.

Sigurskjold, B.W., Berland, C.R. and Svensson, B. (1994) Thermodynamics of inhibitor binding to the catalytic site of glucoamylase from *Aspergillus niger* determined by displacement titration calorimetry. *Biochemistry*, **33**, 10191–10199.

Sorimachi, K., Jacks, A.J., Le Gal-Coeffet, M.F., Williamson, G., Archer, D.B. and Williamson, M.P. (1996) Solution structure of the granular starch binding domain of glucoamylase from *Aspergillus niger* by nuclear magnetic resonance spectroscopy. *J. Mol. Biol.*, **259**, 970–987.

Stoffer, B.B., Dupont, C., Frandsen, T.P., Lehmbeck, J. and Svensson, B. (1997) Glucoamylase mutants in the conserved active-site segment Trp170-Tyr175 located at a distance from the site of catalysis. *Prot. Engng.*, **10**, 81–87.

Stoffer, B., Aleshin, A.E., Firsov, L.M., Svensson, B. and Honzatko, R.B. (1995) Refined structure for the complex of D-*gluco*-dihydroacarbose with glucoamylase from *Aspergillus awamori* var. *X100* to 2.2 Å resolution: dual conformations for extended inhibitors bound to the active site of glucoamylase. *FEBS Lett.*, **358**, 57–61.

Svensson, B., Larsen, K., Svendsen, I. and Boel, E. (1983) The complete amino acid sequence of the glycoprotein, glucoamylase G1, from *Aspergillus niger*. *Carlsberg Res. Commun.*, **48**, 529–544.

Svensson, B., Larsen, K. and Gunnarson, A. (1986) Characterization of a glucoamylase G2 from *Aspergillus niger*. *Eur. J. Biochem.*, **154**, 497–502.

Svensson, B., Stoffer, B., Frandsen, T.P., Søgaard, M., Sierks, M.R., Rodenburg, K.W., Sigurskjold, B.W. and Dupont, C. (1994) Basic molecular features, mechanism and specificity of protein-carbohydrate interactions in amylolytic enzymes. In *Proceedings of 36th Alfred Benzon Symposium*, edited by K. Bock and H. Clausen, pp. 202–213. Copenhagen: Munksgaard.

Svensson, B., Frandsen, T.P., Matsui, I., Juge, N., Fierobe, H.-P., Stoffer, B. and Rodenburg, K.W. (1995) Mutational analysis of catalytic mechanism and specificity in amylolytic enzymes. In *Carbohydrate Bioengineering*, edited by S.B. Petersen, B. Svensson and S. Pedersen, pp. 125–145, Progress in Biotechnology 10. Amsterdam: Elsevier.

12.   PECTIN MODIFYING ENZYMES AND DESIGNER PECTINS

OLGA MAYANS, JOHN JENKINS, DRUMMOND SMITH,
KATHRYN WORBOYS and RICHARD PICKERSGILL

*Department of Food Macromolecular Science, Institute of Food Research, Earley Gate,
Whiteknights Road, Reading RG6 6BZ*

OVERVIEW

Pectins are chemically heterogeneous plant cell wall polysaccharides localised in particular cell wall domains. They are found in the middle lamella, the region of interface between two walls from neighbouring cells, and in the primary plant cell wall as an independent but coextensive network associated with cellulose and xyloglucan. Pectin consists of a linear homogalacturonan backbone, the so-called 'smooth region', interrupted by rhamnogalacturonan which is branched and constitutes the 'hairy region'. These pectic polysaccharides can be found as a wide range of acidic polymers, both in composition and in distribution of the constituent blocks. Homogalacturonans, mostly homogeneous in composition, are formed by $\alpha(1\text{-}4)$ linked D-galacturonic acid residues in varying degrees of methylation, while rhamnogalacturonans are rich in rhamnose and commonly include several neutral saccharides such as arabinans, galactans and arabinogalactans as side chains. The industrial processing of pectic polysaccharides has been traditionally accomplished by chemical means, they are extracted, purified and modified using inorganic acids and bases, however the increasing availability of specific pectin modifying enzymes is already enabling a more effective and precise treatment of these substrates. Enzymatic extraction and modification may enhance the traditional uses of pectins as gelling and stabilising agents and extend their range of applications in both foods and pharmaceuticals. An excellent compendium of the advances in this area has recently been published, edited by Visser and Voragen, 'Pectins and Pectinases' compiles the scientific contributions to the meeting organised by Wageningen Agricultural University in 1995. The focus of the current paper is on the more recent structural studies of enzymes and the implications for molecular evolution and substrate binding.

PECTINS

Pectins and the Plant Cell Wall

The composition of pectin, the structure and, by implication, its role vary among the different plant tissues and even within the different domains of the cell wall. This diversity is also dependent on the stage of development of the tissue. Pectin has been implicated in determining the porosity and development of the cell wall.

Loosening of the pectin network concomitant with wall expansion may involve changes in pectin gel rheology which may influence metabolism of the cellulose/xyloglucan network. This fine control of the local cell wall properties is achieved by: pectin methylesterification, calcium ion concentration and pH-variation (Steele, McCann and Roberts, 1997). In some species, the middle lamella and the cell corners, are rich in relatively unesterified pectins which may function in cell-cell adhesion and play an important role in tissue integrity. Cell corners may act as joists in the scaffolding function of the wall and as pathogens tend to attack these areas, the complex mixture of pectins found there may have a role in signalling defence mechanisms operating through the release of small pectic fragments (McCann and Roberts, 1996).

Pectic Polysaccharides

Pectins are a complex class of plant cell wall polysaccharides that comprise a family of acidic polymers such as homogalacturonans and rhamnogalacturonans with several neutral polymers attached, including arabinans, galactans and arabino-galactans (Schols and Voragen, 1996). Pectin consists of a backbone, in which the 'smooth regions' formed by $\alpha(1\text{-}4)$ linked D-galacturonan are interrupted by ramified rhamnogalacturonan regions which are highly substituted by side chains rich in neutral sugars. The properties of the homogalacturonan polymers are profoundly influenced by the degree of methylation or acetylation of the polymer and by the distribution of the esterification which is usually considered to be either random or block. High methoxy pectins have a degree of methylation above 50%. The presence of xylogalacturonans with the terminal xylose covalently linked to the galacturonsyl backbone also occurs in the 'smooth region' (Schols and Voragen, 1996).

The major subgroups of the rhamnogalacturonan region of pectin are rhamno-galacturonan-I and -II (RG-I and RG-II). The RG-I polymer consists of alternating $\alpha(1\text{-}4)$ linked rhamnose and D-galacturonic acid residues with arabinofuranosyl, galactopyranosyl and minor quantities of fucopyranosyl residues. RG-II is charac-terised by the presence of rare sugars such as 2-O-methyl-fucose, apiose and 3-deoxy-D-lyxo-2-heptusaric acid next to the more common sugar residues rhamnose, fucose, arabinose, galactose, galacturonic acid and glucuronic acid. See Rinaudo (1996) and Axelos *et al.* (1996) for reviews on pectin functionality.

Chemical Stability of Homogalacturonans

In alkaline conditions, homogalalacturonans are degraded by two competing reactions: β-elimination which cleaves the homogalacturonan backbone and creates a double bond next to a methoxylated galacturonic moiety and demethylation by saponification. Both β-elimination and demethylation rates increase at higher pH with the β-elimination rate reducing before demethylation because of the topological requirements for β-elimination. The pH-stability of pectins must be considered in all enzymatic modification steps and the pH-activity profile of enzymes matched to avoid unwanted chemical modification.

(a) Action of pectate lyase and polygalacturonase on demethylated pectin

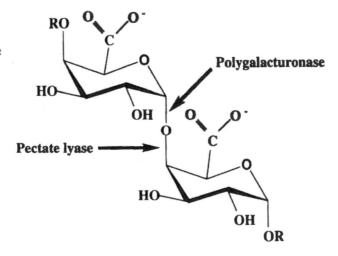

(b) Action of pecin lyase and pectin methylesterase on methylated pectin

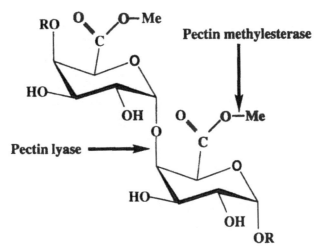

Figure 1 A schematic representation of (a) demethylated pectin and (b) methylated pectin showing the bonds cleaved by the pectin modifying enzymes: polygalacturonase, pectate lyase, pectin methylesterase and pectin lyase.

PECTIN MODIFYING ENZYMES

Introduction to Pectin Modifying Enzymes

Microorganisms produce an array of enzymes that can modify both the homogalacturonan backbone and the branched rhamnogalacturonan regions of the pectin molecule. The enzymes involved in homogalacturonan degradation include polygalacturonase, pectin lyases, pectate lyase and pectin methylesterase (Figure 1). Highly methoxylated homogalacturonans are the substrates of pectin lyases and pectin methylesterases. Pectin lyases cleave the glycosidic linkage adjacent to an esterified residue by β-elimination while pectin methylesterases release methanol from the esterified carboxyl groups and transform methylated polygalacturonan regions into polygalacturonate. This can then be degraded by polygalacturonases and pectate lyases, both cleaving the glycosidic linkage next to a free carboxyl group, the former by hydrolysis and the latter by β-elimination. The rhamnogalacturonan component of pectins is modified by rhamnogalacturonases, rhamnogalacturonan lyases and by the accessory enzymes such as rhamnogalacturonan acetylesterase.

The production of pectin methylesterases and polygalacturonases in plants is well known and reduction of polygalacturonase activity by anti-sense techniques has created genetically-modified tomatoes of commercial relevance (Smith *et al.*, 1988; Sheehy *et al.*, 1988; Kramer and Redenbaugh, 1994). The action of pectin methylesterases from plants tend to produce blocks of demethylated pectin and thus differ in specificity from the generally more random acting microbial enzymes. This may be correlated with the strong product inhibition observed for the plant methylesterases. The reversal of the inhibition on raising the concentration of calcium or other cations (Moustacas *et al.*, 1991; Charnay *et al.*, 1992), allows the activity to be modulated, changing the physical properties of the pectin and rendering it susceptible to degradation by polygalacturonase (Nari *et al.*, 1991). Pectin lyases have not yet been found in plants, so degradation may require pectin methylesterases. However, DNA sequencing techniques have revealed the presence of genes in plants with sequence similarity to the microbial pectate lyases. Recently, two of these genes have been expressed in bacteria and shown to have activity on polygalacturonate, proving that they correspond to plant pectate lyases (Domingo *et al.*, 1997, Taniguchi *et al.*, 1995). These pectic enzymes, naturally occurring in plants, seem to play a role in physiological processes such as the abcission phenomena, and can also produce important textural changes in fruit and vegetables during storage and processing.

Plant Pathogenesis

Extracellular secretion of pectinolytic enzymes by microorganisms has been established to play a major role in phytopathogenesis (Crawford and Kolattukudy, 1987; Barras *et al.*, 1994; Collmer and Keen, 1986). Plant pathogens produce an array of enzymes capable of attacking the plant cell wall components, however a convincing role in virulence has only been established for those enzymes that attack the pectic fraction. Depolimerization of the pectic components of the cell wall leads to tissue maceration and cell death, as occurs in the soft-rotting of plant living tissue and the decay of harvested crops. *Erwinias* (*E. chrysanthemi*, *E. carotovora* spp *carotovora* and *E. carotovora* spp *atroseptica*) have been identified as the main pathogens in primary infection while *Bacillus* (*B. subtilis*, *B. megatherium* and *B. polimyxa*) and pectinolytic strains of *Pseudomonas* seem to act in secondary infection. Understanding the molecular basis for substrate specificity and mode of action of pectin modifying enzymes would aid the design of inhibitors for preventing plant disease.

STRUCTURES OF PECTIN MODIFYING ENZYMES

Known Structures of Pectin Modifying Enzymes

The recent structure determination of several pectin modifying enzymes has stimulated new interest in the biochemistry of these enzymes. The structures of three pectate lyases have been determined by X-ray crystallography: pectate lyase C (PelC) (Yoder *et al.*, 1993), pectate lyase E (PelE) (Lietzke *et al.*, 1994) both from *Erwinia chrysanthemi* and pectate lyase from *Bacillus subtilis* (BsPel) complexed with Ca^{2+} (Pickersgill *et al.*, 1994); recently the crystal structures of rhamnogalacturonase A (RGase A) from *Aspergillus aculeatus* (Petersen *et al.*, 1997) and pectin lyase A (PnlA) from *Aspergillus niger* (Mayans *et al.*, 1997) have also been reported.

The majority of pectate and pectin lyases can be classified as part of the 'extracellular pectate lyase superfamily' identified by Henrissat *et al.* (1995), which includes PelC, PelE, BsPel and PnlA. The lowest pairwise sequence identity for alignment of these structures is 15% (PnlA and PelC). The sequence motif vWiDH (val-TRP-ile-ASP-HIS with upper case: absolutely conserved residues, lower case: conservatively substituted residues) is a characteristic of aligned sequences in this family.

Both *Aspergillus niger* polygalacturonase and *Aspergillus aculeatus* rhamnogalacturonase belong to family 28 of the glycosyl hydrolases family (Henrissat and Bairoch, 1996). It is therefore expected that rhamnogalacturonase will have an inverting mechanism in common with polygalacturonase (Biely *et al.*, 1996).

The Parallel β-helix Architecture of Pectate and Pectin Lyase

The overall architecture of these enzymes comprises a parallel β-helix domain and a major loop region (Figure 2). The parallel β-helix domain is formed by parallel β-strands which fold into a large, right-handed helix, dimpled in cross-section. The β-strands of consecutive turns line up to form three parallel β-sheets called PB1, PB2 and PB3. PB1 and PB2 form an antiparallel β-sandwich, while PB3 lies approximately perpendicular to PB2, giving the cross-section of each coil an L-shape rather than triangular appearance (Figure 3). The β-strands are generally short, typically 3 amino acids for PB1, 4 or 5 amino acids for PB2 and 3 to 5 for PB3 and they form eight complete turns of parallel β-helix. In all structures, both ends of the parallel β-domain are capped, the N-terminal end of the cylinder by a short α-helix which is structurally conserved in the structures determined to date, and the carboxy-terminal end by a polypeptide chain in extended conformation.

The turns between β-sheets have been referred to as T1 (between PB1 and PB2), T2 (between PB2 and PB3) and T3 (between PB3 and PB1). T1 is of the β-arch type and of variable but moderate length. The T2 turn, short and regular in conformation, introduces a change of 90° in the direction of the polypeptide backbone and can be classified as a distorted $\gamma\beta_E$ elbow turn following the nomenclature by Wilmot and Thornton (1990). The turn is composed of two residues, C_i and C_{i+1}, of which Ci is in α_L conformation, typically a glycine or asparagine, and C_{i+1} tend to be an asparagine which participates in the formation of the so-called asparagine ladder described below. The duplet Asn-Asn is characteristic of the T2 turn in pectin and pectate lyases. The secondary structure of this T2 turn relates to the observation by Pickersgill *et al.* (1994) that for some of the β-strands, particularly at the PB2 sheet, the amino acids immediately before and after the β-strand have backbone dihedrals angles in an α_L conformation. Residues in this conformation had been found previously preceding and terminating β-strands in protein structures, and had been classified and termed as β-breakers by Colloc'h and Cohen (1991). The right-handed parallel β-helix of pectic lyases is rich in β-breakers, specially BsPel which shows an unusually high number of occurrences of these structures.

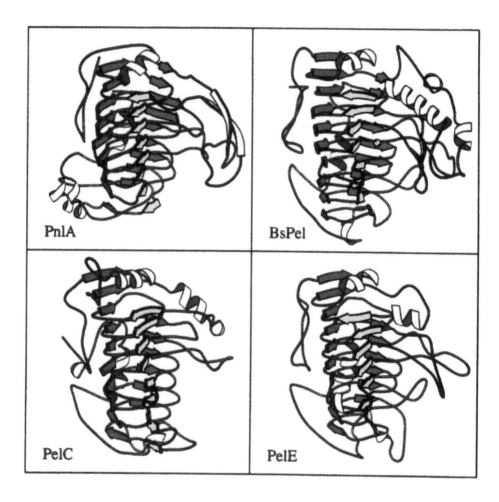

Figure 2 The overall architecture of the pectate and pectin lyases. Top left: pectin lyase (PnlA), top right *Bacillus subtilis* pectate lyase (BsPel), bottom left *Erwinia chrysanthemi* Pel C (PelC) and bottom right *Erwinia chrysanthemi* Pel E (PelE). In this schematic representation, produced using MOLSCRIPT (Kraulis, 1991), arrows represent β-strands with the arrow head to the C-terminal end of the strand and coils are represented by β-helices. Parallel β-sheet 1 (PB1) is shown in yellow, PB2 in green and PB3 in red. In BsPel the T3 loop region (non-coloured region to the right of the parallel β-helix domain) is dominated by a 67 amino acid long loop which comprises a long α-helix. An equivalent but much truncated α-helix is seen in the structure of PelE. The PnlA T3 loop region is dominated by a single T3 loop that protrudes prior to the first turn of the parallel β-helix domain. This T3 loop begins and ends with two β-strands, residues 57-63 and 91-97, which form an antiparallel β-sheet. The middle part of the loop is markedly convoluted, its structure being maintained by two disulphide bridges, Cys 63-Cys 82 and Cys 72-Cys 206. Three of the cysteines involved are part of this same loop Cys 63, Cys 72 and Cys 82; while a forth, Cys 206, is situated in the fourth T3 loop. The latter, of medium length, forms part of the bottom of the substrate binding cleft. The calcium bound to the active site of BsPel is shown as a yellow circle.

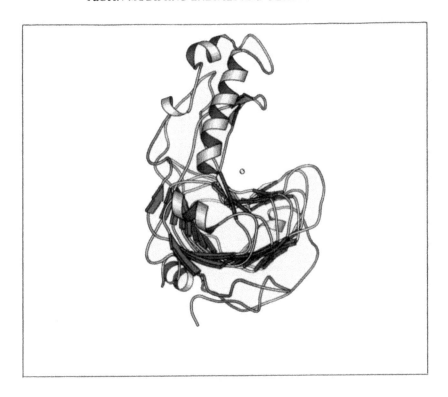

Figure 3   Schematic view of BsPel viewed down the axis of the parallel β-helix. The substrate binding cleft contains the essential calcium (yellow circle).

The T3 loops are commonly lengthy and of more complex conformation, constituting the major loop region. They protrude from the central parallel β-helix, packing against PB1 and forming the substrate binding cleft. The T3 loop region is the area of greatest variability in these parallel β-helices (Figure 2), it forms an independent structural domain as in BsPel, where three long consecutive T3 loops pack against each other to form a hydrophobic core. In PnlA the region is again dominated by a single T3 loop that protrudes prior to the first turn of the parallel β-helix domain. The secondary structure of the T3 loops is also diverse; the BsPel loop contains a long α-helix whilst the corresponding region in PnlA comprises an antiparallel β-sheet. Although the parallel β-domain in these structures is very similar, all four differ in the size and conformation of the T1 and T3 loops. Note that even within the pectate lyases there is substantial variability in the T3 loop regions.

A characteristic of the interior of the parallel β-helix domain, is that the side chains of residues at corresponding positions in consecutive β-strands stack directly upon each other (Figure 4a). The most significant example is the presence of a distinct asparagine ladder in the inner position of the T2 turn in the core of the helix, these asparagines form a characteristic network of hydrogen bonds involving both side chain groups and main chain amides. The hydrogen bonding network links consecutive coils of the parallel β-helix (Figure 4b). In the PnlA structure, this

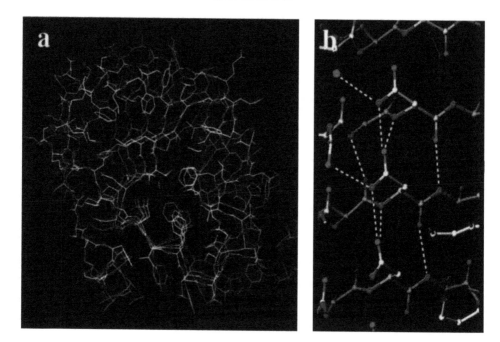

Figure 4 (a) Stacking of amino acid residues in the interior of pectin lyase A (strain 4M-147) viewed from the C-terminal end of the parallel β-helix. The view is such that consecutive T2 turns approximately superimpose and the conserved amphipathic C-terminal α-helix is shown in cross-section, packing against the T2 turn region. Attention is drawn to the internal asparagine ladder and the internal aromatic stack. Note also the tyrosine stack which packs against the C-terminal α-helix. (b) Close up of the hydrogen bonding pattern of the asparagine ladder of PnlA. The carbonyl oxygen of an asparagine side chain hydrogen bonds to both the main chain amide and to the amino-group of the aspartate in the previous turn of parallel β-helix. Hydrogen bonds are shown as dashed lines.

ladder is composed of five residues, while four asparagines are present in the ladder of BsPel, six in PelC and only three in PelE. An internal aromatic stack is also present in these structures, composed of a variable number of residues and commonly located towards the C-terminal end of the parallel β-helix. Aliphatic stacks can also be found within the core. Some side chains that are oriented outward into the solvent also stack, as in the clustering of five external tyrosines in the PnlA structure.

Structural Conservation in Pectate and Pectin Lyases

Within the 'extracellular pectic lyase superfamily' the main functional division occurs between pectate and pectin lyases. The sequences of the different members of the family are distant enough so that structural alignment of the three pectate lyases and the pectin lyase determined, reveals the fundamental structural features of the family. The superimposition of the structures reveals a pairwise sequence identity of approximately 15%. The conserved residues are concentrated far from the active site, along the T2 turns where the N-terminal tail packs against PB2 and including the vWiDH sequence motif. The asparagine ladder and the aromatic stack inside and towards the carboxy-terminal end of the parallel β-helix domain, are also relatively well conserved amongst the members of the family. In PnlA, it is this T2 region of the experimentally determined structures that has the lowest temperature factors and is therefore the most rigid, which may imply that this area is critical for the stability of the structure and may be a nucleus for folding. This would account for the reported difficulties in producing soluble folded protein from mutants of *E. chrysanthemi* PelC with altered sequences in this region (Kita *et al*, 1996).

The conformation of the N- and C-terminal tails is closely related in all the pectic lyase structures. The N-terminal tail packs against PB2 while the C-terminal tail lies across PB3 comprising a highly structurally conserved, amphipathic α-helix, which packs against the T2 turn (Figure 4a). While the N-terminal extensions have some detectable sequence identity, the α-helix capping the end of the parallel β-helix and the C-terminal tail are structurally conserved but have little sequence similarity. In summary, PB2 and PB3 constitute the area of highest structural conservation, which is at a maximum in the asparagine ladder in T2 and the vWiDH pattern in PB2. The high structural conservation occurs in a region distant from the active site while the binding cleft and surroundings are the most divergent part of the molecule. Hence, it could be argued that the N- and C-terminal tails, the asparagine ladder, and the relatively well conserved sequences on PB2 contribute to folding and stability of the parallel β-helix domain, while diversity around the active site allows variation in substrate specificity.

SUBSTRATE RECOGNITION, CATALYSIS AND DESIGNER PECTINS

Pectate and Pectin Lyase have Different Carbohydrate Recognition Strategies

The structures of BsPel and PnlA have allowed the identification of the substrate binding site and several putative catalytic residues and have contributed to a better understanding of the different specificity and catalysis of the distant but related pectate and pectin lyases. Pectate lyases act preferentially on demethylated and therefore highly charged polygalacturonate, require calcium for catalysis and their pH optima are around 8.0. By contrast, pectin lyases show higher activity on highly methylated, relatively hydrophobic homogalacturonans, do not require calcium for catalysis and their pH optima are around 5.5. Prior to the structural studies, it was believed that calcium simply cross-linked the strands of the substrate into a structure that could be recognised by the enzyme. However, the Ca^{2+}-BsPel complex provides direct evidence that calcium binds to the enzyme. In this structure, calcium binds at the bottom of a pronounced cleft to three aspartates, generating a ribbon of positive electrostatic potential complementary to the negatively charged, polygalacturonate substrate. The affinity of BsPel for calcium has been measured at $2.3\,\mu M$ by titration calorimetry. Given that pectate lyases are inactive in the absence of calcium, the calcium binding site is accepted to be the active site and the cleft between the long T3 loops and the parallel sheet PB1 in which it is located, the substrate binding site.

The three pectate lyase structures determined to date present a substrate binding region rich in charged acidic and basic residues, suggesting an electrostatic basis for substrate recognition. In sharp contrast, the substrate binding cleft of pectin

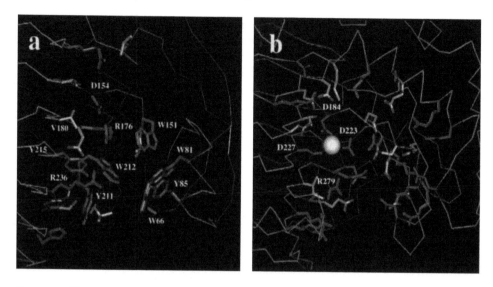

Figure 5 The substrate binding clefts of pectin and pectate lyase. (a) PnlA: residues Arg 154, 176 and Arg 236 are expected to play a role in catalysis and adjacent aromatics are shown in green. (b) BsPel: Calcium is shown as a yellow sphere, note Asp 184 is equivalent to Asp 154 and Arg 279 is equivalent to Arg 236 in PnlA. Amino acids in red are negatively charged and those in blue positively charged at the pH of catalysis. No role in binding or catalysis is implied, the labelled side chains have been identified purely on the basis of their proximity to the essential calcium-ion.

lyase A is dominated by aromatic residues, the only charged amino acids being Asp 154, Arg 176 and Arg 236 (Figures 5a and b) which correspond to Asp 184, Asp 223 and Arg 279 in BsPel. The aromatic residues in the substrate binding cleft are four tryptophans: Trp 66, Trp 81, Trp 151 and Trp 212; and three tyrosines Tyr 85, Tyr 211 and Tyr 215. They constitute the following edge-to-face pairs: Trp 81-Trp 151, Trp 66-Trp 212 and Tyr 211-Trp 212. These residues are conserved in the other *Aspergillus niger* pectin lyases while tyrosines Tyr 85 and Tyr 215, not involved in edge-to-face pairs, are not conserved, therefore, it is possible that the three aromatic pairs are important for the stability of the architecture of the substrate binding cleft. Fluorescence quenching studies using the highly homologous pectin lyase D (Van Houdenhoven, 1975) suggests that one or more of these aromatic residues are involved in substrate binding. It is expected that the aromatics contribute to the affinity of pectin lyase A for the methylated homogalacturonan substrate.

Alignment of the pectin lyase A structure with the pectate lyases confirms the suggestion of Henrissat *et al.* (1995) that only Asp 184 and Arg 279 (BsPel numbering) are invariant in the active site. In the Ca^{2+}-BsPel complex, calcium binds to three aspartates: Asp 184, Asp 223 and Asp 227 (BsPel numbering); within the family of extracellular pectate lyases, Asp184 and Asp 227 are absolutely conserved, while Asp 223 is conservatively substituted for a glutamate. In pectin lyase A, Asp 223 corresponds to Arg 176, and Asp 227 is equivalent to Val 180, Asp 184 (Asp 154 in pectin lyase numbering) being the only conserved carboxylate. The only other absolutely conserved residue in the binding cleft across the members of this family is Arg 236

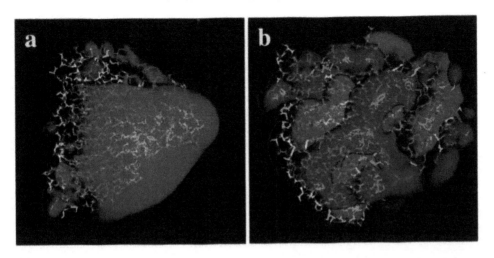

Figure 6 Electrostatic potential for PnlA and BsPel. The calculation using GRASP (Nicholls, 1993) used standard pK values and corresponds to pH 7.0. (a) Electrostatic potential surrounding PnlA contoured at +/− 12 kT/e (blue positive/red negative). (b) Electrostatic potential surrounding BsPel contoured at +/− 2 kT/e (blue positive/red negative).

(pectin lyase numbering). Assuming that extracellular pectate and pectin lyases evolved from a common ancestral lyase, it might be expected that the only two conserved residues, Asp 154 and Arg 236, play equivalent roles in related mechanisms.

Pectin lyase A is rich in acidic residues, containing a total of 48 aspartates and glutamates. Its experimental pI is 3.5 (Kusters-van Someren *et al.*, 1991) and will therefore have an overall negative charge at the pH of optimum catalysis, pH 6.0 (Van Houdenhoven, 1975). The study of the surroundings of the substrate binding site, has revealed that acidic residues cluster around it in both the parallel β-helix domain and the apex of the main T3 loop. The global effect is a significant negative electrostatic potential, enveloping the substrate binding site (Figure 6a). By contrast, the electrostatic field arising in pectate lyase with calcium bound, is an elongated ribbon of positive potential, centered at the calcium, which complements the negatively charged, polymeric substrate (Figure 6b). The different charge distribution of these enzymes may contribute to specificity, polygalacturonate being repelled by the negative charge on pectin lyase but attracted to the pectate lyase cleft. Relatively high ionic strength (0.2 M NaCl) is needed for pectin lyase to express its full activity and for optimal stability (Van Houdenhoven, 1975). In these conditions, the substantial negative potential will be attenuated and the negative partial charges of the methylated homogalacturonan shielded, permitting the binding of this substrate but still contributing significantly to the repulsion of polygalacturonate. On the other hand, the pectate lyase BsPel shows diminishing activity with increasing salt concentration, this being the effect of disrupting the long- and short-range electrostatic interaction of binding. It can be concluded that, substrate specificity is not only a consequence of the hydrophobicity of the binding cleft but also of long-range electrostatic effects.

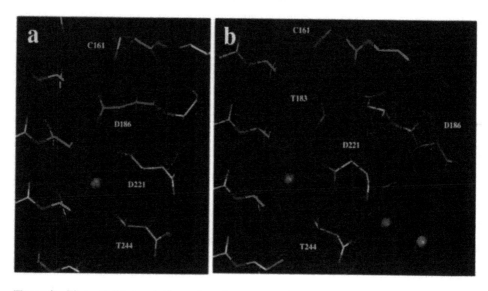

Figure 7 The pH-driven conformation change in PnlA. (a) The fully active enzyme at pH 6.5 with two aspartates (Asp 186 and Asp 221) buried in the core of the parallel β-helix domain (b) Reduced activity enzyme at pH 8.5 with Asp 186 pointing out and into solvent.

A pH Driven Conformational Change in Pectin Lyase A

Pectin lyase A is optimally active within a limited pH range, between 5.5 and 7.0 depending on substrate concentration and ionic strength. This sharp pH activity profile cannot be easily explained in terms of ionisation of one or more of the putative catalytic residues: Asp 154, Arg 176 and Arg 236; the only candidates given the aromatic character of the binding cleft. Instead, the ionisation occurs in an aspartate in a T1 loop which forms part of a lateral wall of the binding cleft. Two pectin lyase structures have been reported, one corresponding to the enzyme at pH 6.5, at which the enzyme is optimally active and a second structure corresponding to pH 8.5, at which the enzyme shows reduced activity. The structure at lower pH, pH 6.5, contains two aspartates buried in the core, Asp 186 in the fourth T1 loop and Asp 221 in the consecutive turn. The distance between the OD2 atoms is 2.7Å, this would not be energetically favourable if both residues were deprotonated, instead a hydrogen bond must be formed between them, suggesting that one of them is deprotonated but not the other. The pectin lyase A structure solved at pH 8.5, shows Asp 221 still inside the core, but Asp 186 has sprung out, and is now exposed to the solvent (Figures 7a and b). It appears that residue Asp 221 is charged, permanently inside the core, and the change observed can be attributed to the deprotonation of residue Asp 186.

At pH 6.5, Asp 186 contributes to stabilise the charged Asp 221 in the inside of the parallel β-helix. At pH 8.5, Asp 186 is not available for this role and the orientation of the side chains of neighbouring residues, residues 182-187, have changed to assure the stability of Asp 221. At this pH, Thr 183 interacts with Asp 221 to form a short, strong hydrogen bond, the involved oxygen atoms being only 2.5 Å apart. Since at both pH values residues from the same T1 loop contribute to

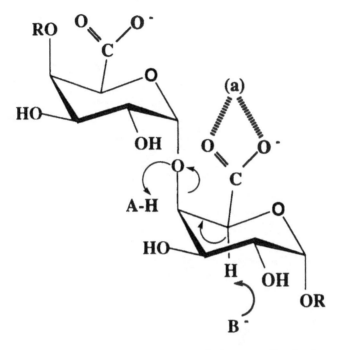

Figure 8 Pectate lyase may accelerate the rate of cleavage of polygalacturonate by: 1. Facilitating abstraction of the C5 proton by increasing its pKa via interaction of the carboxylate with calcium and/or arginine (the calcium and/or arginine are labelled '(a)'). 2. Maintaining favourable stereochemistry for reaction which is almost optimal for a galacturonic acid adopting the chair conformation with the carboxylate equatorial and both the leaving group, O4, and the proton at C5 axial. 3. Providing a base for proton abstraction (B-) and 4. Providing an acid (AH) to protonate the glycosidic oxygen. 'R' represents the continuing chain of galacturonsyl residues in the polygalacturonate polymer.

neutralise the inner Asp 221, structural changes mainly affect one single T1 loop. The structure of neighbouring T1 loops is only affected to a minor extent, and although shifts occur, the relative position of groups is maintained. At pH 8.5, a water molecule appears positioned where the carbonyl group of Thr 183 is at pH 6.5, preserving the local hydrogen bonding pattern of the T1 area.

It is not clear how the structural changes caused by pH relate to the biochemical observation of reduction in activity. The exposure of Asp 186 may cause a direct, through space effect on the catalytic residues. However, in the active form (pH 6.5), Thr 183 is oriented towards the binding cleft at approximately 8.9 Å from the catalytic Arg 176, while at pH 8.5 that space is unoccupied and the surrounding area is somewhat altered. Therefore, it is more likely that the effect on catalysis is a steric effect, in which binding of the substrate is affected either directly or by perturbing the water structure of the active site.

A pH-dependent reduction in activity has also been reported for the other pectin lyases of *Aspergillus niger* (Kester and Visser, 1994) and the pectin lyase of *Aspergillus japonicus* (Ishii and Yokotsuka, 1975). It is possible that the effect is due to a similar structural trigger.

How do Pectin and Pectate Lyase Work?

In a general β-elimination reaction two ionic groups are removed from the substrate: an α-proton commonly assisted by a base, and a leaving group (a β-nucleophile in the reverse reaction) often assisted by an acid, this resulting in the formation of a double bond. This is illustrated for the cleavage of the $\alpha(1\text{-}4)$ glycosidic linkage by pectate lyase in Figure 8. A β-elimination reaction can occur by either a multi-step or a concerted mechanism and the protein is expected to assist catalysis by ensuring optimal stereochemistry of the substrate (transition state), lowering the pKa of the α-proton, providing a base and improving the leaving group by protonation. For a concerted reaction, the rate is critically dependent on the relative orientation of the proton and the leaving group, this is close to optimal in the most stable substrate conformation of the pectin substrate. Pectate lyase but not pectin lyase requires calcium for catalysis, this may reflect the increased stability of the pectate substrate towards depolymerization. In pectin, the proton at C5 is activated by the electron withdrawing carbonyl group of the carboxymethyl at C6, however pectate contains carboxylate anions at the C6 position, and is more stable due to the reduced electron withdrawing power of this group. Calcium in pectate lyase could aid substrate binding by interacting with the carboxylate at C6, increasing the electron withdrawing effect at C5 and facilitating the abstraction of the α-proton by lowering its pKa, which may explain the absolute requirement of a metal cation for the β-elimination catalysis by pectate lyase. The guanidinium group of Arg 176 in PnlA occupies a position close to that of the calcium-ion in pectate lyases. Thus, it could be suggested that the roles of calcium in pectate lyase and the arginine in pectin lyase are related.

Two residues, Lys 247 and Arg 284, are conserved in all pectate lyases but not in pectin lyases and these may have a role in binding and cleaving the pectate substrate.

Site directed mutagenesis of PelC and BsPel have given apparently inconsistent results with the residual activity of mutants of BsPel often higher than in PelC. The results from BsPel suggest that the three calcium binding ligands and the arginine are all important but no one is absolutely essential. Substitution of two of the calcium ligands Asp 223 and Asp 227 by alanine reduces the activity to approximately 0.005% and 0.05% of wild type respectively, the effect of the same substitution for Asp 184 is less striking with 3% of wild type activity remaining. Substituting alanine for Arg 279 results in an activity of 0.2% wild type whilst the conservative substitution of lysine leaves a residual activity of 0.8%. It is reasonable to assume that the position of both the arginine and the calcium are important for catalysis and the surprising result is therefore that the Asp 184 Ala mutation does not also result in an activity of less than 1%, especially as Asp 184 makes bidentate ligands to the calcium-ion. It is unlikely that the conserved carboxylate of Asp 184 acts as the base because of the remaining activity of the Asp 184 Ala mutant and because it is hidden from substrate behind the calcium-ion. Thus its conserved role may be that of positioning the cation (arginine or calcium) in pectin and pectate lyase.

For pectate, but not pectin, lyase there are three other carboxylates, Asp 223, Asp 227 and that of a neighbouring galacturonic acid on the substrate which could act as the base. However, the first two are less plausible nucleophiles due to binding to the calcium while involvement of another galacturonic acid might predict a more rapid lowering of activity as the percentage of methylation increased than is seen in BsPel at least and might not explain exo-pectate lyases. Of course, on substrate binding Asp 223 or Asp 227 may be reoriented and be able to assist proton abstraction. The only plausible alternative is that a water molecule or a hydroxyl anion acts as the base. This mechanism could be conserved across the family with the water binding either calcium or Arg 176. If the water or hydroxide position were critical, it might also account for the observation that calcium cannot be replaced with magnesium, strontium or barium (Tardy et al., 1997). The structure of BsPel with barium in two crystal forms shows that barium binds at the calcium site but also shows that the water structure is altered. Other explanations for the loss of activity when calcium is replaced by barium are that a bound galacturonic acid would be positioned further from the cation site and displacement of Asp 223 or Asp 227 to generate the catalytic base would be less likely.

Given the variability in the active sites of pectin and pectate lyases it is not possible to generate an active site fingerprint to use in a search for homologous lyases. Disulfide bridges are also not conserved within the extracellular family, leaving vWiDH as the only obvious sequence motif defining the extracellular family. No structure has yet been determined for a bacterial pectin lyase. Examination of the sequences suggests that the mechanism may again differ from both the pectate lyases and from the fungal pectin lyases as Arg176 of pectin lyase A is not conserved (Henrissat et al., 1995). Two of the three aspartates of the pectate lyase calcium site may be conserved, suggesting that some cationic group is bound at that site although this might be a lysine or arginine from a neighbouring loop or strand. The limited active site data suggests that the bacterial pectin lyase diverged from pectate lyases rather than from the fungal pectin lyases.

Rhamnogalacturonase and the Other Right Handed Parallel β-Helix Proteins

The backbone of RG-I is composed of rhamnose (Rha) and galacturonic acid (GalUA) residues linked into a polymer which is a repeating dimer unit (1-2)-α-LRha(1-4)-α-DGalUA. The C4 position of Rha is often decorated by arabinan, galactan or arabinogalactan (Schols et al., 1994). Rhamnogalacturonase A (RGase A) is a hydrolase that cleaves the (1-2) linkage. The core of RGase A is composed almost entirely of parallel β-strands, which are rather short and separated by sharp bends. There are 13 complete turns of β-strands forming a larger parallel β-helix than those of pectate and pectin lyases (Petersen et al., 1997). Further differences are the presence of an additional parallel β-sheet, designated PB1a, in the T1 turn and stacks of identical side chains are much less prominent in RGase A. The active site of RGase is again formed by PB1 and the loops at both ends. Structural alignment of RGase A with PelC aligns Asp 180 with Asp 170 of PelC (Asp 184 of BsPel), suggesting that the cluster of carboxylates Asp 156, Asp 177, Asp 180, Asp 197 and Glu 198 is the active site of this family of enzymes, although as yet no substrate-complex structure is available. Asp 156 and Asp 180 are predicted to be the general acid and base involved in the inverting catalytic mechanism.

There are two other known structures, representing three families of proteins, which display the right handed parallel β-helix fold. These are the carboxy-terminal fragment of the phage P22 tailspike protein (Steinbacher et al., 1994, Steinbacher et al., 1997) and pertactin from Bordetella pertussis (Emsley et al., 1996). The carboxy-terminal fragment of the tailspike protein has been shown to be an endorhamnosidase and an inverting mechanism has been suggested with Asp 392 as general acid while Glu 359 and Asp 395 activate a catalytic water molecule as general bases. These residues lie on two neighbouring turns near the middle of the long parallel β-helix, just before or at the start of sheet C, which is equivalent to PB1 of pectate lyases.

Petersen *et al.* (1997) also report that the structure of RGase A can be superposed on that of the tailspike fragment with an rmsd of 2.04Å but did not report if Asp 156, Asp 177 and Asp 180 superpose with the active site residues of tailspike Glu 359, Asp 392 and Asp 395. However, even without the possibility of a closely similar mechanism, the distant relationship between the hydrolases, RGase A and tailspike, seems closer than that with the lyases, because their parallel β-helices are longer and contain almost exclusively hydrophobic residues in their interior. Stacking of residues is also much less pronounced than in the lyases and the asparagine ladder is absent. However, RGase A does share with the lyases an β-helix at the amino-terminus of the parallel β-helix.

Pertactin is a virulence factor of *Bordetella pertussis* and mediates adhesion to mammalian target cells via an RGD sequence. This is the longest parallel β-helix structure yet determined with 16 turns of helix. No enzymatic activity has yet been described nor has a polysaccharide binding been characterised, although poly-saccharide binding is frequently involved in cellular adhesion. Pertactin seems to be a structural intermediate between the lyases and the hydrolyses as the core of the helix is hydrophobic but there is extensive stacking of similar residues.

New Families of Pectate Lyases

In addition to the extracellular pectate lyase family, there are at least five further families of pectate or oligogalacturonate lyases but no other known families of pectin lyases. Firstly there is the periplasmic pectate lyase family (Manulis *et al.*, 1988; Hinton *et al.*, 1989), secondly enzymes from *E. chrysanthemi*, PelI (Hugouvieux-Cotte-Pattat, personal communication) and *E. carotivora* (Heikinheimo *et al.*, 1994), which are homologous to the pectate lyases from *Fusarium solani* (Gonzalez-Candelas and Kolattukudy, 1992, Guo *et al.*, 1995a, 1995b, 1996), third PelX (Brooks *et al.*, 1990) and PelL (Lojkowska *et al.*, 1995) *from Erwinia chrysanthemi*, fourth PelZ from *E. chrysanthemi* (Pissavin *et al.*, 1996) and finally the oligogalacturonate lyases (Reverchon *et al.*, 1989). Despite the absence of any obvious relationship between the sequences, Lojkowska *et al.* (1995) point out a string of similarities between PelL and the extracellular pectate lyase superfamily. PelL is a pectate lyase of similar size (i.e. 400 residues compared to 399 for BsPel), requires calcium for activity and has a similar pH optimum of 8.5, a sequence which suggests an all β-fold and has a high content of asparagines including four pairs of consecutive asparagines and one triple. Pairs of asparagines are common in the extracellular lyase family at the T2 turns. Thus PelL and PelX may well have the parallel β-helix fold despite the lack of any definite evidence for the structure of PelL. The alternative is the convergent evolution of a closely related catalytic mechanism. Similarly, the limited data published on the mechanisms of the family of *Fusarium* pectate lyases (Rao *et al.*, 1996) (which are homologous with PelI and the enzyme from *E. carotivora*) have not yet shown any differences from those of the extracellular family. It is possible to speculate that sequences such as VWWADVCEDA in *Fusarium* PelA are distantly related to HIWIDHCTFN of BsPel. However, in other members of this family the first alanine is substituted by a glutamate, while the equivalent residue in the extracellular family is buried.

The question of the relationship between the various families of lyases will probably not be resolved until a three-dimensional structure is available for a member of each family. However, it is clear that these structures are a challenge for techniques of structure prediction such as 'threading' (Sippl and Weitckus, 1992; Jones *et al.*, 1992). If the enzymes such as PelL, for which crystals have already been obtained, are not distantly homologous to the extracellular family, the convergence to a calcium dependent mechanism would suggest that this is one of very few possible mechanisms for such lyases, including other polysaccharide lyases.

Designer Pectins

Application of the pectin modifying enzymes rhamnogalacturonase and rhamno-galacturonan acetylesterase from *Aspergillus aculeatus* have lead to the improved, if far from complete, characterisation of rhamnogalacturonan structure (Schols *et al.*, 1994; Beldman *et al.*, 1996). These enzymes can also be used to prepare homo-galacturonan regions whose degree and pattern of esterification can then be manipulated by pectin methylesterase and whose molecular weight can be manipulated using pectin lyase, pectate lyase and polygalacturonase. The microbial pectin modifying enzymes are believed to be far less specific than the plant enzymes. It is known, for instance that the microbial pectin methylesterases randomly demethylate whilst the plant enzymes demethylate in blocks. Block demethylation confers profound changes in pectin functionality via interaction with calcium and enzymes with specific patterns of demethylation may give pectins with useful functionalities. For instance some modified pectins have been shown to have anti-cancer activity in mammals (Jiang *et al.*, 1996). Engineering new specificities into the pectin modifying enzymes may offer new routes to enhance pectin functionality.

REFERENCES

Axelos, M.A.V., Garnier, C., Renard, C.M.G.C. and Thibault, J.-F. (1996) Interactions of pectins with multivalent cations: Phase diagrams and structural aspects. In *Pectins and Pectinases*, edited by J. Visser and A.G.J. Vorgen, pp. 35–45. Amsterdam: Elsevier.

Barras, F., van-Gijsegen, F. and Chatterjee, A.K. (1994) Extracellular enzymes and pathogenesis of soft-rot *Erwinia. Ann. Rev. Phytopathol.*, **32**, 201–234.

Beldman, G., Mutter, M., Searle-van Leeuwen, M.J.F., van den Broek, L.A.M., Schols, H.A. and Voragen, A.G.J. (1996) New enzymes active towards pectic structures. In *Pectins and Pectinases*, edited by J. Visser and A.G.J. Vorgen, pp. 231–245. Amsterdam: Elsevier.

Biely, P, Benen, J., Heinrichova, K, Kester, H.C.M. and Visser, J. (1996) Inversion of configuration during hydrolysis of α-1,4-galacturonidic linkage by 3 *Aspergillus* polygalacturonases. *FEBS Letters*, **382**, 249–255.

Brooks, A.D., Sheng, Y.H., Gold, S., Keen, N.T., Collmer, A. and Hutcheson, S.W. (1990) Molecular cloning of the structural gene for exopolygalacturonate lyase from Erwinia chrysanthemi ec16 and characterization of the enzyme product. *J. Bacteriol.*, **172**, 6950–6958.

Charnay, D., Nari, J. and Noat, G. (1992) Regulation of plant cell-wall pectin methyl esterase by polyamines-Interactions with the effects of metal-ions. *Eur. J. Biochem.*, **205**, 711–714.

Collmer, A. and Keen, N.T. (1986) The role of pectic enzymes in plant pathogenesis. *Ann. Rev. Phytopathol.*, **24**, 383–409.

Colloc'h, N. and Cohen, F.E. (1991) β-Breakers: An aperiodic Secondary Structure. *J. Mol. Biol.*, **221**, 603–613,

Crawford, M. and Kolattukudy, P.E. (1987). Pectate lyase from *Fusarium solani* f. sp. *pisi*: purification, characterization, *in vitro* translation of the mRNA, and involvement in pathogenicity. *Arch. Biochem. Biophys.*, **258**, 196–205.

Domingo, C., Roberts, K., Stacey, N.J., Connerton, I., Ruiz-Teran, F. and McCann, M.C. (1997) A pectate lyase from *Zinnia elegans* is Auxin inducible. *Plant J.*, (in press).

Emsley, P., Charles, I.G., Fairweather, N.F. and Isaacs, N.W. (1996) Structure of *Bordetella pertussis* virulence factor p.69 pertactin. *Nature*, **381**, 90–92.

Gonzalez-Candelas., L. and Kolattukudy, P.E. (1992) Isolation and analysis of a novel inducible pectate lyase gene from the phytopathogenic fungus *Fusarium solani f sp pisi* (nectria-haematococca, mating population-vi). *J. Bacteriol.*, **174**, 6343–6349.

Guo, W.J., Gonzalez Candelas, L. and Kolattukudy, P.E. (1995a) Cloning of a new pectate lyase gene *pelC Fusarium-solani f-sp pisi*(*Nectria-haematococca*, mating-type-vi) and characterization of the gene product expressed in *Pichia pastoris. Arch. Biochem. Biophys*, **323**, 352–360.

Guo, W.J., Gonzalez Candelas, L. and Kolattukudy, P.E. (1995b) Cloning of a novel constitutively expressed pectate lyase gene *pelB* from *Fusarium-solani f-sp pisi* (*Nectria-haematococca*, mating-type-vi) and characterization of the gene-product expressed in *Pichia pastoris. J. Bacteriology*, **177**, 7070–7077.

Guo, W.J., Gonzalez Candelas, L. and Kolattukudy, P.E. (1996) Identification of a novel *pelD* gene expressed uniquely in planta by *Fusarium-solani f-sp pisi*(*Nectria-haematococca*, mating-type-vi) and characterization of its protein product as an endo-pectate lyase. *Arch. Biochem. Biophys.*, **332**, 305–312.

Heikinheimo, R., Flego, D., Pirhonen, M., Karlsson, M.B., Eriksson, A., Mae, A., Koiv, V. and Palva, E.T. EMBL Data Library, May 1994 — Accession: S44995.

Henrissat, B. and Bairoch, A. (1996) Updating the sequence-based classification of glycosyl hydrolases. *Biochem. J.*, **16**, 695–696.

Henrissat, B., Heffron, S.E., Yoder, M.D., Lietzke, S.E. and Jurnak, F. (1995). Functional implications of structure-based sequence alignment of proteins in the extracellular pectate lyase superfamily. *Plant Physiol.*, **107**, 963–976.

Hinton, J.C.D., Sidebotham, J.M., Gill, D.R. and Salmond, G.P.C. (1989) Extracellular and periplasmic isoenzymes of pectate lyase from *Erwinia carotovora* subspecies *carotovora* belong to different gene families. *Mol. Microbiol.*, **3**, 1785–1795.

Ishii, S. and Yokotsuka. T. (1975) Purification and properties of pectin lyase from *Aspergillus japonicus. Agr. Biol. Chem.*, **39**, 313–321.

Jiang, Y-H., Lupton, J.R., Chang, W-C.L., Jolly, C.A., Aukema, H.M. and Chapkin, R.S. (1996) Dietry fat and fibre differentially alter intracellular second messengers during tumor development in rat colon. *Carcinogenesis*, **17**, 1227–1233.

Jones, D.T., Taylor, W.R, and Thornton, J.M. (1992) A new approach to protein fold recognition. *Nature*, **358**, 86–89.

Kester, H.C.M. and Visser, J. (1994) Purification and characterization of pectin lyase B, a novel pectinolytic enzyme from *Aspergillus niger. FEMS Micro. Letts*, **120**, 63–68.

Kita, N., Boyd, C.M., Garrett, M.R., Jurnak, F. and Keen, N.T. (1996). Differential effect of site-directed mutations in PelC on pectate lyase activity, plant-tissue maceration, and elicitor activity. *J. Biol. Chem.*, **43**, 26529–26535.

Kramer, M.G. and Redenbaugh, K. (1994) Commercialization of a tomato with an antisense polygalacturonase gene — the flavr savr(tm) tomato story. *Euphytica*, **79**, 293–297.

Kraulis, P.J. (1991). MOLSCRIPT; a program to produce both detailed and schematic plots of proteins. *J. Appl. Cryst.*, **24**, 946–950.

Kusters-van Someren, M.A., Harmsen, J.A.M., Kester, H.C.M. and Visser, J. (1991). Structure of the *Aspergillus niger pelA* gene and its expression in *Aspergillus niger* and *Aspergillus nidulans. Curr. Genet.*, **20**, 293–299.

Lietzke, S.E., Yoder, M.D., Keen, N.T. and Jurnak, F. (1994). The three-dimensional structure of pectate lyase E, a plant virulence factor from *E. chrysanthemi. Plant Physiology*, **106**, 849–862.

Lojkowska, E., Masclaux, C., Boccara, M., Robert-Baudouy, J. and Hugouvieux-Cotte-Pattat, N. (1995) Characterization of the *pelL* gene encoding a novel pectate lyase of *Erwinia chrysanthemi*-3937. *Mol. microbiol.*, **16**, 1183–1195.

Manulis, S., Kobayashi, D.Y. and Keen, N.T. (1988) Molecular cloning and sequencing of a pectate lyase gene from *Yersinia pseudotuberculosis. J. Bacteriol.*, **170**, 1825–1830.

Mayans, O., Scott, M., Connerton, I., Benen, J., Pickersgill, R. and Jenkins, J. (1997). Two crystal structures of pectin lyase A from *Aspergillus* reveals a pH driven conformational change and striking divergence in the substrate binding clefts of pectin and pectate lyases. *Structure*, 5, 677–689.

McCann, M.C. and Roberts, K. (1996) Plant cell wall architecture: the role of pectins. In *Pectins and Pectinases*, edited by J. Visser and A.G.J. Vorgen, pp. 91–107. Amsterdam: Elsevier.

Moustacas, A.M., Nari, J., Borel, M., Noat, G. and Ricard, J. (1991) Pectin methylesterase, metal-ions and plant cell-wall extension — the role of metal-ions in plant cell-wall extension. *Biochem. J.*, 279, 351–354.

Nari, J., Noat, G. and Ricard, J. (1991) Pectin methylesterase, metal-ions and plant cell-wall extension — hydrolysis of pectin by plant cell-wall pectin methylesterase *Biochem. J.*, 279, 343–350.

Nicholls, A., Bharadwaj, R. and Honig, B. (1993) GRASP: graphical representation and analysis of surface-properties. *Biophysical J.*, 64, Pt2, p. A 166.

Petersen, T.N., Kauppinen, S. and Larsen, S. (1997) The crystal structure of rhamnogalacturonase A from *Aspergillus aculeatus*: a right-handed parallel β-helix. *Structure*, 5, 533–544.

Pickersgill, R., Jenkins, J., Harris, G., Nasser, W. and Robert-Baudouy, J. (1994) The structure of *Bacillus subtilis* pectate lyase in complex with calcium. *Nature Struc. Biol.*, 1, 717–723.

Pissavin, C., Robert-Baudouy, J. and Hugouvieux-cotte-pattat, N. (1996) Regulation of *pelZ*, a gene of the *pelB-pelC* cluster encoding a new pectate lyase of *Erwinia chrysanthemi* 3937. *J. Bacteriology*, 178, 7187–7196.

Rao, M.N., Kembhavi, A.A. and Pant, A. (1996) Role of lysine, tryptophan and calcium in the β-elimination activity of a low-molecular-mass pectate lyase *from Fusarium moniliformae*. *Biochem. J.*, 319, 159–164.

Reverchon, S., Huang, Y., Bourson, C. and Robert-Baudouy, J. (1989) Nucleotide sequences of the *Erwinia chrysanthemi ogl* and *pelE* genes negatively regulated by the *kdgR* gene product. *Gene*, 85, 125–134.

Rinaudo, M. (1996) Physiochemical properties of pectins in solution and gel states. In *Pectins and Pectinases*, edited by J. Visser and A.G.J. Vorgen, pp. 21–34. Amsterdam: Elsevier.

Schols, H.A. and Voragen, A.G.J. (1996) Complex Pectins: Structure elucidation using enzymes. In *Pectins and Pectinases*, edited by J. Visser and A.G.J. Vorgen, pp. 3–19. Amsterdam: Elsevier.

Schols, H.A., Voragen, A.G.J. and Colquhoun, I.J. (1994) Hairy (ramified) regions of pectins. Isolation and characterization of rhamnogalacturonan oligomers, liberated during degradation of pectic hairy regions by rhamnogalacturonase. *Carbohydr. Res.*, 256, 97–111.

Sheehy, R.E., Kramer, M. and Hiatt, W.R. (1988) Reduction of polygalacturonase activity in tomato fruit by antisense RNA. *Proc. Natl. Sci.*, 85, 8805–8809.

Sippl, M.J. and Weitckus, S. (1992) Detection of native-like models for amino-acid-sequences of unknown 3-dimensional structure in a data-base of known protein conformations. *Proteins*, 13, 258–271.

Smith, C.J.S., Watson, C.F., Ray, J., Bird, C.R., Morris, P.C., Schuch, W. and Grierson, D. (1988) Antisense RNA inhibition of polygalacturonase gene-expression in transgenic tomatoes. *Nature*, 334, 724–726.

Steele, N.M., McCann, M.C. and Roberts, K. (1997) Pectin modification in cell walls of ripening tomatoes occurs in distinct domains. *Plant physiology*, 114, 373–381.

Steinbacher, S., Baxa, U., Miller, S., Wintraub, A., Seckler, R. and Huber, R. (1997) Crystal-structure of phage P22 tailspike protein complexed with *Salmonella* sp o-antigen receptors. *Proc. Natl Acad. Sci. U.S.A.*, 93, 10584–10588.

Steinbacher, S., Seckler, R., Miller, S., Steipe, B., Huber, R. and Reinemer, P. (1994) Crystal structure of P22 tailspike protein: interdigitated subunits in a thermostable trimer. *Science*, 265, 383–386.

Taniguchi, T., Ono, A., Sawatani, M., Nanba, M., Kohno, K., Usui, M., Kurimoto, M. and Matuhasi, T. (1995) *Cry j*, a major allergen of Japanese ceder pollen, has pectate lyase activity. *Allergy*, 50, 90–93.

Tardy, F., Nasser, W., Robert-Baudouy, J. and Hugouvieux-Cotte-Pattat, N. (1997) Comparative analysis of the five major *Erwinia chrysanthemi* pectate lyases: Enzyme characteristics and potential inhibitors. *J. Bacteriology*, 179, 2503–2511.

Van Houdenhoven, F.E.A. (1975). Studies on pectin lyase. Thesis. Agricultural University, Wageningen, The Netherlands.

Visser, J. and Voragen, A.G.J. (Eds) (1996) 'Pectins and Pectinases' Proceedings of an International Symposium, Wageningen, The Netherlands, December 3–7, 1995. Amsterdam: Elsevier.

Wilmot, C.M. and Thornton, J.M. (1990) β-Turns and their distorsions: a proposed new nomenclature. *Protein Eng.*, 3, 479–493.

Yoder, M.D., Keen, N.T. and Jurnak, F. (1993). New domain motif: the structure of pectate lyase C, a secreted plant virulence factor. *Science*, 260, 1503–1507.

13. ENGINEERING PROTEINS FOR THE DEGRADATION OF RECALCITRANT COMPOUNDS

DICK B. JANSSEN*, JIŘÍ DAMBORSKY[†], RICK RINK
and GEJA H. KROOSHOF

*Department of Biochemistry, Groningen Biotechnology and Biomolecular Sciences Institute,
University of Groningen, 9747 AG Groningen, The Netherlands*

INTRODUCTION

The application of biotechnological processes for the treatment of effluents has a long tradition. Activated sludge systems are used for the removal of easily degradable organics from municipal and industrial waste water, and composting is a standard technique for the treatment of solid organic waste. During the last 15 years, there has been an increasing interest in the development of similar processes for the treatment of waste streams or the cleanup of sites that are contaminated with chemicals of synthetic origin (xenobiotics). However, in many cases these chemicals appear to be resistant to microbial degradation, a problem that was first encountered in the sixties with the environmental persistence of organochlorine insecticides but holds as well for many other agrochemicals, various solvents, waste products and chemical intermediates. Such chemicals are often chlorinated, for example short-chain chloroethylenes, highly chlorinated alkanes and alkenes, PCB's, chlorinated dioxins and dibenzofurans, and chlorobenzenes.

Recalcitrance to biodegradation is usually caused by biochemical blocks in catabolic pathways of microorganisms (for reviews see Janssen, 1994). This means that initial degradation or conversion of a catabolic intermediate does not occur because one or more essential enzymes are absent or their activity is too low for growth. The result of a block may be that a pollutant is inert, that it is partially degraded to products which accumulate, or that toxic effects may occur due to the reactivity of intermediates being produced. Thus, recalcitrance in such a case can be regarded as a problem of enzyme activity, specificity and kinetics. This raises the possibility that the microbial degradation of recalcitrant chemicals can be improved by protein engineering.

The application of engineered proteins in environmental biotechnology is, however, far less straightforward than some other techniques described in this book. First, the chemical composition of the substrate usually is very complex. In many cases, waste streams contain mixtures of chemicals, and it hardly occurs that only a single compound is recalcitrant. Second, for many xenobiotic chemicals the problem is not only a biochemical one. The physical state of a compound greatly influences its biodegradability under practical conditions. For example, even well

†Laboratory of Structure and Dynamics of Biomolecules, Faculty of Science, Masaryk University, Czech Republic
*Corresponding author

degradable chemicals such as linear alkanes can show persistent behavior if absorbed to soil particles. Third, the technologies used in environmental biotechnology are often more similar to the techniques of civil technology than to those of biochemical engineering. Thus, reactors will seldom be run under aseptic conditions, mono-cultures are not used (in fact, the composition of the cultures used is unknown), and control of process conditions is not as sophisticated as in industrial fermentation.

Based on the above, engineered proteins or microorganisms will only find applications in some very specific areas. This may be the removal or valorization of compounds which are formed as major byproducts during industrial synthesis, the cleanup of contaminated groundwater or waste gas that contains only one or very few water-soluble recalcitrant compounds, or the use of broad-specificity enzymes for the conversion of polymeric waste materials.

We have focussed our engineering studies on enzymes catalyzing hydrolytic dehalogenation. This reaction is important for a broad range of chlorinated compounds which are relatively soluble in water. It is an attractive conversion since the carbon-chlorine bond is directly cleaved by water, which causes usually a reduction or loss of toxicity. Haloalkane dehalogenase is one of the very few enzymes that can degrade xenobiotic chemicals and for which detailed insight in the catalytic mechanism and the kinetics has been obtained from studies on wild-type and mutant enzymes.

HALOALKANE DEHALOGENASE

Properties, Structure, and Mechanism

Haloalkane dehalogenases catalyze the hydrolytic cleavage of carbon-halogen bonds in a wide range of halogenated alkanes. A typical reaction is the conversion of 1,2-dichloroethane to 2-chloroethanol. The X-ray structure and catalytic mechanism of the haloalkane dehalogenase (DhlA, 310 amino acids) from *Xanthobacter autotrophicus* GJ10 have been solved (Verschueren *et al.*, 1993a,b; Pries *et al.*, 1994a). The enzyme belongs to a group of hydrolytic proteins called α/β-hydrolase fold enzymes (Ollis *et al.*, 1992, Figure 1). This group includes various other bacterial proteins involved in the biodegradation of natural and xenobiotic compounds, i.e. muconic semi-aldehyde hydrolases, 6-substituted 2-hydroxy-6-oxohexadienoate hydrolases, lipases, and epoxide hydrolases (Figure 2).

Catalysis in haloalkane dehalogenase proceeds via a covalent intermediate (Figure 3), which is formed in DhlA by nucleophilic attack of a carboxylate oxygen of Asp124 on the carbon atom which carries the halogen. The ester is cleaved by a water molecule that is activated by His289 and Asp260. Mutation of His289 to Gln leads to an enzyme in which the covalent intermediate is trapped. Most α/β-hydrolase fold enzymes, such as lipases and esterases, attack an sp^2-hybridized carbonyl carbon atom in the substrate, and have a serine as the nucleophile. The presence of an aspartate in haloalkane dehalogenase and epoxide hydrolase is explained by the fact that these enzymes attack an sp^3-hybridized carbon atom to which no carbonyl moiety is bound. In this way, the enzyme supplies the group required for formation of a tetrahedral intermediate (Figure 2; Verschueren *et al.*, 1993a; Pries *et al.*, 1994a).

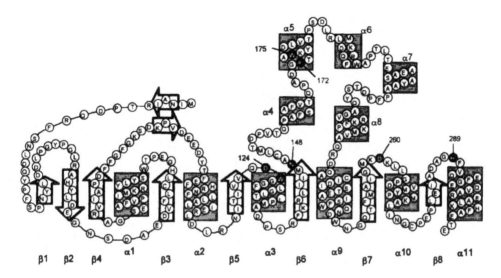

Figure 1 Topology of haloalkane dehalogenase. The α-helices and β-sheets are shown in boxes and arrows, respectively, and the secondary-structure elements that form the α/β-hydrolase fold are indicated with numbers (bottom). The cap domain is formed by helices 4–8 and connecting loops (top). The filled circles indicate catalytic residues, which are also numbered (see Figure 3). The 'charge relay' carboxylate may be present at a topological position corresponding to that of N148 or D260.

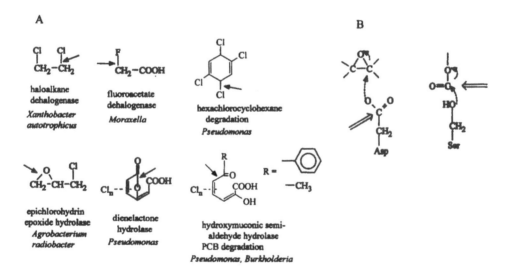

Figure 2 Role of α/β-hydrolase fold enzymes in biodegradation. A, structures of xenobiotic compounds or intermediates converted by α/β-hydrolase fold enzymes. The arrows indicate the bonds that are cleaved. B, covalent intermediates from enzymes with a nucleophilic aspartate (left) or serine (right). The double arrow gives the site of nucleophilic attack of water.

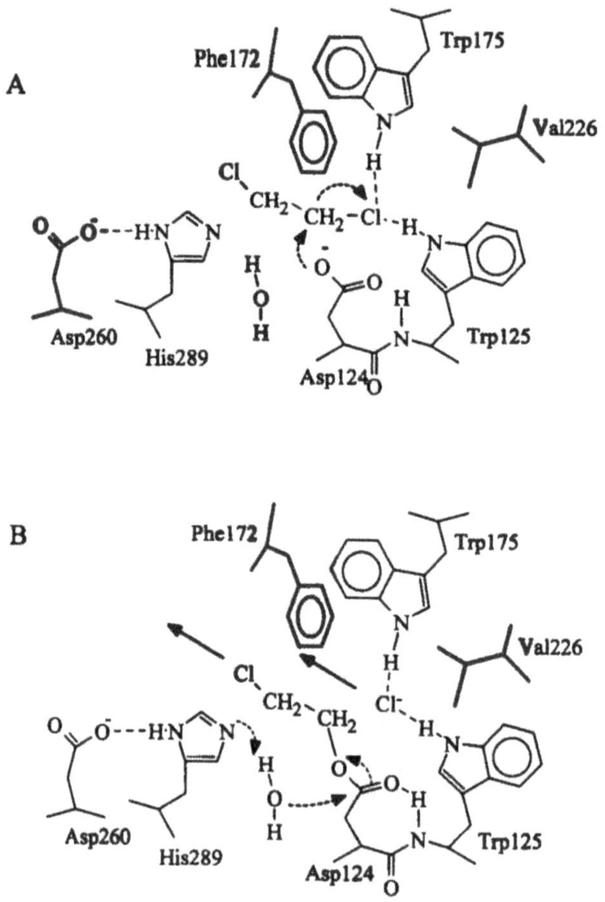

Figure 3 Catalytic mechanism of haloalkane dehalogenase. A, step 2, cleavage of the C-X bond; B, step 3, hydrolysis of the covalent intermediate.

For different reasons, haloalkane dehalogenase is an interesting target for protein engineering studies aimed at improved transformation of environmental pollutants. First, the substrate range is broad, but does not include important chemicals such as 1,2-dichloropropane and 1,1,2-trichloroethane. These compounds are structurally very similar to the 'natural' substrate 1,2-dichloroethane. Their environmental recalcitrance may be explained by the lack of dehalogenases that can convert these compounds. The products of hydrolysis, such as 1-chloro-2-propanol, can be utilized by various organisms, indicating that the critical step for detoxification mainly is the initial dehalogenation. Second, haloalkane dehalogenase converts a class of compounds for which a process technology for large scale application has been developed. The enzyme is used at full scale for groundwater remediation in its natural host *X. autotrophicus* GJ10 (Stucki *et al.*, 1995). The bottleneck for the removal of various other chlorinated chemicals really is the availability of good organisms, i.e. organisms producing the required enzymes.

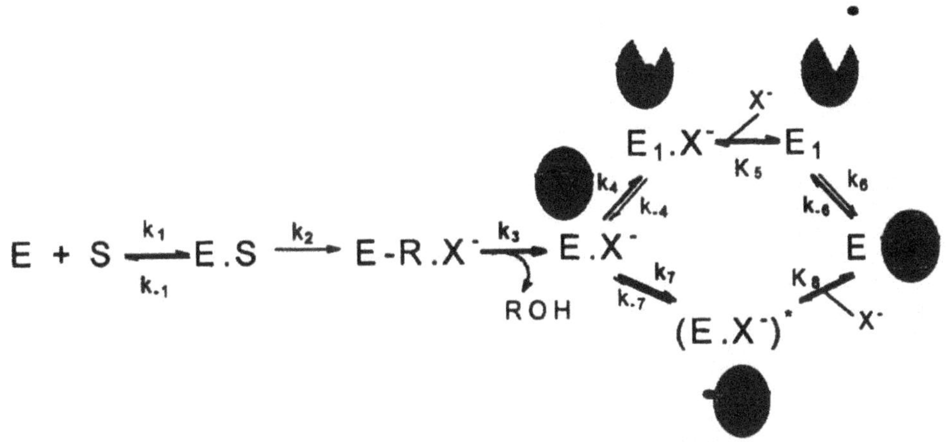

Figure 4 Kinetic mechanism of haloalkane dehalogenase. Step 1, substrate binding; step 2, carbon-halogen bond cleavage; step 3, cleavage of covalent intermediate and alcohol release; step 4, conformational change preceding halide release.

Transient Kinetic Studies

We have spent a lot of effort on deriving a kinetic mechanism and establishing the relevant rate constants of conversion by haloalkane dehalogenase, since the specificity and kinetics of an enzyme-catalyzed reaction is determined by the individual rate constants at the active site, including rates of substrate binding, chemical steps, and product dissociation (Figure 4). The kinetic studies are facilitated by the fact that the intrinsic fluorescence of the dehalogenase is quenched by binding of halides or substrates in the active-site cavity. Two tryptophans (Trp125 and Trp175) are involved in halogen or halide binding due to interaction with the hydrogens bound to the indole nitrogens (Verschueren *et al.*, 1993b). The fluorescence quenching caused by halogen and halide binding makes it possible to determine dissociation constants and to perform stopped-flow studies for measuring the rates of chemical steps in the active site. In such experiments, enzyme and substrate are mixed and the change in fluorescence is followed on a msec to sec time scale. As outlined below, the most complex part of the kinetic mechanism is what seems to be the simplest step of the catalytic cycle: release of the halide ion from the buried active-site cavity.

Additional kinetic studies have been carried out with rapid-quench experiments, in which the reaction of enzyme with substrate is chemically stopped at variable times (msec – sec) after mixing. Non-covalently bound substrate and product (alcohol) are then analyzed by gas chromatography with a sensitivity in the micromolar range. The data are analyzed by direct numerical fitting to reaction schemes or by finding analytical solutions of rate equations. The values of the rate constants are optimized during the fitting process.

From these experiments, a kinetic scheme could be derived which includes all the different reaction rate constants (Schanstra *et al.*, 1996a). We have found these experiments to be essential for understanding the kinetics and specificity of the wild-type enzyme and the properties of mutants. It has allowed us to identify why some substrates are poorly converted and what should be done to improve the activities.

The Kinetics of Halide Export

Initial steady-state and pre-steady-state kinetic measurements indicated that halide release might be the rate-limiting step in the overall catalytic mechanism. For example, rapid-quench experiments performed with excess substrate over enzyme showed a significant burst of 2-bromoethanol production before the steady-state rate is reached. Therefore, halide release was directly analyzed using stopped-flow fluorescence (Schanstra and Janssen, 1996). The results showed that halide release follows a complex mechanism (Figure 4). The essential features of this mechanism and its kinetics are:

- there are two parallel routes, one of which is only kinetically significant for halide import at high concentrations (lower route in Figure 4);
- the major route for halide export is a three-step process, starting with a slow conformational change which is the actual rate-limiting step, and is required to allow halide to leave the active site. After reaching an open conformation, halide rapidly diffuses out. Subsequently, there is a rapid conformational reversal to a state in which the enzyme cannot readily bind halide (upper route in Figure 4);
- halide import at low concentration follows the reversal of the halide export route, with a slow conformational change being required before halide can bind;
- halide export is indeed a slow process, with overall rates of 4 and 8 sec^{-1} for bromide and chloride, respectively. These rates are close to the k_{cat} of the enzyme for the corresponding dihaloethanes and they are limited by the rate of the conformational change preceding actual release. The kinetics of halide release also explain the similar rates (k_{cat}) of 1,2-dibromoethane and 1,2-dichloroethane conversion, although the rate of cleavage of the C-Cl bond is much slower than that of the C-Br bond.

Overall Kinetic Mechanism of Haloalkane Dehalogenase

The complete kinetic mechanism of the enzyme was solved by a combination of stopped-flow fluorescence (both single and multiple turnover experiments) of substrate conversion and halide binding, rapid quench, and steady-state experiments (Schanstra *et al.*, 1996a). The most important conclusions are:

- both substrate binding and C-X cleavage are faster for bromoalkanes than for chloroalkanes. This explains that the K_m for 1,2-dibromoethane is much lower than for 1,2-dichloroethane;
- the conformational change required for halide release is the slowest step for good substrates, such as 1,2-dichloroethane and 1,2-dibromoethane;
- cleavage of the covalent intermediate is not much faster than halide release and occurs at a similar rate for 1,2-dichloroethane and 1,2-dibromoethane;
- long-chain chloroalkanes and polar substrates are poorly converted (low k_{cat} and high K_m, respectively) due to poor binding and/or a low rate of carbon-halogen bond cleavage.

Table 1 Steady-state kinetic parameters and rate constants of haloalkane dehalogenase.

Enzyme/substrate	K_m (mM)	k_{cat} (s^{-1})	k_{cat}/K_m ($M \cdot s^{-1}$)	k_2[a] (s^{-1})	k_4[a] (s^{-1})
1,2-dibromoethane					
Wild type	0.01	3.0	$3.0 \cdot 10^5$	>130	9 ± 1
Phe172Trp	0.025	5.9	$2.4 \cdot 10^5$	30 ± 5	100 ± 30
Val226Ala	0.033	8.2	$2.5 \cdot 10^5$	60 ± 20	58 ± 6
Asp260Asn	n.a.[b]	n.a	<0.5	n.d.[b]	n.d.
Asp260Asn + Asn148Glu	0.43	0.35	600	0.55 ± 0.05	50 ± 5
1,2-dichloroethane					
Wild type	0.53	3.3	$6.2 \cdot 10^3$	50 ± 10	14.5 ± 0.5
Phe172Trp	5.13	2.9	$5.6 \cdot 10^2$	4.5 ± 1	100 ± 10
Val226Ala	1.5	3.8	$2.5 \cdot 10^3$	14 ± 1	65 ± 5
Asp260Asn + Asn148Glu	n.a.	n.a.	<0.5	n.d.	n.d.

[a] k_2, rate constant for carbon-halogen bond cleavage, k_4, rate constant for the conformational change preceeding halide release (Figure 4).

[b] n.a., no activity; n.d., not determined.

Mutants with Faster Conversion of 1,2-Dichloroethane and 1,2-Dibromoethane

Several mutants were constructed that have a significantly improved k_{cat} with 1,2-dibromoethane or 1,2-dichloroethane compared to the wild type. Their properties have given some insight in the structural correlate of the kinetically observed steps during halide release and also revealed some of the bottlenecks associated with improving the catalytic performance of the enzyme for specific substrates.

First, a set of Phe172 mutants was studied (Schanstra *et al.*, 1996b). Phe172 is part of the substrate binding site, but it was not proposed to be directly involved in catalysis. A Phe172Trp mutant enzyme was purified and subjected to kinetic and structural analysis (Table 1). The mutant was found to have a higher k_{cat} for 1,2-dibromoethane, which was due to a faster conformational change required for halide release. The k_{cat} for 1,2-dichloroethane was not increased since the higher rate of halide export was accompanied by a lower rate of cleavage of the carbon-chlorine bond. For 1,2-dichloroethane, C-X cleavage had become rate-limiting in the mutant enzyme. For 1,2-dibromoethane, C-X cleavage is much faster than for 1,2-dichloroethane, and halide export remained the slowest step in the mutant, although it occurs at a higher rate than in the wild type.

The X-ray structure of the Phe172Trp mutant was solved by I. Ridder, B.W. Dijkstra and colleagues (University of Groningen). A small conformational difference was found in the cap domain that shields the buried active site cavity from the bulk solvent. This might have caused destabilization of the region around residues 170-174 and suggests that the conformational change required for halide export occurs in the cap domain.

Similar properties were observed for mutant Val226Ala (Schanstra *et al.*, 1997). This mutant was constructed because it was expected to show improved conversion of β-substituted haloalkanes, which appeared not to be the case, however. Rather, the enzyme showed faster conversion of 1,2-dibromoethane and 1,2-dichloroethane resulting from a higher rate of halide release (Table 1).

The Nature of the Slow Conformational Change

The structural basis of the kinetically observed conformational changes is not yet clear. Molecular dynamics simulations carried out by T. Linssen in the group of H. Berendsen (Biophysical Chemistry, University of Groningen) indicated significant motions in the cap domain that covers the active-site cavity and in helix 10 of the main domain (T. Linssen *et al.*, manuscript in preparation). We suspected that the conformational change takes place in the helix4-loop-helix5 region of the cap domain (Schanstra *et al.*, 1996b). This could partially open up the active-site and allow water to enter and solvate the halide ion. The solvated halide could then leave the active site, according to a rapid equilibrium (Figure 4)

The identity and role of amino acids of which the position is changed during this process and the significance of residues that could compensate the charge on the halide during its release are currently further studied by site-directed mutagenesis and molecular dynamics simulations.

Changing the Position of the Active Site 'Charge Relay' Aspartate

Asp260 of haloalkane dehalogenase was proposed to play a role in stabilizing the positive charge which develops on His289 during activation of water to hydrolyze the covalent intermediate (Figure 3). Its role was tested by mutating it to an Asn (Krooshof *et al.*, 1997). This mutant showed no catalytic activity. Asp260 is probably also required for correct folding of the protein, since the Asp260Asn mutant was unstable and formed inclusion bodies during expression, even at reduced growth temperature.

The effect of mutating Asp260 to Asn could be compensated by replacing Asn148 by an aspartate or glutamate (Figure 1). In this way, the position of the residue involved in stabilizing His289 was shifted from the end of β-strand 7 to the end of β-strand 6. These residues occupy similar positions in the structure, and at both sites a carboxylate could interact with His289. The double mutant showed a much higher activity for 1,2-dibromoethane than the single mutant Asp260Asn, which was also unstable.

Pre-steady-state kinetic measurements have shown that the cleavage of the covalent intermediate is significantly slower in this double Asp260Asn + Asn148Glu mutant enzyme than in the wild type, as expected. Yet, the rate of C-X bond cleavage was reduced even more in the mutant, making this the rate-limiting step. These results indicate that Asp260 is important for correct active-site geometry, in addition to a role in cleavage of the covalent intermediate.

Molecular Modelling of Halide Ion Stabilization

Experimental determination of the 3D structure of haloalkane dehalogenase stimulated a number of computer modelling experiments. The mechanism of the dehalogenation reaction and the role of the catalytic residues for the reaction kinetics and thermodynamics were studied using semi-empirical (AM1) quantum mechanics (Damborsky *et al.*, 1997a; Kuty *et al.*, 1997), and energetically favourable binding sites for the halide ions were proposed from calculation of the molecular interaction fields (Damborsky *et al.*, 1997b). Furthermore, docking experiments were applied to identify different binding modes for saturated and unsaturated substrates.

The quantum mechanical studies suggested that hydrolysis of the alkyl-enzyme intermediate is the slowest chemical reaction step, which is in agreement with the kinetic measurements (Schanstra *et al.*, 1996b). Monitoring of the partial atomic charges on catalytic residues during the reaction course showed a development of positive charges on the nitrogen-bound hydrogens of the two active-site tryptophans (Trp125 and Trp175), which confirmed that these residues contribute to product (halide ion) stabilization, as already proposed from fluorescence quenching (Verschueren *et al.*, 1993b) and site-directed mutagenesis experiments (Kennes *et al.*, 1995).

Residue Phe172 was identified as a third contributor to the stabilization of the developing halide ion. Therefore, quantum mechanical calculations of the carbon-halogen bond cleavage reaction were performed with simple models of all 15 mutants of Phe172 for which activities were reported (Schanstra *et al.*, 1996b). Changes in the charge on the stabilizing atom of residue 172 qualitatively correlated with experimentally determined activities (Figure 5).

QSFR Analysis of Dehalogenase Mutants

Molecular modelling techniques are especially useful in cases where the 3D structure of the protein is known. We have tested whether statistical analysis can be used to establish structure-function relationships in mutants for which no structure is available. A quantitative structure-function relationship (QSFR) analysis was performed with wild type and 15 dehalogenase variants point-mutated at position 172 (Damborsky, 1997c). It was assumed that the mutant proteins are correctly folded. Initially, the substituted amino acids were characterized by 33 indexes quantifying various physico-chemical and structural properties, e.g. size, accessibility to solvent, hydrophobicity, conformational preferences and electronic properties. Four indexes carrying information about the amino acids in the mutant proteins were identified using a projection to latent structures (PLS) analysis, and a mathematical model was constructed which relates these indexes to the activity. The indexes fixed in the model (bulkiness, flexibility, refractivity and aromaticity) are in agreement with mechanistic knowledge about the role of residue 172 in catalysis. These results indicate that QSFR models can be helpful for the interpretation of results from site-directed mutagenesis experiments and that they might be used for predictive purposes.

Comparison of Specificities and 3D Structures of Haloalkane Dehalogenases

An understanding of the structural causes of the differences in substrate specificity among various dehalogenases can be important for efforts to broaden the specificity of from *X. autotrophicus*, which is at present the only haloalkane dehalogenase

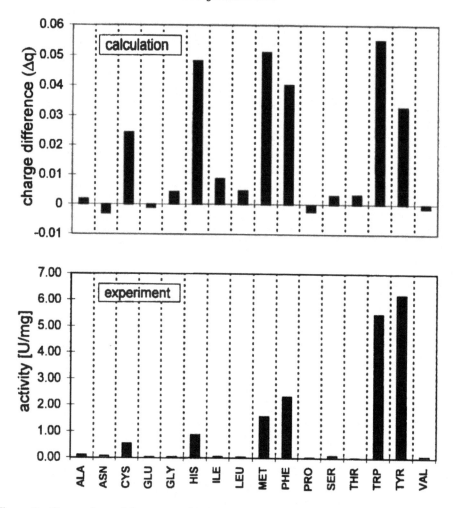

Figure 5   Comparison of the extent of halide ion stabilization and dehalogenase activity for wild type and 15 mutants of DhlA. Only the mutant proteins with significant stabilization provided by residue 172 ($\Delta q > 0.02$ units) showed good activity (Schanstra *et al.*, 1996b). The charge differences ($\Delta q = |q^{educt} - q^{product}|$) refer to the increase in charge on the hydrogen atom that was positioned nearest to the released halide ion (Damborsky, 1997).

for which an X-ray structure is known. The sequences of two other haloalkane dehalogenases were recently reported: LinB from *Sphingomonas paucimobilis* UT26 (Nagata *et al.*, 1993) and DhaA from *Rhodococcus rhodochrous* NCIMB13064 (Kulakova *et al.*, 1997). These enzymes have related sequences: ~50% similarity and ~30% identity.

A comparison of the specificities of these and some other dehalogenases has been carried out using statistical methods. A principal component analysis revealed clustering of the dehalogenases into three distinct substrate-specificity classes (Damborsky *et al.*, 1997b): (1), a group around DhlA of *X. autotrophicus* and *Ancylobacter aquaticus* (van den Wijngaard *et al.*, 1992); (2) a group including DhaA and the

A

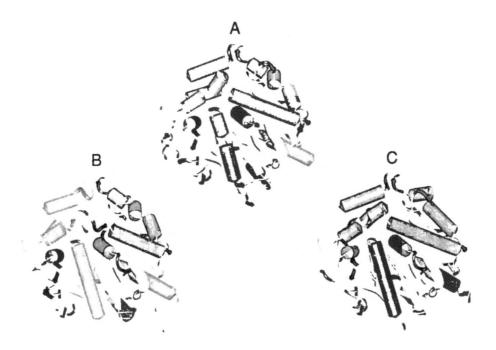

B   C

Figure 6 Comparison of the overall fold of three haloalkane dehalogenases. (A), experimentally determined structure of DhlA from *X. autotrophicus* GJ10; (B) homology model of LinB from *S. paucimobilis* UT26; C, idem, DhaA of *Rhodococcus* NCIMB13064. Cylinders represent α-helices and arrows represent β-sheets. The secondary structure elements were assigned according to Kabsch *et al.* (1983). Homology models were built using the procedure described previously (Sali *et al.*, 1993; Damborsky *et al.*, 1995).

dehalogenases from *Rhodococcus* HA1 (Scholtz *et al.*, 1987), *Rhodococcus erythropolis* Y2 (Sallis *et al.*, 1990), *Corynebacterium* m15-3 (Yokota *et al.*, 1987), and strain GJ70 (Janssen *et al.*, 1987, 1988); (3), the LinB protein, involved in hexachlorohexane metabolism. The first class differs from the others mainly by the high activity with 1,2-dichloroethane and the inability to dehalogenate long-chain chlorinated substrates such as 1-chlorohexane.

Homology modelling was used to construct the hypothetical 3D structures of LinB and DhaA using the structure of DhlA as a template (Damborsky *et al.*, 1995; 1997c; unpublished results). All three dehalogenases share the same tertiary structure (Figure 6). They all carry a nucleophile elbow, the extremely sharp turn which allows the catalytic nucleophile (aspartic acid) to stand proud of the rest of the active site, suggesting the same reaction mechanism for all enzymes. Another residue of the catalytic triad, histidine, is conserved in the dehalogenases. The third residue of the catalytic triad (Asp260 in DhlA) is not conserved in the modelled proteins, but the most likely candidates are Glu130 in DhaA and Glu132 in LinB (Krooshof *et al.*, 1997).

The role of two tryptophans, Trp125 and Trp175, for the stabilization of the halide ion and transition state of the first reaction step in DhlA has been discussed above. In LinB and DhaA there is no aromatic amino acid in the model at a position corresponding to that of Trp175 in DhlA. A missing tryptophan significantly increases the energy barrier of the first reaction step, and therefore can be the reason for the inability of these enzymes to dehalogenate 1,2-dichloroethane (Kennes *et al.*, 1995).

Another important structural difference among the studied dehalogenases is the entrance to and size of the active site cavity. In DhlA, the active site is completely buried inside the protein and is connected to the protein surface via a narrow tunnel. This may be important for degradation of small molecules which have to be tightly bound in the hydrophobic active site during catalysis. The models of LinB and DhaA suggest that larger molecules can be more easily bound in the active site, in agreement with the higher activities of these enzymes for long-chain chloroalkanes. It also can be proposed that these enzymes will have less complex reaction kinetics since no extensive conformation change will be required for release of the halide ion from their active sites, since it is not a buried cavity as in DhlA.

Possibilities for Further Improvement

A number of important conclusions can be drawn from the site-directed mutagenesis studies and the kinetic analysis. First, the enzyme is evolutionary optimized for the conversion of 1,2-dichloroethane, in the sense of maximizing k_{cat}/K_m. For enzymes which have to convert toxic pollutants, k_{cat}/K_m is the most important kinetic parameter since the substrate conversion will be (or should become) very low. This specificity constant is given by:

This parameter was not improved for 1,2-dichloroethane in all the mutants that we studied, although the k_{cat} was elevated in some of them (e.g. Val226Ala (Schanstra *et al.*, 1997), see Table 1). This seems to be caused by the fact that k_2 (the rate of C-X cleavage) and k_4 (the rate of the conformational change that allows halide release, Figure 4) are in some way inversely correlated. Structural or dynamic properties which increase k_4 apparently always lead to a decrease of k_2 for 1,2-dichloroethane. One may see this as follows. Halide release is faster if the conformational changes required for it occur at a higher rate, i.e. if the regions in the cap domain of the enzyme become more flexible. On the other hand, cleavage of the carbon-halogen bond in a small substrate such as 1,2-dichloroethane probably requires a very precise positioning (or, in terms of dynamics, a high frequency with which reactive configurations are reached) of the substrate in the active site, which would be negatively influenced by increased flexibility. The fact that the chloroalkanes used are small hydrophobic molecules leaves little opportunity for creating a substrate-binding site with more specific interactions. Furthermore, the halide ion released during C-Cl bond cleavage needs to be well stabilized for good kinetics of this chemical reaction step, which is achieved by electrostatic interaction with partially positively charged atoms of Trp125, Trp175, and Phe172. On the other hand, the same non-bonding interactions make release of the halide ion difficult. Any structural change that makes electrostatic contributions of Trp125, Trp175, and Phe172 weaker will lead to faster halide release but slower C-X cleavage.

Another important point is that haloalkane dehalogenase is evolutionary better optimized for 1,2-dichloroethane than for 1,2-dibromoethane, although the latter has a low K_m and higher k_{cat}/K_m. The balance between carbon-halogen bond cleavage and halide release is not as well optimized, however, and several mutants with a higher k_{cat} for 1,2-dibromoethane have been obtained. Again, increased halide export is accompanied by a decrease in k_2. The k_{cat}/K_m is less sensitive to this decrease of k_2 than in the case of 1,2-dichloroethane, probably because k_2 is much higher for the brominated analog. 1,2-Dibromoethane is an important soil fumigant that has been extensively used in citrus crops. Bacterial growth on this compound has never been observed, but the wild-type and engineered dehalogenases are capable of rapidly hydrolyzing it to 2-bromoethanol, which may support growth.

Currently, we are investigating the possibility for improving dehalogenase activity towards compounds which are poorly converted due to slow carbon-halogen bond cleavage. This is reflected kinetically in a low k_{cat}, a high K_m, and low k_{cat}/K_m. Interesting substrates, both from a biochemical and biotechnological point of view, are 1,1,2-trichloroethane, which is a competitive inhibitor, and 2-haloalkanes, which are optically active. The possibility of increasing the conversion of substrates that display slow C-X cleavage has already been demonstrated by *in vivo* selection experiments (Pries *et al.*, 1994b). Mutants that show faster conversion of 1-chlorohexane were easily obtained, and all of them appeared to carry mutations in the cap domain (6 different mutations in total). This also shows that the cap domain is the target for adaptive mutations. In fact, we have proposed that the short sequence repeats in the cap domain of haloalkane dehalogenase are the result of recent mutations that are responsible for adaptation of the enzyme to industrially produced 1,2-dichloroethane (Pries *et al.*, 1994b).

DISCUSSION AND FUTURE PROSPECTS

Our studies on haloalkane dehalogenase have demonstrated that the specificity of a dehalogenation reaction may be influenced by selection for spontaneous mutations or by site-directed mutagenesis. We obtained several mutants with increased conversion (k_{cat}) of long-chain chloroalkanes or 1,2-dibromoethane, but the inverse relationship between the rates of carbon-halogen bond cleavage and halide release made it difficult to improve the enzyme for 1,2-dichloroethane and some related environmental pollutants. Furthermore, from the nucleophilic reaction mechanism it may be expected that haloalkane dehalogenase will never be suitable for the hydrolysis of halogens bound to an unsaturated carbon atom, as in chloroethenes or chlorinated aromatics. This will also hold for the L-2-chloropropionic acid dehalogenase of which the structure was recently solved by Soda and coworkers (Hisano *et al.*, 1996). The reaction mechanism has some similarties to that of haloalkane dehalogenase. A conserved Asp close to the N-terminus is the nucleophile that is expected to form a covalent intermediate.

The only hydrolytic dehalogenase that is known to convert a chloroaromatic compound is the 4-chlorobenzoyl CoA dehalogenase from *P. putida* CBS3, which is involved in the degradation of 4-chlorobenzoate to 4-hydroxybenzoate. Dehalogenation occurs after activation of the carboxyl group to a coenzyme A derivative, which makes the aromatic ring sensitive to nucleophilic attack (Scholten *et al.*, 1991). From the X-ray structure, the carboxylate of Asp145 was concluded to be the nucleophile that leads to formation of a Meisenheimer complex as the intermediate, and a histidine was proposed to be involved in the hydrolytic cleavage of the intermediate, just as His289 of haloalkane dehalogenase (Benning *et al.,* 1996). No other enzymes that directly attack vinylic or arenic halogens have been studied in detail so far.

Further studies on the conversion of small chlorinated substrates of high environmental importance by dehalogenase mutants are being continued. In the last section of this review, we summarize some inherent limitations of engineering better biocatalysts for the transformation of environmental pollutants.

Specificity of Binding of Small Hydrophobic Molecules

Many important environmental pollutants, such as the chlorinated solvents, are small hydrophobic molecules with few 'handles' for an enzyme to achieve specific binding. Since substrate binding is not very precise (in contrast to the case of substrates which form hydrogen bonds), it is difficult to predict the catalytic properties of mutants from modelling studies looking only at the complementarity between the protein and the substrate molecules. This problem may possibly be circumvented by:

– the use of the modelling techniques studying the dynamic and catalytic properties of the protein, e.g., molecular dynamics and quantum mechanics;
– the use of combinatorial methods, such as PCR based mutagenesis or gene shuffling, followed by screening on the basis of growth or enzyme activity. This encompasses the need for detailed structure-function insight;
– the use of sequence-activity relationships, i.e. combining activities of different enzymes by splicing different parts of homologous proteins together;
– the use of enzymes which have a very broad activity or a different reaction mechanism, e.g. by formation of radicals, as is the case with some monooxygenases and cytochrome P450. These enzymes generate strong electrophiles by the activation of molecular oxygen and react aspecifically with a wide range of chlorinated substrates.

The Use of Molecular Dynamics: Cytochrome P450

The use of molecular dynamics to serve as a basis for improving the substrate specificity of an enzyme with a small hydrophobic substrate was demonstrated by Ornstein, Sligar, and coworkers (Loida *et al.,* 1995). Based on MD simulations, they constructed mutants of cytochrome P450$_{cam}$ with increased coupling during the conversion of camphor, and with a modified regioselectivity. MD trajectories were analyzed for potentially reactive configurations, and in this way the properties of a mutant enzyme could be predicted even for a hydrophobic substrate such as camphor. Mutants with improved reductive conversion of chloroethanes were identified and constructed using a similar approach (Manchester and Ornstein, 1995a,b). A three-fold increase was observed in the activity of reductive dehalogenation of pentachloroethane with a Phe to Trp mutant which increase was due to an increased frequency of contacts between the substrate and the haem group due to the smaller accessible space in the active site.

Engineering on Basis of Homologous Sequences

If no information on the structure-function relationship is available, it may be attempted to engineer better enzymes on basis of sequence-activity relationships, i.e. by combining properties which are specified by parts of different polypeptide chains. For example, Furukawa and coworkers studied the possibilities to enhance the activity of biphenyl dioxygenases, which are the initial enzymes in the bacterial catabolic pathway of biphenyl and chlorobiphenyls. The activity of these enzymes involves four polypeptides and the catalytic site is a [2Fe-2S] cluster located in the $\alpha_2\beta_2$ terminal dioxygenase. The substrate range of these enzymes mainly determines the range of polychlorinated biphenyls that can be converted by *Burkholderia cepacia* LB400 and *Pseudomonas pseudoalcaligenes* KF707. Although the hydroxylase components are more than 95% identical, the LB400 enzyme has a much wider substrate range. This was recently attributed to the C-terminal half of the large hydroxylase subunit (Kimura *et al.*, 1997). It was shown that a single amino acid replacement (Thr376 to Asn) accounts for most of the differences in substrate selectivity. Introducing this mutation in the KF707 enzyme yielded an enzyme with enhanced substrate range and the acquisition of 3,4-dioxygenase activity towards 2,5,4'-trichlorobiphenyl, a regioselectivity and activity which is absent in the normal KF707 enzyme (Kimura *et al.*, 1997). Similar studies were carried out by Erickson and Mondello (1993), who replaced residues of the LB400 enzyme with amino acids of KF707. This also led to an expansion of the substrate range. Unfortunately, no detailed kinetic analysis or structural information is available about these enzymes.

Another interesting property of these aromatic dioxygenases is their capacity to degrade trichloroethylene. Better activities for trichloroethylene in whole cells could be obtained using hybrid enzymes in which the hydroxylase subunits of the *Pseudomonas putida* enzyme were combined with polypeptides for electron transfer from biphenyl degraders, (Furukawa *et al.*, 1994).

Conversion by Aspecific Oxidative Enzymes

Due to the nucleophilic character of the reaction mechanism, hydrolytic dehalogenases do not usually convert chloroethenes and other low-molecular weight compounds which are not sensitive to SN_2 attack. Conversion is possible, however, by oxygenases which have a completely different reaction mechanism, as they activate molecular oxygen. In this respect, an interesting class of enzymes are the monooxygenases of hydrocarbon-utilizing bacteria. These enzymes have a broad substrate range, which includes trichloroethylene, one of the most important groundwater contaminants. The best studied examples are soluble methane monooxygenases, of which X-ray structures are known (Rosenzweig *et al.*, 1993; Elango *et al.*, 1997), toluene-4-monooxygenase, and toluene-2-monooxygenase. These enzymes are three-component oxygenases composed of an $\alpha_2\beta_2\gamma_2$ hydroxylase that contains a two-iron active site (component A), a small modifying protein that influences hydroxylase coupling (component B), and a flavo-iron-sulfur protein that transfers electrons from NADH to the hydroxylase (C). Based on sequence analysis, toluene-4-mono-oxygenase, methane monooxygenase, and other phenol hydroxylases have similar overall structures, with conserved positions of the active-site residues.

In general, goals for engineering studies with these enzymes may be: an improved activity for trichloroethylene (the conversion rate is rather low), a reduced toxicity of chloroethylene degradation products, and an improved activity with other substrates, such as chloromethanes and chloroethanes or even chloroaromatics. Trichloroethene-degrading monooxygenases are subject to covalent modification by reactive products which leads to inactivation (Oldenhuis *et al.*, 1991; Newman and Wackett, 1995; 1996) and their applicability would improve if it were possible to construct mutants with decreased inactivation.

Natural Adaptation

Perhaps the most important 'competitor' to the efforts of protein engineers to obtain new biocatalysts for the degradation of xenobiotic compounds is nature itself. There may be much better starting points for obtaining a novel or improved enzymate activity than the enzymes that are currently used by the microbiologists that study catabolic pathways. Nature may also use shortcuts that are not commonly included in random or site-directed mutagenesis studies, such as gene exchange and generation of insertions or deletions. Since the acquisition of novel catabolic traits is based on the growth-related selection of rare genetic events, it also should be noted that the size of engineering efforts will be small relative to the magnitude of the evolutionary processes that occur in nature, unless the targets for engineering studies are very carefully selected.

REFERENCES

Benning, M.M., Taylor, K.L., Liu, R.-Q., Yang, G., Xiang, H., Wesenberg, G., Dunaway-Mariano, D. and Holden, H.M. (1996) Structure of 4-chlorobenzoyl coenzyme A dehalogenase determined to 1.8 Å resolution: an enzyme catalyst generated via adaptive mutation. *Biochemistry*, **35**, 8103–8109.

Damborsky, J. (1997) Quantitative structure-function relationships of the single-point mutants of haloalkane dehalogenase: a multivariate approach. *Quant. Struct.-Act. Relat.*, **16**, 126–135.

Damborsky, J., Bull, A.T. and Hardman, D.J. (1995) Homology modelling of the haloalkane dehalogenase of *Sphingomonas paucimobilis* UT26. *Biologia*, **50**, 523–528.

Damborsky, J., Kuty, M., Nemec, M. and Koca, J. (1997a) A molecular modelling study of the catalytic mechanism of haloalkane dehalogenase: I. quantum chemical study of the first reaction step. *J. Chem. Inf. Comp. Sci.*, **37**, 562–568.

Damborsky, J., Kuty, M., Nemec, M. and Koca, J. (1997b) Molecular modelling to understand the mechanisms of microbial degradation — application to hydrolytic dehalogenation with haloalkane dehalogenases. In *QSAR in Environmental Sciences — VII*, edited by F. Chen and G. Schuurmann. Pensacola: SETAC Press.

Damborsky, J., Nyandoroh, M.G., Nemec, M., Holoubek, I., Bull, A.T. and Hardman, D.J. (1997c) Some biochemical properties and classification of a range of bacterial haloalkane dehalogenases, *Biotech. Appl. Biochem.*, **26**, 19–25.

Elango, N., Radhakrishnan, R., Froland, W.A., Wallar, B.J., Earhart, C.A., Lipscomb, J.D. and Ohlendorf, D.H. (1997) Crystal structure of the hydroxylase component of methane monooxygenase from *Methylosinus trichosporium* OB3b. *Protein Sci.*, **6**, 556–568.

Erickson, B.D. and Mondello, F.J. (1993) Enhanced biodegradation of polychlorinated biphenyls after site-directed mutagenesis of a biphenyl dioxygenase gene. *Appl. Environ. Microbiol.*, **59**, 3858–3862.

Furukawa, K., Hirose, J., Hayashida, S. and Nakamura, K. (1994) Efficient degradation of trichloroethylene by a hybrid aromatic ring dioxygenase. *J. Bacteriol.*, **176**, 2121–2123.

Hisano, T., Hata, Y., Fujii, T., Liu, J.Q., Kurihara, T., Esaki, N. and Soda, K. (1996) Crystal structure of L-2-haloacid dehalogenase from *Pseudomonas* sp. YL. An α/β hydrolase structure that is different from the α/β hydrolase fold. *J. Biol. Chem.*, **271**, 20322–20330.

Janssen, D.B. (Ed.) (1994) Genetics of biodegradation of xenobiotic compounds. *Biodegradation*, 5, Nos 3–4.

Janssen, D.B., Gerritse, J., Brackman, J., Kalk, C., Jager, D. and Witholt, B. (1998) Purification and characterization of a bacterial dehalogenase with activity toward halogenated alkanes, alcohols and ethers. *Eur. J. Biochem.*, 171, 67–72.

Janssen, D.B., Jager, D. and Witholt, B. (1987) Degradation of n-haloalkanes and α,ω-dihaloalkanes by wild-type and mutants of *Acinetobacter* sp. strain GJ70. *Appl. Environ. Microbiol.*, 53, 561–566.

Kabsch, W. and Sander, C. (1983) Dictionary of protein secondary structure: pattern recognition of hydrogen-bonded and geometrical features. *Biopolymers*, 22, 2577–2637.

Kennes, C., Pries, F., Krooshof, G.H., Bokma, E., Kingma, J. and Janssen, D.B. (1995) Replacement of tryptophan residues in haloalkane dehalogenase reduces halide binding and catalytic activity. *Eur. J. Biochem.*, 228, 403–407.

Kimura, N., Nishi, A., Goto, M. and Furukawa, K. (1997) Functional analysis of a variety of chimeric dioxygenases constructed from two biphenyl dioxygenases that are similar structurally but different functionally. *J. Bacteriol.*, 179, 3936–3943.

Krooshof, G.H., Kwant, E.M., Damborsky, J. Koca, J. and Janssen D.B. (1997) Repositioning the catalytic triad aspartic acid of haloalkane dehalogenase: effects on stability, kinetics, and structure. *Biochemistry*, in press.

Kulakova, A.N., Larkin, M.J. and Kulakov, L.A. (1997) The plasmid-located haloalkane dehalogenase gene from *Rhodococcus rhodochrous* NCIMB 13064. *Microbiology*, 143, 109–115.

Kuty, M., Damborsky, J., Prokop, M. and Koca, J. (1997) A molecular modelling study of the catalytic mechanism of haloalkane dehalogenase: 2. quantum chemical study of complete reaction mechanism. Submitted for publication.

Loida, P.J., Sligar, S.G., Paulsen, M.D., Arnold, G.E. and Ornstein, R.L. (1995) Stereoselective hydroxylation of norcamphor by cytochrome P450cam. Experimental verification of molecular dynamics simulations. *J. Biol. Chem.*, 270, 5326–5330.

Manchester, J.I. and Ornstein, R.L. (1995) Molecular dynamics simulations indicate that F87W,T185F-cytochrome P450cam may reductively dehalogenate 1,1,1-trichloroethane. *J. Biomol. Struct. Dyn.*, 13, 413–422.

Manchester, J.I. and Ornstein, R.L. (1995) Enzyme-catalyzed dehalogenation of pentachloroethane: why F87W-cytochrome P450cam is faster than wild type. *Protein Eng.*, 8, 801–807.

Nagata, Y., Nariya, T., Ohtomo, R., Fukuda, M., Yano, K. and Takagi, M. (1993) Cloning and sequencing of a dehalogenase gene encoding an enzyme with hydrolase activity involved in the degradation of gamma-hexachlorocyclohexane in *Pseudomonas paucimobilis*. *J. Bacteriol.*, 175, 6403–6410.

Newman, L.M. and Wackett, L.P. (1995) Purification and characterization of toluene 2-monooxygenase from *Burkholderia cepacia* G4. *Biochemistry*, 34, 14066–14076.

Newman, L.M. and Wackett, L.P. (1996) Trichloroethylene oxidation by purified toluene 2-monooxygenase: products, kinetics, and turnover-dependent inactivation. *J. Bacteriol.*, 179, 90–96.

Oldenhuis, R., Oedzes, J.Y., Van der Waarde J.J. and Janssen, D.B. (1991) Kinetics of chlorinated hydrocarbon degradation by *Methylosinus trichosporium* OB3b and toxicity of trichloroethylene. *Appl. Environm. Microbiol.*, 57, 7–14.

Ollis, D.L., Cheah, E., Cygler, M., Dijkstra, B.W., Frolow, F., Franken, S.M., Haral, M., Remington, S.J., Silman, I., Schrag, J., Sussman, J.L., Verschueren, K.H.G. and Goldman, A. (1992) The α/β-hydrolase fold. *Protein Eng.*, 5, 197–211.

Pries, F., Kingma, J., Pentenga, M., Van Pouderoyen, G., Jeronimus-Stratingh, C.M., Bruins, A.P. and Janssen, D.B. (1994a) Site-directed mutagenesis and oxygen isotope incorporation studies of the nucleophilic aspartate of haloalkane dehalogenase. *Biochemistry*, 33, 1242–1247.

Pries, F., van den Wijngaard, A.J., Bos, R., Pentenga, M. and Janssen, D.B. (1994b) The role of spontaneous cap domain mutations in haloalkane dehalogenase specificity and evolution. *J. Biol. Chem.*, 269, 17490–17494.

Rosenzweig, A.C., Frederick, C.A., Lippard, S.J. and Nordlund, P. (1993) Crystal structure of a bacterial non-haem iron hydroxylase that catalyses the biological oxidation of methane. *Nature*, 366, 537–543.

Sali, A. and Blundell, T.L. (1993) Comparative protein modelling by satisfaction of spatial restraints. *J. Mol. Biol.*, 234, 779–815.

Sallis, P.J., Armfield, S.J., Bull, A.T. and Hardman, D.J. (1990) Isolation and characterization of a haloalkane halidohydrolase from *Rhodococcus erythropolis* Y2. *J. Gen. Microbiol.*, 136, 115–120.

Schanstra, J.P. and Janssen D.B. (1996) Kinetics of halide release of haloalkane dehalogenase: evidence for a slow conformational change. *Biochemistry*, 35, 5624–5632.

Schanstra, J.P., Kingma, J. and Janssen, D.B. (1996a) Specificity and kinetics of haloalkane dehalogenase. *J. Biol. Chem.*, **271**, 14747–14753.

Schanstra, J.P., Ridder, I.S., Heimeriks, G.J., Rink, R., Poelarends, G.J., Kalk, K.H., Dijkstra, B.W. and Janssen, D.B. (1996b) Kinetic characterization and X-ray structure of a mutant of haloalkane dehalogenase with higher catalytic activity and modified substrate range. *Biochemistry*, **35**, 13186–13195.

Schanstra, J.P., Ridder, A., Kingma, J. and Janssen, D.B. (1997) Influence of mutations of Val226 on the catalytic rate of haloalkane dehalogenase. *Protein Engin.*, **10**, 53–61.

Scholten, J.D., Chang, K.H., Babbitt, P.C., Charest, H., Sylvestre, M. and Dunaway-Mariano, D. (1991) Novel enzymic hydrolytic dehalogenation of a chlorinated aromatic. *Science*, **253**, 182–185.

Scholtz, R., Leisinger, T., Suter, F. and Cook, A.M. (1987) Characterization of 1-chlorohexane halidohydrolase, a dehalogenase of wide substrate range from an *Arthrobacter* sp. *J. Bacteriol.*, **169**, 5016–5021.

Stucki, G. and Thüer, M. (1995) Experiences of a large-scale application of 1,2-dichloroethane degrading microorganisms for groundwater treatment. *Environ. Sci. Technol.*, **29**, 2339–2345.

Van den Wijngaard, A.J, Van der Kamp, K.W.H.J, Van der Ploeg, J.R., Pries F., Kazemier, B. and Janssen, D.B. (1992) Degradation of 1,2-dichloroethane by *Ancylobacter aquaticus* and other facultative methylotrophs. *Appl. Environ. Microbiol.*, **58**, 976–983.

Verschueren, K.H.G., Seljée, F., Rozeboom, H.J., Kalk, K.H. and Dijkstra, B.W. (1993a) Crystallographic analysis of the catalytic mechanism of haloalkane dehalogenase. *Nature*, **363**, 693–698.

Verschueren, K.H.G., Kingma, J., Rozeboom, H.J., Kalk, K.H., Janssen, D.B. and Dijkstra, B.W. (1993b) Crystallographic and fluorescence studies of the interaction of haloalkane dehalogenase with halide ions. Studies with halide compounds reveal a halide binding site in the active site. *Biochemistry*, **32**, 9031–9037.

Yokota, T., Omori, T. and Kodama, T. (1987) Purification and properties of haloalkane dehalogenase from *Corynebacterium* sp. strain m15-3. *J. Bacteriol.*, **169**, 4049–4054.

14. PROTEIN ENGINEERING FOR BIOSENSORS

MASUO AIZAWA, YASUKO YANAGIDA, TETSUYA HARUYAMA
and EIRY KOBATAKE

*Faculty of Bioscience and Biotechnology, Tokyo Institute of Technology,
Nagatsuta, Midori-ku, Yokohama 226, Japan*

INTRODUCTION

A keen interest has been focused on the information networks in biological systems, where information is transduced, transfered, processed, and stored by a molecule or molecular assemblies. The bioinformation networks may be classified into 1) gene information network, 2) intracellular information network, 3) intercellular information network, 4) sensory information network, and 5) brain information network. Of these bioinformation networks, both intracellular and intercellular information networks are characterized by molecular communication in which information molecules such as hormones, cytokines, and cAMP serve as messenger to transfer some specific information from one site to a target. Extensive researches have been concentrated on the mechanisms of information transduction at the target cells. It is generally recognized that molecular information can be transduced by receptor molecular assemblies on the cellular membranes.

Various types of receptors, which are embedded in the plasma membrane of a target cell, may recognize selectively the corresponding information molecule that is released from a specific cell, and transduce the information into another type of information molecule within the target cell. These receptors are the characteristic protein assemblies for the molecular information transduction across the cellular membranes.

We have long worked on molecular assembling of receptor proteins on the solid matrix surfaces for information transduction from molecular to electronic information (Aizawa *et al.*, 1989a and b; Aizawa, 1994a and b; Aizawa *et al.*, 1995). The protein assemblies for molecular/electronic information transduction has found a wider application in constructing biosensors, because the design concept of biosensors has been derived from the information transduction at the supramolecular systems of receptors embedded in biomembranes as schematically illustrated in Figure 1. In contrast to cellular receptor proteins, enzymes and antibodies have been implemented in biosensors primarily due to an easier transduction from molecular to electronic information on the solid matrix.

Several enzymes have been implemented in electrochemical devices to construct biosensors that have been commercially available and evaluated in various applications. In most enzyme sensors, an enzyme reaction is coupled with an electrochemical reactions for information transduction. An ordered molecular assembly of enzymes on the electrode surface is not necessarily required, if the products of the enzyme reaction can access the electrode surface. However, for an electron transfer type of enzyme sensor, in which enzyme molecules can communicate with the electrode surface through electron transfer, enzyme molecules should be

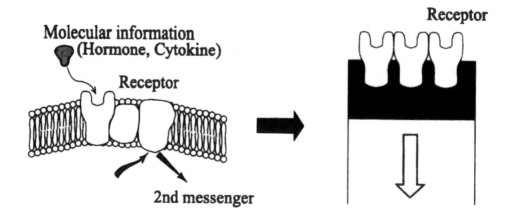

Figure 1 Schematic illustration of biosensor design principle based on the molecular information transduction at the cellular receptor supramolecular systems

properly assembled on the electrode surface so that the active site of the enzyme may be accessible to the electrode. It is extremely necessary that enzyme molecules should be engineered to fulfill these requirements, although enzymes have been used simply in native form for biosensor technology.

Antibodies have also been used for biosensors. Due to a lack of catalytic activity, antibody molecules are in general assembled in an ordered manner on the solid surface to induce a change in the physical properties after the antigen binding. Although these have been proposed so many principles of immunosensors, only a few have been on market. Major problems remained unsolved should be molecular engineering for modifying antibody molecules to get assembled in proper manner on the solid surface with retaining selective binding affinity.

A rapid progress has been made in protein engineering with the support of advanced gene engineering. It should be the next generation to obtain a totally artificially designed protein by protein engineering. However, protein engineering has been successful in rearranging the active center vicinity of a functional protein with modifying the activity and fusing two different functional proteins with retaining their native activities by synthesizing recombinant DNA.

An intensive research and development has been concentrated on implementing native enzymes and antibodies as they are into biosensors, although these proteins suffer from poor stability and processability. Biosensor research and development has been enforced to leap up the next generation of perspectives through the protein engineering of biosensor material.

Protein engineering is expected to offer biosensor materials with various advantages including enhanced resistivities against environmental conditions, superior process performances, and designed functionality. As listed in Figure 2, some of

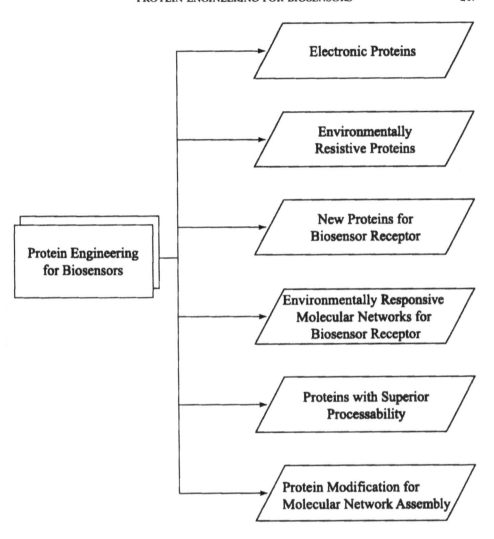

Figure 2   Protein engineering targets relating with biosensor research and development

these advantages have been realized by engineering proteins such as enzymes and antibodies. An expanding application of protein engineering will appear in biosensor research and development.

CHEMICALLY ENGINEERED ELECTRONIC PROTEINS

Potential Assisted Self-Assembly and Molecularly Wired Redox Enzymes on the Electrode Surface

One of the key technologies required for fabricating biomolecular electronic devices concerns the molecular assembly of electronic proteins such as redox enzymes in

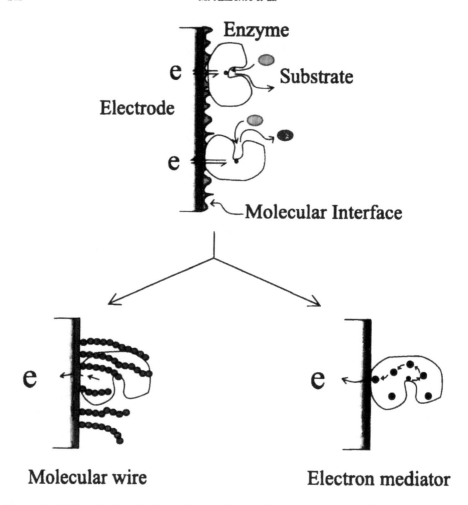

Figure 3   Molecular interfacing technology for making electronic proteins to communicate with an electrode

monolayers on electrode surfaces. Furthermore, the molecularly assembled electronic proteins are required to be electronically communicated with the electrode (Cardosi and Turner, 1991; Degani and Heller, 1987; Gleria and Hill, 1992; Khan *et al.*, 1992). Individual proteins molecules on electrode surfaces should be electronically accessed through the electrode. To fulfill these requirements, two fabrication processes have been proposed by us. One is a potential assisted self-assembly of redox enzymes on the electrode surface, which is followed by an electrochemical fabrication of a monolayer-scale conducting polymer on the electrode surface for molecular interfacing. The other is a self-assembly of mediator-modified redox enzymes on the porous gold electrode surface through the thiol-gold interaction. These two processes are shematically illustrated in Figure 3.

The potential-assisted self-assembly is carried out in an electrolytic cell equipped with a platinum or gold electrode (working electrode) on which a protein monomolecular layer is formed, a platinum counter electrode, and a Ag/AgCl reference electrode. The potential of the working electrode is precisely controlled with a potentiostat with reference to the Ag/AgCl electrode. The protein solution should be prepared, taking into account the isoelectric point because proteins are negatively charged in pH ranges above the isoelectric point.

Fructose dehydrogenase (FDH) is a redox enzyme which has pyrrole-quinoline quinone (PQQ) as a prosthetic group. Upon enzymatic oxidation of D-fructose, the prosthetic group (PQQ) is reduced to $PQQH_2$, and an electron acceptor reoxidizes $PQQH_2$ to PQQ with the liberation of two electrons. FDH is a requisite element of a biosensor for fructose because it can selectively recognize D-fructose as a result of electron transfer from the D-fructose to an electron acceptor in solution. It is, however, difficult for FDH to make the electron transfer from fructose directly to an electrode in place of an electron acceptor in solution due to steric hindrance. Fructose dehydrogenase is, therefore, one of the typical redox enzymes that have demanding conditions for molecular assembly resulting in electronic communication on the electrode surface.

A monolayer of FDH was formed on platinum or gold electrode surfaces by potential-assisted self-assembly (Khan *et al.*, 1992). FDH was dissolved in phosphate buffer (pH 6.0) to make it's net charge negative. FDH molecules instantly adsorb on the electrode surface due primarily to electrostatic interaction. Under controlled electrode potential, FDH adsorption increased with time and reached a steady state. In the potential range from 0 to +0.5 V, adsorption rate of FDH sharply increased with electrode potential. FDH molecules may be self-assembled on electrode surfaces in such a manner that the negatively charged site of the FDH molecule faces to the positively charged surface of the electrode. Enzyme assay clearly showed that electrode-bound FDH retained its enzyme activity without appreciable inactivation.

Adsorption isotherms were obtained at various electrode potentials for potential-assisted self-assembly of FDH on the electrode surface. From the isotherms, the amount of self-assembled FDH can be precisely regulated by electrode potential and potential-controlled time. One can easily obtain a monolayer of electrode-bound protein as either full surface coverage or as less surface coverage with its biological function.

In the next step, a molecular wire of molecular interface was prepared for the electrode-bound FDH, and was made by potential-assisted self-assembly. Polypyrrole was used as a molecular wire of molecular interface for the electrode-bound FDH and was synthesized by electrochemical oxidative polymerization of pyrrole. Electrochemical oxidative polymerization of pyrrole was performed on the FDH-adsorbed electrode in a solution containing 0.1 M pyrrole and 0.1 M KCl under anaerobic conditions at a potential of 0.7 V. The thickness of the polypyrrole membrane was controlled by polymerization electricity. The electrochemical polymerization was stopped when the monolayer of FDH on the electrode surface was presumably covered by the polypyrrole membrane. The total electricity of electrochemical polymerization was controlled at 4 mC. The molecular interfaced FDH was thus prepared on the electrode surface.

Electronic communication between electrode-bound FDH and an electrode has been confirmed by differential pulse voltammetry. Differential pulse voltammetry of the molecular interfaced FDH was conducted in a pH 4.5 buffered solution. A pair of anodic and cathodic peaks were observed for the molecular interfaced FDH. The peaks are attributed to the electrochemical oxidation and reduction of the

PQQ enzyme at redox potentials of 0.08 and 0.07 V vs. Ag/AgCl, respectively. In addition, the anodic and cathodic peak shapes and peak currents of the molecular interfaced FDH were identical, which suggests reversibility of the electron transfer process. On the other hand, FDH exhibited no appreciable peaks in differential pulse voltammetry on the electrode surface without the polypyrrole molecular interface. These results indicate that polypyrrole works as an effective molecular interface for electronic communication between FDH and an electrode.

In addition to FDH, the potential-assisted self-assembly has successfully been applied to several redox enzymes including glucose oxidase and alcohol dehydrogenase. The self-assembled redox enzymes have also been molecularly interfaced with the electrode surface via a conducting polymer.

Self-Assembly of Mediator-Modified Redox Enzymes on the Porous Gold Electrode Surface

In contrast to the molecular wire of molecular interfaces, electron mediators are covalently bound to a redox enzyme in such a manner that an electron tunneling pathway is formed within the enzyme molecule. Therefore, enzyme-bound mediators work as a molecular interface between an enzyme and an electrode. Degani *et al.* proposed the intramolecular electron pathway of ferrocene molecules were covalently bound to glucose oxidase (Degani and Heller, 1987). However, few fabrication methods have been developed to form a monolayer of mediator-modified enzymes on such electrode surfaces. We have succeeded in development of a novel preparation of the electron transfer system of mediator-modified enzyme by self-assembly in a porous gold-black electrode.

Glucose oxidase (from *Asperigilus niger*) and ferrocene carboxyaldehyde were covalently conjugated by the Schiff base reaction, which was followed by $NaBH_4$ reduction. The conjugates were dialyzed against phosphate buffer with three changes of buffer and assayed for their protein and ion contents. Porous gold-black was electrodeposited on a micro-gold electrode by cathodic electrolysis with chloroauric acid and lead acetate. Aminoethane thiol was self-assembled on a smooth gold disk electrode and a gold-black electrode. Ferrocene-modified glucose oxidase was covalently linked to either a modified plain gold or gold-black electrode by glutaraldehyde.

The ferrocene/glucose oxidase conjugates were characterized by the molar ratios of ferrocene to enzyme in the range from 6 to 11. All the oxidase conjugates retained enzyme activity. In cyclicvoltammograms of ferrocene-glucose oxidase in a solution on a smooth gold electrode with and without glucose, a pair of redox peaks of a solid line indicate a reversible electron transfer of ferrocene-modified glucose oxidase on the electrode. A dashed line indicating an increase in catalytic current shows that the enzymatic oxidation of the conjugate is efficiently coupled with the electrochemical oxidation of modified enzyme. Cyclicvoltammetry was also carried out for ferrocene-glucose oxidase conjugate in self-assembled form on the smooth gold electrode. Self-assembled ferrocene-glucose oxidase conjugate on the gold disk electrode showed reversible electron transfer. The anodic peak currents in the cyclicvoltammograms were independent on the molar ratio of ferrocene to enzyme in the range from 6 to 11. It is noted that the anodic peak current prominently increases with an increase in the molar ratio of ferrocene to glucose oxidase whilst the amount of enzyme self-assembled on the electrode surface is fixed. This indicates that each modified ferrocene may contribute to electron transfer between the enzyme and the electrode in the case of gold-black electrode, the ferrocene-modified enzyme could form multi-electron transfer paths on the porous gold-black electrode.

Substrate concentration dependence of response current of the gold-black electrode was compared with that of gold disk electrode. The ferrocene-modified glucose oxidase which was used in this measurement had 11 ferrocenes per glucose oxidase. The electrode potential was controlled at 0.4 V vs. Ag/AgCl. The response current was recorded when the output reached at a steady state. The response current was enhanced when ferrocene-modified glucose oxidase was self-assembled on a porous gold-black electrode.

The porous matrix of gold-black electrode has enabled ferrocene-modified glucose oxidase to perform the smooth electron transfer by means of easy access between self-assembled molecules and electrode surface.

GENETICALLY FUSED PROTEINS

Genetic Engineering for Synthesis of Novel Functional Proteins

Bifunctional proteins, which are synthesized by fusing two different functional proteins, have found promising applications in wider areas including biotechnological processes and clinical anaslysis. A frequently used technique for preparation of such bifunctional proteins is covalent bonding of the proteins with bifunctional reagents. However, covalent preparation of conjugates with bifunctional reagents generally results in the production of complicated conjugate conglomerates, which may cause an inactivation of the protein.

Another approach for the preparation of bifunctional proteins has been achieved by utilizing a recent development in genetic engineering. By using this technique, bifunctional proteins have prepared without any loss of their activity. The structural genes of two proteins are fused in frame, and the resulting protein carries the active site of each protein. Genetically fused bifunctional proteins have been used extensively in biotechnological processes and clinical analyses. Several groups have demonstrated the gene fusion systems for the expression and purification of recombinant proteins. In these systems, the structural gene of interest is connected in frame to a gene encoding a protein which has binding affinity to a ligand. The resulting fusion protein can be purified with high yield through the use of affinity chromatography. A specific cleavage site of protease can be inserted at the junction region between the two proteins to facilitate generation of a desired protein.

For the practical application of such genetically fused proteins, much attention has been focused on the protein for enzyme immunoassay (EIA), because of the ease in stoichiometric control and site-to-site binding of the constituent proteins. As a binding protein for EIA, staphylococcal protein A (SpA) is ideal because of its versatile capability to bindito the Fc region of immunoglobulin of several mammalian species. The gene for SpA has been determined completely, and the gene fusion system of SpA has been available to purify fusion protein by binding to an IgG-resin. So far, we have constructed a genetically fused metapyrocatechase-protein A (Kobatake *et al.*, 1990) and firefly luciferase-protein A (Kobatake *et al.*, 1993), and demonstrated their applicability in EIA.

A chimeric-binding protein which has two different binding regions has also attracted some keen interest as a molecular designed adhesive material. A chimeric-binding protein by fusing an IgG binding domain of protein A and antigen binding site of anti-digoxin antibody, which adhere antibodies to biological materials containing digoxin has been constructed, and a streptavidin-protein A chimeric protein has been designed for enzyme labeling to antibodies. Few reports, however, have appeared on designing a chimeric-binding protein capable of adhering to a solid matrix surface at one terminal. We attempted to construct a chimeric-binding protein as an adhesive molecule for binding of antibody to a solid-phase matrix.

For this purpose, fusion protein between maltose-binding protein (MBP) and SpA was designed. MBP, which is a periplasmic protein of *Escherichia coli (E.coli)*, is an essential element in maltose transport as well as in maltose chemotaxis. This protein binds maltose with a dissociation constant of 1.9 mM. A gene fusion vector plasmid of MBP was constructed by Guan *et al.*, who showed that the fusion proteins can be isolated in a single chromatography step from a cross-linked amylose matrix. On the other hand, SpA from *Staphylococcus aureus* can bind to the Fc region of the immunoglobulin of several mammalian species as mentioned above. Therefore, the fusion protein between MBP and SpA would be expected to be able to bind antibodies to amylose or amylose resin without chemical coupling. The chimeric-binding protein was expressed in *E. coli* and evaluated its applicability especially in EIA.

Protein A-Luciferase Fusion Protein for Bioluminescent Enzyme Immunoassay

The authors have reported sensitive chemiluminescence and bioluminescence techniques to monitor markers in protein binding reactions. The markers were quantitated by counting the photons emitted after the addition of suitable reagents in the peroxidase-catalyzed luminol reaction, or hemin-catalyzed luminol reaction.

A bioluminescent immunoassay has been performed with a genetically fused protein to demonstrate the effectiveness of specifically, stoichiometrically controlled conjugation of the binding protein and the firefly luciferase. Firefly luciferase of *photinus pyralis* is one of the attractive marker enzymes which can be detected with high sensitivity by quantitative analysis of bioluminescence.

In the luminescent reaction of firefly luciferase, trace amount of the enzyme may be determined by luminescence measurement according to

$$\text{Luciferin} + \text{ATP} + O_2 \text{------------} \text{Oxyluciferin} + CO_2 + \text{AMP} + \text{PPi} + \text{ht}$$

As a binding protein for EIA, staphyrococcal protein A is an ideal protein because of its versatile capability of binding to the Fc region of immunoglobulins of several mammalian species. Lindbladh *et al.* constructed a fusion protein between protein A and b-subunit of bacterial luciferase. One drawback of the fusion protein was the instability, which lies in the enzyme itself rather than the fusing of the two proteins. Campbell *et al.* have recently clarified that a few C-terminal amino acids are essential for the photon-generating reaction. On the other hand, several N-terminal amino acids are not necessarily required for the luciferase reaction.

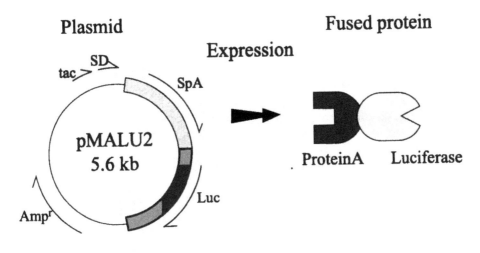

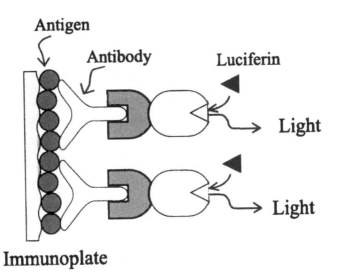

Figure 4 Protein A — luciferase fusion protein for bioluminescent enzyme immunoassay

In addition, the authors aim the development of versatile reagents which can be utilized as universal markers in enzyme immunoassays, because the protein A moiety of the conjugate can bind to the Fc portion of immunoglobulin. The principle of this system is schematically shown in Figure 4.

The protein A expression vector, pMPRA1, was previously constructed in our laboratory. The plasmid has a structural gene of Fc-binding domain of staphylococcal protein A (from pRIT5) under the control of *tac* promoter, with possessing a multicloning site after the protein A gene, the *rrn*BT$_1$T$_2$ terminator, and b-lactamase gene. The plasmid, pT3/T7-1 luc, was digested with *Bam*HI to separate a 1.78 kbp fragment containing a structural gene of firefly luciferase. The fragment was further digested with *Nar*I and then the both terminal ends were blunted by T4 DNA polymerase. By digesting with *Nar*I, N-terminal 12 amino acids of corresponding protein were considered to be deleted. Despite the deletion of these amino acids, the expressed protein retained sufficient luciferase activity. The fragment encoding the luciferase gene was ligated into the multicloning site f pMPRA1. The resulting plasmid, protein A-luciferase fusion gene expression vector, was designated as pMALU2 (Kobatake *et al.*, 1993).

The resulting fusion protein was designed by forming a site-to-site binding in a ratio of 1:1, i.e., the C-terminal site of protein A was fused to the N-terminal site of the deleted luciferase.

The binding capability of the fusion protein of luciferase and protein A to IgG was confirmed by Western blotting analysis.

The luciferase activity of the fusion protein was also confirmed.

Bioluminescent immunoassay for human IgG was performed. A series of human IgG solutions were prepared, and each solution was incubated in a cuvette whose surface was previously coated by $F(ab)_2$ fragment of anti-human IgG antibody. The protein A-luciferase fusion protein was then incubated in the cuvette to carry out a sandwich-format immunoassay. After removing unbound fusion protein, the luciferase bound on the cuvette surface was determined by adding the solution of substrates. Human IgG was determined in the range from 10^{-3} to 10^{-7} g/ml by the bioluminescent immunoassay using the genetically fused protein.

In summary, he protein A-luciferase fusion protein is very useful for the determination of trace amount of antigen or antibody in a very simplified and rapid manner using a photon-counting device. In addition, the fusion protein can be used as a universal marker in sandwich-format immunoassays, since the protein A moiety binds to the major subclasses of immunoglobulin.

One further point should be emphasized. This was the first example of constructing a protein A-firefly luciferase conjugate by recombinant gene technology.

RNA Binding Protein-Luciferase Fusion Protein

Reverse transcription-polymerase chain reaction (RT-PCR) has successfully been applied and evaluated in highly sensitive analysis of RNA with sequence specificity. The method is receiving increasing attention in DNA/RNA analysis as an alternative to standard methods including Northern blot analysis, nuclease protection assay, and *in situ* hybridization using radioisotopic labeling probes. It is noted that this method is specifically powerful for RNA having a low copy number. Reverse-transcription and amplification of a sample RNA using a specific primer are effectively coupled with hybridization analysis. One hybridization analysis is based on the binding of protein along with the labels. Avidin and antibody have been commonly used in conjunction with the corresponding partners. In addition, it is efficient to use enzymes to enhance sensitivity and specificity. Biotinylated or digoxigenin-labeled DNA or RNA probes are detected by alkaline phosphatase (AP) or horseradish peroxidase (HRP) conjugated protein such as avidin or anti-digoxigenin antibody. The specific sites of nucleotide are also detected by the anti-DNA:RNA antibody-AP or HRP conjugate and Lac repressor-b-galactosidase fusion protein, which binds to *E. coli lac* operator DNA sequence. In this work, using the RNA binding protein, we have been developed the novel bifunctional bioanalytical reagent for detection of the specific RNA.

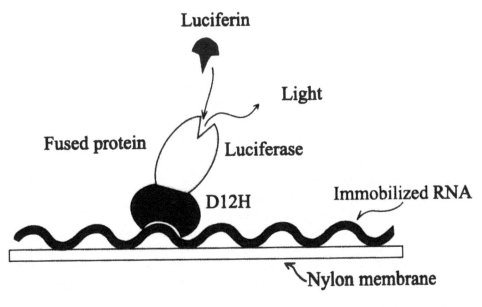

Figure 5   Plasmid construction for RNA binding protein — luciferase fusion protein

Recently, several RNA binding proteins have been isolated and characterized by sequence specific binding activity. Human heterogeneous nuclear ribonucleoprotein (hnRNP) A1 protein bind tightly to RNA containing the sequence UAGGGA/U, and hnRNP C protein to poly(U) 5 containing RNA. *Drosophila* Sex-lethal (Sxl) protein recognizes the alternative spliced site of its RNA and related pre-messenger RNA. In many viruses, such as the human immunodeficiency virus, the viral protein shows the binding specificity to the viral genomic DNA and RNA.

We have isolated RNA binding protein from HeLa cell nuclear extract that bind specifically to single-stranded d(TTAGGG) 4 deoxyribonucleotide oligomer and more tightly to r(UUAGGG) 4 oligoribonucleotides, rH4. The protein exhibited no binding activity to rECGF oligoribonucleotide which has an unrelated sequence having the same length as rH4. The cDNA for the rH4 binding protein has been cloned and sequenced (Kajita *et al.*, 1995). The primary structure predicted by the cDNA sequence reveals that this protein possesses two tandemly arranged RNA binding domains (RBDs) at its amino-terminus, and that the carboxyl-terminal region is particularly rich in glycine which appears to contribute to RNA binding. We have constructed the full-length recombinant protein and deleted mutant including two tandemly arranged RBDs (designated as D12H) to investigate the molecular mechanism of the sequence discrimination. The results indicated that all of these proteins could retain the RNA binding activity to recognize the rH4 with sequence specificity.

By fusing the structural genes of RBDs of r(UUAGGG) 4 binding protein, D12H, and firefly luciferase, a novel bifunctional reagent has been constructed for the bioluminescent detection of rH4. The principle of this system is schematically shown in Figure 5. The reagent could be used to detect the membrane-immobilized rH4 rapidly and specifically. Moreover, when D12H of this protein is replaced by another RNA binding protein, the resulted fusion protein will be effective on the specific detection of different sequence RNAs. This work would be one approach to the rapid and non-radioactive RNA detection of different sequences.

The fusion protein was expressed in *E. coli* strain BL21 (DE3) pLysS which contained a correctly constructed plasmid pEDL8 and carried a T7 RNA polymerase gene under lacUV5 control in the host chromosome. T7 RNA polymerase was expressed by the induction of 1 mM IPTG, and the gene of the fusion protein under the control of the T7 promoter was then transcribed.

The D12H-luciferase fusion protein was composed of three functional structures. From the amino-terminal, they were; the 12 amino acids T7 tag, D12H, and the luciferase at the carboxyl-terminal end. Predicted molecular mass of the fusion protein was 86 kDa, since the molecular masses of the T7 tag, D12H and luciferase are 1 kDa, 22 kDa and 63 kDa, respectively.

In order to investigate whether the fusion protein was expressed correctly, bioluminescent analysis, SDS-PAGE and immunoblotting analysis of the crude extract from homogenized *E. coli* were carried out. The crude lysate of *E. coli* harboring pEDL8 exhibited the bioluminescent activity of luciferase. This indicated that the fusion protein was translated until the carboxyl-terminal end of luciferase in-frame. However, SDS-PAGE and immunoblotting analysis using anti-T7 tag antibody (Novagen) showed many protein bands of smaller molecular weight. We supposed that these were the result of degradation of the fusion protein by proteinase in the host *E. coli* cells. To prevent such degradation, several steps for purification of the fusion protein, ammonium sulfate precipitation, ion-exchange chromatography and size-exclusion chromatography, were carried out. The 86 kDa fusion protein was eluted with 200 mM NaCl, which was confirmed by immunoblotting analysis.

In order to characterize the RNA binding property of the fusion protein, a gel retardation assay was carried out. As described above, RNA binding domains used in this study have the sequence specific binding activity to an oligoribonucleotide. It could bind specifically to rH4, but not to rECGF. Both the fractionated fusion protein and purified D12H were incubated independently with the ^{32}P-labeled rH4 and rECGF for 10 min, then loaded onto a 10% non-denaturing polyacrylamide gel which had been previously run. The position of the migrated complex of fusion protein extract was higher than that of the purified D12H, because of the difference between the molecular mass of the RNA binding protein, 86 kDa fusion protein, and 22 kDa D12H. On the other hand, in the lanes of D12H and fusion protein with rECGF, no slowly migrating complexes were observed. These results indicated that the produced fusion protein, as well as D12H, had the sequence specific binding activity.

In order to evaluate the feasibility of the fusion protein for the detection of rH4 RNA, the bioluminescent RNA detection assay was carried out. Nylon membranes were cut into 6 mm × 6 mm squares to match the size of the cell in the photon counting system. rH4 and rECGF RNA were dropped on to separate membranes, dried in air and immobilized to the surface by UV crosslinking. The membranes were then equilibrated in the buffer, containing 5% RNase free bovine serum albumin, to block the free binding site for nonspecific adsorption.

Then, the unique RNA detection by the fusion protein was carried out. The nylon membranes, on which each series of rH4 and rECGF were immobilized, were soaked in 60 μl of the fusion protein solution for 5 min. After thoroughly washing the membrane with the binding buffer, the bioluminescence activity on the membrane was measured. When the immobilized RNA was less than 50 pmol, for each rH4 and rECGF, the bioluminescence was almost equal to that of the background signal. The bioluminescence then increased by increasing of the amount of rH4 immobilized on the membrane to 100 pmol. Conversely, when rECGF was immobilized on the membrane, increasing the immobilized RNA did not result in an increase in bioluminescence. From these results, it was demonstrated that the D12H-luciferase fusion protein could be applied for the rapid detection of the unique RNA sequence.

GENE ENGINEERING FOR MOLECULAR NETWORKING

Gene Expression Networking Responding Environmentally Hazardous Compounds

Recent development of molecular-based detection techniques has greatly increased the possibility to develop new materials in biosensing system. The insertion of marker genes such as the genes for enzymes and binding protein allows tracking various substances of biological importance. Although these fusion proteins have enormous implications for the study of protein engineering, the main impetus has been the potential technological benefits of genetically engineered microorganisms capable of demonstrating novel functions such as monitoring of environmentally hazardous substances.

The introduction of luminescent enzymes enables specific microorganisms to detect various pollutants in situ, without extracting the marker enzymes. Bioluminescence-based techniques offer several advantages such as nondestructive detection of marked substances in sewage water and in soil samples. These techniques involve introduction of genes for the recognition of the marked substances and for luminescence generation originally cloned from *Pseudomonas putida* and firefly, respectively. TOL-plasmid carries a eries of genes which are required in the digestion of benzene-derivatives. In the plasmid, the gene product of stimulated transcription of the following genes by activating the promotor. We have taken a strategy of fusing to firefly luciferase gene, because we need not the induction of all the enzyme for the digestion of benzene derivative, but the induction of the reporter enzyme.

For this purpose, we have tried to develop a novel strategy for biosensing of benzene-derivatives in environment through the use of recombinant microorganisms harboring gene fusion plasmids between a control region of TOL plasmid and firefly luciferase (Ikariyama *et al.*, 1993) (Figure 6).

The TOL plasmid from *Pseudomonas putida* encodes a series of enzymes which degrade m-xylene to the metabolic pathway (Nakazawa *et al.*, 1980). The expression of these enzymes was controlled with regulating proteins, xylR and xylS (Inouye *et al.*, 1981; Inouye *et al.*, 1983). The complex between m-xylene and xylR protein activates the Ps promoter which controls the expression of xylS. Over the past few years, some researchers have utilized a recombinant microorganism which was introduced a luciferase gene for monitoring of environmental pollutants. King *et al.* reported a recombinant microorganism which induced a light emission of bacterial luciferase when exposed to naphthalene vapor (King *et al.*, 1990). Selifonova *et al.* measured inorganic mercury in contaminated waters through the use of a

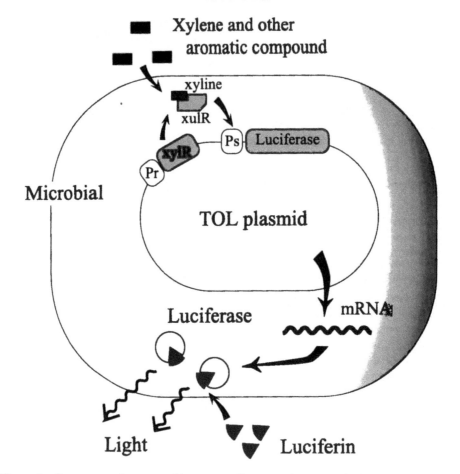

Figure 6   Gene expression networking responding to environmentally hazardous chemicals

microorganism which possesses a fusion gene between luciferase and mercury resistance operon (Selifonova *et al.*, 1993).

In the first step of our investigation, the gene fusion plasmid, pTSN316, for biosensing of benzene derivatives was constructed. As a reporter enzyme, the structural gene of firefly luciferase was inserted under the control of the Ps promoter. The *Escherichia coli* (*E. coli*) cells transformed with pTSN316 generated a luminescence in the presence of aromatic compounds. However, the *E. coli* cells must be treated with EDTA in order to obtain sufficient luminescence, as the permeability of luciferin through the *E. coli* membrane was extremely poor. This process was time-consuming and troublesome for the construction of a convenient and rapid sensing system.

In the further investigation, to make a more convenient and rapid biosensing system for benzene derivatives, pH dependence of a bioluminescence was investigated.

After induction of luciferase expression with m-xylene (0.5 mM), 200 ml of the cell suspension was removed and mixed with 1700 ml of the series of citric buffer solutions with various pH. Following which a luciferin solution was injected and generating bioluminescence was measured with a photon counting system.

Bioluminescence was increased with the increase of pH from 4.0 up to 4.6, and then decreased. When the pH value was higher than 6.0 and lower than 4.0, the luminescence was scarcely observed. On the contrary, the optimum pH for the luciferase reaction was around pH 7.4, and no luminescence was observed when pH was lower than 5.0. These results suggest that *E. coli* cells treated with acidic buffer solution may be changed in the permeability of luciferin through the cell membrane, although the inside of the cell membrane retained neutral pH. Therefore, the citric buffer solution of pH 4.8 was utilized for mixing with the cell suspension before measuring of bioluminescence. Through the use of this method, it was possible to complete the luminescent measurement within 5 min, as the cells could be used directly after culture with an inducer.

In order to investigate the dependence of the concentration of m-xylene on bioluminescence, m-xylene dissolved in ethanol was dissolved in the culture medium of recombinant *E. coli*. m-Xylene was dissolved in ethanol with adequate concentration, and 100 ml of this solution was added to a 10 ml of culture medium of cells. This volume (1/100 to the culture medium) of ethanol had no effect to the luminescence of *E. coli* (data not shown). The intensity of bioluminescence increases with increase of the concentration of m-xylene dissolving in culture medium. The concentration of m-xylene is able to be determined with high sensitivity, the detection limit of m-xylene is about 5 mM. In this stage of experiment, the sensitivity was improved by 10^2 order comparing with our previous results (detection limit was 100 mM).

A relative luminescence to a series of aromatic compound was studied by incubating the *E. coli* in culture medium containing the corresponding chemicals. The final concentration of each benzene derivative was 0.1 mM. The relative luminescence of every bacterial sample incubated with various kinds of aromatic compounds was compared with that of non-induced bacteria. Not only xylene but benzene, toluene, and other derivatives also induced the luminescence from *E. coli*, however, ethyltoluene produced less luminescence. Moreover, it is obvious that meta compound is the most effective inducer in each isomer, in the case of xylene, ethyltoluene, and chlorotoluene. These results suggest that the structure of aromatic compounds, especially the type and location of side-chain group, affects the affinity of these compounds to the xylR protein, which is responsible for the activation of the Ps promoter when it binds to the respective benzene-derivative.

GENE ENGINEERING FOR PROTEIN ASSEMBLY

Self-Assembled Antibody Protein Array on Protein a Monolayer

Biosensors may be classified into two categories; biocatalytic biosensors and bioaffinity biosensors. Biocatalytic biosensors contain a biocatalyst such as an enzyme to recognize the analyte selectively. Bioaffinity biosensors may involve an antibody, a binding protein, or a receptor protein which a form stable complexes with the corresponding ligand. An immunosensor, in which an antibody is used as the receptor, may represent a bioaffinity biosensor.

Advanced biotechnology and monoclonal antibody production have provided strong support for bioaffinity biosensors, and various new principles of electrochemical and optical immunosensors have been proposed. Concentrated efforts have been sharply focused on the development of homogeneous immunosensors, which are free from separation of bound form from free one. Examples include an optical immunosensor based on surface plasmon resonance (SPR), an optical fiber immunosensor based on fluorescence determination using an evanescent wave and an optical fiber electrode immunosensor based on electrochemical luminescence determination. These immunosensors are characterized by a single step of determination and high selectivity as well as high sensitivity. The responses of these immunosensors, however, result from averaging the physicochemical properties of the antibody-bound solid surface. We have succeeded in fabricating an ordered array of antibody molecules on a solid surface and in quantitating individual antigen molecules which are complexed with the antibody array.

Protein A is a cell wall protein from *Staphylococcus aureus* and has a molecular weight of 42 kDa. Since protein A binds specifically to the Fc region of IgG from various animals, it has been widely used in immunoassays and affinity chromatography. We found that protein A could be spread over the water surface to form a monolayer membrane using LB methods (Owaku *et al.*, 1993). On the basis of this finding an antibody array on a solid surface can be obtained by the following two steps. The first step is fabrication of an ordered protein A array on the solid surface by the LB method. The second step is self-assembling of antibody molecules on the protein A array by biospecific affinity between protein A and the Fc region of IgG.

A Fromhertz type of LB trough was used for fabrication of protein A arrays on a highly oriented pyrolytic graphite (HOPG) plate ($15 \times 15 \times 2$ mm). Protein A was dissolved in ultrapure water to make a 0.1×10^{-6} g ml^{-1} solution. With a micropipet, 0.2 ml of a protein A solution was dropped on 150 cm^2 of the air/water interface of the compartment which contained ultrapure water as subphase. The protein A layer was compressed at a rate of 10 mm^2 s^{-1} with a barrier. Compression was stopped at a surface pressure of 11 mN m^{-1} and the monomolecular layer of protein A was transferred to an adjacent compartment containing 0.5% glutaraldehyde solution at a rate of 10 mm^2 s^{-1}. The protein A layer was incubated for 1 hr to be crosslinked by glutaraldehyde, which was followed by transfer to a compartment containing ultrapure water for rinsing.

For preparation of an antibody protein array, a monolayer of protein A which was compressed at a surface pressure of 11 mN m^{-1} was transferred to a compartment containing anti-ferritin antibody in 10 mM phosphate buffer (pH 7.0). The antibody molecules were self-assembled onto the protein A layer. The protein A/antibody molecular membrane was transferred to a compartment containing ultrapure water for rinsing, and was transferred onto the surface of a HOPG plate by the horizontal method.

The antibody array which was self-assembled on the protein A array was also visualized in molecular alignment by AFM. The AFM measurement was conducted at a controlled force of 1.8×10^{-11} N and a scanning rate of 0.5 Hz. Molecular size of antibody was estimated as 7 nm in diameter.

The antibody array was soaked in different concentrations of ferritin solutions for 1 hr, and was assayed for AFM imaging in solution. The ferritin molecules, which were recognized and fixed by the antibody array, were molecularly imaged by AFM. Individual ferritin molecules on the antibody array can be selectively quantitated by AFM.

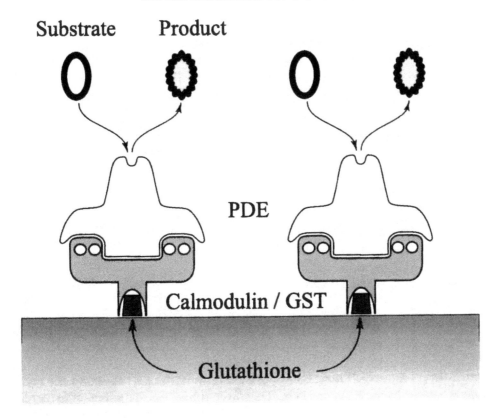

Figure 7 Genetically engineered calmodulin self-assembled on solid matrix and coupled with PDE in solution

Genetically Engineered Calmodulin Self-Assembled on Solid Matrix and Coupled with PDE in Solution

It is of great interest that molecular information is recognized by the binding site of the receptor protein, which may induce a conformational change of the whole receptor protein with a resulting conformational change of the adjacent supramolecular protein. The receptor protein molecular assembly processes molecular information through an intermolecular transfer of conformational change within a supramolecular system. Such an information processing is realized by supramolecular protein systems because of their flexibility in the intermolecular transfer of conformational change.

We have sought for a methodology to construct a protein supramolecular system on solid matrix which can make an information processing through the intermolecular transfer of conformational change (Miwa *et al.*, 1991a and b; Damrongchai *et al.*, 1995). Self-assembling of protein molecules on solid-matrix should be appropriate enough to keep the flexibility of conformational change with retaining the activity. However, calmodulin finds it difficult to be self-assembled on solid matrix as native structure. Therefore, it was concluded that calmodulin should be genetically engineered to be provided with self-assembling function.

We prepared a recombinant calmodulin-glutathione S-transferase fusion protein for the purpose of assembling a modulating protein on a solid-phase matrix (Kobatake et al., 1990) (Figure 7). The recombinant calmodulin may be self-assembled on a glutathione-immobilized matrix through the glutathione S-transferase (GST), because GST has strong affinity to matrix-bound glutathione.

The GST-calmodulin gene fusion vector, pGEX-CaM, was constructed as follows. The coding region of human calmodulin cDNA was amplified by polymerase chain reaction using synthetic oligonucleotides, 5'-AGGGAAGGATTTCAGCTGACCAAC TGACTGAAG and 5'-AGGTCGACTCACTTTGCTGTCATCATTTG. The amplified fragment was digested with XmnI and SalI, subsequently, both terminal ends were blunted with T4DNA polymerase. The isolating gene fragment of calmodulin was ligated to the SmaI site of the pGEX-3X with T4DNA ligase. In the resulting plasmid, the gene of calmodulin was fused to the gene of GST in frame.

The E. coli strain JM105 cells transformed with plasmid, pGEX-CaM, were cultured to produce the fusion protein. The fusion protein was purified through the use of Glutathione-Sepharose beads.

In the fusion protein, there is a recognition site of factor Xa, which is a site-specific protease, between GST and calmodulin. So the GST moiety and calmodulin moieties could be separated easily from the fusion protein by cutting off with factor Xa.

The activity of PDE in the presence of Ca^{2+} is about 4-fold higher than the activity in the absence of Ca^{2+}. Moreover, the PDE activity with calmodulin (not fusing with GST) which separated from the fusion protein by factor Xa was almost the same as the activity by the fusion protein. These results indicate that calmodulin moiety in the fusion protein functioned as an activator of PDE without loosing its activity by fusing with GST.

The fusion protein was self-assembled on a solid-phase matrix. The Glutathione-Sepharose beads were used as matrix for immobilizing the fusion protein through the GST moiety. Matrix-bound calmodulin was coupled with PDE in solution.

In the presence of calcium ion, the PDE activity was almost same as that activated by free calmodulin. Then, the beads were washed three times with 100 mM glycil-glycine buffer (pH 7.5) containing 3 mM EGTA for chelating calcium ion. After centrifugation, the beads were resuspended in 970 mL of the reaction buffer containing EGTA instead of $CaCl_2$, following a solution of cAMP was added. In this case, the PDE activity was restricted at lower level. Then the beads were washed again with 100 mM glycilglycine buffer (pH 7.5) and suspended in a reaction buffer containing 2 mM $CaCl_2$. The PDE activity was recovered to the initial level. These experiments were repeated three times. Almost the same PDE activity could be obtained in each stage of the experiment.

These results suggest that the genetically engineered calmodulin self-assembled on solid matrix could reversibly modulate the PDE activity in responding to calcium ion through intermolecular transfer of conformational change.

Genetically Engineered Lipid-Tagged Antibody

Immunoliposomes bearing antibody molecules on their surface have been used in several biotechnological applications such as drug delivery systems transfection of cells and immunoassays. In order to incorporate soluble antibody molecules stably on the surface of liposomes, it is necessary to introduce hydrophobic moieties to antibody molecules, e. g. by directly coupling antibody molecules to lipids. So far, incorporation of antibody molecules to the surface of liposomes has been performed by chemical coupling. In this procedure, fatty acyl groups in lipid are coupled to appropriately exposed sulphydryl and amino groups in the protein molecule with a bifunctional reagent. However, in such chemical coupling procedures, the conjugate often forms heterogeneous complex in terms of number and location of lipid moieties, and as a result, this treatment may lead to loss or decrease in antigen-binding properties.

Although various fusion proteins have successfully been synthesized by gene engineering, few genetic methods have been applied to modify a protein molecule with lipid at a specific site. A novel method has been developed on the basis of the specific lipid modification of proteins at the bacterial membrane. The new method has been applied to conjugate antibody with lipid for the construction of stable and functional immunoliposomes.

Recent techniques in bacterial expression of functional antibodies (Ward *et al.*, 1989) have also prompted us to use genetic engineering to convert antibodies into membrane-bound molecules for immunoliposome applications. Recombinant Fv fragments, which are the smallest functional unit of an antibody, have been successfully produced in *Escherichia coli* (Huston *et al.*, 1988; Field *et al.*, 1989). For stabilization of Fv fragments, V_H and V_L domains have been linked together with linker peptide, and expressed as a single-chain antibody (Bird *et al.*, 1988; Skerra and Plückthun, 1988). This form of antibody has many advantages for genetic modification because of its simplicity of handling.

In order to construct a stable and functional conjugate between antibody and lipid molecules by gene fusion, the major lipoprotein (lpp) of *Escherichia coli* which contains a specific lipid modification at its aminoterminus to anchor the bacterial membrane has been exploited. The determinants for the biosynthetic lipid modification are contained within a signal peptide of 20 amino acid residues and nine aminoterminal amino acid residues of the lpp (Ghrayeb and Inouye, 1984). Lipid-tagged single-chain antibody has been produced by fusion of genes for a single-chain anti 2-phenyloxazolone antibody and the essential part of lpp of *Escherichia coli* required for lipid-modification (Laukkanen *et al.*, 1993). The resulting lipid-tagged antibody carries a single covalently bound glycerolipid anchor at the aminoterminal cysteinyl residue which is separated from the variable region of the immunoglobulin heavy chain by a linker peptide as shown in Figure 8. The genetically prepared single-chain antibody modified with lipid molecules retained its antigen binding activity. The antibodies were expected to be incorporated stably to liposomes with high orientation. The immunoliposome consisting of the lipid-tagged antibody which was prepared by a detergent dialysis method could demonstrate a possibility of the application for immunoassay by surface plasmon resonance, quartz crystal microbalance (QCM), fluoroimmunoassay, and time-resolved fluoroimmunoassay (Laukkanen *et al.*, 1994 and 1995).

This approach should be applicable to immunoassays for any antigen only replacing the part of antibody. Especially, the recent technology of phage display has dramatically simplified the isolation and modification of antibodies. Not only application to immunoassays, the use of lipid-tagged antibody will open a new field for the development of biosensors, because oriented and high-density assembly of protein molecules to lipid bilayer will be possible by this method.

264	M. AIZAWA *et al.*

REFERENCES

Aizawa, M., Yabuki, S., Shinohara, H., Ikariyama, Y. (1989a) Protein Molecular Assemblies and Molecular Interface for Bioelectronic Devices. In *Molecular Electronics-Science and Technology*, edited by A. Aviram, pp. 301–308. New York: Engineering Found.

Aizawa, M. (1994a) Immunosensors for clinical analysis. *Adv. Clin. Chem.*, **31**, 247–275.

Aizawa, M. (1994b) Molecular Interfacing for Protein Molecular Devices and Neurodevices. *IEEE Eng. Med. Biol.*, Feb/Mar, 94–102.

Aizawa, M., Yabuki, S. and Shinohara, H. (1989b) Biomolecular Interface, in *Molecular Electronics*, edited by F.T. Hong, pp. 269–276. New York: Plenum Press.

Aizawa, M., Nishiguchi, K., Imamura, M., Kobatake, E., Haruyama, T. and Ikariyama, Y. (1995) Integrated Molecular System for Biosensors. *Sensors Actuators*, **9**, B24–25, 1–5.

Bird, R.E., Hardman, K.D., Jacobson, J.W., Johnson, S., Kaufman, B.M., Lee, S.-M., Lee, T., Pope, S.H., Riordan, G.S., Whitlow, M. (1988) Single-chain antigen-binding proteins. *Science*, **242**, 423–426.

Cardosi, M.F. and Turner, A.P.F. (1991) Mediated electrochemistry: A practical approach to biosensing. In *Advances in Biosensors*, edited by A.P.F. Turner, pp. 125–170. London: JAI Press.

Damrongchai, N., Kobatake, E., Haruyama, T., Ikariyama, Y. and Aizawa, M. (1995) Calcium responsive two-dimesional molecular assembling of lipid-conjugated calmodulin. *Bioconjugate Chem.*, **6**, 264–268.

Degani, Y. and Heller, A. (1987) *J. Phys. Chem.*, **91**, 6–12.

Field, H., Yarranton, G.T., Rees, A.R. (1989) Expression of Mouse Immunoglobulin Light and Heavy Chain Variable Regions in *Escherichia coli* and Reconstitution of Antigen-Binding Activity. *Protein Eng.*, **3**, 641–647.

Ghrayeb, J., Inouye, M. (1984) Nine amino acid residues at the NH₂-terminal of lipoprotein are sufficient for its modification, processing, and localization in the outer membrane of *Escherichia coli*. *J. Biol. Chem.*, **259**, 463–467.

Gleria, K.D. and Hill, A.O. (1992) New developments in bioelectrochemistry. In *Advances in Biosensors*, edited by A.P.F. Turner, pp. 53–78. London: JAI Press.

Huston, J.S., Levison, D., Mudgett-Hunter, M., Tai, M.-S., Novotny, J., Margolies, M.N., Ridge, R.J., Bruccoleri, R.E., Haber, E., Crea, R., Oppermann, H. (1988) Protein Engineering of Antibody Binding Sites: Recovery of Specific Activity in an Anti-Digoxin Single-Chain Fv Analogue Produced in *Escherichia coli. Proc. Natl. Acad. Sci. U.S.A.*, **85**, 5879–5883.

Ikariyama, Y., Nishiguchi, S., Kobatake, E., Aizawa, M., Tsuda, M. and Nakazawa, T. (1993) Luminescent Biomonitoring of Benzene Derivatives in the Environment Using Recombinant *Escherichia coli*. *Sensors and Actuators, B*, **13–14**, 169–172.

Inouye, S., Nakazawa, A. and Nakazawa, T. (1981) Molecular Cloning of Gene xylS of the TOL Plasmid: Evidence for Positive Regulation of the xylDEGF Operon by xylS. *J. Bacteriol.*, **148**, 413–418.

Inouye, S., Nakazawa, A. and Nakazawa, T. (1983) Molecular Cloning of Regulatory Gene xylR and Operator-Promotor Regions of the xylABC and xylDEGF Operons of the TOL Plasmid. *J. Bacteriol.*, **155**, 1192–1199.

Kajita, Y., Nakayama, J., Aizawa, M. and Ishikawa, F. (1995) The UUAG-Specific RNA Binding Protein. Heterogeneous Nuclear Ribonucleoprotein D0. Common Modular Structure and Binding Properties of the 2xRBD-Gly Family. *J. Biol. Chem.*, **270**(38), 22167–22175

Khan, G.F., Kobatake, E., Shinohara, H., Ikariyama, Y. and Aizawa M. (1992) Molecular interface for an Activity Controlled Enzyme Electrode and Its Application for the Determiniation of Fructose. *Anal. Chem.*, **64**, 1254–1258.

King, J.M. H., DiGrazia, P.M., Applegate, B., Burlage, R., Sanseverino, J., Dunbar, P., Larimer, F. and Sayler, G.S. (1990) Rapid, Sensitive Bioluminescent Reporter Technology for Naphthalene Exposure and Biodegradation. *Science*, **249**, 778–780.

Kobatake, E., Ikariyama, Y., Aizawa, M., Miwa, K. and Kato, S. (1990) Hyperproduction of a Biofunctional Hybrid Protein, Metapyrocatechase-Protein A, by Gene Fusion. *J. Biotechnol.*, **16**, 87–96.

Kobatake, E., Iwai, T., Ikariyama, Y. and Aizawa, M. (1993) Application of a Fusion Protein, Metapyrocatechase/Protein A, to an Enzyme Immunoassay. *Anal. Biochem.*, **208**, 300–305.

Laukkanen, M.-L., Teeri, T.T., Keinänen, K. (1993) Lipid-tagged Antibodies: Bacterial Expression and Characterization of a Lipoprotein-Single-Chain Antibody Fusion Protein. *Protein Eng.*, **6**, 449–454.

Laukkanen, M.-L., Alfthan, K., Keinänen, K. (1994) Functional Immunoliposomes Harboring a Biosynthetically Lipid-Tagged Single Chain Antibody. *Biochemistry*, **33**, 11664–11670.

Laukkanen, M.-L., Orellana, A., Keinänen, K. (1995) Use of Genetically Engineered Lipid-tagged Antibody to Generate Functional Europium chelate-Loaded Liposomes. Application in Fluoroimmunoassay. *J. Immunol. Methods*, **185**, 95–102.

Miwa, T., Damrongchai, N., Shinohara, H., Ikariyama, Y., Aizawa, M. (1991a) Activity Self-Controllable Enzyme-Calmodulin Hybrid by Ca^{2+} Information — An Approach Towards the Regulation of a Molecular Devices Which Breaks Chemical Structure. *J. Biotechnol.*, **20**, 141–146.

Miwa, T., Kobatake, E., Ikariyama, Y. and Aizawa, M. (1991b) Ca^{2+}-Responsive Extensible Monolayer Membrane of Calmodulin-Albumin Conjugate. *Bioconjugate Chem.*, **2**, 270–274. (1995) *Bioconjugate Chem.*, **6**, 264–268.

Nakazawa, T., Inouye, S. and Nakazawa, A. (1980) Physical and Functional Mapping of RP4-TOL Plasmid Recombinants: Analysis of Insertion and Deletion Mutants. *J. Bacteriol.*, **144**, 222–231.

Owaku, K., Goto, M., Ikariyama, Y. and Aizawa, M. (1993) Optical Immunosensing for IgG. *Sensors and Actuators*, **B13–14**, 723–724.

Selifonova, O., Burlage, R. and Barkay, T. (1993) Bioluminescent Sensors for Bioavailable Hg(II) in the Environment. *Appl. Environ. Microbiol.*, **59**, 3083–3090.

Skerra, A., Plückthun, A. (1988) Assembly of a Functional Immunoglobulin Fv Fragment in *Escherichia coli. Science*, **240**, 1038–1043.

Ward, E.S., Gussow, D., Griffiths, A.D., Jones, P.T., Winter, G. (1989) Binding Activities of a Repertoire of Single Immunoglobin Variable Domains Secreted from *Escherichia coli. Nature*, **341**, 544–546.

15. PROTEIN ENGINEERING IN VACCINE DEVELOPMENT

GUIDO GRANDI*, MARIA GRAZIA PIZZA, ENZO SCARLATO
and RINO RAPPUOLI

Chiron Vaccines, Via Fiorentina, 1, 53100 Siena ITALY

INTRODUCTION

In the early twenties Glenny and Ramon discovered that if the culture supernatant of *Corynebacterium diphtheriae* was treated with formaldehyde it became incapable of killing guinea pigs and the treated animals acquired resistance to subsequent challenges with diphtheria toxin (Glenny and Hopkins, 1923; Ramon, 1924). This observation, destined to remain one of the most relevant in the history of vaccinology, opened the way to the large scale detoxification of bacterial toxins, such as tetanus and diphtheria toxins, and to the introduction of mass vaccination against *C. tetanus* and *C. diphtheriae* infections. Furthermore, the use of formaldehyde and other chemical agents was extended to the inactivation of whole bacteria and viruses and today many vaccines are still produced using exactly the same procedure described by Ramon and Glenny at the beginning of the century (Woodrow, 1997). Widely used vaccines based on chemical detoxification are diphtheria, tetanus, whole cell pertussis, polio, influenza and many others.

Although generally very efficacious and inexpensive, the chemically detoxified vaccines suffer three major drawbacks:

1. The chemical treatment, causing covalent cross-linking among the free NH_2 groups of the amino acid side chains, inevitably perturbs the 3D structure of the vaccine components. It is therefore necessary to identify conditions which from on the one hand fully inactivate molecules and/or microorganisms which would otherwise be dangerous, and on the other do not completely destroy the structural epitopes relevant for inducing protective immunity.
2. The chemical reaction may be reversible and toxicity may be restored by long term storage.
3. The production process requires handling large quantities of very dangerous material.

Protein Engineering turns out to be a very efficacious way to design vaccine components with superior immunogenicity and none of the drawbacks of the chemically detoxified counterparts.

In the following chapter we present two examples of how protein engineering has been exploited in our laboratories for the production of new vaccines. Two toxins, the pertussis toxin (PT) and the *E. coli* heat-labile toxin (LT) have been

---

*Corresponding author

genetically detoxified and used as antigens of vaccine components and as potent adjuvants for vaccines to be delivered at mucosal level.

In the following sessions the structural features of the two toxins will be briefly described together with the strategies used for their detoxification. Finally, the immunological properties and the clinical applications of the detoxified toxins will be presented.

PERTUSSIS, CHOLERA AND E. COLI TOXINS: ADP-RIBOSYLTRANSFERASES WITH STRUCTURAL FEATURES IN COMMON

Pertussis toxin (PT), Cholera toxin (CT) and E. coli heat labile toxin (LT) are bacterial toxins with an ADP-ribosylating activity (Domenighini et al., 1995). In the ADP-ribosylation reaction, the toxins bind NAD and transfer the ADP-ribose moiety to the major target proteins. In the case of PT, the target proteins are the GTP-binding proteins G_i and G_0, both involved in signal transduction of eukaryotic cells (Katada and Ui, 1982; Bokoch et al., 1983; West et al., 1985; Fong et al., 1988). On the contrary, LT and CT transfer the ADP-ribose mostly to G_s, a protein that activates the adenylate cyclase, thus inducing the synthesis of the cAMP second messenger (Van Dop et al., 1984; Iiri et al., 1989).

The toxins are organised in two functionally different subunits: the enzymatically active A subunit, responsible for toxicity, and the binding B subunit which interacts with the specific receptors located on the surface of the eukaryotic cells (Rappuoli and Pizza, 1991).

The A subunits of the three toxins, having similar enzymatic activities, share similarities in size as well as in the primary and tertiary structures. They are organised in two domains, A1 and A2. The A1 domain carries the NAD binding sites and the catalytic site. In particular, NAD is housed in two cavities, one binding the nicotinamide ring and the other the adenine moiety. Interestingly, in PT, a toxin with three orders of magnitude higher affinity for NAD as compared to CT and LT, the nicotinamide ring nicely interacts with two tyrosines, residues missing in both CT and LT.

The two amino acids essential for catalysis, a glutamic acid and an arginine, are both located in the nicotinamide binding pocket (Domenighini et al., 1995).

As far as the A2 domain is concerned, it has the function to fix the A subunit to the B subunit and in all three toxins it has the common feature to be organised in a long alpha helix which penetrates into and interacts with a cavity present in the B subunit.

Reflecting the difference in target cell specificity and in the mechanism used to translocate the A subunit into the target cell, the B subunit of CT and LT differs substantially from the B subunit of PT. In CT and LT, the B subunit is a pentamer of five identical monomers (Sixma et al., 1991), whereas PT is composed of four different monomers (S2, S3, S4 and S5) present in the 1:1:2:1 ratio (Tamura et al., 1982).

Pertussis Toxin and Anti-Pertussis Vaccine

Pertussis, or whooping cough, is a disease caused by Bordetella pertussis, a gram-negative bacterium which adheres to the cilia of the upper respiratory tract of humans and induces local and systemic damages by releasing a number of toxic substances, including PT (Weiss and Hewlett, 1986). The disease is characterised by long lasting paroxysmal cough, accompanied by whoops, vomiting, cyanosis and apnoea. The most common complications are pneumonia, seizures, encephalopathy and death. It has been estimated that worldwide pertussis causes over 500,000 deaths each year, with mortality being most frequent under 1 year of age when it can be as high as 1/100.

Antibiotic treatment is very effective in clearing *Bordetella pertussis* but has little consequence on the disease, mostly because the disease can only be diagnosed when the bacteria have already released the toxins that are responsible for local and systemic damages.

Vaccination is the only way to control pertussis. Mass vaccination using killed bacteria was introduced in the 1950s and reduced the incidence of disease in infants by 99% (Farizo *et al.*, 1992). However, in spite of its efficacy, the inactivated, whole cell vaccine is not widely used mostly because of the side effects which have been associated with it. Most of these are mild reactions such as redness, swelling and fever and occur in 30 to 70% of the vaccinees. However, the vaccine has also been associated with more severe reactions such as neurological damages and death, which have been reported with a frequency of 1/100,000 and 1/300,000, respectively.

In 1981, Yuji and Sato developed a new pertussis vaccine composed of partially purified protein components from *B. pertussis*, the most effective of which is PT (Sato *et al.*, 1984). For PT to be included in the vaccine, it needs to be detoxified. Sato's group used formalin treatment to make the toxin inactive, the same procedure developed by Ramon. However, this procedure was not entirely satisfactory for pertussis toxin and in some cases it showed reversion to toxicity.

Our approach to detoxify PT stems from the assumption that a genetically modified toxin in which the amino acid residues involved in catalysis are replaced, becomes fully inactive and maintains the same immunogenicity properties as the wild type toxin. Such a genetically modified toxin is expected to be safer than the chemically detoxified PT and superior to it in terms of immunogenicity, thus allowing the use of a lower amount of antigen in the final vaccine formulation.

Starting from the cloning and sequencing of the 3.5 kb DNA fragment encoding the five PT subunits (Locht and Keith, 1986; Nicosia *et al.*, 1986), the two catalytic residues Arg9 and Glu129 of S1 (A subunit) were replaced with Lys and Gly, respectively (Figure 1). The modified S1 gene was then introduced back into *B. pertussis* using classical genetic approaches (Stibitz *et al.*, 1986), thus obtaining a genetically modified strain which produced a fully detoxified PT (Pizza *et al.*, 1989).

Extensive studies were performed to prove that PT9K/129G was identical in structure to the wild type molecule but had lost its original toxicity. The molecule showed identical behaviour on SDS-PAGE and a large panel of monoclonal antibodies recognised the mutated and the wild type molecules with identical affinities. The absence of toxicity was confirmed *in vitro* by the CHO cell assay and *in vivo* by a number of assays such as the histamine sensitivity assay, the enhancement of insulin secretion, the induction of lymphocytosis and the potentiation of anaphylaxis.

The ability of PT9K/129G to induce appropriate immune responses was tested in mice, guinea pigs and rabbits. In all instances, the recombinant molecule was found to be from 10 to 20-fold more immunogenic than any toxin form that had been detoxified by chemical treatment (Nencioni *et al.*, 1991). Furthermore, and most important, the molecule induced strong protection against bacterial infection in mice in a dose-dependent manner.

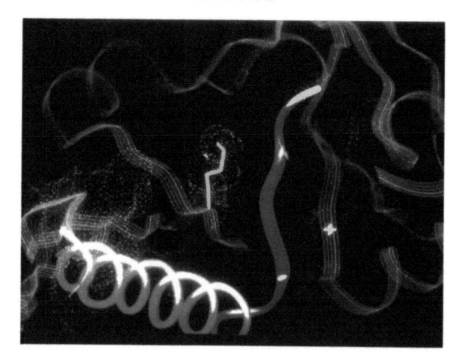

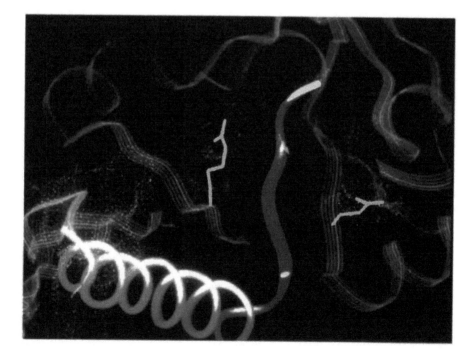

Figure 1 3D model structure of the catalytic pocket of wild-type PT (left) and PT9K/129G mutant (right). Highlighted are the two residues, Arg9 and Glu129, which have been replaced with Lys and Gly to create a fully detoxified molecule.

Today, PT9K/129G is the component of a trivalent vaccine available on the Italian market against diphtheria, tetanus and pertussis (DTaP), the anti pertussis part of the vaccine being composed of three antigens, PT9K/129G, FHA and 69K (5, 2.5 and 2.5 μg each, respectively).

The high efficacy of the Chiron DTaP vaccine has been recently demonstrated by a comparative efficacy trial conducted in Italy (Greco et al., 1996). In this trial, the Chiron vaccine was compared with the whole cell vaccine and with another commercially available acellular vaccine differing from the Chiron formulation in two important aspects. Firstly, PT is inactivated by the conventional chemical treatment, and secondly, the amount of PT, FHA and 69K is 25 μg, 25 μg and 8 μg, respectively. The results of the study showed that the genetically detoxified PT was the most immunogenic, inducing a superior ELISA and toxin neutralising antibody titre. Furthermore, the Chiron formulation was the only one able to protect vaccinees starting from the first vaccination dose and showed a longer lasting protective immunity. In addition, it showed a lower reactogenicity. The early protection observed with the Chiron vaccine is most likely due to the superior immunogenicity of PT, while the low reactogenicity to the low amount of antigens used.

E. coli Heat Labile Toxin (LT) and Mucosal Delivery of Vaccines

Different immune responses may be required to protect against different pathogens. Since many pathogens enter the host through mucosal surfaces, the induction of mucosal immune response may be of paramount importance for protection. It has been shown that one possible way to induce mucosal immunity is to deliver antigen to the mucosal surfaces rather than systemically by injection. Furthermore, mucosal delivery of antigens can elicit systemic immunity as well, offering a way to activate the host immune system in a more general manner. Unfortunately, very few of the currently licensed vaccines are suitable for mucosal delivery and these are limited to live vaccines such as Polio and Bacille Calmette-Guèrin (anti-tuberculosis vaccine).

The key factor that makes mucosal vaccine development particularly difficult is that most pure antigens delivered mucosally are either poor immunogens or even induce a state of immune tolerance that inhibits subsequent responses to the same antigen delivered systemically. Needless to say that inducing immune tolerance would be extremely dangerous for the host, in that he becomes exposed to subsequent infection by the microorganism from which the antigen has been derived.

However, not all antigens induce tolerance. De-Aizpurua and Russell-Jones (1988) were perhaps the first to study the ability of different proteins to elicit local and systemic antibody responses following oral feeding. They found that the antigens that could elicit systemic antibody responses following oral feeding were those that had the ability to bind to receptors on the host cells. For example, bacterial toxins such as CT and LT turned out to be very potent mucosal immunogens. Interestingly, if the toxins were inactivated using chemical denaturation or formalin treatment, they lost their mucosal immunogenicity, in line with the involvement of cell binding in mucosal immunogenicity (Pierce and Gowan, 1975; Pierce, 1978).

In addition to their ability to act as mucosal immunogens, both CT and LT are potent mucosal adjuvants. In fact, administration of CT and LT to mucosal surfaces can activate immune responses to coadministered poorly immunogenic antigens provided that they are delivered simultaneously to the same mucosal surface

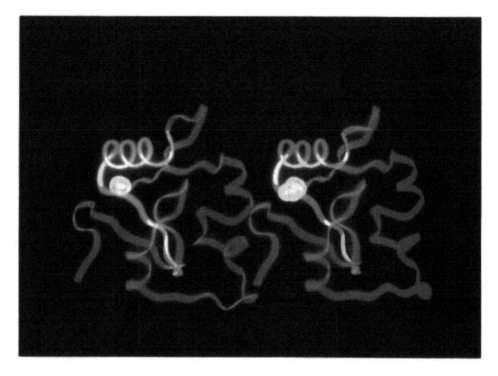

Figure 2 3D model structure of the A2 domain of wild-type LT (left) and LT-K63 mutant (right). Highlighted is the Ser63 residue which in the mutant was replaced with Lys to fill the catalytic pocket, thus preventing substrate binding.

(Holmgren and Czerkinsky, 1992; Walker, 1994). Therefore, the use of CT and LT in combination with any selected antigen appears to be a possible solution to the development of effective mucosal vaccines.

However, both CT and LT are highly toxic to humans. Clinical trials in volunteers showed that oral ingestion of a few micrograms of these toxins can induce severe diarrhoea (Levine *et al.*, 1983).

Encouraged by the results obtained with PT, we reasoned that a detoxified LT mutant could have maintained the mucosal adjuvanticity properties and therefore could have been safely used for mucosal vaccine formulations.

As mentioned above, the catalytic residues Glu129 and Arg9 which were successfully replaced to obtain a detoxified PT are strictly conserved in both CT and LT (present at position 112 and 7, respectively). Therefore, the first substitutions attempted were Glu with Ala and Arg with His. Indeed, these substitutions turned out to give fully inactive molecules. However, in contrast with the finding observed with pertussis toxin, the mutant toxins were very sensitive to protease digestion and very unstable to storage and manipulation (Burnette *et al.*, 1991; Hase *et al.*, 1994). Therefore, new amino acid substitutions were designed by computer modelling using the available coordinates of the LT structure. In particular, the Ser63>Lys substitution was selected, with the idea to fill the catalytic pocket of the enzymatic subunit with the bulky side chain of the Lys residue (Figure 2). LTK63 mutant was found to be enzymatically inactive, nontoxic both *in vivo* and *in vitro*, and very stable to protease

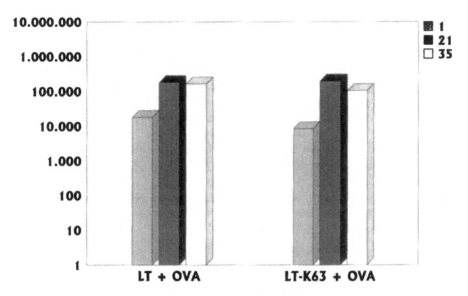

Figure 3 IgG-specific responses to wild type LT and to LT-K63 mutant. Two groups of four mice each were immunised intranasally with 1 µg of either LT or LT-K63 on days 1, 21 and 35 and blood samples were collected on days 0 (pre-immune sera), 20, 34 and 49.The titers represent the mean and were calculated as the reciprocal of the dilution of sera that gave an OD of 0.3 above the response shown in nonimmunised controls.

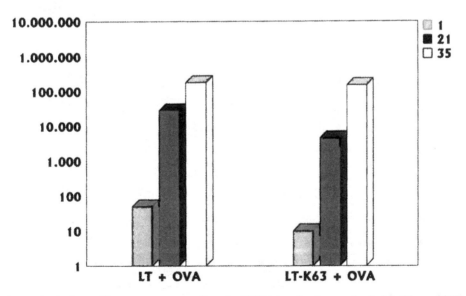

Figure 4 IgG-specific responses to ovalbumin (OVA) in mice immunised with 10 µg of OVA and 1 µg of either wild-type LT and LT-K63. Two groups of four mice each were immunised intranasally on days 1, 21 and 35 and blood samples were collected on days 0 (pre-immune sera), 20, 34 and 49.The values represent the mean titers and were calculated as the reciprocal of the dilution of sera that gave an OD of 0.3 above the response shown in nonimmunised controls.

treatment. Therefore, the molecule was purified and tested in immunogenicity and adjuvanticity studies (Pizza *et al.*, 1994).

When administered intranasally, the mutant was found to be as immunogenic as the wild type LT (Figure 3). LTK63 was also used to immunise mice intranasally in combination with the poorly immunogenic ovalbumin to test its adjuvanticity property. As shown in Figure 4, although at the first and second dose LTK63 is less efficient than the wild type LT to induce antibody response against ovalbumin, its adjuvanticity effect equals the wild type after the third dose.

LTK63 is currently tested in a few oral vaccine formulations. For example, we have recently shown that the mutant protected mice against *Helicobacter pylori* infection when orally delivered with the bystander antigens of *H. pylori* vacA and cagA. In addition, nasal delivery of LT mutants with inactivated influenza virus fully protects mice from the challenge with the virus. Human clinical trials with this anti-flu vaccine formulation are in progress.

REFERENCES

Bokoch, G.M., Katada, T., Northup, J.K., Hewlett, E.L. and Gilman, A.G. (1983) Identification of the predominant substrate for ADP-ribosylation by islet activating protein. *J. Biol. Chem.*, 258, 2072–2075.
Burnette, W.N., Mar, V.L., Platler, B.W., Schlotterbeck, J.D., McGinley, M.D., Stoney, K.S., Rhode, M.F. and Kaslow, H.R. (1991) Site-specific mutagenesis of the catalytic subunit of cholera toxin; substituting lysine for arginine 7 causes loss of activity. *Infect. Immun.*, 59, 4266–4270.
De-Aizpurua, D.H. and Russel-Jones, I. (1988) Oral vaccination: Identification of classes of proteins that provoke an immune response upon oral feeding. *J. Exp. Med.*, 167, 440–451.
Domenighini, M., Pizza, M. and Rappuoli, R. (1995) Bacterial ADP-ribosyltransferases. In *Bacterial Toxins and Virulence Factors in Disease*, edited by J. Moss, B. Iglewski, M. Vaughan and A.T. Tu. New York: Marcel Dekker, Inc.
Farizo, K.M., Cochi, S.L., Zell, E.R., Brink, E.W., Wassilak, S.G. and Patriarca, P.A. (1992) Epidemiological features of pertussis in United States, 1980–1989. *Clin. Infect. Dis.*, 14, 708–719.
Fong, H.K., Yoshimoto, K.K., Eversole-Cire, P. and Simon, M.I. (1988) Identification of a GTP-binding protein alpha subunit that lacks an apparent ADP-ribosylation site for pertussis toxin. *Proc. Natl. Acad. Sci. U.S.A.*, 85, 3066–3070.
Glenny, A.T. and Hopkins, B.E. (1923) Diphtheria toxoid as an immunizing agent. *Br. J. Exp. Pathol.*, 4, 283–288.
Greco, D., Salmaso, S., Mastrantonio P., Giuliano, M., Tozzi, A.E., Ciofi, M.L., Giammarco, A., Panei, P., Blackwelder, W.C., Klein D.L. and Wassilak, S.G.F. (1996) A controlled trial of two acellular vaccines and one whole-cell vaccine against pertussis. Progetto Pertosse Working Group. *N. Engl. J. Med.*, 334(6), 341–348.
Haze, C.C., Thai, L.S., Boesmanfinkelstein, M., Mar, V.L., Burnette, W.N., Kaslow, H.R., Stevens, L.A., Moss, J. and Finkelstein, R.A. (1994) Construction and characterisation of recombinant *Vibrio cholerae* inactive cholera toxin analogs. *Infect. Immun.*, 62, 3051–3057.
Holmgren, J. and Czerkinsky, C. (1992) Cholera as a model for research on mucosal immunity and development of oral vaccines. *Curr. Opin. Immunol.*, 4, 387–392.
Iiri, T., Tohkin, M., Morishima, N., Ohoka, Y., Ui, M. and Katada, T. (1989) Chemotactic peptide receptor-supported ADP-ribosylation of a pertussis toxin substrate GTP-binding protein by cholera toxin in neutrophil-type HL-60 cells. *J. Biol. Chem.*, 264, 21394–21400.
Katada, T. and Ui, M. (1982) ADP ribosylation of the specific membrane protein of C6 cells by islet-activating protein associated with modification of adenylate cyclase activity. *J. Biol. Chem.*, 257, 7210–7216.
Levine, M.M., Kaper, J.B., Black, R.E. and Clements, M.L. (1983) New knowledge on pathogenesis of bacterial enteric infections as applied to vaccine development. *Microbiol. Rev.*, 47(4), 510–550.
Locht, C. and Keith, J.M. (1986) Pertussis toxin gene: nucleotide sequence and genetic organization. *Science.*, 232, 1258–1264.

Nencioni, L., Volpini, G., Peppoloni, S., De Magistris, M.T., Marsili, I. and Rappuoli, R. (1991) Properties of the pertussis toxin mutant PT-9K/129G after formaldehyde treatment. *Infect. Immun.*, 59, 625–630.

Nicosia, A., Perugini, M., Franzini, C., Casagli, M.C., Borri, M.G., Antoni, G., Almoni, M., Neri, P., Ratti, G. and Rappuoli, R. (1986) Cloning and sequencing of the pertussis toxin genes. opernon structures and gene duplication. *Proc. Natl. Acad. Sci. USA.*, 83, 4631–4635.

Pierce, N.F. (1978) The role of antigen form and function in the primary and secondary intestinal immune response to cholera toxin and toxoid in rats. *J. Exp. Med.*, 148, 195–206.

Pierce N.F. and Gowan, J.L. (1975) Cellular kinetics of the intestinal immune response to cholera toxoid in rats. *J. Exp. Med.*, 142, 1550–1563.

Pizza, M., Covacci, A., Bartoloni, A., Perugini, M., Nencioni, L., De Magistris, M.T., Villa, L., Nucci, D., Manetti, R., Bugnoli, M., Giovannoni, F., Olivieri, R., Barbieri, J.T., Sato, H. and Rappuoli, R. (1989) Mutants of pertussis toxin suitable for vaccine development. *Science*, 246, 497–500.

Pizza, M., Domenighini, M., Hol, W., Giannelli, V., Fontana, M.R., Giuliani, M.M., Magagnoli, C., Peppoloni, S., Manetti, R. and Rappuoli, R. (1994) Probing the structure-activity relationship of *Escherichia coli* LT-A by site-directed mutagenesis. *Molec. Microbiol.*, 14, 51–60.

Ramon, G. (1924) Sur la toxine et sur l'anatoxine diphtheriques. *Ann. Inst. Pasteur*, 38, 1–10.

Rappuoli, R. and Pizza, M. (1991) Structure and evolutionary aspects of ADP-ribosylating toxins. In *Sourcebook of bacterial protein toxins*, edited by J. Alouf and J. Freer, 1–20. Academic Press.

Sato, Y., Kimura, M. and Fukumi, H. (1984) Development of a pertussis component vaccine in Japan. *Lancet*, i, 122–126.

Sixma, T.K., Pronk, S.E., Kalk, K.H., Wartna, E.S., van Zanten, B.A., Witholt, B. and Hol, W.G. (1991) Crystal structure of a cholera toxin-related heat-labile enterotoxin from *E. coli*. *Nature*, 351, 371–377.

Stibitz, S., Black, W. and Falkow, S. (1986) The construction of a cloning vector designed for gene replacement in *Bordetella pertussis*. *Gene*, 50, 133–140.

Tamura, M., Nogimori, K., Murai, S., Yajima, M., Ito, K., Katada, T., Ui, M. and Ishii, S. (1982) Subunit structure of the islet-activating protein, pertussis toxin, in conformity with the A-B- model. *Biochemistry.*, 21, 5516–5522.

Van Dop, C., Tsubokawa, M., Bourne, H.R. and Ramachandran, J. (1984) Amino acid sequence of retinal transducin at the site ADP-ribosylated by cholera toxin. *J. Biol. Chem.*, 259, 696–698.

Walker, R.I. (1994) New strategies for using mucosal vaccination to achieve more effective immunisation. *Vaccine.*, 12, 387–399.

Weiss, A. and Hewlett, E.L. (1986) Virulence factors of *Bordetella pertussis*. *Ann. Rev. Microbiol.*, 40, 661–686.

West, R.E., Jr., Moss, J., Vaughan, M., Liu, T. and Liu, T.Y. (1985) Persussis toxin-catalyzed ADP-ribosylation of transducin. Cysteine 347 is the ADP-ribose acceptor site. *J. Biol. Chem.*, 260, 14428–14430.

Woodrow, G.C. (1997) An overview of Biotechnology as applied to vaccine development. In *New Generation Vaccines*, edited by M.M. Levine, G.C. Woodrow, J.B. Kaper and G.S. Cobon. New York: Marcel Dekker, Inc.

16. EXPLORING MOLECULAR RECOGNITION BY COMBINATORIAL AND RATIONAL APPROACHES

ANNA TRAMONTANO[1] and FRANCO FELICI[2]

[1]IRBM P. Angeletti, Via Pontina Km 30.600, I-00045 Pomezia, Rome
[2]Kenton Labs, IRCCS S. Lucia, Via Azdeotina 306, I-00179 Rome

INTRODUCTION

The problem of identifying molecules with a given biological activity often coincides with that of engineering a specific ligand for a given molecule. This can be accomplished through procedures based on approaches utilising more or less 'rational' information components, mainly depending on the amount of structural information available for the target molecule. When this type of information is available, small molecules interacting with specific ligands (for example enzyme inhibitors or receptor agonists and antagonists) can be developed by rational design.

But the information available is often insufficient for design, and combinatorial techniques offer a powerful method for the production and screening of very large collections of molecules, thus constituting a "combinatorial" approach, which complements the lack of enough structural information to adopt a purely "rational" methodology. For example, drug discovery has traditionally relied on the combinatorial approach, screening large numbers of organic molecules to identify novel lead compounds.

We describe here a series of case studies of various approaches utilising different rational/combinatorial content.

A procedure based on a combinatorial approach is *a-priori* more or less powerful, depending on the diversity of the random collection, but it is also true that the greater the complexity, the more difficult the identification of the positive molecules from the population becomes. In recent years, the generation of large "repertoires" of chemically (Geysen *et al.*, 1986; Fodor *et al.*, 1991; Lam *et al.*, 1991; Houghten *et al.*, 1991) or biologically (Scott *et al.*, 1990; Devlin *et al.*, 1990; Cwirla *et al.*, 1990; Felici *et al.*, 1991) synthesised molecules has provided a novel, powerful approach to the identification of unknown ligands or variants of already known molecules, with the addition of new and more favourable properties. Most of these libraries readily encompass $>10^7$ different peptide sequences, thereby creating a rich source of structural diversity. Different methods have been devised to identify members of the library with specific binding properties; among the molecular biology tools being used to construct ligand libraries, those exploiting filamentous bacteriophage as molecular vectors constitute the majority and have been successfully used in many different cases (for reviews see Felici *et al.*, 1995; Daniels and Lane, 1996; Cortese *et al.*, 1996).

The closely related filamentous bacteriophage M13, f1, fd (Model and Russel, 1988) are made up of a proteic envelope formed by several copies of five different proteins, and contain a single-stranded DNA molecule carrying the genetic

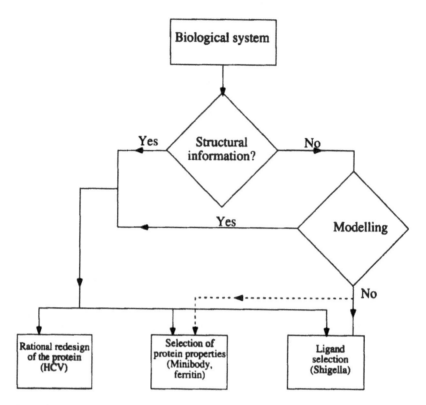

Figure 1   Schematic flowchart of the strategy to identify molecules with the desired biological activity.

information. When, through site-directed mutagenesis, foreign sequences are inserted into the gene encoding for one of the capsid proteins, the corresponding fusion product is displayed on the surface of the viral particle. By using appropriate strategies it is possible to enrich phage displaying peptide sequences with the desired properties from a very heterogeneous mixture of recombinant phage particles. Such phage clones, containing specific peptide products physically associated with their genetic information, can be rescued and propagated by infection of competent bacteria, and the sequence of the selected peptides can be easily deduced from the sequence of the corresponding DNA molecule. In this way the individual selected leads can be identified and characterised, even when only tiny amounts are recovered in the selection procedure.

This selection strategy has been applied to many different ligand/ligate systems and has led to the identification of new peptide sequences which do not necessarily resemble the natural ones, but display analogous binding specificity.

It is also feasible to adopt solely rational strategies to modify and/or extend the recognition ability of a given protein in order to enhance or modify its natural function, but this approach requires some prior knowledge of the protein structure and the determinants of the recognition process (Figure 1). The difficulties encountered for this approach depend on the quality of the structural information, as will be illustrated by one of our case studies.

SELECTION OF PEPTIDES MIMICKING *SHIGELLA FLEXNERI* LIPOPOLYSACCHARIDE

In order to develop effective vaccines against several important bacterial pathogens, it would be extremely useful to be able to mimic carbohydrate antigens, as these antigens are known to be T-cell independent, inducing weak immune responses associated with lack of a memory response.

Until now, two approaches have been attempted to overcome these difficulties: polysaccharide-protein conjugated vaccines and anti-idiotype vaccines. The main drawbacks presented by the former are the purification of the polysaccharide (since it must be totally devoid of any endotoxic activity due to residual lipid A component), and the carbohydrate moiety's loss of immunogenicity related to the type of coupling with the protein carrier. Carbohydrate synthesis may lessen the problems associated with antigen purification, but remains a rather limited solution, due to the overall difficulty found with carbohydrate chemistry. The second approach, based on the mimicry of sugar antigens by anti-idiotype antibodies, is not a feasible alternative, since manipulation of the idiotypic network by inducing anti-idiotype antibodies directed against a monoclonal antibody specific for a carbohydrate structure is difficult and time-consuming, and the use of such material in humans is still a matter of debate.

Shigella flexneri is a mucosal bacterial pathogen which is responsible for most of the endemic forms of shigellosis, a dysenteric syndrome that causes a high mortality rate among infants, particularly in developing countries. Secretory IgA antibodies constitute a first line of defence against such pathogens and many experimental and clinical studies have shown that resistance to infection correlates with specific titres of these antibodies (Mazanec *et al.*, 1993).

Systemic and mucosal immune responses to *S. flexneri* infection are mainly directed against lipopolysaccharide (LPS) and plasmid-encoded proteins. The protection provided by natural infection is considered to be serotype-specific, pointing to the LPS as a primary target antigen for protective immunity.

In a recent paper (Phalipon *et al.*, 1997), two monoclonal dimeric IgA antibodies which have been shown to be protective against infection in a mouse animal model (Phalipon *et al.*, 1995) were used to select specific phage-displayed peptides through affinity-selection and immunological screening. Several different positive clones were identified, and all of these phage-displayed peptides were able to compete in ELISA with the specific LPS for binding to the antibody, thus demonstrating that they are effective antigenic mimics of the LPS antigen and therefore represent one of the few examples of peptide sequences able to mimic sugars.

The molecular basis of this mimicry is that antigen-specific antibodies, being ligates for the antigen's epitopes, contain implicit structural information about these epitopes. A peptide specifically binding to the antigen-binding site of an antibody might be a positive image of the antigen, thus mimicking part of its structural and biological properties. If this mimicry is sufficiently accurate, the selected peptides can be suitable substitute immunogens, completely distinct from the natural antigen.

In fact, all the nineteen clones identified using the two anti-LPS monoclonal IgA antibodies were separately used to immunise mice and two of them were indeed able to induce a LPS-specific immune response. Thus in this case, immunogenic mimics of protective carbohydrate epitopes were selected by a straightforward approach using only phage-displayed peptide libraries and protective monoclonal antibodies specific for the pathogen of interest, allowing a significant step forward in the search for potential leads for the development of peptide-based vaccines against polysaccharides.

It is therefore clear that mimicking non-protein ligands (i.e. carbohydrates) can be achieved through screening random peptide libraries where the specific sequences were identified using a purely combinatorial approach, without using any structural information on the original target. This constitutes a unique and essential feature for those projects where the nature of the ligands is either known to contain non-protein elements which are important for ligate recognition, or unknown.

MAPPING OF A DISCONTINUOUS EPITOPE OF HUMAN H FERRITIN

Monoclonal antibodies to antigens are powerful reagents for structural and functional analysis, particularly when the epitope recognised by them can be identified. A detailed picture of the amino acids that comprise a given epitope (structural epitope) has been described only in a few cases, where antigen-antibody complexes were analyzed by direct physical methods (x-ray crystallography or NMR hydrogen-deuterium exchange), because these methods are time consuming and technically difficult.

The utilisation of overlapping synthetic peptides matching the antigen sequence, or the construction of random peptide libraries, constitute alternative techniques for mapping the folded antigenic determinants of proteins. Screening a random peptide library with a monoclonal antibody usually reveals a number of peptides that bind to the selector molecule and often display a common consensus amino acid sequence. When this sequence shows similarity with a portion of the primary structure of the protein antigen, the linear epitope recognised by the antibody can be mapped (for a review see Cortese et al., 1995).

Nevertheless, all the protein structural epitopes so far reported are discontinuous (Davies et al., 1990; MacCallum et al., 1996) so it would hence appear that the conformation recognised by the majority of antibodies is composed of amino acids from two or more segments which are separate from each other in the primary structure of the polypeptide chain, but brought together on the surface during folding. In such cases, comparison between the consensus sequence of the selected peptides and the antigen primary structure does not usually permit unequivocal matching, in the absence of data on the 3D structure.

The first successful attempt at defining a discontinuous epitope by screening a random peptide library was reported by Luzzago et al. in 1993. In this paper, a cysteine-constrained nonapeptide library was screened using a monoclonal antibody (H107) raised against human H ferritin (H-Fer). The 3D structure of the assembled molecule of H-Fer had been already solved at high resolution by X-ray crystallography (Lawson et al., 1991) and a previous study had shown that H107 binding is sensitive to antigen conformational changes, as extensive alterations of three separate regions of H-Fer can affect the binding of this antibody, but not that of other monoclonal antibodies (Arosio et al., 1990). From the analysis of the specific phage clones selected, two consensus sequences were derived (YDxxxxW and GSxF, in one-letter amino acid code, where x represents any amino acid residue), neither of which were

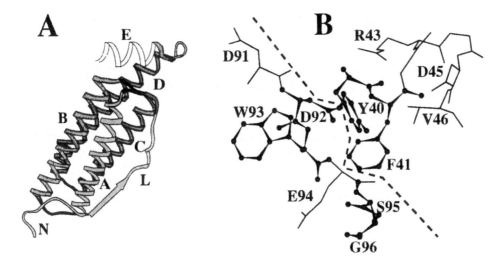

Figure 2 A) Structure of a monomer of human H ferritin. The two darker loops contain the residues involved in binding to MAb H107. B) Structure of the regions 40-46 and 91-96 of H ferritin. Only the side chains of residues exposed to solvent are shown. Residues in ball and stick representation are candidates for mimicry by the selected peptides.

found in the sequence of H-Fer. However, a clearer picture emerged by examining the 3D structure of the protein, and a tentative match between the identified residues and protein regions exposed in the assembled complex was found. When the putative binding residues were mutated, the resulting proteins no longer reacted with H107, thus confirming their involvement in the binding to the antibody.

This study led to the identification of two distinct H-Fer regions constituting the H107 epitope. Each of the two families of phage-displayed peptides bearing the two different consensus sequences appear to mimic part of the antibody's binding surface, in both cases spanning the "discontinuity" between these two regions (Figure 2).

The above example is particularly useful to illustrate how the synergy between computational and experimental approaches can be effective to gain a better understanding of the system in hand.

Two independent computational methods were in fact used:

In one approach the problem can be stated as follows: given the structure (or a model of the structure) of an antigen and the structure (or a model of the structure) of an antibody raised against the antigen, can we construct a model of the complex between the two structures in order to identify the site of interaction? In order to answer this question a docking procedure can be used to derive a set of possible orientations of the antigen with respect to the antibody in the complex (Helmer-Citterich et al., 1995).

The second approach uses knowing which peptide sequences are able to mimic the surface of the antigen. In this case the question can be formulated as follows: is it possible to identify the region of the antigen mimicked by a peptide of known sequence? A protein surface can be viewed as an ensemble of overlapping 'pseudo-peptides', i.e. peptides which could assume a conformation such that their side chains are in the same position of those exposed on the protein surface (Pizzi *et al.*, 1995). A peptide library can be viewed as a collection from which one or more of the mimicking peptides can be selected. It is possible to enumerate the sequences of all 'pseudo-peptides' that can putatively mimic the surface of a protein and organise them in a data base. The resulting sequence data base can then be searched for peptides resembling those selected in the phage library screening experiment.

Both approaches were used for the case of ferritin. For the first, the hypervariable regions of the H107 antibody genes were sequenced and a model was built and 'docked' to the known ferritin structure, using PUZZLE, a docking method based on surface complementarity (Helmer-Citterich and Tramontano, 1994). For the second, the 'pseudo-peptide' data base constructed on the basis of the ferritin surface was searched using the sequences selected in the phage library experiment described above. The conclusions reached with both approaches identify the same epitope of ferritin, and are in perfect agreement with the experimental mutation data.

This represents an excellent example where data deriving from a "combinatorial" approach are "rationally" analyzed utilising structural information and computing methods to design "ad hoc" experiments (in this case site-directed mutagenesis) verifying the hypothesis.

MINIBODY

It is conceivable that complex epitopes could be more effectively mimicked by ligands which are more composite than just short peptide sequences. Ideally, one would like to select structurally constrained regions of a ligand analogously to what happens in the immune system, where antibodies can recognise a wide variety of antigens using a recognition site formed by the combination of six exposed loops mounted on a stable structural scaffold. Immunoglobulins contain four polypeptide chains, each one built of domains with a common fold: two disulphide linked beta-sheets packed face to face. The antigen binding site is contained in the V domains of the light (L) and heavy (H) chains. These domains pack to form a structurally conserved framework onto which three loops from the V_L and three from the V_H domains are arranged to form the antigen binding site (Alzari *et al.*, 1988). The loops from V_L and V_H are denoted L1, L2 and L3, and H1, H2 and H3 according to Chothia and Lesk (1987). The loops L1 and H1 connect strands belonging to different sheets within V_L and V_H domains; L2, L3, H2 and H3 are hairpin loops connecting adjacent strands in the same beta-sheet. Variations in length and amino acid sequence of these loops generate differences in specificity and affinity of immunoglobulins (Kabat *et al.*, 1991). Importantly, the conformational determinants of antigen-binding loops of immunoglobulins have been identified and it has been shown that they can have a discrete repertoire of conformations called "canonical structures" (Chothia and Lesk, 1987; Chothia *et al.*, 1989; Morea *et al.*, 1998). These canonical structures are stabilised by a small number of residues through packing or hydrogen bonding interactions, or their ability to assume unusual main chain conformations. Other residues are relatively free to vary, to modulate surface topography and charge distribution of the antigen-binding site.

Using a complete antibody (or its variable Fv fragment) as a scaffold to expose randomised regions in a conformationally constrained fashion has the drawback

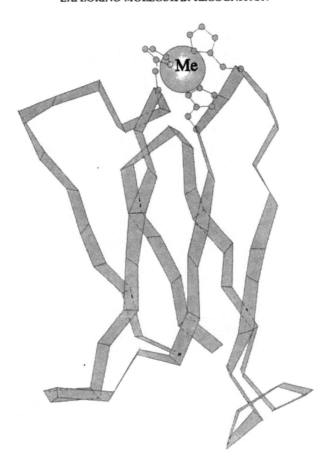

Figure 3   Cartoon of the Minibody structure.

that the antibody is multimeric and the six antigen binding loops come from two different domains. Possible solutions to this problem are the construction of a 'single chain' antibody (where the V_H and V_L fragments are covalently linked using a long flexible linker) or to redesign a new protein, small, monomeric but still retaining some of the immunoglobulin's antigen binding loops, whose sequences can then be randomised and used to select ligands. This second approach has indeed met with remarkable success in the case of the Minibody (Pessi *et al.*, 1993).

The Minibody is a designed protein of 61-residues and its predicted structure includes three beta-strands from each of the two beta-sheets of the variable (V) heavy chain domain of the mouse antibody McPC603 (Satow *et al.*, 1986), along with the segments corresponding to the exposed hypervariable H1 and H2 loops of the immunoglobulin as defined by Chothia and Lesk (1987) (Figure 3). The sequence of the beta-sheet portion of the Minibody obviously differs from that of the McPC603 immunoglobulin, since regions that would be buried either within a domain or at the interface with the V_L domain in the complete immunoglobulin find a completely different environment in the Minibody, so that appropriate amino acid changes had to be designed. In principle, the Minibody loops should have the desirable property of being predictable, as is the case for their counterparts in the complete immunoglobulin molecule (Chothia and Lesk, 1987; Chothia *et al.*, 1989; Tramontano *et al.*, 1990).

Our understanding of the rules relating sequence to structure in proteins is still fairly rudimentary, consequently redesigning a known protein fold as a framework for the insertion of functional sites is extremely challenging and it is necessary to demonstrate that the desired structure has been preserved. As a result, the Minibody was extensively characterised using a variety of methods.

The findings obtained indicated that the design was successful. In particular, the minibody is a compact globular monomer in solution, its secondary structure content is in excellent agreement with the expected one and denaturation studies with urea showed that the molecule unfolds with a co-operative transition, and its $\Delta G_D^{H_2O}$ value, although lower than the values (5–15 Kcal/mol) observed for natural proteins (Creighton, 1990), is very similar to that of the reduced immunoglobulin C_L domain. To test that the loops were in the predicted position in the designed structure, the authors also engineered a metal binding site formed by residues belonging to both loops (Figure 3) and demonstrated that the molecule did indeed bind metal. The design and synthesis of an all beta-protein is a remarkable achievement because producing such a structure is inherently more difficult than making a helical protein (Richardson and Richardson, 1989).

The minibody serves as an excellent presentation scaffold for the construction of an "all-purpose" constrained library because of its properties. In fact, it has been possible to construct a fusion phage and show that the minibody is effectively displayed onto the f1 phage surface when fused to the N-terminus of pIII and to construct a library of constrained loops by randomisation of the minibody sequence corresponding to the hypervariable regions. The recombinant phage was affinity-selected to search for molecules with novel functions and it has indeed been possible to select from such a library minibody mutants with high affinity for human interleukin 6 (Martin *et al.*, 1996).

Peptide sequences and conformations generated by the phage-displayed minibody might be conceivably utilised in the drug development process. However, peptide-based pharmaceuticals have a number of potential disadvantages, such as poor oral activity and short circulating half-lives, hence diminishing their appeal as drug candidates (Plattner and Norbeck, 1990). The conversion of peptides into peptidomimetics is a challenging task, daunted by modelling difficulties, synthetic barriers, etc. (Hruby *et al.*, 1990). However, by conformationally-constraining peptide loops onto a "presentation" framework such as the minibody, their three-dimensional conformation can then be elucidated, or even predicted as for immunoglobulin hypervariable loops (Chothia *et al.*, 1989) thus making chemical synthesis of biologically active mimetics based on the affinity selected peptide ligands practicable (Saragovi *et al.*, 1992; Taub and Greene, 1992).

HCV PROTEASE

So far we have described combinatorial approaches, even if based on structural insights, to select molecules able to recognise a protein. Rational approaches can be exploited when more structural information is available, as illustrated by the case of the Hepatitis C Virus (HCV) protease.

HCV is the major etiological agent of both parenterally-transmitted and sporadic non-A, non-B hepatitis (NANB-H). The HCV virion has a positive-strand RNA genome of about 9.5 kilobases which contains a single open reading frame encoding a polyprotein of about 3000 amino acids. The polyprotein is subsequently cleaved and gives rise to the E1, E2, NS3, NS4A, NS4B, NS5A and NS5B proteins. It has been demonstrated that the proteolytic cleavage at the NS3/NS4A, NS4A/NS4B, NS4B/ NS5A, and NS5A/NS5B junctions is mediated by a virus-encoded serine protease, contained within NS3. Its sequence contains the characteristic sequence pattern of the small cellular proteases of the trypsin superfamily (Hijikata et al., 1993; Bartenschlager et al., 1993; Grakoui et al., 1993; Tomei et al., 1993). In addition to the NS3 serine-protease, the NS4A protein is required for cleavage at the NS3/NS4A and NS4B/NS5A sites and it increases the efficiency of cleavage at the NS5A/NS5B and NS4A/NS4B junctions (Bartenschlager et al., 1994; Failla et al., 1994; Lin et al., 1994; Tanji et al., 1994).

The structure of the HCV NS3 protease was only recently solved (Love et al., 1996; Kim et al., 1996), but well before this the need arose to understand the substrate specificity of this enzyme, since it would be of great help in designing substrate-based inhibitors.

There is no protease of known structure sharing more than a 13% sequence identity with the HCV NS3 protease, so that the construction of a reliable homology model is basically impossible. Nevertheless, one can take advantage of the wealth of structural data on the serine protease protein family to try and obtain an approximate model.

The major obstacle to constructing a model at such a low level of sequence identity is the quality of the sequence alignment. There are, however, many known structures belonging to the same folding family, which allow a sequence profile based on a multiple structural superposition to be constructed, thus facilitating the alignment of the protein sequence to be modelled (target sequence) to the profile.

There are several methods to obtain the optimal structural superposition of a family of protein structures and their description goes beyond the scope of this chapter. Here it suffices to say that, given some distance threshold, it is possible to obtain a list of all residues in the family which correspond to each other in three-dimensional space within that threshold (structure based sequence profile). Some regions will diverge structurally and no sequence profile will be derived for them.

The structure based profile will give information about which amino acids are allowed in each of the aligned positions (all hydrophobic, only large, always charged,..) This will facilitate the alignment of the target sequence to the other proteins of the family but, since the profile does not include the complete sequences, the alignment will only be partial. Luckily however, the best conserved regions of homologous proteins always include the important functional regions and the alignment will usually include the active site region and most of the important structural and functional features of the family.

The alignment can be used to construct a homology model of the parts corresponding to the conserved regions. This relies on the assumption (usually correct) that regions structurally conserved in the family will also be conserved in the target protein. The most important problem is that the model will be as wrong as inaccurate the alignment is, i.e. if the target sequence has been aligned incorrectly to the structure based profile in a region, the model will be wrong in that region.

For practical purposes it is usually convenient to complete the model by joining up the modelled regions with 'reasonable' protein substructures, but it should be kept in mind that they are not the results of an accurate prediction.

Following the above protocol, it was possible to construct a partial model of the NS3 protease which could be used to predict the specificity of the protease.

In the model, the physico-chemical environment and the shape of the specificity pocket are primarily determined by phenylalanine 213 (according to the chymotrypsin numbering scheme). Visual inspection of the modelled pocket suggested that the P1 residue of the substrate could have been a cysteine or a serine. In addition, the pocket appeared to be very hydrophobic and closed by the aromatic ring of the phenylalanine. Since the sulphydryl group of cysteine has been shown to interact favourably with the aromatic ring of phenylalanine, cysteine was considered the most reasonable P1 residue. This prediction was subsequently confirmed by N-terminal sequencing of the processed NS4B, NS5A and NS5B proteins (Grakoui et al., 1993; Pizzi et al., 1993).

The success of this test was certainly encouraging, but was not sufficient to guarantee the correctness of the alignment of the residues forming the specificity pocket of the enzyme. Consequently, before using the modelling results as a starting point for further experiments and to attempt the design of substrate based inhibitors, it was necessary to gain more confidence in the model. If a model is correct, one should be able to use it to rationally modify some properties of the protein and this should increase confidence in the model itself.

The model was thus used as a guide for site-specific mutagenesis aimed at modifying the specificity of the protease. The rationale was to replace the residues predicted to form the specificity pocket of NS3 with the corresponding ones of Streptomyces griseus protease B (SGPB) (Read et al., 1983), a protease of known structure able to recognise a phenylalanine residue in the P1 position of the substrate.

The optimal P1 residues for the substrate of SGPB are phenylalanine, tyrosine and, to a lesser extent, leucine. The specificity pocket of this protease is formed by the side chains of alanine 190, threonine 213, glycine 216 and threonine 226. Both proteins have a threonine in position 226, which was consequently left unchanged. In positions 213, 216 and 190, the NS3 protease has a phenylalanine, an alanine and a tyrosine. In the model the latter does not contribute to the specificity of the enzyme (Pizzi et al., 1993), so only the phenylalanine and the alanine were mutated to the corresponding residues of the SGPB sequence (threonine and glycine, respectively) (Figure 4).

The experiment was successful: the redesigned enzyme acquired, through only two mutations, the ability to cleave the substrate recognised by Streptomyces griseus protease B, an enzyme of known structure used as template, and this strongly supported the reliability of the model (Failla et al., 1996).

The recent resolution of the structure of the NS3 enzyme has fully supported the model of the specificity pocket.

This example shows that even when only limited structural information is available on a protein, the careful analysis of a low reliability model can still single out those regions that are more likely to be predicted correctly and these can be used to modify the properties of the target protein.

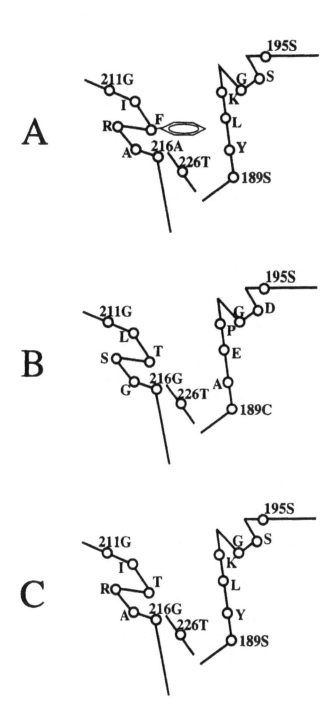

Figure 4 Schematic view of the specificity pocket of A) the model of HCV protease; B) Streptomyces Griseus protease B and C) the mutant HCV protease that acquires the specificity of SGPB.

Other examples of changes in specificity of serine proteases have been reported in the literature (Hedstrom *et al.*, 1992; Caputo *et al.*, 1994), but the HCV protease example is interesting because it illustrates that even in a case as difficult as this, it is conceivable to rationally modify the ability of a protein to recognise its cognate molecule, thus opening up the road to a number of very interesting experiments.

CONCLUSIONS

Although molecular recognition has been among the most studied fields in biology for the last twenty years, our comprehension of the rules regulating the subtle and complex balance between affinity and specificity is far from being satisfactory.

We believe that the use of large collections of random variants and the ever more elaborate and powerful methods for selection and screening exploited in combinatorial approaches, will prove to be an essential tool for increasing our basic understanding of the recognition processes. Components of naturally observed complexes have evolved and co-evolved and have to satisfy far more biological requirements than recognition alone, so that the information of interest is embedded in a variety of other complex phenomena. Conversely, selection from random collections of sequences and/or structures is likely to more easily reveal the factors that are essential for binding and thus help us understand molecular recognition.

The case studies that we have described here were chosen to represent a variety of different approaches, ranging from those where essentially no structural information about the system is available, to those where this information is already known, or can be gained using several different methodologies together. However, it is clear that in all cases a combination of approaches (combinatorial, rational, experimental, computational) can be nothing but beneficial to achieve the desired properties of the biological system (Figure 1).

Are combinatorial approaches more powerful than rational ones? The answer we propose is that a combination of the two is obviously the right way to go, but requires careful balancing.

To be more precise, we do not believe that 'the best system' exists, but rather that the system is dictated by each specific case, according to the information already available and the goal of the project. In some cases we may be interested in just replacing one ligand with another, as is the case for vaccine development, and here the power of combinatorial selection strategies is enormous since not even prior knowledge of the pathologic agent is required.

In other cases we might need to design an all-purpose system and the more we know about it in advance the easier it will be to design optimal experiments and interpret our results. For example, an even partial knowledge of the structural features of the natural components may help in designing combinatorial libraries having structural constraints or utilising molecular scaffolds that would be best suited to mimic the authentic interactions, and will also lower the diversity needed to achieve a fairly complete representation of all the possible variants, that is often a stumbling block.

We might also need to modify the properties of our system just to gain a better understanding of it, or a receive confirmation of our hypotheses.

The few cases described here were not selected on the basis of their success or of the scope of their application, but rather to provide an overview of the wide variety of possible approaches, and to stress the great power deriving from their optimal combination.

ACKNOWLEDGEMENTS

The authors are grateful to Janet Clench for carefully reading the manuscript and to Brenda McManus for excellent assistance.

REFERENCES

Alzari, P.M., Lascombe, M.B. and Poljak, R.J. (1988) Three-dimensional structure of antibodies. *Annu. Rev. Immunol.*, 6, 555–580.

Arosio, P., Cozzi, A., Ingrassia, R., Luzzago, A., Ruggeri, G., Iacobello, C., Santanbrogio, P. and Albertini, A. (1990) A mutational analysis of the epitopes of recombinant human H-ferritin. *Biochim. Biophis. Acta*, 1039, 197–203.

Bartenschlager, R., Ahlborn-Laake, L., Mous, J. and Jacobsen, H. (1993) Nonstructural protein 3 of the hepatitis C virus encodes a serine-type proteinase required for cleavage at the NS3/4 and NS4/5 junctions. *J. Virol.*, 68, 1045–5055.

Bartenschlager, R., Ahlborn-Laake, L., Mous, J. and Jacobsen, H. (1994) Kinetic and structural analysis of hepatitis C virus polyprotein processing. *J. Virol.*, 68, 5045–5055.

Caputo, A., James, M.N.G., Powers, J.C., Hudig, D. and Bleackley, R.C. (1994) Conversion of the substrate specificity of mouse proteinase granzyme B. *Nature Structural Biology*, 6, 364–367.

Chothia, C. and Lesk, A.M. (1987) Canonical structures for the hypervariable regions of immunoglobulins. *J. Mol. Biol.*, 196, 901–918.

Chothia, C., Lesk, A.M., Tramontano, A., Levitt, M., Smith-Gill, S.J., Air, G., Sheriff, S., Padlan, E.A., Davies, D., Tulip, W.R., Colman, P.M., Spinelli, S., Alzari, P.M. and Poljak, J. (1989) Conformations of immunoglobulin hypervariable regions. *Nature*, 342, 877–883.

Cortese, R., Monaci, P., Luzzago, A., Santini, C., Bartoli, F., Cortese, I., Fortugno, P., Galfrè, G., Nicosia A. and Felici, F. (1996) Selection of biologically active peptides by phage display of random peptide libraries. *Curr. Op. Biotechnol.*, 7, 616–621.

Cortese R., Monaci, P., Nicosia, A., Luzzago, A., Felici, F., Galfrè G., Pessi, A., Tramontano, A. and Sollazzo, M. (1995) Identification of biologically active peptides using random libraries displayed on phage. *Curr. Opin. Biotechnol.*, 6, 73–80.

Creighton, T.E. (1990) Protein folding. *Biochem. J.*, 270, 1–16.

Cwirla, S.E., Peters, E.A., Barrett, R.W. and Dower, W.J. (1990) Peptides on phage: A vast library of peptides for identifying ligands. *Proc. Natl. Acad. Sci. USA*, 87, 6378–6382.

Daniels, D.A. and Lane, D.P. (1996) Phage peptide libraries. *Methods: A Companion to Methods in Enzymology*, 9, 494–507.

Davies, R.D., Padlan, E.A. and Sheriff, S. (1990) Antibody-antigen complexes. *Annu. Rev. Biochem.*, 59, 439–473.

Devlin, J.J., Panganiban, L.C. and Devlin, P.E. (1990) Random Peptide Libraries: A Source of Specific Protein Binding Molecules. *Science*, 249, 404–406.

Failla, C., Pizzi, E., De Francesco, R. and Tramontano, A. (1996) Redesigning the substrate specificity of the hepatitis C virus NS3 protease Fold. *Des.* 1, 35–42.

Failla, C., Tomei, L. and De Francesco, R. (1994) Both NS3 and NS4A are required for processing of hepatitis C virus polyprotein. *J. Virol.*, 68, 4017–4026.

Felici, F., Castagnoli, L., Musacchio, A., Jappelli, R. and Cesareni, G. (1991) Selection of antibody ligands from a large library of oligopeptides expressed on a multivalent exposition vector. *J. Mol. Biol.*, 222, 301–310.

Felici, F., Luzzago, A., Monaci, P., Nicosia, A., Sollazzo, M. and Traboni, C. (1995) Peptide and protein display on the surface of filamentous bacteriophage. In *Biotechnology Annual Review*, edited by M.R. El-Gewely, Vol. 1, pp. 149–183. Amsterdam, The Netherlands: Elsevier Science.

Fodor, S.P.A., Leighton, R.J., Pirrung, M.C., Stryer, L., Lu, A.T. and Solas D. (1991) Light-directed, spatially addressable parallel chemical synthesis. *Science*, 251, 767–773.

Geysen, H.M., Rodda, S.J. and Mason, T.J. (1986) A priori delineation of a peptide which mimics a discontinuous antigenic determinant. *Mol. Immunol.*, 23, 709–715.

Grakoui, A., McCourt, D.W., Wychowski, C., Feinstone, S.M. and Rice, C.M. (1993) Characterization of the hepatitis C virus-encoded serine proteinase: determination of proteinase-dependent polyprotein cleavage sites. *J. Virol.*, 67, 2832–2843.

Grakoui, A., Wychowski, C., Lin, C., Feinstone, S.M. and Rice, C.M. (1993) Expression and identification of hepatitis C virus polyprotein cleavage products. *J. Virol.*, 67, 1385–1395.

Hedstrom L., Szilagyi, L. and Rutter, W.J. (1992) Converting trypsin to chymotrypsin: the role of surface loops. *Science*, 255, 1249–1253.

Helmer-Citterich, M., Rovida, E., Luzzago, A. and Tramontano, A. (1995) Modelling antibody antigen interactions: ferritin as a case study. *Molecular Immunology*, 32, 1001–1010.

Helmer Citterich, M. and Tramontano, A. (1994) PUZZLE: A new method for automated protein docking based on surface shape complementarity. *J. Mol. Biol.*, 235, 1021–1031.

Hijikata, M., Kato, N., Ootsuyama, Y., Nakagawa, M. and Shimotohno, K. (1993) Two distinct proteinase activities required for the processing of a putative nonstructural precursor protein of hepatitis C virus. *J. Virol.*, 67, 4665–4675.

Houghten, R.A., Pinilla, C., Blondelle, S.E., Appel, J.R., Dooley C.T. and Cuervo, J.H. (1991) Generation and use of synthetic peptide combinatorial libraries for basic research and drug discovery. *Nature*, 354, 84–86.

Hruby, V.J., Al-Obeidi, F. and Kazmierski, W. (1990) Emerging approaches in the molecular design of receptor-selective peptide ligands: conformational topographical and dynamic consideration. *Biochem. J.*, 268, 249–262.

Kabat, E.A., Wu, E.T., Perry, H.M., Gottesman, K.S. and Foeller, C. (1991) In: *Proteins of immunological interest*. U.S. Department of Health and Human Service NIH.

Kim, J.L., Morgenstern, K.A., Lin, C., Fox, T., Dwyer, M.D., Landro, J.A., Chambers, S.P., Markland, W., Lepre, C.A., O'Malley, E.T., Harbeson, S.L., Rice, C.M., Murchko, M.A., Caron, P.R. and Thomson, J.A. (1996) Crystal structure of the hepatitis C virus NS3 protease domain complexed with a synthetic NS4A cofactor peptide. *Cell*, 87, 343–355.

Lam, K.S., Salmon, S.E., Hersh, E.M., Hruby, V.J., Kazmierski W.M. and Knapp, R.J. (1991) A new type of synthetic peptide library for identifying ligand-binding activity. *Nature*, 354, 82–84.

Lawson, D.M., Artymiuk, P.J., Yewdall, S.J., Smith, J.M.A., Livingstone, J.C., Treffry, A., Luzzago, A., Levi, S., Arosio, P., Cesareni, G., Thomas, C.D., Shaw, W.V. and Harrison, P.M. (1991) Solving the structure of human H ferritin by genetically engineering intermolecular crystal contacts. *Nature*, 349, 541–544.

Lin, C., Amberg, S.M., Chambers, T.J. and Rice, C.M. (1994) Cleavage at a novel site in the NS4A region by the yellow fever virus NS2B-3 proteinase is a prerequisite for processing at the downstream 4A/4B signalase site. *J. Virol.*, 4, 297–304.

Love, R.A., Parge, H.E., Wickersham, J.A., Hostomsky, Z., Habuka, N., Moomaw, E.W., Adachi, T. and Hostomska, Z. (1996) The crystal structure of Hepatitis C virus NS3 proteinase reveals a trypsin like fold and a structural zinc binding site. *Cell*, 87, 331–342.

Luzzago, A., Felici, F., Tramontano, A., Pessi, A. and Cortese, R. (1993) Mimicking of discontinuous epitopes by phage-displayed peptides, I. Epitope mapping of human H ferritin using a phage library of constrained peptides. *Gene*, 128, 51–57.

MacCallum R.M., Martin A.C.R. and Thornton J.M. (1996) Antibody-antigen Interactions: Contact Analysis and Binding Site Topography. *J. Mol. Biol.*, 262, 732–745.

Martin F., Toniatti C., Salvati A.L., Ciliberto G., Cortese R. and Sollazzo M. (1996) Coupling protein design and in vitro selection strategies: improving specificity and affinity of a designed betaprotein IL-6 antagonist. *J. Mol. Biol.*, 255, 86–97.

Mazanec M.B., Nedrud J.G., Kaetzel C.S. and Lamm M.E. (1993) A Three-tiered view of the role of IgA in mucosal defence. *Immunology Today*, 14, 430–435.

Model, P. and Russel, M. (1988) Filamentous Bacteriophage. In *The bacteriophages*, edited by R. Calendar, pp. 375–456. New York: Plenum Press.

Morea, V., Tzamontano, A., Rustici, M., Chothia, C. and Lesk, A.M. (1998) Conformations of the third hypervariable region in the VH domain of immunoglobins. *J. Mol. Biol.*, **275**, 269–294.

Phalipon, A., Folgori A., Arondel J., Sgaramella G., Fortugno P., Cortese R., Sansonetti P.J. and Felici, F. (1997) Induction of anti-carbohydrate antibodies by phage library-selected peptide mimics *Eur. J. Immunol.*, **27**, 2620–2625.

Phalipon, A., Kaufmann, M., Michetti, P., Cavaillon, J.M., Huerre, M., Sansonetti, P.J. and Kraehenbuhl, J.P. (1995) Monoclonal IgA antibody directed against serotype-specific epitope of *Shigella flexneri* lipopolysaccharide protects against murine experimental shigellosis. *J. Exp. Med.*, **182**, 769–778.

Pessi, A., Bianchi, E., Crameri, A., Venturini, S., Tramontano, A. and Sollazzo, M. (1993) A designed metal binding protein with a novel fold. *Nature*, **362**, 367–369.

Pizzi, E., Cortese, R. and Tramontano, A. (1995) Mapping epitopes on protein surfaces. *Biopolymers*, **36**, 675–680.

Pizzi, E., Tramontano, A., Tomei, L., La Monica, N., Failla, C., Sardana M., Wood, T. and De Francesco, R. (1993) Molecular model of the specificity pocket of the hepatitis C virus protease: implications for substrate recognition. *Proc. Natl. Acad. Sci.*, **91**, 888–892.

Plattner, J.J. and Norberk, D.W. (1990) In *Drug Discovery Technologies*, edited by C.R. Clark and W.H. Moos, pp. 92–131. Chichester, England: Horwood Press.

Read, R.J., Fujinaga, M., Sielecki, A.R. and James, M.N.G. (1983) Structure of the complex of Streptomyces griseus protease B, a serine protease, with the third domain of the ovomucoid inhibitor at 1.8 Å resolution. *Biochemistry*, **22**, 4420–4433.

Richardson, J.S. and Richardson, D.C. (1989) The *de novo* design of protein structures. *Trends Biochem. Sci.*, **14**, 304–309.

Saragovi, H.U., Greene, M.I., Chrusciel, R.A. and Kahn, M. (1992) Loops and secondary structure mimetics: development and applications in basic science and rational drug design. *BioTech.*, **10**, 773–778.

Satow, Y., Cohen, G.H., Padlan, E.A. and Davies, D.R. (1986) Phosphocholine binding immunoglobulin Fab McPC603: an X-ray diffraction study at 2.7 Å. *J. Mol. Biol.*, **190**, 593–604.

Scott, J.K. and Smith, G.P. (1990) Searching for peptide ligands with an epitope library. *Science*, **249**, 386–390.

Tanji, Y., Hijikata, M., Hirowatari, Y. and Shimotono, K. (1994) Identification of the domain required for *trans*-cleavage activity of hepatitis C viral serine protease. *Gene*, **145**, 215–219.

Taub, R. and Greene, M.I. (1992) Functional validation of ligand mimicry by anti-receptor antibodies: structural and therapeutics implications. *Biochemistry*, **31**, 7431–7435.

Tomei, L., Failla, C., Santolini, E., De Francesco, R. and La Monica, N. (1993) NS3 is a serine protease required for processing of hepatitis C virus polyprotein. *J. Virol*, **67**, 4017–4026.

Tramontano, A., Chothia, C. and Lesk, A.M. (1990) Framework residue 71 is a major determinant of the position and conformation of the second hypervariable region in the VH domains of immunoglobulins. *J. Mol. Biol.*, **215**, 175–182.

17. PRODUCTION AND POTENTIAL APPLICATIONS OF RECOMBINANT GASTRIC LIPASES IN BIOTECHNOLOGY

STEPHANE CANAAN*, LILIANE DUPUIS*, MIREILLE RIVIÈRE*,
ANNE BELAÏCH†, ROBERT VERGER*+, JEAN L. ROMETTE‡
and CATHERINE WICKER-PLANQUART*

*Laboratoire de Lipolyse Enzymatique, UPR 9025, de l'IFR-1 du CNRS,
31 Chemin Joseph-Aiguier, 13402 Marseille Cedex 20, France
†Laboratoire de Bioénergétique et Ingénierie des Protéines, UPR 9036,
de l'IFR-1 du CNRS, 31 Chemin Joseph-Aiguier, 13402 Marseille Cedex 20, France
‡AFMB/DISP, UPR 9039 du CNRS, ESIL Case 925,
163 avenue de Luminy, 13288 Marseille Cedex 9, France

INTRODUCTION

For a long time, the occurrence of gastric lipolysis was attributed to contamination of the gastric content by pancreatic lipase as the result of duodeno-gastric reflux. Volhard (1901) established, however, that there exists a gastric enzyme capable of hydrolysing triacylglycerols in humans. In 1917, Hull and Keaton (1917) observed the presence of a dog gastric lipase under conditions precluding contamination by pancreatic lipase. Nevertheless, the real importance of gastric lipolysis was first understood very belatedly in studies on patients with cystic fibrosis as well as on chronic pancreatitis, where the amount of fat absorption amounted to 70% of the total ingested lipids, whereas the pancreatic lipase and bile salt levels were particularly low (Hamosh, 1984, Ross and Sammons, 1955, Roulet et al., 1980). In other patients, where pancreatic lipase is completely lacking due to a congenital deficiency, 50% of the ingested dietary fat was absorbed (Muller et al., 1975).

One particularity of all preduodenal lipases is that they are stable and hydrolyse their substrates under acidic conditions. Acidic lipases have been screened in various species (rat, mouse, calf, sheep, humans, rabbit, dog, cat, monkey, horse, hog, guinea pig) from the tongue to the pylorus (Moreau et al., 1988a, De Nigris et al., 1988). Up to now, six preduodenal lipases have been purified and characterized biochemically: rat lingual lipase (RLL, Hamosh et al., 1979, Field and Scow, 1983, Roberts et al., 1984), human gastric lipase (HGL) (Tiruppathi and Balasubramanian, 1982), calf pharyngeal lipase (CPL, Bernbäck et al., 1985), lamb pharyngeal lipase (LPL, de Caro et al., 1995), dog gastric lipase (DGL, Carrière et al., 1991), rabbit gastric lipase (RGL, Moreau et al., 1988b). These enzymes form a new family of lipases: the acidic lipases. The present review focuses on two enzymes belonging to this family: HGL and DGL. Their biochemical properties have by now been clearly established and HGL has been expressed in vitro in active form, using the baculovirus/ insect cells system (Wicker-Planquart et al., 1996).

+ to whom correspondance should be sent

Physiological Aspects

Gastric lipases have been purified either from gastric juice or from fundic gastric mucosae and characterized *in vitro*. HGL was first purified by Tiruppathi and Balasubramanian (1982) and its secretion was stimulated by injecting pentagastrin (Szafran *et al.*, 1978, Moreau *et al.*, 1988). Nowadays, the enzyme is purified using cation-exchange (Mono S) chromatography followed by an immunoaffinity column procedure using monoclonal antibodies. Lipase activity can be measured potentiometrically on short-chain (tributyroylglycerol, TC4), medium-chain (trioctanoylglycerol, TC8) and long-chain (Intralipid™) triacylglycerols at 37°C using a pH-stat (Gargouri *et al.*, 1986). The maximum specific activities are 1160 units/mg at pH 6, 1110 units/mg at pH 6 and 600 units/mg at pH 5, on TC4, TC8 and Intralipid™, respectively.

DGL (Carrière *et al.*, 1991) was purified from the gastric juice of dogs with chronic gastric fistulae. Little if any DGL activity was observed in basal or stimulated gastric secretion, whereas the gastric mucosa was found to contain substantial amounts of lipolytic activity. This paradox can be explained by the occurence of an irreversible inactivation of DGL at low pH levels, which are present in the stomach under fasting conditions. By buffering the acidic secretion *in vivo*, or by using an anti-acid secretion drug such as omeprazol during the stimulation, the effects of several gastric secretagogues on the secretion of DGL (Carrière *et al.*, 1992) were determined. Unlike HGL secretion, DGL secretion was poorly stimulated by pentagastrin, while urecholine, 16,16-dimethyl prostaglandin E_2 and secretin were potent secretagogues. DGL was extracted from soaked dog gastric mucosa and purified after cation exchange, anion exchange and gel filtration chromatographies. The maximum specific activities were 550 units/mg on tributyroylglycerol at pH 5.5, 750 units/mg on trioctanoylglycerol at pH 6.5 and 950 units/mg on Intralipid™ at pH 4 (Carrière *et al.*, 1991). DGL is nearly twice as active on long chain triacylglycerol as on short chain triacylglycerol.

HGL (Bodmer *et al.*, 1987) and DGL (Bénicourt *et al.*, 1993) cDNAs were synthesized by RT-PCR amplification using stomach biopsies and reverse transcriptase was performed for 60 min at 37°C. The HGL and DGL coding regions 1.2 kb pairs and 1.55 kb pairs, respectively, were amplified by performing polymerase chain reaction (PCR) procedures. The cDNA coding for HGL and DGL were ligated in a pUC 18 plasmid. The deduced amino acid sequences of HGL and DGL were those of a 379 amino acid polypeptide, and the homology between the two proteins was about 85%.

The molecular masses of the HGL and DGL are 50 kDa. These enzymes are glycoproteins, containing around 15% carbohydrate. Endoglycosidase F (Moreau *et al.*, 1992) treatment of HGL results in the appearance of three major products, which can be separated. The glycan part of HGL consists of asparagine linked carbohydrates. HGL possesses four potential N-glycosylation sites (asparagine 15, 80, 252 and 308) with the consensus sequence Asn-X-Ser or Asn-X-Thr, but three of these were found to be occupied.

Several attempts at crystallising purified native preparations of HGL and DGL were unsuccessful. The most successful crystals obtained from native DGL diffracted at a resolution of 4,5 Å (Moreau *et al.*, 1988c). Charge homogeneity seemed to be essential to obtaining properly diffracting crystals. These crystals contained four to eight molecules per asymmetric unit in the case of DGL. This pattern of cristallisation in probably due to the large amounts of carbohydrates present in the protein.

Biochemical Aspects

Acidic lipases are specifically inhibited by micellar diethyl p-nitrophenyl phosphate (E600) (Moreau *et al.*, 1991), which is a well-known serine esterase inhibitor. Acidic lipases show the consensus sequence "(LIV)-X-(LIVFY)-(LIVST)-G-(HYWV)-S-X-G-(GSTAC)" (Ollis *et al.*, 1992) located around serine 153, characteristic of serine enzymes. In pancreatic lipases, the active serine residue (Ser 152) is located at the hinge of the nucleophilic elbow: this residue has been found in other lipases to participate in the charge relay system, together with a histidine and an aspartic (or glutamic) residue. Tetrahydrolipstatin, a chemical substance derived from lipstatin produced by *Streptomyces toxitricini*, inhibited pancreatic and gastric lipases *in vitro* (Hadvàry *et al.*, 1988, Borgström, 1988) when emulsified tributyrin (Gargouri *et al.*, 1991) or a dicaprin monolayer (Ransac *et al.*, 1991) were used as substrates. This inhibitor is linked to serine 152 in the case of pancreatic lipase (Peng *et al.*, 1991).

Three cysteine residues are present in both HGL and DGL. Titration of HGL (Gargouri *et al.*, 1988) and DGL (Carrière *et al.*, 1991) was performed by incubating the lipase with classical sulfhydryl reagents such as dithio-nitrobenzoic acid (DTNB), lauroyl thio-nitrobenzoic acid (C12-TNB) or di(thiopyridine) (4-PDS), and one titrable sulfhydryl group was found to exist per molecule of enzyme. Further addition of a denaturing agent such as SDS or urea did not increase the number of modified SH groups. As only the free sulfhydryl group was titrated, it can be concluded that gastric lipases contain a single disulphide bridge. The modification of the sulfhydryl group was accompanied by a concomitant loss of gastric lipase activity. It can be concluded that HGL and DGL possess a sulfhydryl group that seems to be either directly or indirectly involved in their catalytic activity. Using the monomolecular film technique, it was established that after the modification of one sulfhydryl group, gastric lipase still binds to films of phosphatidylcholine or dicaprin. The modification of the sulfhydryl group therefore seems to prevent the access of the substrate to the catalytic site and not to interfere with the interfacial binding step.

EXPRESSION OF GASTRIC LIPASES

Expression of Gastric Lipases in *E. coli*

In a preliminary study, HGL cDNA was fused with the maltose binding protein cDNA and expressed in *E. coli*. A soluble fusion protein of about 80 kDa was produced, which was recognized by rabbit polyclonal antibodies, but showed no activity towards triglycerides. Other attempts to produce soluble HGL were made (Wicker-Planquart *et al.*, unpublished data), using pET plasmids such as pET 22b(+) plasmid (from Novagen). This vector carries an N-terminal pel B signal sequence for potential periplasmic localisation. It also contains the natural promoter and coding sequence of the *lac* repressor (*lac I*), oriented so that the T7 *lac* and *lacI* promoters diverge. This vector was used to express HGL in *E. coli* strains: BL21(DE3) and BL21(DE3)pLysS, which is lysogenic for λDE3, and carries the T7 RNA polymerase gene under *lac UV5* control. In the absence of IPTG, the *lac* repressor acts both at the *lac UV5* promoter by repressing the transcription of the T7 RNA polymerase gene by the host polymerase and at the T7 *lac* promoter, by blocking the transcription of the target gene by any T7 RNA polymerase which occurs. HGL expression in this system was modulated by modifying IPTG addition as well as the *E. coli* cell culture temperature during the induction phase. When *E. coli* cells were grown at 37°C, HGL was mostly produced in the form of insoluble material (inclusion bodies), whereas when the induction was performed at 15 or 20°C, a soluble protein was produced, that was recognized by anti-HGL antibodies with an apparent molecular mass of 43 kDa on SDS/PAGE gels. In all the cases tested, however, using the pH-stat technique, the protein was inactive towards tributyroylglycerol. Similar results have been obtained with the cDNA coding for DGL (Bénicourt *et al.*, 1993). These data seem to indicate that the expression of a catalytically active lipase requires glycosylation. Expression experiments were therefore performed with HGL in eukaryotic systems, with which post transcriptional modifications of the enzyme such as glycosylation and disulphide bond formation can be performed.

Expression of Human Gastric Lipase in Insect Cells

The baculovirus/insect cell system

Baculoviruses belong to a group of diverse large double stranded DNA viruses that infect many different species of insects and arthropods, which are their natural hosts. Baculovirus strains are highly species-specific and are not known to infect any vertebrate hosts. The most extensively studied baculovirus strain so far is the *Autographa californica* nuclear polyhedrosis virus (AcNPV). The baculovirus infection cycle can be divided in two phases: the early phase in which extracellular virus particles (ECV) sprout from the cell membrane of the infected insect cells, and the later phase, during which occluded virus particles (OV) are assembled inside the nucleus. The protein polyhedrin is the main component of the occlusion bodies. This protein, which is expressed in large amounts, is not essential to the baculovirus life cycle and its synthesis is directed by a strong promoter. It can therefore be replaced by another gene, such as a gastric lipase gene, and expressed in insect cells. The gene of interest is transferred to the baculovirus by means of a transfer vector which contains an *E. coli* origin of replication as well as an ampicillin resistance gene and identical sequence stretches to those flanking the polyhedrin gene in the wild type genome. During cotransfection of insect cells by the baculovirus and the transferred gene, homologous recombination between these sequences occurs, resulting in a recombinant virus containing the gene of interest (Figure 1).

Expression of recombinant HGL (r-HGL)

Construction and expression of the recombinant baculovirus in insect cells
The gene from HGL was cloned into the baculovirus transfer vector PVL 1392 downstream of the strong polyhedrin promoter. The recombinant transfer vector was cotransfected with the DNA baculovirus, BaculoGold virus from Pharmingen. BaculoGold™ DNA contains a lethal deletion and does not code for viable viruses

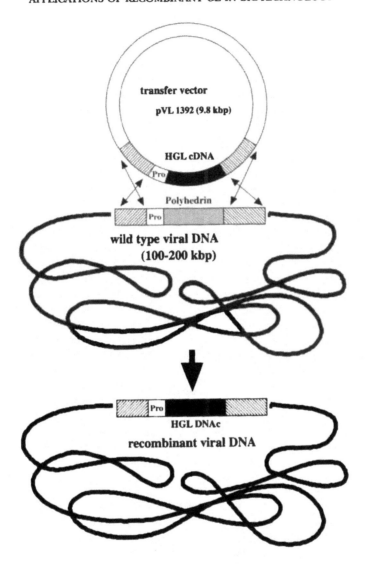

Figure 1 Construction of HGL recombinant baculovirus

by itself, but cotransfection with the recombinant transfer vector restores the deletion and yields a viable virus with a recombination efficiency of pratically 100%. The recombinant virus stock was amplified by performing several rounds of growth and a high titer virus stock solution was harvested. For the expression of recombinant protein, BTI-TN-5B1-4 (or High Five™) insect cells were used, which were grown in a serum-free Ex-cell™ 400 medium.

Production and purification of r-HGL in tissue culture plates
The insect cells were infected with a multiplicity of infection (MOI) of between 1 and 5. The amount of HGL accumulated in the supernatant of the insect cells was monitored directly at various post-infection times by measuring the lipase activity.

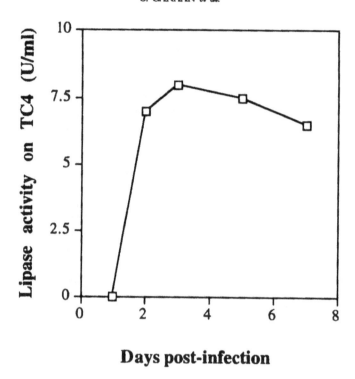

Figure 2   Kinetics of the onset of r-HGL activity in the supernatant of insect cells grown in monolayer flasks infected with HGL recombinant baculovirus. Monolayers of BTI-TN-5B 1-4 cells were grown in serum free medium (Ex-Cell™ 400) in 75-ml flasks prior to infection at a MOI = 3.1 ml of supernatant corresponds to 5×10^6 cells

The highest HGL activity, 8 units/ml of cell culture, was recorded 3 days post-infection. r-HGL was secreted in an almost pure form in the supernatant of insect cell culture media during the first three days post-infection. From day 3 onwards, the activity decreased slightly with time (Figure 2). This decrease was consistent with the data available on cell lysis and the proteolytic degradation of r-HGL. The release of insect cell lysosomal proteases from lysed cells may attributable to the degradation of r-HGL in the culture medium.

r-HGL was purified in a single step by performing cation-exchange chromatography on SP-Sepharose (Wicker-Planquart *et al.*, 1996).

Production and purification of r-HGL in bioreactor
Insect cells can be grown in large-volume suspension bioreactors. The possibility of producing the recombinant protein in a biorector is a great improvement over the use of flasks with monolayer cell cultures, since the handling of the insect cells is facilitated and large amounts of recombinant protein can be obtained in this way. The number of insect cells increases two-fold every 18–24 hours in shaker cultures, as compared with the production time of 28–30 hours in the case of monolayer cultures. Agitation can easily damage insect cells (Tramper *et al.*, 1986) however, and the productivity of the system depends greatly on a number of factors, including the cell line (Hink *et al.*, 1991), the age and condition of the cells at time of infection

(Goodwin *et al.*, 1980, Vaugn *et al.*, 1991), the MOI (Yang *et al.*, 1996) and the number of passages of the virus inoculum (Kool *et al.*, 1991, Wickham *et al.*, 1991). We have used High Five™ cells, which were reported to be an excellent host for the expression of recombinant proteins (Grossmann *et al.*, 1997), and several cell infection procedures with the MOI and the host cell density at the time of infection as variables. As the post-infection addition of glucose can facilitate enhanced recombinant protein production (Wang *et al.*, 1993a, b), we kept glucose concentration at 3 mg/ml by adding a concentrated solution every two-days. Two procedures are described below, the one based on a low MOI and a low cell mass, with which all the viral particles become attached to the cells (which are in great excess), and the other based on a high cell mass and a high MOI.

Insect cells obtained from *Trichoplusia ni*, BTI-TN-5B1-4 were grown in suspension and in serum free medium Ex-Cell™ 400. An inoculum was produced in spinner flasks (INTEGRA Biosciences) by infecting 600 ml of host cells ($2.5 \ 10^6$ cells per ml) with a low viral titer solution (MOI = 0.1) of the recombinant baculovirus, and incubated for 3 hrs before being transferred to a 4-litre working volume bioreactor (Cytoflow, INCELTECH-France). The cells were then diluted down to a density of $6 \ 10^5$ cells per ml with fresh medium. In order to prevent the cells from aggregating, 0.1% (v/v) pluronic F-68 acid (from Sigma) was added to the medium. A constant glucose concentration of 3 mg/ml was maintained (Figure 3a). The r-HGL production was monitored on-line by measuring the enzyme activity in 3-ml samples collected from the bulk solution. The cell density and cell viability were determined using the tryptan blue exclusion method with a Malassez haemocytometer. Centrifugation was performed at 5000 rpm for 10 min and cell supernatant was collected when cellular lysis reached 60–70% (Figure 4). With the second procedure, a large cell mass was obtained after growing the host cells in the 4-litre bioreactor for four days until a cell density of $3 \ 10^6$ cells/ml was reached and then infecting the cells with a high viral titer solution (MOI = 5, Figure 3b).

The recombinant protein yield differs considerably between those two procedures, since the method which consists of using a low cell density and a low MOI is about 4 times more efficient than that involving a large cell mass and a high MOI (Figure 4). Under the low conditions, about 8–20 mg of recombinant HGL (depending on the runs) were harvested in 1.5 l culture medium on day 6–7 post-infection. The two procedures actually differ not only the initial cell mass and the yield, but also in the dynamic interactions occurring among the propagating viruses, the cell division processes, the availability of essential nutrients, and the accumulation of waste metabolites. With the low procedure, the cell population is doubled at least once after the infection (infected cells do not multiply). The high productivity obtained under these conditions was directly due to the cells being infected in a relatively early phase of growth (*i.e.* in an exponential phase). In addition, with this method, the virus replication and subsequent horizontal spread of the infection take place in a better nutritional environment. The low procedure is the more efficient of the two, probably because in this case the recombinant protein expression is completed before the depletion of nutrients, as observed by other authors (Caron *et al.*, 1990, Reuveny *et al.*, 1993). On the other hand, with the high MOI-high cell mass protocol, the infection is performed into an already worked-out medium (three to four days of culture at high cell density), a rapid arrest of cell growth occurs and the cell viability decreases. Similar results have been obtained elsewhere

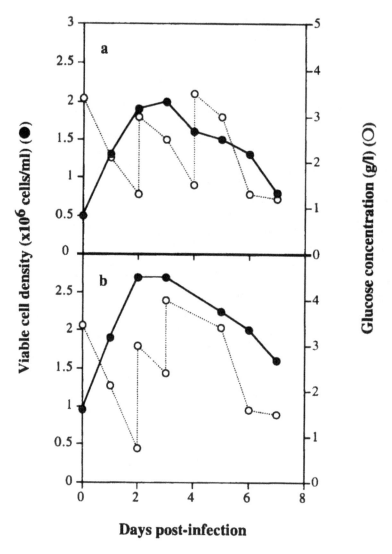

Figure 3 Growth curve of BTI-TN-5B 1-4 insect cells grown in Ex-Cell™ 400 medium using a 4-litre bioreactor. (a) The cell culture (2.2×10^6 cells/ml) was infected at a MOI of 0.1 in a spinner flask for 3 hours, transferred to a bioreactor and diluted. The cell concentration and residual glucose concentration were quantified once a day during the bioreactor run. Glucose was added to the bioreactor culture medium on post-infection days 2 and 4. (b) The cell culture (3×10^6 cells/ml) was infected at a MOI of 5 in a bioreactor. The cell concentration and residual glucose concentration were quantified once a day during the bioreactor run. Glucose was added to the bioreactor culture medium on post-infection days 2 and 3

by Yang *et al.* (1996): these authors developed a serum-free cell culture process using a recombinant baculovirus to infect *Trichloplusia ni* insect cells to produce human lysosomal glucocerebrosidase. They developed a model for predicting the recombinant protein yield based on the key assumption that maximum protein production was limited by the nutritional value of the medium. Based on this model, it was

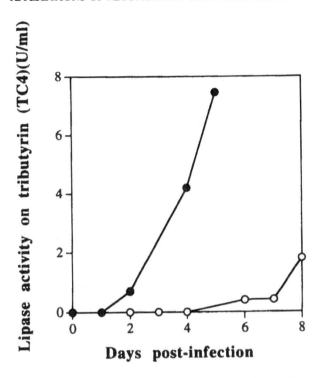

Figure 4   Production curves of rHGL in Ex-Cell™ 400 medium. The cell culture and infection conditions were those described in the legend of Figure 3. The activity of r-HGL was measured using a pH stat method in the supernatants of the culture media of BTI-TN-5B-1-4 cells infected with the recombinant baculovirus at a MOI of 0.1 (●) or 5 (O)

predicted that optimum protein expression would be achieved with a 4-day batch process at a low host cell density at the time of infection (1/10 of the maximum cell density at the end of the cell growth) and a MOI of 0.09. Experimental data confirmed the results obtained with this predictive model.

One of the drawbacks of using of suspension cultures for this purpose is that a high rate of cell lysis occurs and intracellular proteins are discharged from the insect cells into the culture medium. During the r-HGL purification procedure, some problems arose due to the strong tendency of these lipases to adsorb on hydrophobic surfaces. The decrease in the HGL glycosylation level in the baculovirus/insect cells system may account for the decrease in solubility of the enzyme. We observed in some cases that the specific activity of r-HGL decreased in the culture medium after post-infection day 6. This result suggests that r-HGL might aggregate, since no degradation fragments were detected in the immunoblotting experiments.

We nevertheless succeeded in purifying r-HGL by performing two chromatographic steps (SP-Sepharose followed by an anti HGL immunoaffinity column, data not shown).

Biochemical characterization of r-HGL

Recombinant HGL was identified using the Western blotting procedure with rabbit polyclonal antibodies. The fact that the protein migrated with an apparent molecular mass of 45 kDa under SDS-PAGE analysis (compared with 50 kDa in the case of natural HGL) indicates that insect cells have only a limited HGL glycosylation capacity.

The amino acid sequence of r-HGL has been found to resemble that of native HGL (Wicker-Planquart *et al.*, 1996). The N-terminal protein sequence (LFGKL...) determined from the r-HGL secreted by insect cells infected with the recombinant baculovirus containing its natural signal peptide nucleotide sequence indicates that the cleavage site occurs at the expected position, between Gly -1 and Leu +1, as observed with native HGL. The amino acid sequence of r-HGL was determined up to residue 16. Asn 15 was detected, indicating that this residue is probably not glycosylated. We were unable to detect any residue at position 15 of native HGL, however, which suggests the presence of a glycan chain. Lamb pregastric lipase has also been found to be glycosylated at that position (De Caro *et al.*, 1995). This result confirms that heterologous recombinant proteins are less glycosylated by insect cells than by mammal cells.

The pI pattern of r-HGL displays three isoforms, with calculated pI of 7.0 (minor band), 7.5 and 8.1. Some heterogeneity therefore existed as the result of glycosylation, although it was less marked than in the natural form (Wicker-Planquart *et al.*, 1996).

r-HGL was assayed on short-, medium- and long-chain triglycerides. The maximum specific activities of the r-HGL were 434 (Intralipid™), 730 (trioctanoylglycerol) and 562 (tributyroylglycerol) units/mg (Figure 5, Wicker-Planquart *et al.*, 1996).

POTENTIAL APPLICATIONS

Substitutive Enzyme Therapy

Exocrine pancreatic insufficiency, which occurs, for example, in cystic fibrosis (1/2500 children) always gives rise to two serious problems: malnutrition and steatorrhea. These problems can be partly solved by the administration of porcine pancreatic extracts (Bénicourt *et al.*, 1993, Lankisch, 1993) as a replacement therapy. To reduce the malabsorption of fat, the enzymes delivered into the duodenum must amount to 5–10% of the quantities usually present after maximum stimulation of the pancreas (DiMagno *et al.*, 1973). Under optimum conditions, if no inactivation of the supplement enzymes occurs in the stomach or duodenum, approximately 30,000 IU of lipase must be taken with each meal (DiMagno *et al.*, 1982), which corresponds to about 5–10 g of lyophilised pancreatic powder per day. In the past, preparations of this kind were far from satisfactory, since a large proportion of the enzymes administered were denatured in the stomach due to the extreme acidity of the gastric juice. Pancreatic lipase is irreversibly inactivated at pH levels downward of 4.0. This problem can be partly solved by using enteric-coated microspheric pancreatic preparations. The lipase activity is released from the microspheres at pH levels of around 5.5. It is necessary to bear in mind, however, that a considerable loss of activity occurs along the gastrointestinal tract, since only 8% of the exogeneous lipase was recovered in the above study. The lipase was degraded at the duodenal level by proteases present in the pancreatic extracts.

The use of acidic-resistant lipases should in principle yield more satisfactory results than the pancreatic preparations currently in use. The co-administration of acidic lipases, which hydrolyse dietary lipids under acidic conditions, should help

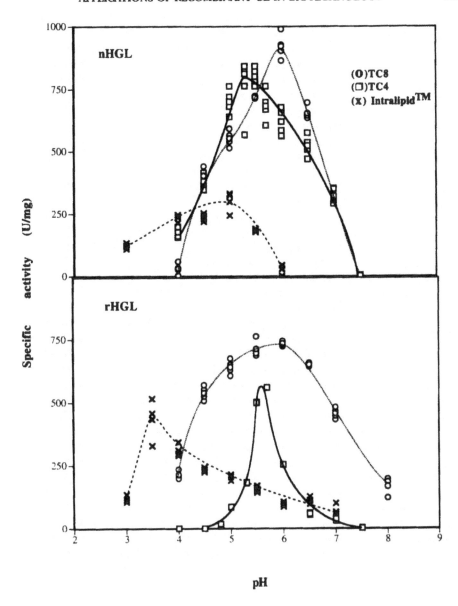

Figure 5 Effects of pH on the specific activity of natural HGL (upper panel) and r-HGL (lower panel) on tributyroylglycerol (TC4), trioctanoylglycerol (TC8) and long-chain (Intralipid™) triacylglycerols

to treat patients with various forms of pancreatic deficiency. Physiological studies have shown that preduodenal lipases are capable of acting not only in the stomach but also in the duodenum in synergy with a pancreatic lipase (Carrière *et al.*, 1993a, 1993b). Various clinical studies have been conducted on both animals and humans to assess the efficacy of enzymatic replacement therapies using acid-resistant lipases of bacterial origin to treat exocrine pancreatic insufficiency (Suzuki *et al.*, 1997). Although this treatment significantly increased the weight and reduced the steatorrhea in dogs, the use of these enzymes is limited because of their sensitivity to the proteolytic action of gastric pepsin, as well as by legal requirements.

Unlike acidic lipases of bacterial origin, gastric lipases are resistant to digestive enzymes. Furthermore, they are active on long-chain triacylglycerols, which account for most of the dietary fat in humans (specific activity of HGL and DGL on Intralipid™: 600 U/mg (Gargouri *et al.*, 1986) and 950 U/mg (Carrière *et al.*, 1991), respectively). r-HGL displays interesting properties, since its pH optimum was found to be 3.5 on long-chain triglycerides and the recombinant enzyme still shows 100% activity after being incubated at pH 1 for 4 h (Wicker-Planquart *et al.*, 1996).

Acid Lipases in the Dairy Industry

Preduodenal lipases are widely used in the dairy industry: crude pharyngeal extracts are able to impart typical flavours to all sorts of cheeses and cheese-like products (Bech, 1992, Birschback, 1992), and to accelerate cheese ripening. The traditional sources of pregastric lipases are animal tissues of young ruminants (kid, lamb, calf) which are processed to yield partially purified samples (liquid extracts, pastes and vacuum- or freeze-dried powders). Each cheese has its own characteristic flavor, which is mainly due to the amount of a particular type of free fatty acids that are generated by the action of lipases on milk fat. Thus, free short-chain fatty acids (butyric acid C4:0 and caproic acid C6:0) are responsible for a sharp, tangy flavour while free medium-chain (lauric acid C12:0 and myristic acid C14:0) impart a soapy taste to the product. The free fatty acid profile in each cheese depends to the source of preduodenal lipase used, which has its own particular fatty acid selectivity (Bech, 1992, Birschback, 1992 and 1994, Ha *et al.*, 1993). For example, lamb (de Caro *et al.*, 1995) and kid (Lai *et al.*, 1997) pregastric lipase have a very marked specificity for short-chain lipids. Using synthetic monoacid triglycerides, Lai *et al.* (1997) showed that the Kcat value of partially purified goat and kid lipases decreased by at least twofold relative to the value for tributyroylglycerol hydrolysis for each additional pair of carbon atoms in the carboxylic acid chain, whereas the Km values for the goat and kid extracts raised by a 16-fold factor when hydrolysing a C12:0 versus a C4:0 or C6:0 substrate (Km = 0.15 mM for the short-chain substrates and 8.2 mM for the C12:0 substrate). A similar reactivity with natural milk fat triglycerides was also observed (Lai *et al.*, 1997). Thus, each preduodenal lipase impart its own characteristic flavour profile: a buttery and slighly peppery flavour (calf), a sharp "piccante" flavour (kid), a strong "pecorino" flavour (lamb) (Bech, 1992). Kid or lamb pregastric extracts are used in manufacturing Romano, Domiati and Feta cheeses, calf or kid pre-gastric lipases for Mozarella, Parmesan and Povolone cheese while a combination of calf gastric lipase and goat pregastric lipase can produce Cheddar or Provolone cheese (Chaudhari *et al.*, 1971). Cheese flavour is also function, among other factors, on the composition of milk fat: for example, ovine milk is richer in short- and medium- chain triglycerides, bovine milk fat is richer in long-chain and unsaturated triglycerides, and caprine milk fat is the richest in polyunsaturated medium-chain triglycerides (Ruiz-Sala *et al.*, 1996). It is nevertheless possible to use preduodenal lipases for imitation of cheese made from ewe's or goat's milk. Addition of pre-gastric lipase from goat or kid to bovine milk fat leds to free fatty acid profile for hydrolysis of the bovine milk fat by goat and kid extract very similar to the one of Parmesan cheese (Lai *et al.*, 1997). It should be noticed that short-chain fatty acids are available in the highest proportions at the sn-3 position of bovine milk fat (28.4% C4:0). Goat and kid pregastric lipases selectively hydrolyze triglycerides at this position. Positional selectivity alone cannot, however, account for the high concentration of butyric acid observed after hydrolysis by goat and kid lipases (40 and 42.5%, respectively). One can imagine that recombinant technologies will made available in the future numerous acid lipases with finely tuned fatty acid specificities.

ACKNOWLEDGEMENTS

We are grateful to Takoua Debèche for her assistance in setting up the *E. coli* expression experiments, to Karine Faessel for performing the bioreactor experiments, and to Dr Jessica Blanc for revising the English manuscript.

REFERENCES

Bech, A.M. (1992) Enzymes for the acceleration of cheese-ripening. In Fermentation-produced Enzymes and Accelerate Ripening in Cheesemaking. *IDF Bull.*, **269**, 24–28. Brussels, Belgium: Int. Dairy Fed.

Bénicourt, C., Blanchard, C., Carrière, F., Verger, R. and Junien, J.-L. (1993) Potential use of a recombinant dog gastric lipase as an enzymatic supplement to pancreatic extracts in cystic fibrosis. In *Clinical ecology of cystic fibrosis*, edited by H. Escobar, C.F. Baquero and L. Suárez, pp. 291–295. Elsevier Science Publishers B.V.

Bernbäck, S., Hernell, O. and Blackberg, L. (1985) Purification and molecular characterization of bovine pregastric lipase. *Eur. J. Biochem.*, **148**, 233.

Birschback, P. (1992) Pregastric lipases. In Fermentation-produced Enzymes and Accelerate Ripening in Cheesemaking. *IDF Bull.*, **269**, 36–39. Brussels, Belgium: Int. Dairy Fed.

Birschback, P. (1994) Origins of lipases and their characteristics. In The Use of Lipases in Cheesemaking. *IDF Bull.*, **294**, 7. Brussels, Belgium: Int. Dairy Fed.

Bodmer, M.W., Angal, S., Yarranton, G.T., Harris, T.J.R., Lyons, A., King, D.J., Piéroni, G., Rivière, C., Verger, R. and Lowe, P.A. (1987) Molecular cloning of a human gastric lipase and expression of the enzyme in yeast. *Biochim. Biophys. Acta*, **909**, 237–244.

Borgström, B. (1988) Mode of action of tetrahydrolipstatin: A derivative of the naturally occurring lipase inhibitor lipstatin. *Biochim. Biophys. Acta*, **962**, 308–316.

Caron, A.W., Archambault, L. and Massie, B. (1990) High level recombinant protein production in bioreactors using the baculovirus-insect cell expression system. *Biotechnol. Bioeng.*, **36**, 1133–1140.

Carrière, F., Moreau, H., Raphel, V., Laugier, R., Bénicourt, C., Junien, J.-L. and Verger, R. (1991) Purification and biochemical characterization of dog gastric lipase. *Eur. J. Biochem.*, **202**, 75–83.

Carrière, F., Raphel, V., Moreau, H., Bernadac, A., Devaux, M.-A., Grimaud, R., Barrowman, J.A., Bénicourt, C., Junien, J.-L., Laugier, R. and Verger, R. (1992) Dog gastric lipase: Stimulation of its secretion in vivo and cytolocalization in mucous pit cells. *Gastroenterology*, **102**, 1535–1545.

Carrière, F., Laugier, R., Barrowman, J.A., Douchet, I., Priymenko, N. and Verger, R. (1993a) Gastric and pancreatic lipase levels during a test meal in dogs. *Scand. J. Gastroenterol*, **28**, 443–454.

Carrière, F., Barrowman, J.A., Verger, R. and Laugier, R. (1993b) Secretion and contribution to lipolysis of digestive lipase during a test meal in humans: Gastric lipase remains active in the duodenum. *Gastroenterology*, **105**, 876–888.

Chaudhari, R.V., and Richardson, G.H. (1971) Lamb gastric lipase and protease in cheese manufacture. *J. Dairy Sci.*, **54**, 467–471.

De Caro, J., Ferrato, F., Verger, R. and de Caro, A. (1995) Purification and molecular characterization of lamb pregastric lipase. *Biochim. Biophys. Acta*, **1252**, 321–329.

DeNigris, S.J., Hamosh, M., Kasbekar, D.K., Lee, T.C. and Hamosh, P. (1988) Lingual and gastric lipases: species differences in the origin of prepancreatic digestive lipases and in the localisation of gastric lipase. *Biochim. Biophys. Acta*, **959**, 38–45.

DiMagno, E.P., Go, V.L.W. and Summerskill, W.H.J. (1973) Relations between pancreatic enzyme outputs and malabsorption in severe pancreatic insufficiency. *N. Engl. J. Med.*, **288**, 813–815.

DiMagno, E.P. (1982) Controversies in the treatment of exocrine pancreatic insufficiency. *Dig. Dis. Sci.*, **27**, 481–484.

Field, R. and Scow, R.O. (1983) Purification and characterization of rat lingual lipase. *J. Biol. Chem.*, **258**, 14563–14569.

Gargouri, Y., Piéroni, G., Rivière, C., Saunière, J.-F., Lowe, P.A., Sarda, L. and Verger, R. (1986) Kinetic assay of human gastric lipase on short- and long-chain triacylglycerol emulsions. *Gastroenterology*, **91**, 919–925.

Gargouri, Y., Moreau, H., Piéroni, G. and Verger, R. (1988) Human gastric lipase: A sulfhydryl enzyme. *J. Biol. Chem.*, **263**, 2159–2162.

Gargouri, Y., Chahinian, H., Moreau, H., Ransac, S. and Verger, R. (1991) Inactivation of pancreatic and gastric lipases by THL and $C_{12:0}$-TNB: a kinetic study with emulsified tributyrin. *Biochim. Biophys. Acta*, **1085**, 322–328.

Goodwin, R.H. and Adams, J.R. (1980) Nutrient factors influencing viral replication in serum free insect cell line culture. In *Invertebrate systems in vitro*, edited by Kurstak, Marmamorosch and Dubendorfer, pp. 493–509. Amsterdam: Elsevier.

Grossmann, M., Wong, R., Teh, N.G., Tropea, J.E., East-Pamer, J., Weintraub, B.D. and Szudlinski, M.W. (1997) Expression of biologically active human thyrotropin (hTSH) in a baculovirus system: effect of insect cell glycosylation on hTSH activity *in vitro* and *in vivo*. *Endocrinology*, **138**, 92–100.

Ha, J.K. and Lindsay, R.C. (1993) Release of volatile branched-chain and other fatty acids from ruminants milk fats by various lipases. *J. Dairy Sci.*, **76**, 677–690.

Hadvàry, P., Lengsfeld, H. and Wolfer, H. (1988) Inhibition of pancreatic lipase *in vitro* by the covalent inhibitor tetrahydrolipstatin. *Biochem. J.*, **256**, 357–361.

Hamosh, M., Ganot, D. and Hamosh, P. (1979) Rat lingual lipase: Characteristics of enzyme activity. *J. Biol. Chem.*, **254**, 12121–12125.

Hamosh, M. (1984) Lingual lipase. In *Lipases*, edited by B. Borgström, and H.L. Brockman, pp. 49–81. Amsterdam: Elsevier.

Hink, W.F., Thomsen, D.R., Davidson, D.J., Meyer, A.L. and Castellino, F.J. (1991) Expression of three recombinant proteins using baculovirus vectors in 23 insect cell lines. *Biotechnol. Progr.*, **7**, 9–14.

Hull, M. and Keaton, R.W. (1917) The existence of a gastric lipase. *J. Biol. Chem.*, **32**, 127–140.

Kool, M., Voncken, J.M., van Lier, F.L.J., Tramper, J. and Vlak, J.M. (1991) Detection and analysis of Autographa california nuclear polyhedrosis virus mutants with defective interfering properties. *Virology*, **183**, 739–746.

Lai, D.T., Mackenzie, A.D., O'Connor, C.J. and Turner, K.W. (1997) Hydrolysis characteristics of bovine milk fat and monoacid triglycerides mediated by pregastric lipase from goats and kids. *J. Dairy Sci.*, **80**, 2249–2257.

Lankisch, P.G. (1993) Enzyme treatment of exocrine pancreatic insufficiency in chronic pancreatitis. *Digestion*, **54**, 21–29.

Moreau, H., Gargouri, Y., Lecat, D., Junien, J.-L. and Verger, R. (1988a) Screening of preduodenal lipases in several mammals. *Biochim. Biophys. Acta*, **959**, 247–252.

Moreau, H., Gargouri, Y., Lecat, D., Junien, J.-L. and Verger, R. (1988b) Purification, characterization and kinetic properties of the rabbit gastric lipase. *Biochim. Biophys. Acta*, **960**, 286–293.

Moreau, H., Saunière, J.-F., Gargouri, Y., Piéroni, G., Verger, R. and Sarles, H. (1988c) Human gastric lipase: Variations induced by gastrointestinal hormones and by pathology. *Scand. J. Gastroenterol.*, **23**, 1044–1048.

Moreau, H., Abergel, C., Carrière, F., Ferrato, F., Fontecilla-Camps, J.C., Cambillau, C. and Verger, R. (1992) Isoform purification of gastric lipases. Towards crystallisation. *J. Mol. Biol.*, **225**, 147–153.

Moreau, H., Moulin, A., Gargouri, Y., Noël, J.-P. and Verger, R. (1991) Inactivation of gastric and pancreatic lipases by diethyl p-nitrophenyl phosphate. *Biochemistry*, **30**, 1037–1041.

Muller, D.P.R., McCollum, J.P.K., Tompeter, R.S. and Harries, J.T. (1975) Studies on the mechanism of fat absorption in congenital isolated lipase deficiency. *Gut*, **16**, 838.

Ollis, D.L., Cheah, E., Cygler, M., Dijkstra, B., Frolow, F., Franken, S.M., Harel, M., Remington, S.J., Silman, I., Schrag, J., Sussman, J.L., Verschueren, K.H.G. and Goldman, A. (1992) The α/β hydrolase fold. *Protein Eng.*, **5**, 197–211.

Peng, Q., Hadvary, P. and Maerki, H.P. (1991) Identification of the THL binding site on human pancreatic lipase. *GBF Monogr.*, **16**, 141–144.

Ransac, S., Gargouri, Y., Moreau, H. and Verger, R. (1991) Inactivation of pancreatic and gastric lipases by tetrahydrolipstatin and alkyl-dithio-5-(2-nitrobenzoic acid). A kinetic study with 1, 2-didecanoyl-sn-glycerol monolayers. *Eur. J. Biochem.*, **202**, 395–400.

Reuveny, S., Kim, Y.J., Kemp, C.W. and Shiloach, J. (1993) Production of recombinant proteins in high-density insect cell cultures. *Biotechnol. Bioeng.*, **42**, 235–239.

Roberts, I.M., Montgomery, R.K. and Carey, M.C. (1984) Lingual lipase: Partial purification, hydrolytic properties and comparison with pancreatic lipase. *Am. J. Physiol.*, **247G**, 385–393.

Roulet, M., Weber, A.M., Paradis, Y., Roy, C.C., Chartraud, L., Lasalle, R. and Morin, C.L. (1980) Gastric emptying and lingual lipase activity in cystic fibrosis. *Pediatr. Res.*, **14**, 1360–1362.

Ross, C.A.C. and Sammons, H.C. (1955) Non-pancreatic lipase in children with pancreatic fibrosis. *Arch. Dis. Child.*, **30**, 428–431.

Suzuki, A., Mizumoto, A., Sarr, M.G. and Dimagno, E.P. (1997) Bacterial lipase and high-fat diets in canine exocrine pancreatic insufficiency: a new therapy of steatorrhea? *Gastroenterology*, **112**, 2048–2055.

Szafran, Z., Szafran, H., Popiela, T. and Trompeter, G. (1978) Coupled secretion of gastric lipase and pepsin in man following pentagastrin stimulation. *Digestion*, **18**, 310–318.

Tiruppathi, C. and Balasubramanian, K.A. (1982) Purification and properties of an acid lipase from human gastric juice. *Biochim. Biophys. Acta*, **712**, 692–697.

Tramper, J., Williams, J.B., Jous, T.D. and Vlak, J.M. (1986) Shear sensitivity of insect cells in suspension. *Enzyme Microb. Technol.*, **8**, 3–36.

Vaugn, J.L., Fan, F., Dougherty, E.M., Adams, J.R., Guzo, D. and McClintock, J.T. (1991) The use of commercial serum replacements in media for the *in vitro* replication of nuclear polyhedrosis virus. *J. Invert. Pathol.*, **58**, 279–304.

Volhard, F. (1901) Über das fettspaltende ferment des Magens. *Z. Klind. Med.*, **42**, 414–429.

Wang, M.Y., Vakharia, V. and Bentley, W.E. (1993a) Expression of epoxide hydrolase in insect cells: a focus on the infected cell. *Biotechnol. Bioeng.*, **42**, 240–247.

Wang, M.Y., Simon, K. and Bentley, W.E. (1993b) Effects of oxygen glucose/glutamine feeding on insect cell baculovirus protein expression: a study of epoxide hydrolase production. *Biotechnol. Prog.*, **9**, 355–361.

Wicker-Planquart, C., Canaan, S., Rivière, M., Dupuis, L. and Verger, R. (1996) Expression in insect cells and purification of a catalytically active recombinant human gastric lipase. *Protein Eng.*, **9**, 1225–1232.

Wickham, T.J., Davis, T., Granados, R.R., Hammer, D.A., Shuler, M.L. and Wood, H.A. (1991) Baculovirus defective interfering particles are responsible for variations in recombinant protein production as a function of multiplicity of infection. *Biotechnol. Lett*, **13**, 483–488.

Yang, J.D., Gecik, P., Collins, A., Czarnecki, S., Hsu, H.H., Lasdun, A., Sundaram, R., Muthukumar, G. and Silberklang, M. (1996) Rational scale-up of a baculovirus insect cell batch process based on medium nutritional depth. *Biotechnol. Bioeng.*, **52**, 696–706.

18.   INHIBITION OF RAS-DEPENDENT SIGNAL TRANSDUCTION FOR THERAPEUTIC PURPOSES: PROTEIN ENGINEERING OF A RAS-SPECIFIC GUANINE NUCLEOTIDE EXCHANGE FACTOR

MARCO VANONI, PAOLA BOSSÙ[§], RICCARDO BERTINI[§]
and LILIA ALBERGHINA

Dipartimento di Biotechnologie e Bioscienze, Università degli Studi di Milano-Bicocca via Emanueli 12, 20126 Milano, Italy;
[§]Centro Ricerche Dompé, 67100 L' Aquila Italy

SIGNAL TRANSDUCTION AND THERAPY OF PROLIFERATIVE DISORDERS

The correct functioning of cells and tissues relies on the fidelity and precise coordination of signalling pathways, whose malfunctioning and lack of coordination may originate several diseases, most notably proliferative disorders, such as cancer. This recognition has opened the way to a novel approach to disease therapy, based on the inhibition of specific signal transduction pathways. A potential problem of such an approach concerns its selectivity, since the signal transduction pathways altered in a specific disease are present also in normal cells. A possible explanation for the enhanced sensitivity of cells in pathological state has been proposed, at least in the case of cancer cells that overexpress a specific signal transduction pathway and are often dependent on said pathway for their survival, possibly because they have lost the redundancy in signal transduction pathways typical of their normal counterparts.

In principle both universal and selective targets for signal transduction therapy can be envisaged: the former are targets which are altered in most cancer types, such as protein tyrosine kinases, the latter can be oncogenes or proto-oncogenes whose alteration is specific to a given tumor type. The exceedingly low toxicity in animal models of drugs directed towards universal targets indicates that these can in fact find application for therapy and deserve extensive clinical trials (reviewed in Levitzky, 1996).

Manipulation of signal transduction pathways for disease management should not be limited to small molecules, although so far they have been most successful. Antisense nucleotides and RNAs to signalling molecules and ribozymes can be used as well. An intriguing, novel possibility relies on the use of proteins able to estinguish or to attenuate signal transduction pathways and will be examined in closer detail later. Three major types of mutant proteins may be expected in protein involved in signal transduction pathways: loss of function mutants that are no longer able to exert their role but do not interfere with the pathway; gain of function mutants which become hyperactivated and dominant-negative mutants wich down regulate the pathway to which they belong by sequestering the cognate interacting protein(s).

Protein engineering may aid in the development of new pharmacological lead compounds, not only by allowing the unravelling of the molecular mechanisms of pharmacologically relevant targets, but directly providing potential lead compounds to be used in clinical trials. The major advance of this approach is that it can be used rationally to design variant proteins with pharmacologically useful properties: in other words, proteins can be designed — or screened — to fulfill specific needs in order to alter a given signal transduction pathway. Thus this approach may bypass — or effectively complement — high throuput screenings by directly providing structural scaffolds to be used as basis to rationally design molecules interfering with the desired targets.

THE RAS CYCLE AND ITS REGULATORS

A major target for the therapy of proliferative disorders is represented by the *ras* proto-oncogenes and their encoded proteins, which are small guanine nucleotide binding proteins acting as molecular switches in cell proliferation and signal transduction (Barbacid, 1987). In humans three major *ras* genes exist, H-*ras*, K-*ras* and N-*ras*. Each of the *ras*-encoded protein, which we will refer to collectively as p21^{ras}, can work as an oncoprotein upon mutational activation. Activated *ras* genes have been found in a great variety of human tumors. The highest incidence of *ras* mutations can be found in pancreas (90%), colon (50%) and lung (30%) adonocarcinomas, as well as in thyroid tumors (50%) and myeloid leukemias. Ras proteins undergo an extensive series of post-translational modification at their C-terminal tails which increase protein hydrophobicity and result in ras association with the inner face of the plasma membrane. Such modifications include farnesylation of cys186 (catalyzed by farnesyl transferase) wich is followed by proteolytic removal of three residues downstream of cys186 and methylation of cys186 carboxyl group. These post-translational modifications are required for ras transforming activity.

Within cells, ras proteins exist in either a GTP• or a GDP•bound form. The protein is in its "on" state when it is bound to GTP and in its "off" state when it is bound to GDP. Analysis of the GDP• and GTP•bound forms by X-ray cristallography and — more recently — by NMR has shown that their conformation differs mainly in the region spanning residues 30–38 and 60–76 (switch I and switch II regions). Ras proteins are endowed with a low intrinsic GTPase activity and like all GTPases can go through a cycle of reactions, summarized in Figure 1. Oncogenic version of p21^{ras} are found mostly in the GTP•bound form either because of a reduced GTPase acivity or because of an increased GDP/GTP exchange. The level of the active, GTP•bound form results from the balance of the competing activity of GTPase Activating Proteins (**GAP**) and Guanine nucleotide Exchange Factors (**GEF**) (reviewed in Bollag and McCormick, 1991; Feig, 1994; Lowy and Willumsen, 1993; Boguski and McCormick, 1993).

GAP

P120-GAP isolated in *Xenopus* oocytes and mammalian cells has been the first vertebrate protein shown to regulate ras activation state. It is a ubiquitous, soluble

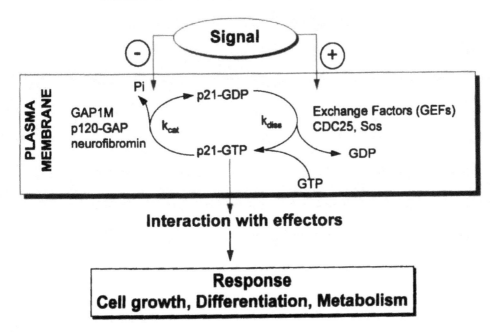

Figure 1   The ras cycle

polypeptide of 1047 residues endowed with a potent catalytic activity able to stimulate ras intrinsic GTPase activity up to 5 orders of magnitude. All ras wild type isotypes are sensitive to GAP action, while some oncogenic forms are no longer affected by GAP binding. Later, other proteins with GAP-like activities have been isolated, such as neurofibromin (NF1), Gap1m, IQ-GAP and GAPIII. Different regulatory mechanisms of GAP activity of these molecules, together with their different structural organization and tissue-specificity may contribute to fine tune ras charging. All GAPs share a ca 350 aminoacid long catalytic domain, often referred to as GRD (GAP Related Domain) whose three dimensional structure has recently been solved by X-ray cristallography (reviewed in Wittinghofer *et al.*, 1997). Among GAP-like molecules NF1 is particularly interesting, since its encoding gene is inactivated in neurofibromatosis, an autosomically dominant disease characterized by an high frequency of both benign and malignant tumors. Cell lines derived from tumor Schwann cells and devoided of NF1 activity, display an elevated ras•GTP level despite the presence of p120-GAP, showing that NF1 acts as an anti-oncogene and that p120-GAP alone is not enough in these cells to down-regulate ras.

GEF

The prototypes of ras-specific GEFs are the products of the yeast *CDC25* gene (Martegani *et al.*, 1986; Camonis *et al.*, 1986) and its close homolog *SDC25* (Damack *et al.*, 1991), the first molecule of this class for which a GEF activity was shown (Crechet *et al.*, 1990). In mammalian cells two ras-specific GEF classes have been identified. One class — whose cDNA was originally isolated by functional

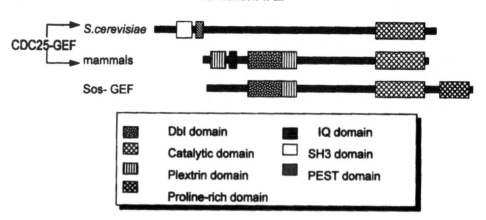

Figure 2   Modular structure of eukaryotic GEF

complementation of a *cdc25* mutant of *Saccharomyces cerevisiae* — has been called CDC25[Mm] and is mostly brain-specific (Martegani *et al.*, 1992; Ferrari *et al.*, 1994: Wei *et al.*, 1993). CDC25[Mm] homologs have also been isolated from rat (Shou *et al.*, 1992) and man (Schweighoffer *et al.*, 1993; Wei *et al.*, 1994). A second related gene has been isolated from mouse (Fam *et al.*, 1997). Immunological and sequence data derived from EST projects indicate that other proteins of the same class may exist, but whose role and specificity still needs to be addressed (see for instance Tung *et al.*, 1997). The other class — Sos-GEF — has been isolated on the basis of homology with the *Son of Sevenless* gene of *Drosophila* (Simon *et al.*, 1991) and is expressed ubiquitously. Both GEFs are large, multidomain proteins, with a widely different structural organization (Figure 2). In the brain-specific GEF CDC25[Mm], the catalytic domain is carboxy-terminal and has been mapped to the last 256 residues (Jacquet *et al.*, 1992; Coccetti *et al.*, 1995), whereas the Sos catalytic domain is located in the central part of the molecule, the tail being a proline-rich domain (Chardin *et al.*, 1993). The two catalytic domains are approximately 35% identical (70% homologous). The differences in structural organization appear to reflect the involvement in different signal transduction pathways.

THE RAS SIGNAL TRANSDUCTION CASCADE

Sos-GEF has been shown to work downstream of tyrosine kinase receptors. After ligand/receptor interaction, Sos binds to the receptor *via* interaction of its carboxyl terminal proline-rich domain with the SH3 domain of the adaptor protein grb2. Receptor binding of the grb2/Sos complex may be direct or mediated by phosphorylated Shc (Buday and Downward, 1993; Li *et al.*, 1993; Rozakis-Adcock *et al.*, 1993; Feig, 1994). Following Sos-GEF recruitment in the membrane, the Mitogen Activated Protein (MAP) kinase cascade is activated, whose first element is raf-1 (reviewed in Downward, 1997). The signal transduction pathway in which CDC25[Mm] is involved is less well defined. Available data suggest that this GEF may interact with trimeric G proteins (Van Biesen *et al.*, 1995; Zippel *et al.*, 1996).

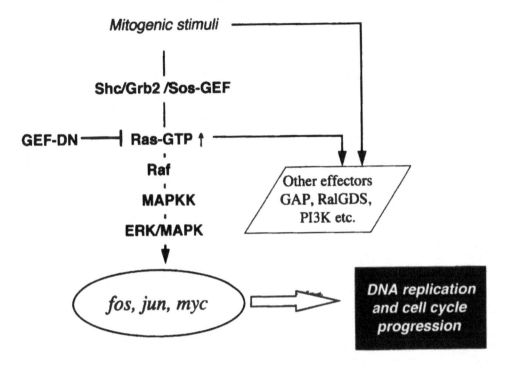

Figure 3 A simplified view of the Ras/MAPK signal transduction cascade and the putative point of action of GEF-DN

Recently it has been shown that upon stimulation of muscarinic receptors by carbacol, phosphorylation-dependent activation of CDC25^{Mm} takes place (Mattingly and Macara, 1996). Both classes appear able to respond to changes in Ca^{++} levels (Farnsworth *et al.*, 1995; Buchsbaum *et al.*, 1996; Freshney *et al.*, 1997). A role for CDC25^{Mm} in synaptic transmission and long-term memory has recently been proposed on the basis of biochemical, physiological and genetic data (Sturani *et al.*, 1997; Brambilla *et al.*, 1997). Deregulated expression of both GEF classes has been shown to be transforming in different experimental systems (Barlat *et al.*, 1993; Zippel *et al.*, 1994; Egan *et al.*, 1993). A simplified view of the current view of the regulation of the ras cycle is reported in Figure 3.

THERAPEUTIC TARGETS IN THE RAS CYCLE

Ras Proteins and their Processing

So far different possible targets in the ras cycle have been explored for therapeutic intervention. Much effort has been directed in altering correct ras subcellular localization, a mandatory prerequisite for ras function. Different classes of farnesyl transferase inhibitors have been isolated and are currently under different stages of pharmacological study. Although ras is not the only important protein to undego farnesylation, some farnesyl transferase inhibitors reduce tumors in animal models down to undetectable size, with little or any toxic side-effect even after months takes place of treatment. Molecular evidence suggests that reversion of ras-transformed cells takes place by a complex mechanism that involves regulation of the actin cytoskeleton, possibly interfering with rho activity (Prendergast *et al.*, 1994; Lebowitz *et al.*, 1995). Interestingly it has been shown that farnesyl transferase inhibitors are able to revert some hyperplastic phenotypes induced by NF1 loss (Kim *et al.*, 1997).

Prenylated proteins can be subjected to palmytoylation, COOH-terminal proteolysis and methylation. Recent studies in yeast have shown that RAS proteins mislocalize to the interior of the cell in strains lacking the two specific processing proteases RCE1 and AFC1 required for processing of farnesylated RAS. Loss of proteolysis reduces but does not eliminate RAS function in the mutant strains, which in fact remain viable. Thus these studies suggest that the prenylprotein-specific protease and methyl transferase may be good targets for the development of anti-tumor drugs (Boyartchuck *et al.*, 1997).

Upstream Effectors: Grb2

As an example of upstream targets, we can mention results obtained with mutant Grb2 proteins, adaptor molecules connecting activated tyrosine kinase receptors to Sos-GEF. A Grb2 isoform lacking its receptor binding SH2 domain has been shown to act as a dominant negative mutant over wild type Grb2 and, by suppressing proliferative signals, to trigger active programmed cell death (Fath *et al.*, 1994) while a form lacking the NH_2-terminal SH3 domain has been shown to induce reversal of the transformed phenotype caused by the point mutation-activated rat *HER-2/Neu* (Xie *et al.*, 1995).

Downstream Effectors: the raf1/MAP Kinase Cascade

A synthethic MAP kinase kinase inhibitor has recently been described (Alessi *et al.*, 1995) which has been shown to reverse the transformed phenotype induced by ras overexpression (Dudley *et al.*, 1995), as well as NGF-induced differentiation of PC12 cells (Pang *et al.*, 1995).

Ras Interacting Proteins: GAP and NF1

The search for therapeutic agents able to down-modulate the ras signallling pathway has also started to explore both upstream and downstream elements. A first series of results shows that peptides containing consensus sequences defining a ras-binding domain within neurofibromin, a GAP-like molecule whose encoding gene is inactivated in neurofibromatosis, an autosomically dominant disease characterized by an high frequency of both benign and malignant tumors, are able to block ras activity *in vitro* and *in vivo* (Clark *et al.*, 1996).

Ras-interacting Proteins: GEF

Deletion mutants

GEF are large multidomain proteins, so that different mechanisms may be used to generate interfering mutants. Tipically deletion mutants lacking the catalytic domain

Table 1 Properties of mutants in the catalytic domain of ras-specific GEF

	in vitro		in vivo		
	dissociation	exchange	cdc25 complementation	fos-luciferase stimulation	tumor[e]
wild type	+[a]	+[a]	+	+	+
loss of function	−	−	−	−	−
gain of function	++	++	+	++	++
dominant negative	+[b]	−	−[c]	−[d]	−−−

[a] May be temperature sensitive
[b] Non catalytic
[c] May reduce growth rate or even abolish growth
[d] May inhibit basal and stimulated level of activity
[e] Overexpression of wild type GEF induce tumor formation (+), which may be increased in gain of function mutants (++), or lost in loss of function mutants (−). Overexpression of a dominant negative mutant in a ras-transformed cell reduce growth rate and restores normal phenotype (−−−).

are expected to downregulate the ras pathway by making non-productive interactions with upstream effectors. Mutants of this kind have been described both for Sos-GEF and CDC25^{Mm}. Sos molecules lacking the catalyic domain have been shown to specifically inhibit insulin-induced ras activation in mammalian fibroblasts (Sakaue et al., 1995). More recently overexpression of Sos PH domain has been shown to have a pronounced dominant-negative effect on serum induced activation of the ras pathway (Chen et al., 1997). In a similar way it could be shown that overexpression of a truncated version of CDC25^{Mm} lacking the COOH-terminal catalytic domain was able to effectively reduce serum- and LPA-induced ras activation, while having no effect on ras basal activity (Zippel et al., 1996).

Point mutations within the GEF catalytic domain

Understanding the mechanism by which GEF activates ras is of paramount importance not only for unravelling the biochemical basis of ras-mediated signal transduction (Polakis and McCormick, 1993), but also as an aid in designing novel molecules able to interfere with ras signalling. In the absence of any three dimensional structure for ras-specific GEF, initial mutagenic efforts of mammalian GEF have been based either on systematic approaches, on sequence homology data or on functional screening in the yeast S. cerevisiae.

The properties of the three major types of mutant proteins may be expected in ras-GEFs (loss of function, gain of function and dominant negative) are summarized in Table 1. Mutants of each class have been obtained by different groups within the catalytic domains of different ras-specific GEFs and will be described below. Particular emphasis will be given to results obtained from our laboratories.

Loss of function and gain of function mutants

By using alanine scanning mutagenesis on the human grf55, the human homolog of CDC25^{Mm}, different loss of function mutants have been isolated. On the basis of two hybrid and functional complementation experiments of *cdc25* mutants different regions of the catalytic possibly involved in GEF function were identified (Camus *et al.*, 1995). Other mutants have been isolated in yeast CDC25 GEF on the basis of allele-specific suppression of mutants in the RAS protein-encoding genes (Park *et al.*, 1994, 1997).

We have so far isolated and characterized nine mutants, mapping at three different positions within CDC25^{Mm} catalytic domain (Vanoni *et al.*, submitted; unpublished results from our laboratory). Although most mutants were catalytically inactive, we could isolate both one gain of function and one dominant negative mutant. Mutants obtained in our laboratory have been analyzed by a combination of techniques. GEF activity was scored *in vivo* in both yeast — using complementation of *S. cerevisiae cdc25-1ts* mutants — and mammalian cells — by scoring *trans*-activation of a ras-responsive *fos*-luciferase reporter gene — and *in vitro*, by testing GEF activity using purified mutant GEF and ras proteins, as well as interaction with ras by means of a competition assay and biomolecular interaction analysis (BIAcore) experiments. The dominant-negative mutant has been characterized further in terms of *focus* and tumor formation. Results obtained with different assays were found to agree in most cases, although a few exceptions were found. These differences may reflect not only different intrinsic sensitivities of each assay, but also depend on several other factors including solubility and stability of the purified proteins, differences in expression levels and presence of stabilizing proteins in the *in vivo* assays. Such differences have to be taken into account when comparing results from different laboratories.

One of our most interesting result was finding both a loss of function and a gain of function mutant by mutating the same amino acid position. We will refer to these mutants as CDC25^{Mm}-LF (for Loss of Function) and CDC25^{Mm}-GF, (for Gain of Function) respectively. Mutant CDC25^{Mm}-LF was inactive in both the yeast complementation assay and in the *fos*-luciferase assay in mammalian fibroblasts. Mutant CDC25^{Mm}-GF complemented the *cdc25* mutation as efficiently as the wild type and displayed a ras-dependent *fos*-luciferase activity significantly higher than the wild type molecule. When this mutation was reintroduced into the full length molecule a similar if not stronger effect was observed. In fact both mutant and wild type proteins show the same pattern of dose-response curves, but maximal activity of the CDC25^{Mm}-GF mutant was obtained at lower doses.

By using a sensitive dissociation assay using fluorescent nucleotides it could be shown that the CDC25^{Mm}-LF mutant was not completely inactive, but rather exceedingly thermolabile. On the contrary mutation of the same residue in mutant CDC25^{Mm}-GF increased the thermostability of the mammalian GEF, since nearly full GEF activity was retained even after pre-incubation for 15 min at 40°C, conditions which almost completely inactivate the wild type protein. GEF activity in the two mutants was affected in opposite ways: CDC25^{Mm}-GF is approximately 2 times more active than wyld type, which on its turn is approximately 3 times more active than CDC25^{Mm}-LF. Biacore experiments showed that the K_D for the dissociation of the ras/GEF complex in the CDC25^{Mm}-GF mutant was slightly reduced (ca 30%) when compared to the wild type, while that of the CDC25^{Mm}-LF mutant was about 2.5–3 times higher. Qualitatively the differences in K_D of the mutant GEF/ras complexes parallel the differences in activity seen both *in vivo* and *in vitro*. The complex catalytic GEF cycle makes it difficult to state whether the observed alterations in K_D represent the only biochemical difference between wild type and mutant GEFs.

Dominant negative mutants

One example of a dominant negative molecule within the GEF catalytic domain of the yeast GEF CDC25 has been reported. The authors showed that the mutant protein showed reduction in growth rate in yeast strain which could be reversed by RAS2 overexpression. They also showed that the protein could bind RAS2 *in vitro* but was catalytically inactive. No data are available regarding its ability to down-regulate ras signalling in mammalian cells (Park *et al.*, 1997).

We have been able to identify a dominant negative mutant in the CDC25^{Mm} molecule (unpublished results from our laboratory). We will refer to this mutant as CDC25^{Mm}-DN (for Dominant Negative). The CDC25^{Mm}-DN mutant was almost completely inactive in the standard exchange and dissociation assays. When added at equimolar concentration with ras it promoted non-catalytic dissociation of bound nucleotide, but no exchange. Once formed the binary complex between the mutant GEF and p21^{ras} remained completely stable and could not be displaced by a 10 fold excess of wild type GEF. Within the current view of the ras/GEF exchange cycle, this finding implies that the mutant DN protein acts efficiently in dissociating the ras-bound nucleotide, "freezing" ras in the nucleotide-free conformation resulting in a very stable ras/GEF binary complex, possibly because of a higher affinity of the mutant GEF for nucleotide-free ras.

On this basis it was predicted that the CDC25^{Mm}-DN protein could titrate out ras *in vivo*, thus efficiently downregulating the ras signal. Preliminary experiments with the *fos*-luciferase assay were consistent with this hypothesis, since overexpression of the full length CDC25^{Mm}-DN mutant in PDGF-stimulated NIH3T3 cells, reduced luciferase activity below control in a statistically significant manner. Later experiments confirmed and extended this observation. A construct expressing the reconstructed full length mutant CDC25^{Mm}-DN gene was stably transfected in K-*ras* transformed NIH3T3 fibroblasts. Expression of the mutant reverted the phenotype of the K-*ras* transformed NIH3T3 fibroblasts back to wild type phenotype. CDC25^{Mm}-DN expressing cells showed in fact flat morphology and regained anchorage dependent growth. Preliminary experiments indicated that these clones also display dramatically attenuated tumor-forming ability in a nude mice model compared to mock-transfected k-*ras* transformed fibroblasts.

FUTURE DIRECTIONS

On the basis of the results summarized in the previous paragraph, the DN mutant GEF may represent the prototype of a new class of pharmaceutical compounds specifically interfering with the ras cascade, the rational being that the mutant GEF molecule can efficiently down-modulate the ras cascade by sequestering intracellular ras in the inactive nucleotide-free state. The most effective molecule so far identified is the 1260 aa-long, whole length molecule: thus gene therapy would be the most obvious choice for delivery of the dominant negative GEF. Recently a report of gene therapy which makes use of a dominant-negative *ras* allele which block the ras cascade to reduce ballon-induced hyperplasia has benn published (Indolfi *et al.* 1995).

The ras/GEF interfaces have not been identified yet. Mutational analysis has identified sites on ras which define a possible interaction surface, but very few informations are yet available regarding the complementary GEF surface. Conventional molecular genetic analyses, as used here and in other reports, are starting to shed some light on the GEF residues involved. Also, powerful new genetic tools are now available to directly select either for mutational events disrupting protein/protein interactions — thus highlighting residues playing major roles in the energetics of binding — or for (macro)molecules interfering with binding (Vidal *et al.*, 1996). It is expected that these studies may help in designing peptides with interfering capability which, once obtained, might be used as templates in the design of peptidomimetics with enhanced pharmacological properties, whose design might be greatly improved by the powerful techniques of combinatorial chemistry.

The availability of the three dimensional structure of a GEF, alone or in complex with ras, would greatly aid in rational design of drugs interfering with the formation of the ras/GEF complex. The ras/GEF interface is likely to be quite large. Nevertheless, it is quite likely that only a subset of actual contacts contribute significantly to the energetics of binding. (see for instance Wells, 1996 and references within). Our own results indicate that minimal perturbation of the GEF structure may have dramatic effects on its functionality. So in this light it should not be impossible to find small molecules able to interfere either with ras/GEF binding or with GEF catalytic action. It is interesting to note in this respect that antibiotics inhibiting guanine nucleotide exchange on protein synthesis elongation factor Tu and the sensitivity of the molecule to its GEF have been reported and characterized (Anborgh *et al.*, 1993).

ACKNOWLEDGEMENTS

The work in the authors' laboratories has been supported by grants from Associazione Italiana Ricerca contro il Cancro (AIRC), European Union (contract BIO2CT-93005), CNR ACRO (grant 94.0163.PF39) and Dompé SpA, Italy.

REFERENCES

Alessi, D.A., Cuenda, A., Cohen, P., Dudley, D.T. and Saltiel, A.R. (1995) PD 098059 is a specific inhibitor of the activation of mitogen-activated protein kinase kinase *in vitro* and *in vivo. J. Biol. Chem.*, **270**, 27489–27494.

Anborgh, P.H. and Parmeggiani, A. (1993) probing the reactivity of the GTP- and GDP-bound conformations of elongation factor Tu in complex with the antibiotic GE2270 A. *J. Biol. Chem.*, **268**, 24622–24628.

Barbacid, M. (1987) *Ras* genes. *Ann. Rev. Biochem.*, **56**, 779–827.

Barlat, I., Shweighoffer, F., Chevallier-Multon, M.C., Duchesne, M., Fath, I., Landais, D., Jacquet, M. and Tocque, B. (1993) The *Saccharomyces cerevisiae* gene product SDC25 C-domain functions as an oncoprotein in NIH3T3 cells. *Oncogene*, **8**, 215–218.

Boguski, M.S. and McCormick, F. (1993) Proteins regulating Ras and its relatives. *Nature*, **366**, 643–654.

Bollag, G. and McCormick, F. (1991) Regulators and effectors of ras proteins. *Ann. Rev. Cell. Biol.*, **7**, 601–632.

Boyartchuk, V.L., Ashby, M.N. and Rine, J. (1997) Modulation of ras and a-factor function by carboxyl-terminal proteolysis. *Science*, **275**, 1796–1800.

Brambilla, R., Gnesutta, N., Minichiello, L., White, G., Roylance, A.J., Herro, C.E., Ramsey, M., Wolfer, D.P., Cestari, V., Rossi-Arnaud, C., Grant, S.G.N., Chapman, P.F., Lipp, H.-P., Sturani, E. and Klein, R. (1997) A role for the ras signalling pathway in synaptic transmission and long-term memory. *Nature*, **390**, 281–286.

Buchsbaum, R., Telliez, J.-B., Goonesekera, S. and L.A. Feig (1996) The N-terminal pleckstrin, coiled-coil and IQ domains of the exchange factor ras-GRF act cooperatively to facilitate activation by calcium. *Mol. Cell. Biol.*, **16**, 4888–4896.

Buday, L. and Downward, J. (1993) Epidermal Growth factor regulates p21^{ras} through the formation of a complex of receptor, grb2 adapter protein and Sos nucleotide exchange factor. *Cell*, **73**, 611–620.

Camonis, J.H., Kalekine, M., Gondrè, B., Garreau, H., Boy-Marcotte, E. and Jacquet, M. (1986) Characterization, cloning and sequence analysis of the *CDC25* gene which controls the cAMP level of *Saccharomyces cerevisiae*. *EMBO J.*, **5**, 375–380.

Camus, C., Hermann-Le Denmat, S. and Jacquet, M. (1995) Identification of guanine exchange factor key residues involved in exchange activity and ras interaction. *Oncogene*, **11**, 951–959.

Chardin, P., Camonis, J.H., Gale, N.W., Van Aelst, L., Schlessinger, J., Wigler, M.H. and Bar-Sagi, D. (1993) Human Sos1: a guanine exchange factor for ras that binds to grb2. *Science.*, **260**, 1338–1343.

Chen, R.-H., Corbalan-Garcia, S. and Bar-Sagi, D. (1997) The role of the PH domain in the signal-dependent membrane targeting of Sos. *EMBO J.*, **16**, 1351–1359.

Clark, G.J., Drugan, J.K., Terrell, R.S., Bradham, C., Der, C.J., Bell, R.M. and Campbell, S. (1996) Peptides containing a consensus ras binding sequence from Raf-1 and the GTPase activating protein NF1 inhibit Ras function. *Proc. Natl. Acad. Sci. USA*, **93**, 1577–1581.

Coccetti, P., Mauri, I., Alberghina, L., Martegani, E. and Parmeggiani, A. (1995) The minimal active domain of the mouse ras exchange factor CDC25^{Mm}. *Biochem. Biophys. Res. Comm.*, **206**, 253–259.

Crechet, J.B., Poullet, P., Mistou, M.Y., Parmeggiani, A., Camonis, J., Boy-Marcotte, E., Damak, F. and Jacquet, M. (1990) Enhancement of the GDP to GTP exchange of ras proteins by the carboxyl terminal domain of SDC25 gene product. *Science*, **248**, 866–868.

Damak F., Boy-Marcotte E., Le-Rosquet D., Guilbaud R. and Jacquet, M. (1991) *SDC25*, a *CDC25*-like gene which contains a RAS-activating domain and is a dispensable gene of *Saccharomyces cerevisiae*. *Mol. Cell. Biol.*, **11**, 202.

Downward, J. (1997) Cell cycle: routine role for Ras. *Current Biology*, **7**, R258–R260.

Dudley, D.T., Pang, L., Decker, S.J., Bridges, A.J. and Saltiel, A.R. (1995) A synthetic inhibitor of the mitogen activate protein kinase cascade. *Proc. Natl. Acad. Sci. USA*, **92**, 7686–7689.

Egan, S.E., Giddins, B.W., Brooks, M.W., Buday, L., Sizeland, A.M. and Weinberg, R.A. (1993) Association of Sos ras exchange protein with grb2 is implicated in tyrosine kinase signal transduction and transformation. *Nature*, **363**, 45–51.

Fam, N.P., Fan, W.-T., Wang, Z., Zhang, L.-J., Chen, L. and Moran, M.F. (1997) Cloning and characterization of ras-GRF2, a novel guanine nucleotide exchange factor for ras. *Mol. Cell. Biol.*, **17**, 1396–1406.

Farnsworth, C.L., Freshney, N.W., Rosen, L., Ghosh, A., Greenberg, M.E. and Feig, L.A. (1995) Calcium activation of ras mediated by the neuronal exchange factor ras-GRF. *Nature*, **376**, 524–527.

Fath, I., Schweighoffer, F., Rey, I., Multon, M.-C., Boiziau, J., Duchesne, M. and Tocque, B. (1994) Cloning of Grb2 isoform with apoptotic properties. *Science*, **264**, 971–974.

Feig, L.A. (1994) Guanine nucleotide exchange factors: a family of positive regulators of ras and related GTPases. *Curr. Opin. Cell Biol.*, **6**, 204–211.

Ferrari, C., Zippel, R., Martegani, E., Gnesutta, N., Carrera, V. and Sturani, E. (1994) Expression of two different products of CDC25^{Mm}, a mammalian ras activator, during development of mouse brain. *Exp. Cell Res.*, **210**, 353–357.

Freshney, N.W., Goonesekera, S.D. and Feig, L.A. (1997) Activation of the exchange factor Ras-GRF by calcium require an intact Dbl homology domain. *FEBS Letters*, **407**, 111–115.

Indolfi, C., Avvedimento, E.V., Rapacciuolo, A., Di Lorenzo, E., Esposito, G., Stabile, E., Feleciello, A., Mele, E., Giuliano, P., Condorelli, G.L. and Chiarello, M. (1995) Inhibition of cellular *ras* prevents smooth muscle cell proliferation after vascular injury *in vivo*. *Nature Medicine*, **1**, 541–545.

Jacquet, E., Vanoni, M., Ferrari, C., Alberghina, L., Martegani, E. and Parmeggiani, A. (1992) A mouse CDC25-like protein enhances the formation of the active GTP complex of human ras p21 and *Saccharomyces cerevisiae* RAS2 proteins. *J. Biol. Chem.*, **267**, 24181–24183.

Kim, H.A., Ling, B. and Ratner, N. (1997) *Nf1*-deficient mouse Schwann cells are angiogenic and invasive and can be induced to hyperproliferate: reversion of some phenotypes by an inhibitor of farnesyl protein transferase. (1997) *Mol. Cell. Biol.*, **17**, 862–872.

Lebowitz, P.F., Davide, J.P. and Prendergast, J.P. (1995) Evidence that farnsyltransferase inhibitors suppress ras transformation by interfering with rho activity. *Mol. Cell. Biol.*, 15, 6613–6622.

Levitzky, A. (1996) Targeting signal transduction for disease therapy. *Curr. Opinion Cell Biol.*, 8, 239–244.

Li, N., Batzer, A., Daly, R., Yajnik, V., Skolnik, E., Chardin, P., Bar-Sagi, D., Margolis, B. and Schlessinger, J. (1993) Guanine nucleotide releasing factor hSos1 binds to Grb2 and links receptor tyrosine kinases to ras signalling. *Nature*, 363, 85–88.

Lowy, D.R. and Willumsen B.M. (1993) Function and regulation of ras. *Ann. Rev. Biochem.*, 62, 851–891.

Martegani, E, Baroni, M.D., Frascotti, G. and Alberghina, L. (1986) Molecular cloning and transcriptional analysis of the start gene *CDC25* of *Saccharomyces cerevisiae*. *EMBO J.*, 5, 2363–2369.

Martegani, E., Vanoni, M., Zippel, R., Coccetti, P., Brambilla, R., Ferrari, C., Sturani, E. and Alberghina, L. (1992) Cloning by functional complementation of a mouse cDNA encoding a homologue of CDC25, a *Saccharomyces cerevisiae* RAS activator. *EMBO J*, 11, 2151–2157.

Mattingly, R.R. and Macara, I.G. (1996) Phosphorylation dependent activation of the ras-GRF/CDC25[Mm] exchange factor by muscarinic receptors and G proteins betagamma subunits. *Nature*, 382, 268–272.

Pang, L., Sawada, T., Decker, S.J. and Saltiel, A.R. (1995) Inhibition of MAP kinase kinase blocks the differentiation of PC-12 cells induced by nerve growth factor. *J. Biol. Chem.*, 270, 13585–13588.

Park,W., Mosteller, R. D. and Broek, D. (1994) Amino acid residues in the CDC25 guanine nucleotide exchange factor critical for interaction with ras. *Mol. Cell. Biol.*, 14, 8117–8122.

Park, W., Mosteller, R.D. and Broek, D. (1997) Identification of a dominant-negative mutation in the yeast CDC25 guanine nucleotide exchange factor for ras. *Oncogene*, 14, 831–836.

Polakis, P. and McCormick, F. (1993) Structural requirements for the interaction of p21[ras] with GAP, exchange factors and its biological effectors. *J. Biol. Chem.*, 268, 9157–9160.

Prendergast, G.C., Davide, J.P., deSolms, J., Giuliani, E.A., Graham, S.L., Gibbs, J.B., Oliff, A. and Kohl, N.E. (1994) Farnsyltransferase inhibition causes morphological revertion of ras-transformed cells by a compex mechanism that involves regulation of the actin cytoskeleton. *Mol. Cell. Biol.*, 14, 4193–4202.

Rozakis-Adcock, M., Fernley, R., Wade, J., Pawson, T. and Bowtell, D. (1993) The SH2 and SH3 domains of mammalian grb2 couple the EGF receptor to the ras activator mSos1. *Nature*, 363, 83–85.

Sakaue, M., Bowtell, D. and Kasuga, M. (1995) A dominat-negative mutant of mSos1 inhibits insulin-induced ras activation and reveals ras-dependent and — independent insulin signalling pathway. *Mol. Cell. Biol.*, 15, 379–388.

Schweigthoffer, F., Faure, M., Fath, I., Chevalier-Multon, M.C., Apiou, F., Dutrillaux, B., Sturani, E., Jacquet, M. and Tocque, B. (1993) Identification of a human nucleotide releasing factor. *Oncogene*, 8, 1477–1485.

Shou, C., Farnsworth, C.L., Neel, B.G. and Feig, L.A. (1992) Molecular cloning of cDNAs encoding a guanine nucleotide releasing factor for ras p21. *Nature*, 358, 351–354.

Simon, M.A., Bowtell, D.D.L., Dodson, G.S., Laverty, T.R. and Rubin, G.M. (1991) Ras1 and a putative guanine nucleotide exchange factor perform crucial steps in signalling by the sevenless protein tyrosin kinase. *Cell*, 67, 701–716.

Sturani, E., Abbondio, A., Branduardi, P., Ferrari, C., Zippel, R., Martegani, E., Vanoni, M. and Denis-Donini, S. (1997) The ras guanine nucleotide exchange factor CDC25[Mm] is present at the synaptic junction. *Exp. Cell Res.*, 235, 117–123.

Tung, P.S., Fam, N.P., Chen, L. and Moran, M.F. (1997) A 54-kDa protein related to ras-guanine nucleotide release factor expressed in rat exocrine pancreas. *Cell Tissue Res.*, 289, 505–515.

van Biesen, T., Hawes, B.E., Luttrell, D.K., Krueger, K.M., Touhara, K., Porfiri, E., Sakaue, M., Luttrel, L.M. and Lefkowitz, R.J. (1995) Receptor-tyrosine kinase- and G beta gamma-mediated MAP kinase activation by a common signalling pathway. *Nature*, 376, 781–784.

Vidal, M., Brachmann, R., Fattaey, A., Harlow, E. and Boecke, J.D. (1996) Genetic selection for the dissociation of protein-protein and DNA-protein interactions. *Proc. Natl. Acad. Sci. USA*, 93, 10315–10320.

Wei, W., Das, B., Park, W. and Broek, D. (1994) Cloning and analysis of human cDNAs encoding a 140-kDa brain guanine nucleotide-exchange factor, Cdc25GEF, which regulates the function of ras. *Gene*, 151, 279–284.

Wei, W., Schreiber, S.S., Baudry, M., Tocco, G. and Broek, D. (1993) Localization of the cellular expression pattern of cdc25NEF and ras in the juvenile brain. *Mol. Brain Res.*, 19, 339–344.

Wells, J.A. (1996) Binding in the growth hormone receptor complex. *Proc. Natl. Acad. Sci. USA*, 93, 1–6.

Wittinghofer, A., Scheffzek, K. and Ahmadian, M.R. (1997) The interaction of Ras with GTPase activating proteins. *FEBS Letts.*, 410, 63–67.

Xie, Y., Pendergast, A.M. and Hung, M.-C. (1995) Dominant-negative mutant of Grb2 induced reversal of the transformed phenotype caused by point mutation-activated rat *HER-2/Neu. J. Biol. Chem.*, **270**, 30717–30724.

Zippel, R., De Maddalena, C., Porro, G., Modena, D. and Vanoni, M. (1994) Increased p21^{ras}-specific guanine nucleotide exchange causes tumor formation in nude mice. *Int. J. Oncol.*, **4**, 175–179.

Zippel, R., Orecchia, S., Sturani, E. and Martegani, E. (1996) The brain specific ras exchange factor CDC25^{Mm}: modulation of its activity through Gi-protein mediated signals. *Oncogene*, **12**, 2697–2703.

19.   EXPRESSION AND USE OF TAGGED CELL SURFACE RECEPTORS FOR DRUG DISCOVERY

KERRY J. KOLLER*, EMILY TATE, ANN M. DAVIS, LESLEY FARRINGTON
and ERIK A. WHITEHORN

Affymax Research Institute, 4001 Miranda Ave, Palo Alto, CA 94304, USA

INTRODUCTION

In the last decade, advances in molecular biology and protein engineering have greatly altered the way pharmaceutical companies screen for new drug candidates. These new tools have allowed for the production of large quantities of proteins to be used as targets in high-throughput screens (HTS) of small organic molecules or of very large biologically generated peptide libraries. Target proteins that have been traditionally of interest can be broadly classified as either enzymes or receptors. Of these, cell surface receptors have proved to be targets that mediate the actions of many established pharmaceuticals. Based on our interests, we have developed new protein engineering techniques to specifically facilitate the HTS of a subset of these proteins, type I membrane and G-protein coupled receptors.

Type I membrane proteins constitute a family of proteins that include receptors for growth factors, cytokines, and cell adhesion molecules. These proteins contain three functional regions: (i) a rather large N-terminal extracellular domain which is responsible for binding the natural ligand or counter-receptor, (ii) a single transmembrane domain, and (iii) an intracellular domain which can be important for the initiation of signal transduction subsequent to ligand activation. The natural agonists for these receptors are large, often multi-subunit, proteins that can be either soluble or membrane bound. These ligands recognize regions within the extracellular domain (ECD) of the receptor and cause dimerization or other physical changes in the receptor to begin a signal transduction cascade. Soluble forms of the ECDs of many of these receptors occur naturally or can be produced by recombinant techniques (Rose-John and Heinrich, 1994). In many cases, these soluble ECDs retain their ability to bind their natural ligand.

G-protein coupled receptors constitute a large family of cell surface proteins that interact with various types of hormones and transduce a cellular signal. These receptors are also known as serpentine or 7-transmembrane receptors (7TMRs) because they transverse the cell membrane seven times. The N-terminal portions of these proteins are expressed on the extracellular surface of the cell and range in lengths from several hundred to as few as seven amino acids. The types of natural ligands that bind these receptors also range from large proteins, such as the heterodimeric glycoprotein hormones with molecular weights of approximately 30,000 Da, to small molecules like the biogenic amines with sizes of about 150 to 200 Da. The binding sites for these ligands can be found on the extracellular

---

* To whom correspondence should be addressed

domain or in regions within the transmembrane domains, and after ligand binding, intracellular portions of the receptors interact with various G-proteins to begin the signal transduction cascade. Some members of this receptor family can be activated by substances that are not typically thought to be receptor ligands. Specifically, changes in extracellular calcium levels are signalled to some cells by binding of Ca^{++} or other polyvalent ions to the calcium-sensing receptor (Brown *et al.*, 1993). Other 7TMRs are activated after proteolytic cleavage of their N-termini. For example, thrombin recognizes and cleaves the first 41 amino acids of a specific 7TMR, and the newly-formed N-terminus binds to and activates other regions within the receptor (Vu *et al.*, 1991). Since the biologically active regions of these receptors are found throughout the protein, the study of 7TMRs has been historically done using membrane or cell preparations from tissues or cell lines naturally expressing these proteins. Recently, G-protein coupled receptors have also been expressed recombinantly in heterologous host cell lines. Most often, the recombinant receptors retain their expected pharmacology and functional activity.

We have recently developed two methods for the production of these two classes of cell surface receptors to specifically increase the efficiency of drug discovery efforts (Whitehorn *et al.*, 1995; Koller *et al.*, 1997). Both of these techniques are similar; novel expression vectors have been designed (Figure 1) to engineer a generic antibody epitope tag into the proteins, and cells expressing large amounts of the receptors of interest can be selected using a fluorescent-activated cell sorter (FACS). Because these procedures are rapid and generic, we have successfully made reagents for a large number of receptor targets for HTS. The purpose of this chapter is to review the two methods and illustrate their uses in drug discovery.

TYPE I MEMBRANE RECEPTORS

Soluble ECDs of type I membrane receptors are often used to study receptor-ligand interactions. In addition, soluble receptors have decided advantages over their native counterparts for drug discovery research. The ability to produce large amounts of homogeneous receptor reagents greatly facilitates the development of efficient drug screening assays. In addition, information derived from structural studies of purified soluble receptors may allow the rational design of drug candidates. Typically, these reagents have been generated using a recombinant, genetic engineering approach by inserting a stop codon into the receptor cDNA sequence immediately upstream from the predicted transmembrane domain sequence. When the gene is expressed in mammalian cells, the ECD is secreted into the extracellular media. We have developed an alternative method of producing soluble ECDs which employs a strategy in which DNA encoding the ECD is fused to DNA encoding a signal sequence of a protein which is normally found on the cell surface in phospha-tidylinositol glycan (PI-G) linked form (Lin *et al.*, 1990; Whitehorn *et al.*, 1995). The signal sequence directs the expression of the ECD to the cell surface where it is found linked through a PI-G moiety. The ECD can then be released in soluble form by treatment of the cells with the enzyme, PI-PLC. Because the sequences that direct the cell surface anchorage are common for every R-ECD fusion, they provide a common epitope that is recognized by a high-affinity antibody. Cells expressing high levels of ECDs can be identified, FACS-sorted and cloned. After cleavage with PI-PLC, the soluble ECDs can be purified, labeled and manipulated in a number

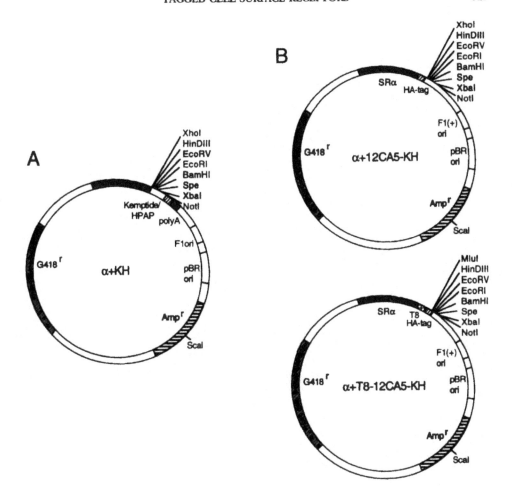

Figure 1 Expression plasmids for cloning epitope tagged receptors

Receptor cDNAs of interest are subcloned into the appropriate vector using restriction sites within the polylinker. The resultant plasmid is linearized by digestion with ScaI and transfected into CHO cells as previously described (Whitehorn *et al.*, 1995). (A) PI-G linked receptor plasmid, α+KH. Using cloning sites of the polylinker, the carboxy terminus of the R-ECD is fused to the kemptide sequence (LRRASALG), a alanine-alanine linker and the carboxy terminal 46 amino acid sequence of human placental alkaline phosphatase to form the following C-terminal sequence (Kemptide/HPAP): LRRASLGAACLEPYTACDLAPPAGTTD* AAH-23 amino acids-TAP. The epitope for mAb179, CLEPTYACD, is retained in the resulting protein fusion after the PI-G linkage is formed at the D* residue. (B) 7TM-receptor plasmids: α+12CA5-KH for expression of 7TMRs with short N-terminal extracellular domains; α+T8-12CA5-KH for expression of 7TMRs with signal sequences. The cDNA sequence of the full length 7TMR is subcloned into the polylinker. For α+12CA5-KH, the endogenous starting methionine is replaced with a methionine from the vector followed by the nine amino acid HA-tag sequence, YPYDVPDYA. For α+T8-12CA5-KH, the endogenous starting methionine and signal sequence are replaced with the methionine and T8 signal sequence followed by the HA-tag. G418^r, gene encoding the resistance to the mammalian antibiotic, G418; SRα, mammalian cell promoter (Takebe *et al.*, 1988); HA-tag, sequence encoding the antibody 12CA5 epitope; T8, sequence encoding the T8 signal sequence; Amp^r, gene encoding the resistance to bacterial antibiotic, ampicillin; F1(+) ori and pBR ori, origins of replication for growth of plasmid in bacteria

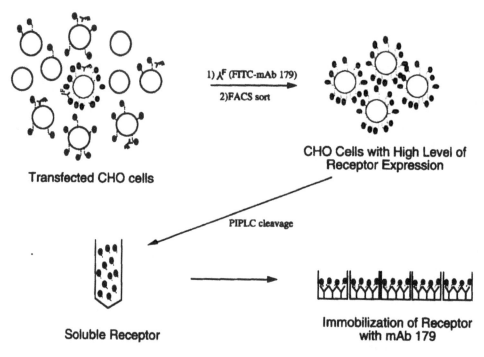

Figure 2 Schematic representation of the production, isolation and immobilization of soluble R-ECDs from type I membrane receptors. G418-resistant population of transfected cells are selected for high receptor expression by labeling with FITC-conjugated mAb179 and sorting with the FACS. The clonal cell line is treated with PI-PLC to remove cell surface R-ECD. The isolated soluble receptor retains the epitope for mAb179; therefore, a high density array of active receptor can be immobilized by capturing on plates passively adsorbed this antibody

of useful ways that enhance the value of these receptors in drug discovery research. Figure 2 shows a schematic representation of the steps involved in isolating a cell line expressing high levels of membrane-bound receptor ECD and enriching for and immobilizing soluble ECD.

Receptor Cloning

A generic expression vector, α+KH (Figure 1A), has been developed for subcloning of R-ECDs sequences upstream from and in-frame with the PI-G signal sequences of human placental alkaline phosphatase (HPAP), a protein that is normally expressed in a PI-G-linked form. The plasmid contains an SRα promoter (Takebe *et al.*, 1988) to drive the synthesis of the recombinant protein and the neo^r gene encoding a protein that confers resistance to the mammalian antibiotic, G418. The plasmid has been engineered such that the expressed protein will also contain the "kemptide" sequence, LRRASLG, immediately before the HPAP-derived amino acids. This sequence can serve as a substrate for protein kinase A (PKA) and therefore, facilitates the site-specific radiolabeling *in vitro* of soluble R-ECDs.

The polymerase chain reaction (PCR) provides a facile route to editing and then cloning the sequences of genes coding for cDNAs of the R-ECDs. Oligonucleotide primers are synthesized which correspond to the 5′-sense and 3′-antisense strands of the cDNA. The 5′ primer includes a unique restriction site at the 5′ end of the oligonucleotide to facilitate directional cloning into the polylinker, which is followed by a -3 purine, the initiation Met codon and approximately 20 additional bases of homologous downstream sequence. The 3′ primer contains a different unique restriction site for cloning, the stop codon and about 20 bases of upstream coding sequence from the R-ECD. Only five cycles of PCR are used to edit cloned cDNAs in plasmids, and in our experience, any sequence errors were located predominately in the primer homologous regions. Primers as described above have also been used to clone R-ECDs from RNA sources using RT-PCR (Kawasaki *et al.*, 1988). Following sequence confirmation of the cloned cDNA, the expression plasmids are transfected into host mammalian cells, typically CHO cells, via electroporation.

Isolation of ECD-expressing Cell Lines and Soluble ECDs

After transfection, the cells are selected for approximately 10 days in G418. The R-ECD is expressed on the cell surface as a chimera of the R-ECD sequence and a small 20 amino acid sequence from HPAP and linked to the membrane via a PI-G lipid anchor. We have produced a high-affinity antibody, mAb179, which recognizes an epitope, CLEPTYACD, within the HPAP-derived sequence on the C-terminus of the R-ECD chimera. As shown schematically in Figure 2, the level of receptor expression varies greatly in the G418-resistant population of transfected cells. To identify the highest receptor-expressing cells, we have used fluorescently-labeled mAb179 and the FACS to clone single cells. These cells are allowed to expand, and the colonies are reanalyzed for receptor expression. After several FACS analyses, the highest expressing clone is chosen for further analysis. In general, the intensity of mAb179 staining of isolated clones correlates well with the number of binding sites for the natural ligand present on the cells.

To isolate soluble R-ECD, the chosen CHO clone is expanded and grown in a large 15 to 36 liter bioreactor. The cells are then collected, washed in serum-free medium, and concentrated. The R-ECD is removed from the CHO membrane by treatment with PI-PLC. The efficiency of this process can be monitored using FACS analysis. Figure 3 shows that treatment of cells expressing the ECD of intracellular cell adhesion molecule (ICAM) with PI-PLC removes greater than 90% of the receptor protein from cell surface. This procedure yields a concentrated preparation of active receptor with only low levels of contaminating proteins from the cells or the medium (see Figure 2). After PI-PLC cleavage from the cells, the mAb179 epitope remains fused to the R-ECD and provides a useful, generic tag. Affinity chromatography has been used for further purification of the receptors; the R-ECDs can be isolated at greater than 90% purity by a single pass over an mAb179 affinity column. Growth of R-ECD-expressing cells in 36 liter stirred-tank bioreactors followed by PI-PLC cleavage and purification has resulted in a typical yield of 1 mg R-ECD per 10^{10} cells (5–10 mg per 36 liter reactor).

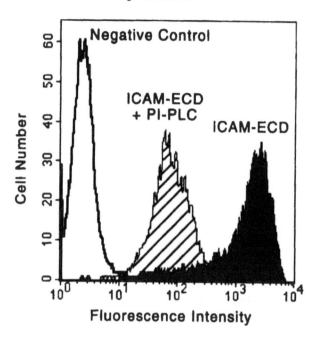

Figure 3 FACS analysis of intracellular cell adhesion molecule (ICAM)-ECD expressing CHO cells. Cells are incubated with FITC-labeled mAb179 for 30 minutes at 4°C and analyzed by FACS. Cells expressing the ECD of ICAM (black) have a mean peak channel fluorescence of 1222. After treatment of the cells with PI-PLC for one hour, the R-ECD is removed from the cell surface and the mean fluorescence intensity of these cells (striped) is reduced to 69. Antibody staining of untransfected CHO cells (white) is shown as a negative control (mean fluorescence = 2.6)

Characterization and Use of Soluble ECDs

Once a soluble preparation of a R-ECD is obtained, it is characterized for biological activity by measuring binding of the natural ligand. The affinity of mAb179 for the epitope is sufficiently high that the receptors are efficiently captured by passively immobilized antibody. In Figure 4, typical results from experiments of this type are shown using the interleukin-2 (IL2) αR-ECD. Competition binding assays were performed with the CHO clonal cell line expressing the IL2αR-ECD on the cell surface via a lipid anchor (Figure 4A) and with the soluble R-ECD immobilized by mAb179 (Figure 4B). These results illustrate that the affinity of IL2 for the R-ECD is approximately 14 nM in both systems. Therefore, cleavage of the recombinant R-ECD from the cell surface does not alter the binding capacity of the protein, and the interaction of the R-ECD with mAb179 does not negatively affect the binding of the natural hormone. We have expressed several dozens different receptor subunit ECDs in this manner, and in all cases the antibody, either in solution or immobilized, binds to the receptor without blocking the binding of the natural ligand.

Receptor ECDs prepared as described here are particularly useful in the screening of libraries of recombinant peptides and synthetic, combinatorially assembled

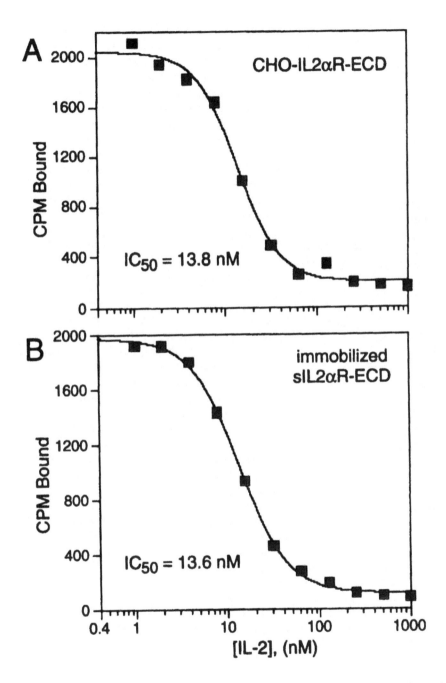

Figure 4   Ligand binding activity of IL2αR-ECD. Binding experiments were performed on CHO cells expressing IL2αR-ECD linked to the cell surface via the PI-G anchor (A) and on soluble IL2αR-ECD immobilized onto an assay plate by capture with mAb179 (B). The receptor preparation was incubated at 4°C for two hours with ^{125}I-IL2 in the presence or absence of various concentrations of unlabeled IL2. Unbound radioligand was removed by aspiration. Non-specific binding was defined by addition of 1 μM IL2

compounds. Screening recombinant peptide display libraries requires that the target molecule be immobilized at a high density and in active form to capture the displayed peptides. The binding characteristics of immobilized mAb179 satisfy this requirement. Over 20 immobilized R-ECDs have been used to screen a large selection of recombinant peptide libraries (Cwirla *et al.*, 1990; Cull *et al.*, 1992). We have successfully identified peptides in the range of 12 to 20 amino acids that can specifically bind to a target receptor and have a desired biological activity. Specifically, peptide antagonists have been identified for the type I interleukin-1 (IL-1) receptor (Yanofsky *et al.*, 1996) and E-selectin (Martens *et al.*, 1995). In addition, small peptide agonists have recently been developed for the erythropoietin (Wrighton *et al.*, 1996) and thrombopoietin (Cwirla *et al.*, 1997) receptors.

Peptide families that interact with the above receptors were first identified by screening RPD libraries on a high density array of immobilized R-ECD. However, peptide ligand optimization was facilitated by screening analog libraries with labeled purified receptor. Specifically, R-ECDs have been phosphorylated on their kemptide sequence with ^{32}P using PKA and γ-^{32}P-ATP used to detect the binding of peptide ligands immobilized on a nitrocellulose filter (Yanofsky *et al.*, 1996). Since this receptor-ligand interaction is monomeric, this screening strategy identifies only the highest affinity ligands. Labeled soluble R-ECDs have also been used to screen synthetic, combinatorial libraries of compounds displayed on beads. MAb179 labeled directly or complexed with secondary reagents has provided a simple means of tagging soluble receptors with fluorescent, chromogenic, chemiluminescent, or other detectable groups. High affinity peptides containing non-natural amino acids have been identified by isolating R-ECD/bead complexes by the FACS. In general, the result of these types of more stringent secondary screening formats is the identification of higher affinity receptor ligands.

Finally, purified soluble R-ECDs bound to hormone or peptide agonists (Vos *et al.*, 1992; Vigers *et al.*, 1997; Livnah *et al.*, 1996) and antagonists (Shreuder *et al.*, 1997) have been sucessfully used in x-ray crystallography studies to elucidate the 3-D structure of the receptors. These results have provided critical information for the rational design of small molecule agonist or antagonists for type I membrane receptors. Proteins produced in the manner described here could potentially serve as a source of soluble R-ECDs for use in these types of studies.

7-TRANSMEMBRANE RECEPTORS

7TMRs are often studied by subcloning the cDNA of the protein of interest into an expression vector which, when transfected into mammalian cell lines, allows for the expression of the receptor on the cell surface. Clonal cell lines expressing 7TMRs are usually identified from an enriched population of transfected cells by screening a relatively small number of randomly isolated clones for receptor expression. This process has a number of drawbacks. It does not allow for the selection of the most stable cell lines or of those with the highest receptor density. In addition, it requires a specific reagent, such as a radioligand or an antibody, for each receptor. To develop a robust and generic expression system for 7TMRs, we have utilized the nine amino acid sequence tag, YPYDVPDYA, derived from influenza virus hemagglutinin (HA), and the commercially available monoclonal antibody, 12CA5, that recognizes

Table 1 Overview of steps in producing tagged 7TMR-expressing cell lines

	Procedure	Time Frame

I	Construct Expression Plasmid
	    PCR from cDNA or RT-PCR from RNA; sequence plasmid for
	    PCR-induced errors
		~one week

II	Transfect Host Mammalian Cells Select for transfected cells in G418	~10 days

III	FACS analyze Cells;
	    FACS Clone top 2% of G418-resistant population expand
	    cloned cells for assay
		~10 days

IV	Analyze clones in 96-well plate assay expand 12 best clones for further analysis	~10 days
V	Select best clone by FACS analysis expand cell line for characterization	~two weeks
VI	Verify biological activity test cell line in binding and functional assays	~one week

this sequence (Wilson *et al.*, 1984). The HA sequence has been successfully used to tag a number of recombinant proteins on both the N- and C-termini, and the resultant proteins have been used for studying, among other things, protein-protein interactions (Murray *et al.*, 1995), receptor phosphorylation and desensitization (Ali *et al.*, 1993), and protein degradation after cell stimulation (Levis and Bourne, 1992). By developing expression plasmids (Figure 1B) that place the HA-tag on the N-terminus of any 7TMR, we have been able to rapidly produce mammalian cell lines stably expressing high levels of a desired receptor (Koller *et al.*, 1997). This generic system has also proved especially useful for producing cell lines expressing orphan receptors where no known ligands are available to facilitate cell line selection. Importantly, the activity of the tagged receptors appears to be identical to their untagged versions in both binding and functional assays. An outline of the steps involved in our technique is shown in Table 1. An important advantage of this method is the rapidity with which clonal cell lines can be made. As shown by the approximate time frame each step of the procedure takes, we can develop a stable, high-expressing clonal cell line in about five weeks from transfection.

Receptor Cloning

To develop a generic system that allows for the production of high-expressing receptor cell lines for all types of 7TMRs, two novel expression vectors were made by modifying α+KH described above (Figure 1). Both of these vectors introduce the HA-tag at the N-terminus of a 7TMR. For many 7TMRs, the starting methionine is the first amino acid of the mature protein. For these receptors, the α+12CA5-KH vector has a set of restriction sites in a polylinker immediately downstream from sequences coding for an initiating methionine and the HA-tag. The protein encoded

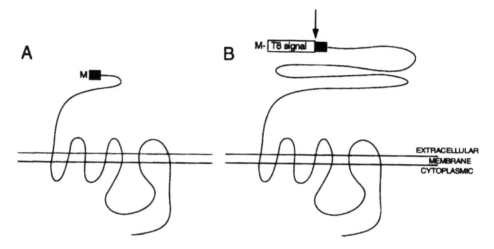

Figure 5 Schematic representation of tagged 7TMRs expressed on CHO cells. (A) For receptors lacking a signal sequence, the HA-tag sequence (black box) is located immediately following the starting methionine of the expressed receptor. (B) For receptors that require a signal sequence for proper expression, the HA-tag sequence is expressed immediately following the signal peptidase cleavage site (indicated by the arrow) of the exogenous T8 signal sequence on the N-terminus of the receptor. The sequence of the pre-processed amino terminus is the following: MALPVTALLLPLALLLHAARP/ DYASYPYDVPDYA. Signal peptidase cleavage (designated by "/") occurs after the proline at residue 21 (Littman *et al.*, 1985) and the N-terminus of the mature receptor contains the nine residue HA-tag preceded by four amino acids derived from the restriction site used to construct the vector. Membrane, transmembrane region of receptor; Extracellular, portion of the receptor expressed on the cell surface; Cytoplasmic, portion of the receptor expressed within the cytoplasm of the cell

by this plasmid will contain the nine amino acid epitope immediately following the starting methionine (Figure 5A). A subset of 7TMRs contains a rather large first ECD. In these proteins, the mature receptor is formed subsequent to cleavage of a signal sequence. Therefore, an additional vector, α+T8-12CA5-KH, was engineered to provide both an initiation methionine and an exogenous signal sequence derived from the human T cell surface protein, T8. These receptors will express the HA-tag on the N-terminus without a methionine (Figure 5B). The 7TMR cDNA of interest is cloned without its endogenous methionine or signal sequence, if present, into these vectors by using chimeric primers in a PCR reaction as was described above for the type I membrane receptors. After receptor sequence verification, the plasmids are transfected into the host mammalian cells.

Isolation of High Receptor-Expressing Cell Lines

After the transfected cells are selected by growth in G418 for ten days, thousands of G418-resistant colonies are generated. As with the type I membrane protein R-ECD-expressing cells, these cells are harvested as a population and analyzed by FACS after incubation of the cells with the anti-HA antibody followed by phycoerythrin (PE)-conjugated goat anti-mouse IgG. Typically, there is a range of receptor expression

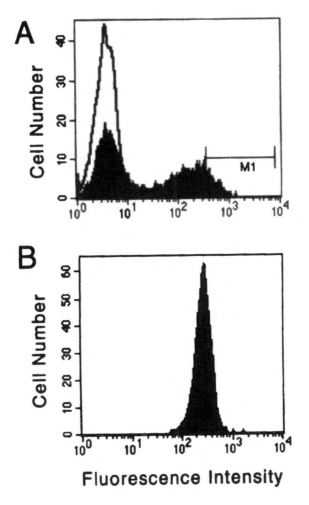

Figure 6 FACS analysis of calcitonin receptor-expressing CHO cells. (A) Ten days after selection, the G418-resistant population of cells transfected with the calcitonin receptor were analyzed by FACS. A histogram of the fluorescence intensity of the analyzed cells is shown in black. The profile of untransfected CHO-K1 cells (white) is shown to illustrate background levels of fluorescence. M1 refers to the top 2% of the population of cells from which individual highly fluorescent cells were cloned. (B) FACS analysis of the clonal cell line isolated from the population shown in (A). Mean peak channel fluorescence for this cell line was 243

in this cell population as demonstrated with calcitonin receptor-expressing cells in Figure 6A. The untransfected CHO-K1 cells have a low mean peak channel fluorescence (3.9), while the population of cells selecting for expression of tagged calcitonin receptor has a range of fluorescence that is as high 10^3 fluorescent intensity units. After the top 2% of the fluorescent cells are cloned (M1 in Figure 6A), the best expressing clone is chosen after several rounds of FACS analysis. The mean peak channel fluorescence for the calcitonin receptor clone chosen from this population was 243 (Figure 6B). As observed for the mAb179 epitope on type I membrane R-ECDs, the intensity of 12CA5 antibody staining of isolated clones correlates well with the number of binding sites for the natural ligand on these cells. We have generated greater than thirty cell lines expressing different 7TMRs by using this protocol, and some receptors tend to express more highly than others. In general, we have isolated clones with mean peak channel fluorescence in range of eight- to 450-fold over the CHO non-transfected controls. When receptor density was measured in a membrane binding assay, we measured a range of expression from 0.5 to 15 pmol/mg protein.

In addition to expressing very high amounts of receptor, the cell lines produced by this procedure are also extremely stable. Unlike the cells expressing the R-ECDs, the 7TMR-expressing cells are capable of intracellular signaling after ligand activation. Some recombinant proteins can be toxic to cells, and, in theory, a functional receptor may be more difficult to express recombinantly than a non-signaling ECD. In fact, during the clone isolation procedure, we often observe what may be loss of cell surface receptor expression in clones that were initially chosen for their high epitope levels. The procedure requires cells to maintain receptor expression through at least three rounds of dilution and expansion, and only the hardiest cells survive the initial cloning procedure. Therefore, in general, this method provides a rapid and generic technique for producing stable, high-expressing cell lines of any type of 7TMR.

Characterization and Use of Tagged Receptor Cell Lines

All of the tagged cell lines that we have produced expressing 7TMRs are biologically active. Figure 7 shows the results of characterizing a cell line expressing the human somatostatin type 4 receptor (SSTR4). Membranes from the cells were isolated and used for receptor binding studies, and the affinity of SSTR4 for SST-14, its natural ligand, is 1.1 nM (Figure 7A), very close to the literature value for untagged SSTR4 of 0.53 nM (Patel and Srikant, 1994). To measure the functional activity of the receptor, we assayed its ability to mediate SST-14's inhibition of forskolin-induced cAMP production in whole cells (Figure 7B) and SST-14's induction of extracellular acidification (McConnell *et al.*, 1992) using the microphysiometer (Figure 7C). In both cases, the activity of the tagged receptor correlated well with its binding affinity and was similar to what is expected for the untagged receptor (Bito *et al.*, 1994). Of the more than 30 HA-tagged 7TMR cell lines we have made using the technique described here, none has failed to bind or respond to their natural ligand as expected. Therefore, the presence of the tag does not adversely affect the ligand binding or signal transduction of these receptors.

The cell lines expressing tagged 7TMRs described here have proved to be essential for our efforts in screening synthetic, combinatorially assembled libraries of small molecules in high-throughput screens (HTS). In general, as the pharmaceutical industry expands its use of HTS, the amount of biological material needed to complete these screens will increase, and high-expressing cell lines may provide a means to fill that need. In addition, the development of sensitive and automatable assays for HTS may be facilitated by the availability of high-expressing cell lines. For example, scintillation proximity assays (SPA) for radioligand binding have greatly increased the efficiency of HTS (Cook, 1996). In our experience, however, the development of a robust assay is dependent on a high level of receptor

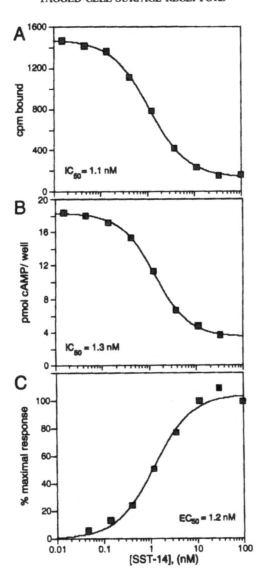

Figure 7 Characterization of the activity of tagged SSTR4 expressed in CHO cells. (A) Binding activity of radioligand. Membranes prepared from SSTR4-CHO cells were incubated with ^{125}I-somatostatin in the presence and absence of varying concentrations of unlabeled somatostatin-14 (SST-14). Bound radioligand was determined using a scintillation proximity assay (Cook 1996). Non-specific binding was defined by 1 μM SST-14. (B) Ligand-induced inhibition of forskolin-induced cAMP production. Cells expressing SSTR4 were stimulated with 10μM forskolin and incubated with varying concentrations of SST-14 for 15 minutes. After extraction with ethanol, intracellular cAMP was measured using a radioimmunoassay. (C) Ligand-induced stimulation of extracellular acidification rates as a measure of cellular activation. Cells were stimulated with various concentrations of SST-14 for one minute as described (Koller *et al.*, 1997) and the acidification rates obtained four minutes post-compound addition were expressed as a percent of maximal response. The IC_{50} and EC_{50} values for these experiments were determined by analyzing the data with the sigmoidal-logistic curve-fitting function on DeltaGraph

expression. In addition, a higher receptor density on cell lines allows one to decrease the use of costly SPA reagents. Since we have been able to produce cell lines with a very high level of receptor expression, we have successfully developed very robust SPA-based assays for nearly all the cell lines we have made. Similarly, the development of non-radioactive HTS assays such as fluorescence polarization (Jolley, 1996) and homogeneous time-resolved fluorescence (Kolb *et al.*, 1996) may be facilitated by high levels of receptor expression.

SUMMARY

The generic aspects of the two systems described here have allowed us to rapidly produce active, soluble R-ECDs of more than 40 type I membrane receptors and functional high-expressing cell lines for more than 30 7TMRs. Using these methods, both of these target classes have been successfully incorporated into our drug discovery programs. Both of these methods utilize a tagging approach that allows us to use FACS technology to identify robust recombinant cell lines. With these clonal lines, we can generate reagents for use in classic and novel assay designs. These reagents will be essential as we begin to implement non-radioactive miniaturized high throughput assays. The engineered soluble R-ECDs have proved to be particularly well suited to screening combinatorial peptide and small molecule libraries. We have successfully identified many potent and active compounds for cytokine receptors and cell adhesion molecules. In addition, these purified receptors could potentially be used in crystallography studies that are crucial for the rational design of small molecule drugs.

REFERENCES

Ali, H., Richardson, R.M., Tomhave, E.D., Didsbury, J.R. and Snyderman, R. (1993) Differences in phosphorylation of formylpeptide and C5a chemoattractant receptors correlate with differences in desensitization. *J. Biol. Chem.*, **268**, 24247–24254.

Bito, H., Mori, M., Sakanaka, C., Takano, T., Honda, Z., Gotoh, Y., Nishida, E. and Shimizu, T. (1994) Functional coupling of SSTR4, a major hippocampal somatostatin receptor, to adenylate cyclase inhibition, arachidonate release and activation of the mitogen-activated protein kinase cascade. *J. Biol. Chem.*, **269**, 12722–12730.

Brown, E.M., Gamba, G., Riccardi, D., Lombardi, M., Butters, R., Kifor, O., Sun.A., Hediger, M.A., Lytton, J. and Hebert, S.C. (1993) Cloning and characterization of an extracellular Ca²⁺-sensing receptor from bovine parathyroid. *Nature*, **366**, 575–580.

Cook, N.D. (1996) Scintillation proximity assay: a versatile high-throughput screening technology *Drug Discovery Technology*, **1**, 287–294.

Cull, M.G., Miller, J.F. and Schatz, P.J. (1992) Screening for receptor ligands using large libraries of peptides linked to the C terminus of the lac repressor. *Proc. Natl. Acad. Sci. USA*, **89**, 1865–1869.

Cwirla, S.E., Peters, E.A., Barrett R.W. and Dower, W.J. (1990) Peptides on phage: a vast library of peptides for identifying ligands. *Proc. Natl. Acad. Sci. USA*, **87**, 6378–6382.

Cwirla, S.E., Balasubramanian, P., Duffin, D.J., Wagstrom, C.R., Gates, C.M., Singer, S.C., Davis, A.M., Tansik, R.L., Mattheakis, L.C., Boytos, C.M., Schatz, P.J., Baccanari, D.P., Wrighton, N.C., Barrett, R.W. and Dower, W.J. (1997) Peptide agonist of the thrombopoietin receptor as potent as the natural cytokine. *Science*, in press.

Jolley, M.E. (1996) Fluorescence polarization assays for the detection of proteases and their inhibitors. *J. Biomol. Screening*, **1**, 33–38.

Kawasaki, E.S., Clark, S.S., Coyne, M.Y., Smith, S.D., Champlin, R., Witte, O.N. and McCormick, F.P. (1988) Diagnosis of chronic myeloid and acute lymphocytic leukemias by detection of leukemia-specific mRNA sequences amplified in vitro. Proc. Natl. Acad. Sci. USA, 85, 5698–5702.

Kolb, J.M., Yamanaka, G. and Manly, S.P. (1996) Use of a novel homogeneous fluorescent technology in high throughput screening. J. Biomol. Screening, 1, 203–210.

Koller, K.J., Whitehorn, E.A., Tate, E., Ries, T., Aguilar, B., Chernov-Rogan, T., Davis, A.M., Dobbs, A., Yen, M. and Barrett, R.W. (1997) A Generic method for the production of cell lines expressing high levels of 7-transmembrane receptors. Anal. Biochem., 250, in press.

Levis, M.J. and Bourne, H.R. (1992) Activation of the alpha subunit of Gs in intact cells alters its abundance, rate of degradation and membrane avidity. J. Cell. Biol., 119, 1297–1307.

Lin, A.Y., Devaux, B., Green, A., Sagerstrom, C., Elliott, J.F. and Davis, M.M. (1990) Expression of T cell antigen receptor heterodimers in a lipid-linked form. Science, 249, 677–679.

Littman, D.R., Thoman, Y., Maddon, P.J., Chess, L. and Axel, R. (1985) The isolation and sequence of the gene encoding T8: a molecule defining functional classes of T lymphocytes. Cell, 40, 237–246.

Livnah, O., Stura, E.A., Johnson, D.L., Middleton, S.A., Mulcahy, L.S., Wrighton, N.C., Dower, W.J., Jolliffe, L.K. and Wilson, I.A. (1996) Functional mimicry of a protein hormone by a peptide agonist: the EPO receptor complex at 2.8 Å. Science, 273, 464–471.

Martens, C.L., Cwirla, S.E., Lee, R., Y.-W., Whitehorn, E., Chen, E., Y.-F., Bakker, A., Martin, E.L., Wagstrom, C., Gopalan, P., Smith, C.W., Tate, E., Koller, K.J., Schatz, P.J., Dower, W.J. and Barrett, R.W. (1995) Peptides which bind to E-selectin and block neutrophil adhesion. J. Biol. Chem., 270, 21129–21136.

McConnell, H.M., Owicki, J.C., Parce, J.W., Miller, D.L., Bazter, G.T., Wada, H.G. and Pitchford, S. (1992) The cytosensor microphysiometer: biological applications of silicon technology. Science, 257, 1906–1912.

Murray, P.J., Watowich, S.S., Lodish, H.F., Young, R.A. and Hilton, D.J. (1995) Epitope tagging of the human endoplasmic reticulum HSP70 protein, BiP, to facilitate analysis of BiP-substrate interactions. Anal. Biochem., 229, 170–179.

Patel, Y.C. and Srikant, C.B. (1994) Subtype selectivity of peptide analogs for all five cloned human somatostatin receptors (hsstr 1-5). Endocrinology, 135, 2814–2817.

Rose-John, S. and Heinrich, P.C. (1994) Soluble receptors for cytokines and growth factors: generation and biological function. Biochem J., 300, 281–290.

Shreuder, H., Tardif, C., Trump-Kallmeyer, S., Soffientini, A., Sarubbi, E., Akeson, A., Bowlin, T., Yanofsky, S. and Barrett, R.W. (1997) A new cytokine-receptor binding mode revealed by the crystal structure of the IL-1 receptor with an antagonist. Nature, 386, 194–200.

Takebe, Y., Seiki, M., Fujisawa, J.-I., Hoy, P., Yokota, K., Arai, K.I., Yoshida, M. and Arai, N. (1988) SR alpha promoter: an efficient and versatile mammalian cDNA expression system composed of the simian virus 40 early promoter and the R-U5 segment of human T-cell leukemia virus type 1 long terminal repeat. Mol. Cell. Biol., 8, 466–472.

Vigers, G.P.A., Anderson, L.J., Caffes, P. and Brandhuber, B.J. (1997) Crystal structure of the type-1 interleukin-1 receptor complexed with interleukin-1β. Nature, 286, 190–194.

Vos, A.M., Ultsch, M. and Kossiakoff, A.A. (1992) Human growth hormone and extracellular domain of its receptor: crystal structure of the complex. Science, 255, 306–312.

Vu, T.-K.H., Hung, D.T., Wheaton, V.I. and Coughlin, S.R. (1991) Molecular cloning of a functional thrombin receptor reveals a novel proteolytic mechanism of receptor activation. Cell, 64, 1057–1068.

Whitehorn, E.A., Tate, E., Yanofsky, S.D., Kochersperger, L., Davis, A., Mortensen, R.B., Yonkovich, S., Bell, K., Dower, W.J. and Barrett, R.W. (1995) A Generic method for expression and use of "tagged" soluble versions of cell surface receptors. Bio/Technology, 13, 1215–1219.

Wilson, I.A., Niman, H.L., Houghten, R.A., Cherenson, A.R., Connolly, M.L. and Lerner, R.A. (1984) The structure of an antigenic determinant in a protein. Cell, 37, 767–778.

Wrighton, N.C., Farrell, F.X., Chang, R., Kashyap, A.K., Barbone, F.P., Mulchay, L.S., Johnson, D.L., Barrett, R.W., Joliffe, L.K. and Dower, W.J. (1996) Small peptides as potent mimetics of the protein hormone erythropoietin. Science, 273, 458–464.

Yanofsky, S.D., Baldwin, D.N., Butler, J.H., Holden, F.R., Jacobs, J.W., Balasubramanian, P., Chinn, J.P., Cwirla, S.E., Peters-Bhatt, E., Whitehorn, E.A., Tate, E.H., Akeson, A., Bowlin, T.L., Dower, W.J. and Barrett, R.W., (1996) High affinity type I interleukin 1 receptor antagonists discovered by screening recombinant peptide libraries. Proc. Natl. Acad. Sci. USA, 93, 7381–7386.

20. ESTROGEN RECEPTOR AS A TARGET FOR NEW DRUG DISCOVERY

CAROLINE E. CONNOR and DONALD P. McDONNELL

Department of Pharmacology and Cancer Biology, Duke University Medical Center, Durham, NC 27710, USA

ESTROGEN ACTION

Homeostasis within the female reproductive system is controlled primarily by estrogens, the principal one being 17β-estradiol (E2). This sex steroid, a member of a class of compounds which includes cortisol, testosterone, and progesterone, is a rigid, lipophilic molecule which circulates in the blood stream and passively diffuses into all cells but only exhibits activity in those cells which contain estrogen receptors (McDonnell *et al.*, 1993; Sutherland *et al.*, 1988). The primary functions of this hormone include regulation of the estrous cycle, development of the reproductive tract and mammary glands, regulation of lactation, and general control of female reproductive behavior (Sutherland *et al.*, 1988).

In addition, studies on the long-term effects of estrogen deprivation have highlighted important non-reproductive functions for this hormone (Horowitz, 1993; Orimo *et al.*, 1993). Specifically, it has been shown that after menopause, women are at increased risk for atherosclerosis and osteoporosis, as well as undesirable climacteric effects such as mood swings and hot flushes (Edman, 1983). The rising incidence of coronary heart disease in postmenopausal women (Murabito, 1995) and the observation that over one-third of all Caucasian women are expected to experience some of the debilitating effects of osteoporosis (Horowitz, 1993) further highlight the importance of this hormone. Reintroduction of estrogen has been associated with more favorable serum lipid levels (Orimo *et al.*, 1993) and decreased debilitating bone resorption associated with the osteoporotic state (Horowitz, 1993). The fact that Estrogen Replacement Therapy is usually sufficient to provide relief from menopausal symptoms indicates that estrogen is the key etiologic agent in this condition.

Recently, links have also been made between estrogen and improvements in cognitive function (Wickelgren, 1997). In addition, estrogen has been shown in some studies to delay the onset of Alzheimer's disease in women (Tang *et al.*, 1996). These latest findings are extremely important and serve to highlight the medical and economic need to develop safe and effective Hormone Replacement Therapy (HRT).

Though the link has not been fully substantiated, HRT has been associated with an increased risk of various reproductive cancers (Stanford *et al.*, 1995). Specific estrogen-related cancers include that of the breast (Sunderland and Osborne, 1991), ovary (Rao and Slotman, 1991), and endometrium (Gottardis *et al.*, 1988b). The role of estrogen in these cancers is clearly important, as ablation often leads to a reversal in the pathology. Tumor regression can be accomplished through medical castration in conjunction with the use of antiestrogens, compounds that oppose the action of estrogen (Dreicer and Wilding, 1992). Consequently, although the benefits of estrogen are largely undisputed, the serious side effects of HRT remain cause for concern. This presents the pharmaceutical community with a challenge to design better and safer estrogens and estrogen mimetics.

THE BIOLOGICAL ACTIONS OF ESTROGEN ARE MEDIATED THROUGH A SPECIFIC INTRACELLULAR RECEPTOR

Estrogen, present physiologically at nanomolar concentrations (McDonnell et al., 1993), affects complex regulatory signaling pathways via interaction with a specific high affinity estrogen receptor (ER) localized within target cell nuclei. This protein is a member of an intracellular receptor superfamily, the nuclear hormone receptor family. Also included in this group are receptors for steroids, thyroid hormone, vitamin D_3, and retinoic acid (Carson-Jurica et al., 1990; Evans, 1988). The physiological concentration of nuclear hormone receptors is quite low, typically less than 0.01% of the total cellular protein (McDonnell et al., 1987); therefore, cloning was essential in obtaining information regarding the structure and function of these conserved family members. All steroid hormones operate by a similar mechanism (depicted in Figure 1) in which the latent receptor is bound in a multiprotein complex comprised of receptor, heat-shock protein 90 (hsp90), hsp70, p59, and other proteins (Smith and Toft, 1993). Upon ligand binding, the receptor undergoes a conformational change, the inhibitory proteins are displaced, and spontaneous dimerization of the receptor occurs (Kumar and Chambon, 1988; McDonnell et al., 1994). In this activated state the receptor can interact with specific DNA sequences within target genes and manifest either a positive or negative effect on transcription.

Structure and Function

The cDNA for ER was cloned initially from MCF-7 breast cancer cells and was found to have an open reading frame (ORF) of 1785 nucleotides and encode for a 595 amino acid, 66 kilodalton polypeptide (Greene et al., 1986). A decade of subsequent studies focused on the molecular biology of estrogen via this receptor. However, a search for novel steroid receptors culminated in a description by Kuiper et al. (1996) of a second receptor for estrogen which was highly related to the original protein. This new receptor isoform, designated ERβ to distinguish it from the former (now called ERα), was described as a 485 amino acid, 54 kilodalton protein (Kuiper et al., 1996). However, recent data suggest that the N-terminus of this protein is longer than previously observed and that this additional sequence is critical for function (T. Willson, personal communication). Northern analysis has revealed a wide tissue distribution of ERα, but ERβ mRNA has been found primarily in ovary, testis, prostate, spleen, thymus, and hypothalamus (Mosselman et al., 1996; Shughrue et al., 1996). The results of these mRNA distribution studies need to be confirmed using antibodies that can specifically recognize ERβ.

The estrogen receptor can be divided into six structural domains, regions A through F (Figure 2). Regions C and E contain the highest degree of homology to other steroid receptor family members, while the N-terminal regions, A and B, and region D, the hinge region, are poorly conserved (Carson-Jurica et al., 1990; Evans,

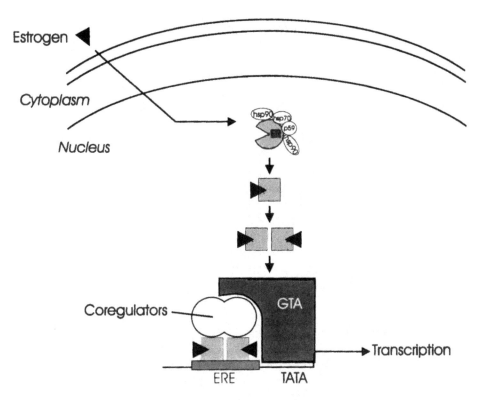

Figure 1 Mechanism of estrogen action. The estrogen receptor (ER) exists in an inactive form in the absence of ligand, bound to several specific inhibitory proteins. Upon ligand binding, the receptor undergoes a conformational change and the complex dissociates. Subsequent dimerization occurs, followed by DNA binding to a specific estrogen response element (ERE). Once bound, the receptor can interact with coregulator proteins and the general transcription apparatus (GTA) and affect gene transcription in a manner dependent upon the cellular environment

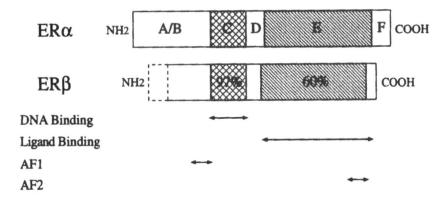

Figure 2 Molecular organization of the functional domains of both ERα and ERβ. The percent homology of ERβ with respect to ERα is indicated for the DBD and HBD. The dotted line at the N-terminus of ERβ reflects the uncertainty regarding its length

1988). Additionally, ERβ shares approximately 97% and 60% homology to ERα in regions C and E, respectively. Region E, found in the C-terminal region of the receptor, is of hypervariable length and functions as the hormone-binding domain (HBD). This region forms a hydrophobic pocket and confers to the receptor specific ligand-binding properties (Green et al., 1986).

The two estrogen receptors manifest their biological activity at target gene promoters by binding with high affinity to specific DNA regulatory elements. Precise DNA recognition by the receptors is controlled by a 66-68 amino acid DNA-binding domain (region C). This domain includes nine perfectly conserved cysteines and forms two zinc fingers, analogous to other transcription factors (Krust et al., 1986). The nature of the sequence-specific regulatory elements was discovered upon the analysis of the Xenopus laevis vitellogenin genes, direct targets of estrogen action. It was found that one to three copies of a specific sequence was found preceding each of the vitellogenin genes and also in the chicken apoVLDLII gene, which is also estrogen-regulated (Klein-Hitpass et al., 1986). This sequence, known as the consensus estrogen response element, or ERE, is a palindromic sequence: 5'-GGTCAnnnTGACC-3' where nnn = any three nucleotides. The consensus ERE is similar to, yet distinct from, the response elements of the progesterone (PR), glucocorticoid (GR), and vitamin D (vitD) receptors (Carson-Jurica et al., 1990). An enhancer element, the ERE can confer estrogen-dependent regulation to a heterologous promoter in a position- and orientation-independent manner (Kumar et al., 1987; Martinez et al., 1987). It is now clear that several half-sites (sequences comprised of half of the ERE palindrome) or, alternatively, multiple imperfect EREs, can act cooperatively or synergistically to permit ER responsiveness, as is seen with the vitellogenin B1 (Martinez et al., 1987) and complement 3 (Norris et al., 1996) genes. The ability of ER and other hormone receptors to act as ligand-inducible transcription factors distinguishes them from classical membrane-bound receptors.

The observation that estrogen's effects are mediated via an interaction between trans-acting factors (ER) and cis-acting elements (EREs) within genes (Klein-Hitpass et al., 1986; Kumar et al., 1987) led to the development of reconstituted ER-responsive transcription systems. Specifically, the ER/ERE interaction system can be reconstituted in both yeast (Metzger et al., 1988) and mammalian (Tzukerman et al., 1994) cells, allowing the dissection and manipulation of the functional domains. While intact DNA- and hormone-binding domains are required for transcriptional activation, studies using chimeric GR/ER proteins suggest that the HBD is not necessary for ERE recognition and binding (Kumar et al., 1987). Region D may be altered in length and composition with little or no effect, substantiating its role as a hinge region (Kumar et al., 1987).

Though transcriptional activation by ER requires intact DNA- and hormone-binding domains, much of this is dictated by two independent, non-acidic activation domains (Tora et al., 1989; Tzukerman et al., 1994). Activation Function 1 (AF1) and AF2 are contained within the receptor's A/B region and HBD, respectively. Estrogen has been shown to induce a tertiary structure within the receptor, thus creating a functional AF2 (Tora et al., 1989). The two activation functions can facilitate the disruption of chromatin structure (Pham et al., 1991), allowing dimerized, ligand-bound ER to contact the general transcription apparatus within the cell. AF1 and AF2 are promoter-dependent and cell-type specific; in some cell contexts either AF is sufficient for maximal transcriptional activity, whereas other cells require one specific AF or in some cases both AFs (Tzukerman et al., 1994). This observation suggested that ER is not operating in the same way in all cells and provided the impetus to develop compounds which could activate one AF but not the other, thereby generating some selectivity (McDonnell et al., 1995).

The inability of AF1 and AF2 to fully explain the actions of ligand-bound ER led to the recent discovery of a third activation domain. A yeast genetic screen identified a region in ER known as AF2a (Pierrat et al., 1994), but this autonomous domain initially appeared unable to transactivate in mammalian cells. New data suggest that this region is able to activate transcription in the mammalian hepatocarcinoma cell line HepG2 (Norris et al., 1997). The precise role of this region in ER pharmacology remains to be determined.

Coactivators

The mechanism by which DNA-bound ER alters the transcription of target genes is only beginning to be elucidated. Though in vitro data suggest that ER directly contacts TFIIB, a component of the basal transcriptional machinery, this is insufficient to explain promoter activation (Ing et al., 1992). Insight into the mechanism of ER transcriptional activity was provided by experiments demonstrating that overexpression of the activation domains of ER could "squelch" the activity of some but not all transcription factors (Meyer et al., 1989). These data implied that ER contacted the transcription machinery through intermediary proteins which were present in limiting amounts within the cell. This prompted an ongoing search for receptor associated proteins which are able to potentiate the transcriptional activity of ligand-bound steroid receptors.

This effort has led to the recent identification of several proteins which interact with the HBD of ER and when introduced into the cell potentiate ER transcriptional activity (Halachmi et al., 1994; Hong et al., 1996; Horwitz et al., 1996; Joyeux et al., 1997). In addition, when these proteins are overexpressed in the cell, they reverse the ER-mediated squelching of transcription (Onate et al., 1995; Voegel et al., 1996). One of these proteins, steroid receptor coactivator-1 (SRC-1), initially identified as a PR coregulator, was determined to be a limiting factor needed for efficient transactivation by both PR and ER (Onate et al., 1995). The discovery of SRC-1 was followed by that of many other proteins, most of which impart some effect on ER transcriptional activity. These proteins were termed coactivators because they are not components of the basal transcription machinery and are limiting factors which are utilized only by ligand-bound receptor (Horwitz et al., 1996). However, most of these proteins are expressed ubiquitously and behave promiscuously with respect to receptors.

The majority of coactivators for ER and other steroid receptors may be general transcription factors and not specific targets for pharmacological intervention of the ER signaling pathway (Horwitz et al., 1996). Specific proteins, however, could have important consequences in ER-dependent cancers. For example, AIB1, a recently cloned SRC-1 family member that enhances estrogen-dependent transcription, is amplified in breast and ovarian cancer cell lines and tumor specimens (Anzick et al., 1997). This suggests that altered gene expression of this coactivator may offer a selective advantage for tumor growth.

THE ROLE OF LIGAND IN ER ACTION; IMPLICATIONS FOR ER PHARMACOLOGY

The classical model of ER action suggests that ligand acts as a switch to convert the receptor into a transcriptionally active form (Clark and Peck, 1979). Antiestrogens, synthetic compounds which oppose the action of estrogen, were once believed to function solely as competitive antagonists, inhibiting agonist access to the receptor. Recent discoveries in the ER field imply that this model is oversimplified and cannot account for the observed pharmacology of known ER agonists and antagonists (Katzenellenbogen *et al.*, 1996).

Much of the information in support of this hypothesis is derived from studies of the antiestrogen tamoxifen. Tamoxifen is used extensively as adjuvant chemotherapy in the treatment of estrogen-dependent breast cancer. Though usually considered an antiestrogen, tamoxifen was shown to function as an agonist in bone, not blocking the actions of estrogen in this organ system but rather mimicking them (Love *et al.*, 1992). This "paradox" was further substantiated when it was demonstrated that the use of tamoxifen in postmenopausal women attenuated cardiovascular risks, suggesting again that the compound functions as an agonist in some circumstances (Love *et al.*, 1991). These findings, incompatible with the definition of tamoxifen as a classical antagonist, support the hypothesis that this ligand alters ER in such a manner that the protein/ligand complex can be recognized differently in different cells (McDonnell *et al.*, 1995).

Properties of ER Ligands

One molecular explanation for the differential pharmacology of E2 and tamoxifen was provided upon biochemical analysis of ER structure. Using an assay to determine sensitivity to proteases, it was shown that both E2 and tamoxifen induce a conformational change within ER upon binding. However, the structure of the ER/tamoxifen complex was different from the ER/E2 complex (McDonnell *et al.*, 1995). More extensive protease digestion analysis has led to the classification of ER antagonists (Figure 3) into three mechanistically distinct groups (McDonnell *et al.*, 1995). Each unique conformation can alter the ability of ER to modulate gene transcription, perhaps due to the selective binding of coactivator proteins (Smith *et al.*, 1997).

The pure antiestrogens, represented by ICI 182,780, compose the first class of ER antagonists. The mechanism of action of this group of compounds is complex in that the conformation adopted by ER leads to receptor activation but does not permit efficient dimerization or DNA binding (Fawell *et al.*, 1990). In addition, this ligand induces turnover of ER by targeting the receptor to the lysosomes for degradation (Dauvois *et al.*, 1993). *In vivo*, pure antiestrogens do not exhibit ER agonist activity and oppose estradiol activity under all circumstances examined (Wakeling and Bowler, 1988). Recent *in vitro* studies, however, have shown that ICI 182,780 can function as an agonist in certain conditions (Paech *et al.*, 1997), implying that even "pure" antiestrogens may manifest ER agonist activity in some cell contexts. Additionally, there is now evidence which suggests that ICI 182,780 may actually act as an inverse agonist, decreasing the basal activity of wild-type ER (Willson *et al.*, 1997).

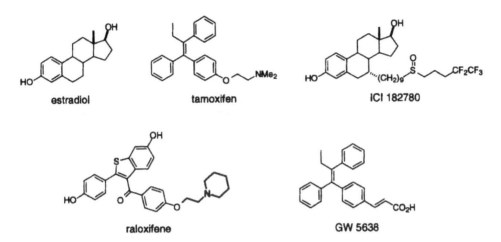

Figure 3 Chemical structures of different estrogen receptor ligands

Additional classes of antagonists, represented by the benzothiophene-derived raloxifene and the triphenylethylenes tamoxifen and GW5638, have more elaborate mechanisms of action. All of these compounds function as cell-specific agonists/antagonists, but each exhibits a unique biology. Because of their distinct activities, these compounds have been called SERMs, or Selective Estrogen Receptor Modulators. The differential ability of each type of SERM to stimulate the transcriptional activity of the receptor in various cell and promoter contexts reflects the physiological activity of these compounds. The partial agonist activity of tamoxifen, for instance, is likely responsible for its uterotrophic effects and may be the reason breast cancer cells "fail" tamoxifen treatment and begin to recognize it as an agonist (Kedar *et al.*, 1994). However, the cell-selective partial agonist activity of this drug may also be responsible for its beneficial effects, including prevention of ovariectomy-induced bone loss in animals and increased bone mineral density in postmenopausal breast cancer patients (Love *et al.*, 1992). Tamoxifen also promotes antiatherogenic effects (Saarto *et al.*, 1996), including a decrease in low-density lipoproteins and Lp(a) in a similar population of women.

Some of the complex biology of these SERMs can be explained by the fact that ER does not work the same way in all cells. The preferential influence of these compounds on AF1 or AF2 may explain the partial agonist activity which they exhibit in certain tissues. For example, tamoxifen functions as a partial agonist (relative to E2) in contexts where AF1 alone is required for ER transcriptional activity (Berry *et al.*, 1990; Tzukerman *et al.*, 1994). In contrast, tamoxifen manifests antagonist properties in tissues or contexts in which AF2 is required for activation. Raloxifene and GW5638 act as antiestrogens on wild-type ER and do not appear to manifest ER agonist activity through AF1 (McDonnell *et al.*, 1995; Willson *et al.*, 1997). It has been suggested that the differential expression of a subset of coactivators and corepressors may partially explain the agonist/antagonist activity of a mixed agonist such as tamoxifen (Smith *et al.*, 1997).

Raloxifene, currently being investigated in clinical trials as an agent for the treatment and prevention of osteoporosis (Black *et al.*, 1994), has less of a stimulatory effect on uterine tissue than tamoxifen (Sato *et al.*, 1996). The partial agonist effects of raloxifene are demonstrated by its ability to decrease serum cholesterol in postmenopausal women (Kauffman *et al.*, 1997). Recent studies have provided insight into a potential mechanism to explain raloxifene's protective effects on bone. It was observed that this compound, which functions as an antagonist in breast and uterine tissue, is able to prevent bone loss in an ovariectomized rat model, possibly by inducing the transcription of transforming growth factor-β3 (TGFβ3) (Yang *et al.*, 1996a). TGFβ3 is a key factor in bone formation, induction, and repair (Linkhart *et al.*, 1996). This isoform potently inhibits osteoclast differentiation in two models, substantiating its role in bone preservation (Yang *et al.*, 1996). A putative raloxifene response element found within the promoter region of TGFβ3 was proposed as a DNA sequence which was activated selectively by raloxifene and did not require the DBD of ER (Yang *et al.*, 1996a; Yang *et al.*, 1996b). However, this was later found to be insufficient to explain the regulation of this promoter (Yang *et al.*, 1997). It remains unclear how members of a wide spectrum of compounds are able to elicit the same effects in one tissue (bone) but opposing effects in others (uterus).

In Search of Improved SERMs

The observation that tamoxifen and other antiestrogens can manifest both agonist and antagonist activity suggests that the classical description of whether a compound is an agonist or an antagonist is obsolescent. The ability of the ligand to induce a distinct conformational change in the receptor alters the receptor's ability to interact with coactivator proteins and thus the general transcription machinery. This phenomenon allows mechanistically distinct compounds to operate through a single receptor type. The introduction of human ERβ (Mosselman *et al.*, 1996) has broadened this belief, suggesting that perhaps certain ligands can selectively activate one receptor subtype over another. However, the ligand affinity profiles of the two receptor subtypes on classical EREs remains indistinguishable (Kuiper *et al.*, 1997).

The ideal SERM would mimic the role of estrogen in its protective effects in bone and the cardiovascular system and oppose estrogen's proliferative actions in the breast and uterus. HRT efficacy and compliance could be vastly enhanced by a compound which could prevent hot flushes and mood swings associated with menopause while still providing adequate bone and cardiovascular protection. In addition, such a compound should lack undesirable proliferative activity on the endometrium and breast. A compound exhibiting this ideal profile has potential use in treating breast cancer patients who have become resistant to tamoxifen therapy. Currently, the pure antiestrogens (such as ICI 182,780) are the only ER modulators that are effective in the treatment of these cancers (Gottardis *et al.*, 1989). However, these compounds have been shown to deplete ER due to increased receptor turnover (Dauvois *et al.*, 1992). Despite this therapeutic potential, however, experiments in ovariectomized rats demonstrate that ICI 182,780 lacks the bone protective capabilities of tamoxifen (Gallagher *et al.*, 1993). Thus, there is an unmet medical need for new antiestrogens which can function as "pure" antagonists in the breast and uterus but agonists in the bone and cardiovascular system.

Novel compounds which exhibit these properties *in vitro* are now being evaluated for their potential in many clinical venues. One example, GW5638, was discovered in a search for improved tamoxifen analogs (Willson *et al.*, 1994). This compound lacks the uterine proliferative effects of tamoxifen and protects against bone loss in the ovariectomized rat model (Willson *et al.*, 1997). GW5638 exhibits a unique pharmacological profile, suggesting that the compound is mechanistically distinct from other types of known ER ligands (Willson *et al.*, 1997). Therefore, GW5638 may be an ideal candidate for the treatment of tamoxifen-failed tumors.

Ligand-Independent Activation of ER

The transcriptional activity of steroid hormone receptors is generally considered to be regulated by specific, high-affinity ligands. However, regulation is actually more complex in that activation can also occur in a ligand-independent manner (Aronica *et al.*, 1994; Chalbos *et al.*, 1994; Power *et al.*, 1991). One example is a phosphorylation-dependent pathway which affects ER through the neurotransmitter dopamine (DA). The actions of DA are mediated through G-protein/adenylyl cyclase interactions and involve the second messenger cyclic AMP. In CV-1 cells, derived from monkey kidney, DA is able to activate ER in the absence of E2 (Power *et al.*, 1991). This suggests a crucial role for a phosphorylation cascade and cross-talk between a variety of cell pathways, with ER as the site of convergence.

Another observation involving ligand-independent activation of ER is that the phosphorylation of Ser118 in the receptor is required for the full activity of AF1 (Ali *et al.*, 1993). Growth factor signaling pathways have been implicated in receptor activation, and this effect may be mediated in part by the Ras-MAPK (Mitogen-Activated Protein Kinase) cascade (Kato *et al.*, 1995). Further evidence supporting ER involvement is noted in the ability of the pure antiestrogen ICI 164,384 to block the response to Ras (Kato *et al.*, 1995). Okadaic acid, a phosphatase inhibitor, can also activate ER (Power *et al.*, 1991). Alterations in the regulation of ER phosphorylation may therefore be involved in the adverse effects of growth factors and oncogenes in breast cancer (Kato *et al.*, 1995).

APPROACHES TO DRUG DISCOVERY AND EVALUATION

Recent advances in both molecular and biochemical techniques have permitted more efficient synthesis and evaluation of ER modulators. Additionally, dissection of the ER signal transduction pathway has yielded new ideas regarding the manipulation of ER action. One of the most significant advances in this respect is the development of mechanism-based high-throughput screens.

Cis-trans Assays

Cell culture has provided a viable medium for the rapid analysis of potential ER ligands. The "cis-trans" assay (Figure 4) involves the transfection of a reporter plasmid and, if necessary, an ER-containing plasmid into mammalian cells (Evans, 1988). The reporter plasmid contains an ERE upstream of the luciferase gene. In the case of ER-positive cell lines, such as MCF-7, no transfection of receptor is necessary. The readout of reporter gene activity (luciferase) is a direct reflection of the transcriptional capability of the ligand-induced receptor.

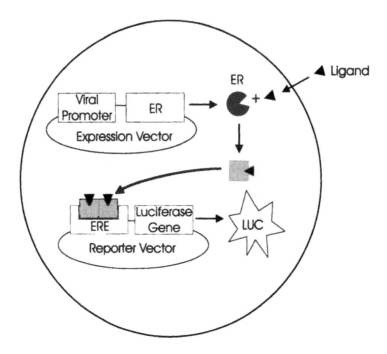

Figure 4   Characterization of estrogen receptor ligands using the cis-trans assay in mammalian cells. The assay involves the cotransfection of plasmids containing the estrogen receptor, driven by a constitutive promoter, and a reporter plasmid which contains an estrogen-responsive element. The latter is upstream of the luciferase gene and therefore the production of luciferase, which can be quantitated, is dependent upon the ability of the ligand to bring the receptor to the reporter element and activate transcription

This type of assay is also useful in the comparison of the relative abilities of antagonists to inhibit the agonist activity of estrogen or other partial agonists. By using constructs expressing specific receptor mutants, the relative AF1 or AF2 activity of the ligand-bound receptor can be assessed (Tzukerman *et al.*, 1994). For example, the removal of the AF1 portion of the receptor results in a protein which is still ligand-inducible and capable of binding DNA, allowing the activity of AF2 alone to be evaluated (Tzukerman *et al.*, 1994). Conversely, a three amino acid mutation in the AF2 domain diminishes its transactivational capability without affecting hormone binding (Danielian *et al.*, 1992). By comparing the ability of ligands to activate these mutants versus the wild-type ER, a "fingerprint" of compounds can be created (Willson *et al.*, 1997). Thus, ER modulators can be classified into different groups, each of which exhibits distinct biological activities.

Animal Models

High-throughput screening for ligands can be extremely useful, but a relevant physiological model is required to assess the validity of a putative pharmaco-therapeutic. Animal models are useful in the appraisal of compounds which show promise in cell culture or other *in vitro* assays. The ovariectomized rat model remains the standard for the evaluation of the effects of compounds on the bone or in the uterus. The ovariectomized animal experiences a loss in bone mineral density, which can be fully restored with estrogen and to a lesser extent with tamoxifen (Love *et al.*, 1992). Conversely, estrogen promotes uterotrophic activity in this model which can be inhibited by "pure" antiestrogens. Novel compounds may be evaluated similarly to establish improved efficacy or potency.

The most well-established animal model for breast cancer is the athymic nude mouse (Soule and McGrath, 1980). These mice, incapable of rejecting a tumor immunologically (Giovenella *et al.*, 1978), provide a good repository for human breast cancer cells. Thus, they are a valuable tool for studying the effects of various endocrine manipulations. This model offers the advantage over an animal carcinoma in that the malignancy is of human descent and retains many characteristics of the original tumor.

The ovariectomized athymic mouse is inoculated with the ER-positive, estrogen-dependent MCF-7 cell line. This type of breast cancer responds to and eventually fails anti-hormone treatment (develops resistance) (Gottardis and Jordan, 1988; Soule and McGrath, 1980). To control estrogen levels, sustained release E2 pellets are implanted in each animal. This model has been further characterized in studies designed to evaluate the effects of tamoxifen on tumor growth *in vivo* (Gottardis *et al.*, 1988a). Tamoxifen inhibits estradiol-stimulated growth but does not cause tumor growth when implanted alone (Gottardis *et al.*, 1988a; Gottardis *et al.*, 1988b), although resistance eventually occurs (Gottardis and Jordan, 1988). A pure antagonist, ICI 164,384, inhibits the growth of the tamoxifen-resistant tumor (Gottardis *et al.*, 1989). This work substantiates the use of the athymic mouse to determine anti-hormone efficacy (Gottardis *et al.*, 1988a), and this model provides an invaluable tool for assessing the therapeutic potential of novel breast cancer treatments.

FUTURE DIRECTIONS

ERβ

The recent cloning of a novel estrogen receptor isoform, ERβ (Kuiper *et al.*, 1996; Mosselman *et al.*, 1996) unveils the possibility that additional subtypes exist. Extensive research has yet to be completed for ERβ, but the recent ER knockout mouse (Lubahn *et al.*, 1993) phenotype indicates a physiologic role for this receptor. Specifically, while these mice have the expected reproductive deficiencies, they lack gross cardiovascular abnormalities anticipated with the loss of estrogen signaling (Iafrati *et al.*, 1997). These data imply that each receptor may regulate distinct sets of genes. Further evaluation of this isoform suggests that clinically useful subtype-specific ligands may be developed.

Crystal Structure

The recent discovery of the three-dimensional crystal structure of the hormone-binding domains for isoforms of the human retinoic acid (Renaud *et al.*, 1995) and retinoid-related (Bourguet *et al.*, 1995) receptors was a significant accomplishment in the receptor field. Because the HBD spans nearly half of ER (Evans, 1988), its structure is expected to be quite complex in order to accommodate the variety of small ligands known to bind the protein. Crystal structures for both receptor isoforms, particularly if they each regulate a unique subset of genes, will facilitate rational drug design at these important hormone targets. The crystal structure of the HBD of ERα in the presence of both agonist (E2) and antagonist (raloxifene) has now been reported (Brzozowski *et al.*, 1997). The authors suggest that the antagonist causes a change in the protein which prevents a functional AF-2 domain from interacting with coactivators. However, because the N- and C-termini of ER are thought to interact (Kraus *et al.*, 1995), the full length crystal structure is necessary for a more complete understanding of the receptor.

Target Genes

There is currently a paucity of genes known to be directly regulated by estrogen through a classical response element. These include PR (Savouret *et al.*, 1991), c-fos (Hyder *et al.*, 1992), pS2 (Berry *et al.*, 1989), and oxytocin (Richard and Zingg, 1990). These genes serve as markers for estrogen and antiestrogen action in a variety of tissues. The continued identification of genes in specific tissues will allow for rapid classification of novel SERMs.

FINAL THOUGHTS

The importance of ER in regulating female endocrine function, in addition to its implications in a multitude of disease states, provides impetus for the creation of specific, selective ligands for this protein. While several selective modulators have been designed, further insights into the mechanism of action of this receptor will allow for the production of therapeutics useful for treating and preventing osteoporosis, cardiovascular disease, and breast cancer. Continued technological advances in the ER field permit a tethering of the fields of descriptive biology and drug discovery.

REFERENCES

Ali, S., Metzger, D., Bornert, J.-M. and Chambon, P. (1993) Modulation of transcriptional activation by ligand-dependent phosphorylation of the human oestrogen receptor A/B region. *EMBO J.*, **12**, 1153–1160.

Anzick, S.L., Kononen, J., Walker, R.L., Azorsa, D.O., Tanner, M.M., Guan, X.-Y., Sauter, G., Kallioniemi, O.-P., Trent, J.M. and Meltzer, P.S. (1997) AIB1, a Steroid Receptor Coactivator Amplified in Breast and Ovarian Cancer. *Science*, **277**, 965–968.

Aronica, S.M., Kraus, W.L. and Katzenellenbogen, B.S. (1994) Estrogen action via the cAMP signaling pathway: Stimulation of adenylate cyclase and cAMP-regulated gene transcription. *Proc. Natl. Acad. Sci. USA*, **91**, 8517–8521.

Berry, M., Metzger, D. and Chambon, P. (1990) Role of the two activating domains of the oestrogen receptor in the cell-type and promoter-context dependent agonistic activity of the anti-oestrogen 4-hydroxytamoxifen. *EMBO J.*, **9**, 2811–2818.

Berry, M., Nunez, A.-M. and Chambon, P. (1989) Estrogen-responsive element of the human pS2 gene is an imperfectly palindromic sequence. *Proc. Natl. Acad. Sci. USA*, **86**, 1218–1222.

Black, L.J., Sato, M., Rowley, E.R., Magee, D.E., Bekele, A., Williams, D.C., Cullinan, G.J., Bendele, R., Kauffman, R.F., Bensch, W.R., Frolik, C.A., Termine, J.D. and Bryant, H.U. (1994) Raloxifene (LY139481 HCl) Prevents Bone Loss and Reduces Serum Cholesterol without Causing Uterine Hypertrophy in Ovariectomized Rats. *J. Clin. Invest.*, **93**, 63–69.

Bourguet, W., Ruff, M., Chambon, P., Gronemeyer, H. and Moras, D. (1995) Crystal structure of the ligand-binding domain of the human nuclear receptor RXR-α. *Nature*, **375**, 377–382.

Brzozowski, A.M., Pike, A.C.W., Dauter, Z., Hubbard, R.E., Bonn, T., Engstrom, O., Ohman, L., Greene, G.L., Gustafsson, J.-A. and Carlquist, M. (1997) Molecular basis of agonism and antagonism in the oestrogen receptor. *Nature*, **389**, 753–758.

Carson-Jurica, M.A., Schrader, W.T. and O'Malley, B.W. (1990) Steroid Receptor Family: Structure and Functions. *Endocr. Rev.*, **11**, 201–220.

Chalbos, D., Philips, A. and Rochefort, H. (1994) Genomic cross-talk between the estrogen receptor and growth factor regulatory pathways in estrogen target tissues. *Semin. Cancer Biol.*, **5**, 361–368.

Clark, J.H. and Peck, Jr., E.J. (1979) *Female Sex Steroids*. Berlin: Springer-Verlag.

Danielian, P.S., White, R., Lees, J.A. and Parker, M.G. (1992) Identification of a conserved region required for hormone dependent transcriptional activation by steroid hormone receptors. *EMBO J.*, **11**, 1025–1033.

Dauvois, S., Danielian, P.S., White, R. and Parker, M.G. (1992) Antiestrogen ICI 164,384 reduces cellular estrogen receptor content by increasing its turnover. *Proc. Natl. Acad. Sci. USA*, **89**, 4037–4041.

Dauvois, S., White, R. and Parker, M.G. (1993) The antiestrogen ICI 182,780 disrupts estrogen receptor nucleocytoplasmic shuttling. *J. Cell Sci.*, **106**, 1377–1388.

Dreicer, R. and Wilding, G. (1992) Steroid Hormone Agonists and Antagonists in the Treatment of Cancer. *Cancer Invest.*, **10**, 27–41.

Edman, C.D. (1983) The Climateric. In *The Menopause*, edited by H.J. Buchsbaum, pp. 23–33. New York: Springer-Verlag.

Evans, R.M. (1988) The Steroid and Thyroid Hormone Receptor Superfamily. *Science*, **240**, 889–895.

Fawell, S.E., White, R., Hoare, S., Sydenham, M., Page, M. and Parker, M.G. (1990) Inhibition of estrogen receptor-DNA binding by the "pure" antiestrogen ICI 164,384 appears to be mediated by impaired receptor dimerization. *Proc. Natl. Acad. Sci. USA*, **87**, 6883–6887.

Gallagher, A., Chambers, T.J. and Tobias, J.H. (1993) The Estrogen Antagonist ICI 182,780 Reduces Cancellous Bone Volume in Female Rats. *Endocrinology*, **133**, 2787–2791.

Giovenella, B.P., Stehlin, J.S., Williams, Jr., L.J., Lee, S.-S. and Shepard, R.C. (1978) Heterotransplantation of Human Cancers into Nude Mice. *Cancer*, **42**, 2269–2281.

Gottardis, M.M., Jiang, S.-Y., Jeng, M.-H. and Jordan, V.C. (1989) Inhibition of Tamoxifen-stimulated Growth of an MCF-7 Tumor Variant in Athymic Mice by Novel Steroidal Antiestrogens. *Cancer Res.*, **49**, 4090–4093.

Gottardis, M.M. and Jordan, V.C. (1988) Development of Tamoxifen-stimulated Growth of MCF-7 Tumors in Athymic Mice after Long-Term Antiestrogen Administration. *Cancer Res.*, **48**, 5183–5187.

Gottardis, M.M., Robinson, S.P. and Jordan, V.C. (1988a) Estradiol-Stimulated Growth of MCF-7 Tumors Implanted in Athymic Mice: A Model to Study the Tumoristatic Action of Tamoxifen. *J. Steroid Biochem.*, **30**, 311–314.

Gottardis, M.M., Robinson, S.P., Satyaswaroop, P.G. and Jordan, V.C. (1988b) Contrasting Actions of Tamoxifen on Endometrial and Breast Tumor Growth in the Athymic Mouse. *Cancer Res.*, **48**, 812–815.

Green, S., Walter, P., Kumar, V., Krust, A., Bornert, J.M., Argos, P. and Chambon, P. (1986) Human oestrogen receptor cDNA: sequence, expression and homology to v-erb-A. *Nature*, **320**, 134–139.

Greene, G.L., Gilna, P., Waterfield, M., Baker, A., Hort, Y. and Shine, J. (1986) Sequence and Expression of Human Estrogen Receptor Complementary DNA. *Science*, **231**, 1150–1154.

Halachmi, S., Marden, E., Martin, G., MacKay, H., Abbondanza, C. and Brown, M. (1994) Estrogen Receptor-Associated Proteins: Possible Mediators of Hormone-Induced Transcription. *Science*, **264**, 1455–1458.

Hong, H., Kohli, K., Trivedi, A., Johnson, D.L. and Stallcup, M.R. (1996) GRIP1, a novel mouse protein that serves as a transcriptional coactivator in yeast for the hormone binding domains of steroid receptors. *Proc. Natl. Acad. Sci. USA*, **93**, 4948–4952.

Horowitz, M.C. (1993) Cytokines and Estrogen in Bone: Anti-Osteoporotic Effects. *Science*, **260**, 626–627.

Horwitz, K.B., Jackson, T.A., Bain, D.L., Richer, J.K., Takimoto, G.S. and Tung, L. (1996) Nuclear Receptor Coactivators and Corepressors. *Mol. Endocrinol.*, **10**, 1167–1177.

352 C.E. CONNOR and D.P. McDONNELL

Hyder, S.M., Stancel, G.M., Nawaz, Z., McDonnell, D.P. and Loose-Mitchell, D.S. (1992) Identification of an Estrogen Response Element in the 3'-Flanking Region of the Murine c-fos Protooncogene. *J. Biol. Chem.*, **267**, 18047–18054.

Iafrati, M.D., Karas, R.H., Aronovitz, M., Kim, S., Sullivan, Jr., T.R., Lubahn, D.B., O'Donnell, Jr., T.F., Korach, K.S. and Mendelsohn, M.E. (1997) Estrogen inhibits the vascular injury response in estrogen receptor α-deficient mice. *Nat. Med.*, **3**, 545–548.

Ing, N.H., Beekman, J.M., Tsai, S.Y., Tsai, M.-J. and O'Malley, B.W. (1992) Members of the Steroid Hormone Receptor Superfamily Interact with TFIIB (S300-II). *J. Biol. Chem.*, **267**, 17617–17623.

Joyeux, A., Cavailles, V., Balaguer, P. and Nicolas, J.C. (1997) RIP 140 Enhances Nuclear Receptor-Dependent Transcription *in Vivo* in Yeast. *Mol. Endocrinol.*, **11**, 193–202.

Kato, S., Endoh, H., Masuhiro, Y., Kitamoto, T., Uchiyama, S., Sasaki, H., Masushige, S., Gotoh, Y., Nishida, E., Kawashima, H., Metzger, D. and Chambon, P. (1995) Activation of the Estrogen Receptor Through Phosphorylation by Mitogen-Activated Protein Kinase. *Science*, **270**, 1491–1494.

Katzenellenbogen, J.A., O'Malley, B.W. and Katzenellenbogen, B.S. (1996) Tripartite Steroid Hormone Receptor Pharmacology: Interaction with Multiple Effector Sites as a Basis for the Cell- and Promoter-Specific Action of These Hormones. *Mol. Endocrinol.*, **10**, 119–131.

Kauffman, R.F., Bensch, W.R., Roudebush, R.E., Cole, H.W., Bean, J.S., Phillips, D.L., Monroe, A., Cullinan, G.J., Glasebrook, A.L. and Bryant, H.U. (1997) Hypocholesterolemic Activity of Raloxifene (LY139481): Pharmacological Characterization as a Selective Estrogen Receptor Modulator. *J. Pharmacol. Exp. Ther.*, **280**, 146–153.

Kedar, R.P., Bourne, T.H., Powles, T.J., Collins, W.P., Ashley, S.E., Cosgrove, D.O. and Campbell, S. (1994) Effects of tamoxifen on uterus and ovaries of postmenopausal women in a randomised breast cancer prevention trial. *Lancet*, **343**, 1318–1321.

Klein-Hitpass, L., Schorpp, M., Wagner, U. and Ryffel, G.U. (1986) An Estrogen-Responsive Element Derived from the 5' Flanking Region of the Xenopus Vitellogenin A2 Gene Functions in Transfected Human Cells. *Cell*, **46**, 1053–1061.

Kraus, W.L., McInerney, E.M. and Katzenellenbogen, B.S. (1995) Ligand-dependent, transcriptionally productive association of the amino- and carboxyl-terminal regions of a steroid hormone nuclear receptor. *Proc. Natl. Acad. Sci. USA*, **92**, 12314–12318.

Krust, A., Green, S., Argos, P., Kumar, V., Walter, P., Bornert, J.M. and Chambon, P. (1986) The chicken oestrogen receptor sequence: homology with v-erbA and the human oestrogen and glucocorticoid receptors. *EMBO J.*, **5**, 891–897.

Kuiper, G.G.J.M., Carlsson, B., Grandien, K., Enmark, E., Haggblad, J., Nilsson, S. and Gustafsson, J.-A. (1997) Comparison of the Ligand Binding Specificity and Transcript Tissue Distribution of Estrogen Receptors α and β. *Endocrinology*, **138**, 863–870.

Kuiper, G.G.J.M., Enmark, E., Pelto-Huikko, M., Nilsson, S. and Gustafsson, J.-A. (1996) Cloning of a novel estrogen receptor expressed in rat prostate and ovary. *Proc. Natl. Acad. Sci. USA*, **93**, 5925–5930.

Kumar, V. and Chambon, P. (1988) The Estrogen Receptor Binds Tightly to Its Responsive Element as a Ligand-Induced Homodimer. *Cell*, **55**, 145–156.

Kumar, V., Green, S., Stack, G., Beery, M., Jin, J.-R. and Chambon, P. (1987) Functional Domains of the Human Estrogen Receptor. *Cell*, **51**, 941–951.

Linkhart, T.A., Mohan, S. and Baylink, D.J. (1996) Growth Factors for Bone Growth and Repair: IGF, TGFβ and BMP. *Bone*, **19**, 1S-12S.

Love, R.R., Mazess, R.B., Barden, H.S., Epstein, S., Newcomb, P.A., Jordan, V.C., Carbone, P.P. and DeMets, D.L. (1992) Effects of tamoxifen on bone mineral density in postmenopausal women with breast cancer. *New Engl. J. Med.*, **326**, 852–856.

Love, R.R., Wiebe, D.A., Newcomb, P.A., Cameron, L., Leventhal, H., Jordan, V.C., Feyzi, J. and DeMets, D.L. (1991) Effects of tamoxifen on cardiovascular risk factors in postmenopausal women. *Ann. Intern. Med.*, **115**, 860–864.

Lubahn, D.B., Moyer, J.S., Golding, T.S., Couse, J.F., Korach, K.S. and Smithies, O. (1993) Alteration of reproductive function but not prenatal sexual development after insertional disruption of the mouse estrogen receptor gene. *Proc. Natl. Acad. Sci. USA*, **90**, 11162–11166.

Martinez, E., Givel, F. and Wahli, W. (1987) The estrogen-responsive element as an inducible enhancer: DNA sequence requirements and conversion to a glucocorticoid-responsive element. *EMBO J.*, **6**, 3719–3727.

McDonnell, D.P., Clemm, D.L., Hermann, T., Goldman, M.E. and Pike, J.W. (1995) Analysis of Estrogen Receptor Function *in Vitro* Reveals Three Distinct Classes of Antiestrogens. *Mol. Endocrinol.*, **9**, 659–669.

McDonnell, D.P., Clemm, D.L. and Imhof, M.O. (1994) Definition of the cellular mechanisms which distinguish between hormone and antihormone activated steroid receptors. *Semin. Cancer Biol.*, 5, 503.1–503.10.

McDonnell, D.P., Clevenger, B., Dana, S., Santiso-Mere, D., Tzukerman, M.T. and Gleeson, M.A.G. (1993) The Mechanism of Action of Steroid Hormones: A New Twist to an Old Tale. *J. Clin. Pharmacol.*, 33, 1165–1172.

McDonnell, D.P., Mangelsdorf, D.J., Pike, J.W., Haussler, M.R. and O'Malley, B.W. (1987) Molecular Cloning of Complementary DNA Encoding the Avian Receptor for Vitamin D. *Science*, 235, 1214–1217.

Metzger, D., White, J.H. and Chambon, P. (1988) The human oestrogen receptor functions in yeast. *Nature*, 334, 31–36.

Meyer, M.-E., Gronemeyer, H., Turcotte, B., Bocquel, M.-T., Tasset, D. and Chambon, P. (1989) Steroid Hormone Receptors Compete for Factors That Mediate Their Enhancer Function. *Cell*, 57, 433–442.

Mosselman, S., Polman, J. and Dijkema, R. (1996) ERβ: identification and characterization of a novel human estrogen receptor. *FEBS Lett.*, 392, 49–53.

Murabito, J.M. (1995) Women and Cardiovascular Disease: Contributions from the Framingham Heart Study. *J. Am. Med. Womens Assoc.*, 50, 35–39, 55.

Norris, J.D., Fan, D., Kerner, S.A. and McDonnell, D.P. (1997) Identification of a Third Autonomous Activation Domain within the Human Estrogen Receptor. *Mol. Endocrinol.*, 11, 747–754.

Norris, J.D., Fan, D., Wagner, B.L. and McDonnell, D.P. (1996) Identification of the Sequences within the Human Complement 3 Promoter Required for Estrogen Responsiveness Provides Insight into the Mechanism of Tamoxifen Mixed Agonist Activity. *Mol. Endocrinol.*, 10, 1605–1616.

Onate, S.A., Tsai, S.Y., Tsai, M.-J. and O'Malley, B.W. (1995) Sequence and Characterization of a Coactivator for the Steroid Hormone Receptor Superfamily. *Science*, 270, 1354–1357.

Orimo, A., Inoue, S., Ikegami, A., Hosoi, T., Akishita, M., Ouchi, Y., Muramatsu, M. and Orimo, H. (1993) Vascular Smooth Muscle Cells as Target for Estrogen. *Biochem. Biophys. Res. Commun.*, 195, 730–736.

Paech, K., Webb, P., Kuiper, G.G.J.M., Nilsson, S., Gustafsson, J.-A., Kushner, P.J. and Scanlan, T.S. (1997) Differential Ligand Activation of Estrogen Receptors ERα and ERβ at AP1 Sites. *Science*, 277, 1508–1510.

Pham, T.A., Hwung, Y.-P., McDonnell, D.P. and O'Malley, B.W. (1991) Transactivation Functions Facilitate the Disruption of Chromatin Structure by Estrogen Receptor Derivatives *in Vivo. J. Biol. Chem.*, 266, 18179–18187.

Pierrat, B., Heery, D.M., Chambon, P. and Losson, R. (1994) A highly conserved region in the hormone-binding domain of the human estrogen receptor functions as an efficient transactivation domain in yeast. *Gene*, 143, 193–200.

Power, R.F., Mani, S.K., Codina, J., Conneely, O.M. and O'Malley, B.W. (1991) Dopaminergic and Ligand-Independent Activation of Steroid Hormone Receptors. *Science*, 254, 1636–1639.

Rao, B.R. and Slotman, B.J. (1991) Endocrine Factors in Common Epithelial Ovarian Cancer. *Endocr. Rev.*, 12, 14–26.

Renaud, J.-P., Rochel, N., Ruff, M., Vivat, V., Chambon, P., Gronemeyer, H. and Moras, D. (1995) Crystal structure of the RAR-γ ligand-binding domain bound to all-*trans* retinoic acid. *Nature*, 378, 681–689.

Richard, S. and Zingg, H.H. (1990) The Human Oxytocin Gene Promoter Is Regulated by Estrogens. *J. Biol. Chem.*, 265, 6098–6103.

Saarto, T., Blomqvist, C., Ehnholm, C., Taskinen, M.-R. and Elomaa, I. (1996) Antiatherogenic Effects of Adjuvant Antiestrogens: A Randomized Trial Comparing the Effects of Tamoxifen and Toremifene on Plasma Lipid Levels in Postmenopausal Women With Node-Positive Breast Cancer. *J. Clin. Oncol.*, 14, 429–433.

Sato, M., Rippy, M.K. and Bryant, H.U. (1996) Raloxifene, tamoxifen, nafoxidine, or estrogen effects on reproductive and nonreproductive tissues in ovariectomized rats. *FASEB J.*, 10, 905–912.

Savouret, J.F., Bailly, A., Misrahi, M., Rauch, C., Redeuilh, G., Chauchereau, A. and Milgrom, E. (1991) Characterization of the hormone responsive element involved in the regulation of the progesterone receptor gene. *EMBO J.*, 10, 1875–1883.

Shughrue, P.J., Komm, B. and Merchenthaler, I. (1996) The distribution of estrogen receptor-β mRNA in the rat hypothalamus. *Steroids*, 61, 678–681.

Smith, C.L., Nawaz, Z. and O'Malley, B.W. (1997) Coactivator and Corepressor Regulation of the Agonist/Antagonist Activity of the Mixed Antiestrogen, 4-Hydroxytamoxifen. *Mol. Endocrinol.*, 11, 657–666.

Smith, D.F. and Toft, D.O. (1993) Steroid Receptors and Their Associated Proteins. *Mol. Endocrinol.*, **7**, 4–11.

Soule, H.D. and McGrath, C.M. (1980) Estrogen Responsive Proliferation of Clonal Human Breast Carcinoma Cells in Athymic Mice. *Cancer Lett.*, **10**, 177–189.

Stanford, J.L., Weiss, N.S., Voigt, L.F., Daling, J.R., Habel, L.A. and Rossing, M.A. (1995) Combined Estrogen and Progestin Hormone Replacement Therapy in Relation to Risk of Breast Cancer in Middle-aged Women. *JAMA*, **274**, 137–142.

Sunderland, M.C. and Osborne, C.K. (1991) Tamoxifen in Premenopausal Patients With Metastatic Breast Cancer: A Review. *J. Clin. Oncol.*, **9**, 1283–1297.

Sutherland, R.L., Watts, C.K.W. and Clarke, C.L. (1988) Oestrogen actions. In *Hormones and their Actions Part I*, edited by B.A. Cooke, R.J.B. King and H.J. van der Molen, pp. 197–215. Amsterdam: Elsevier.

Tang, M.-Y., Jacobs, D., Stern, Y., Marder, K., Schofield, P., Gurland, B., Andrews, H. and Mayeux, R. (1996) Effect of oestrogen during menopause on risk and age at onset of Alzheimer's disease. *Lancet*, **348**, 429–432.

Tora, L., White, J., Brou, C., Tasset, D., Webster, N., Scheer, E. and Chambon, P. (1989) The Human Estrogen Receptor Has Two Independent Nonacidic Transcriptional Activation Functions. *Cell*, **59**, 477–487.

Tzukerman, M.T., Esty, A., Santiso-Mere, D., Danielian, P., Parker, M.G., Stein, R.B., Pike, J.W. and McDonnell, D.P. (1994) Human Estrogen Receptor Transactivational Capacity is Determined by both Cellular and Promoter Context and Mediated by Two Functionally Distinct Intramolecular Regions. *Mol. Endocrinol.*, **8**, 21–30.

Voegel, J.J., Heine, M.J.S., Zechel, C., Chambon, P. and Gronemeyer, H. (1996) TIF2, a 160 kDa transcriptional mediator for the ligand-dependent activation function AF-2 of nuclear receptors. *EMBO J*, **15**, 3667–3675.

Wakeling, A.E. and Bowler, J. (1988) Novel Antioestrogens Without Partial Agonist Activity. *J. Steroid Biochem.*, **31**, 645–653.

Wickelgren, I. (1997) Estrogen Stakes Claim to Cognition. *Science*, **276**, 675–678.

Willson, T.M., Henke, B.R., Momtahen, T.M., Charifson, P.S., Batchelor, K.W., Lubahn, D.B., Moore, L.B., Oliver, B.B., Sauls, H.R., Triantafillou, J.A., Wolfe, S.G. and Baer, P.G. (1994) 3-[4-(1,2-Diphenylbut-1-enyl)phenyl]acrylic Acid: A Non-Steroidal Estrogen with Functional Selectivity for Bone over Uterus in Rats. *J. Med. Chem.*, **37**, 1550–1552.

Willson, T.M., Norris, J.D., Wagner, B.L., Asplin, I., Baer, P., Brown, H.R., Jones, S.A., Henke, B., Sauls, H., Wolfe, S., Morris, D.C. and McDonnell, D.P. (1997) Dissection of the Molecular Mechanism of Action of GW5638, a Novel Estrogen Receptor Ligand, Provides Insight into the Role of Estrogen Receptor in Bone. *Endocrinology*, **138**, 3901–3911.

Yang, N.N., Bryant, H.U., Hardikar, S., Sato, M., Galvin, R.J.S., Glasebrook, A.L. and Termine, J.D. (1996) Estrogen and Raloxifene Stimulate Transforming Growth Factor-β3 Gene Expression in Rat Bone: A Potential Mechanism for Estrogen- or Raloxifene-Mediated Bone Maintenance. *Endocrinology*, **137**, 2075–2084.

Yang, N.N., Venugopalan, M., Hardikar, S. and Glasebrook, A. (1997) Correction: Raloxifene Response Needs More Than an Element. *Science*, **275**, 1249.

Yang, N.N., Venugopalan, M., Hardikar, S. and Glasebrook, A. (1996) Identification of an Estrogen Response Element Activated by Metabolites of 17β-Estradiol and Raloxifene. *Science*, **273**, 1222–1225.

21. MOLECULAR BIOSCREENING IN ONCOLOGY

GIULIO DRAETTA[1] and MICHAEL BOISCLAIR[2]

[1]Department of Experimental Oncology, European Institute of Oncology,
Via Ripamonti 435, 20141 Milan, Italy and [2]Mitotix, Inc.,
One Kendall Square, Building 600, Cambridge, MA 02135, USA

MECHANISM-BASED DRUG DISCOVERY

Modern drug discovery, the process aiming at the generation of novel drugs, starts with the identification of a given molecular target, e.g. a DNA, RNA, or protein molecule within the human body or within a human pathogen that will then be targeted for the identification of "interfering entities", usually small organic molecules. The identification of compounds that show potent and selective activity against the chosen molecular targets requires the combined expertise of biochemists, pharmacologists and medicinal chemists who will work together to improve the ability of said compounds to affect that target function both *in vitro* and most importantly *in vivo*.

Many drugs still used in clinical practice have been discovered through empirical approaches which tested the ability of complex mixtures derived from natural sources (e.g. plant extracts) to cure human diseases. Their biochemical mechanism of action was often elucidated many years after they had been utilized as pharmaceutical agents. This is the case for example of aspirin: already in ancient Egypt extracts of willow bark were used to treat inflammation, yet acetylsalicylic acid was synthesized much later towards the end of the nineteenth century. The elucidation of its mechanism of action as a cyclooxygenase inhibitor had to wait until much later in this century.

The detailed knowledge we now have of the large number of biochemical reactions that control the proper functioning of a human body and of their alterations in disease has allowed the development of molecular pharmacology. Through a detailed knowledge of a particular molecular cascade it is nowadays possible to identify specific inhibitors of a particular biochemical reaction. Together with medicinal chemistry this allows the tailoring of an inhibitor to that particular target. Once an inhibitor is identified the knowledge of its biochemical mechanism of action and of any possible other effects on a multitude of *in vitro* and *in vivo* assays is of paramount importance for the drug discovery project.

The cancer agents currently utilized in therapy have largely been identified through studies in which compounds were tested for efficacy in cells or whole model organisms directly. Only recently the identification of compounds which target selected molecular targets has been attempted. The compounds identified for their ability to block cell proliferation *in vitro* or to control tumor growth in animal models turned out in large part to be potent albeit non specific inhibitors of cell proliferation. In disease areas other than Oncology, the identification of powerful and highly selective inhibitors of biochemical pathways that have then shown incredible selectivity *in vivo* has been achieved with great success. A notable example

Table 1 Steps in drug discovery

1	Target identification	
	1.1	Genetics: identification of a gene based on its ability to affect cell function
	1.2	Biochemistry: identification of a molecule showing activity in a given assay
	1.3	Bioinformatics: screening for homologues in DNA/protein databases
2	Target Validation	
	2.1	Biology: what are the cellular effects of altering a molecule's function
	2.2	Biochemistry: how does a molecule work *in vitro*
3	Screening	
	3.1	Primary: identify and rank inhibitors based on their potency
	3.2	Secondary: assess inhibitor specificity and ability to work on intact cells
	3.3	*In vivo* testing: assess compound ability to show efficacy in animal models

could be offered by drugs that act by inhibiting specific membrane receptors or those that block critical enzyme reactions such as the ones controlling cholesterol biosynthesis.

Molecular bioscreening is a central part of the general strategy for drug discovery. Through this process a large set of quantitative measurements of the molecular interactions of a given compound, or a family of them, with its targets can be generated and evaluated.

From the initial studies made to identify and to validate a target (i.e. to offer a rationale for inhibiting a specific set of molecular reactions), the drug discovery process continues with the set-up of specific molecular assays that will allow to screen for the identification of specific inhibitors. Once inhibitors are identified in the screen and a choice is made on which to focus upon, an iterative process starts that requires a close collaboration between organic chemists and scientists with expertise in screening to generate compounds that are powerful and specific for the selected target(s). The process continues with the testing in both cellular and animal models, for the ability of the identified compounds to cross cell membranes and to reach their target tissues at the appropriate concentration and for the required period of time to exert their effects. The steps involved in preclinical drug discovery research are listed in Table 1.

In this review article we will attempt to describe the criteria used to select a cancer target for bioscreening and to implement screening efforts. In Oncology, it is increasingly clear that multiple alterations are responsible for the occurrence of the disease. In solid tumors, through fluorescence in situ hybridization and comparative genome hybridization it has been found that a large fraction of the DNA in a given patient's tumor can be rearranged and functionally altered. Furthermore, different genetic alterations can be found within individual tumors as cells grow and metastasize. It becomes essential therefore to establish a number of criteria to be followed for the selection of a given molecular target for a drug discovery effort. These criteria could include the demonstration of the frequency of a molecular alteration in a given cancer patient population, the correlation between the presence of such alteration and the patient's prognosis, the time of occurrence of the alteration

Table 2 Validating a potential cancer target

1	Biological criteria	
	1.1	Defined elements in a signal cascade
	1.2	Proven to be required for function *in vivo*
	1.3	Availability of animal model data
	1.4	Proven role in disease maintenance (frequency, correlation with prognosis, time of appearance during tumor progression)
2	Biochemical criteria	
	2.1	Enzymes, receptors
	2.2	Involvement of small molecule (substrates, ligands, allosteric regulators)
	2.3	Information on three dimensional structure (X-ray crystallography, nuclear magnetic resonance)

during tumor progression. In addition to these criteria any experimental evidence able to demonstrate that tumor cell growth can be blocked by affecting the target molecule's function will be a necessary point for decision making. The Ras small GTP binding protein and cyclin D1 CDK4 protein kinase are good examples of molecules that have satisfied all of the above criteria and are now being targeted for inhibitor discovery in a number of pharmaceutical companies.

A list of criteria that should help to select the "ideal" drug target is presented in Table 2. These are divided into two subgroups. Priority of course is given to the identification of a potential target in the context of a specific disease. Once the listed "biological" criteria are satisfied though, it is also important to consider the fact that traditionally certain molecular reactions have been easier to inhibit than others. For example, it is easier to develop an enzyme inhibitor than to block the interaction between two proteins. This is partly because enzymes have defined active sites and mechanism of action, the knowledge of which will help to design specific inhibitors. Within the enzyme category certain proteases and protein kinases have also been easier to inhibit compared to other enzymes. In general, the presence of a binding site for a small co-substrate or allosteric regulator can also help the process. Furthermore, the availability of chemical inhibitors of certain enzyme classes allows their immediate testing for activity against the new target identified. This could help to "jump start" the lead identification process (see below). The knowledge of the pathway within which the biochemical reaction one wants to inhibit lies, can often come to help: it is at least in principle possible to choose to try to inhibit a reaction downstream or upstream of a given target in the pathway if this increases the chance of identifying specific inhibitors.

New technological improvements have recently been introduced to accelerate drug discovery. They span from the availability of huge databases of cDNA sequences created as a result of large scale genome sequencing projects, to techniques that allow the rapid identification of DNA rearrangements (comparative genome hybridization) or alterations in mRNA expression (cDNA and oligonucleotide microarrays, RDA, SAGE etc.). Functional genomics, using large scale antisense or ribozyme expression, or genetic complementation approaches with cDNA expression libraries, aims at accelerating target validation by affecting physiological functions in human cells. Support to drug discovery is also being given by continued efforts in screening miniaturization and automation and in combinatorial chemistry (see below).

CHOOSING AN APPROPRIATE MOLECULAR TARGET

What processes can be targeted to block the growth of a solid tumor? Broadly speaking they are: tumor cell growth (increased proliferation and failure to enter apoptosis), tumor angiogenesis, tumor invasion and metastasis. Presently, all of them are being targeted for drug discovery. Twenty years after the discovery of the first oncogenes, scientists have reached the conclusion that multiple genetic defects underlie tumor growth. The need for validating a target is of paramount importance to demonstrate which of the many defects identified is necessary for tumor cell survival and which instead is secondary to the event and has no bearing on its maintenance. Most often these changes affect selected pathways like the ones controlling the G1/S transition of the cell division cycle, or the p53 checkpoint control cascade.

The identification of genetic alterations in human cancers has received a substantial acceleration during the past five to ten years. Through the study of tumor prone kindred and the use of positional cloning techniques, alterations directly linked to several inherited cancer predisposition patterns have been identified. As of somatic mutations, recent studies using techniques such as comparative genome hybridization and representational difference analysis (RDA), have highlighted the fact that as much as 20 to 30 percent of the human genome can be found rearranged in solid tumors. As a result of large scale genome mapping and sequencing efforts, the identity of more and more of these gene sequences will be known. From this the need of trying to assess by biochemical and genetic means the role played by each of these alterations in the generation of the tumor phenotype and most importantly in sustaining tumor cell growth and survival, prior to developing therapeutic agents. After the initial burst given by the identification of novel cDNA sequences though genomic projects, it has become clear that in the absence of biological validation no potential target would become of use. At the molecular level, a pathway can be affected by distinct events all having the same final effect. One example of this could be the inactivation of p53 which can occur by gene mutation, or by hyper degradation of the protein as a consequence of its interaction with the E6 protein of human papilloma virus or of its interaction with the Mdm2 protein. It is evident that while blocking p53 degradation could have an effect on tumors carrying HPV infection, no effect would be seen in conditions in which p53 is inactivated by genetic mutations.

How to Validate a Target

The term "biological validation" is used in the pharmaceutical industry to mean the process through which the role that a putative target molecule has in controlling a given cellular process can be assessed. Before undergoing a drug discovery program centered around a given target molecule the logical first step is to have actually collected sufficient data form either the literature or *ad hoc* experiments to be able to predict whether a compound that affects that molecule's function will likely have any effects *in vivo*. Of course early during the process and until a specific inhibitor is identified it will be difficult to translate an indirect observation with what can happen when an inhibitor is administered to a living organism or even added to a cell.

The following paragraphs will list a number of approaches that are utilized for target validation, starting from a demonstration of a deregulation of its expression to genetics experiments demonstrating its function *in vivo*. The biological validation of a target, or a family of targets, will have to involve techniques that can reasonably tell what the consequences are in a cell or an animal of inhibiting a given function.

The age of bioinformatics and "virtual" genome cloning

Much information can be gathered from the analysis of genome database sequences. Both full length cDNA clones as well as expressed sequence tags (EST), short segments of cDNAs that are being collected in publicly available databases, can be used for homology searches using consensus amino acid sequences that identify give protein targets. Through this effort it is possible to obtain a listing of cDNA sequences encoding families of proteins the expression of which can be examined at both the RNA and protein levels, in normal and tumor tissues. Such an effort is suitable for automation and can therefore be conducted in relatively short periods of time.

Gene knock outs in mouse

Genetic inactivation in mouse is often pursued to demonstrate the function of a given gene product *in vivo*. The results of such experiments can often be difficult to interpret since they may show that contrary to acute inactivation experiments performed in cell lines, little or no effect is observed when a gene is deleted in a mouse embryo. One can expect that lack of a critical function that could have resulted embryonic lethality is often compensated by epigenetic changes occurring during the embryonic development and tend to obfuscate the observed phenotype. The setting up of inducible systems that allow the acute inactivation of a gene function in the adult mouse should obviate to these problems. Mouse molecular genetics, by providing both gene knock outs and transgenic animals, has nonetheless provided excellent animal models of disease that mimic the molecular lesions detected in humans. This has greatly enhanced our ability to screen for inhibitors of relevant targets *in vivo*. Notable examples could be the cancer phenotypes of the p53 knock out and the cyclin D1 transgenic mice.

Antisense oligonucleotide and ribozyme technology

Antisense oligonucleotides have the ability to block gene expression by interacting with mRNA molecules and causing their degradation. Chemically modified oligonucleotides are now used that show more stability and less non specific toxic effects compared to non-modified oligonucleotides. Ribozymes, RNA enzymes that can block gene expression by cleaving specific mRNA molecules, can also be utilized for validation purposes.

Microinjection of plasmids and antibodies

Cell microinjection although not quite a routine technique has the advantages of being able to demonstrate the effects of acutely administering a given protein, cDNA or antibody molecule to mammalian cells. As with all analytical techniques this technique is time consuming and cannot be rendered fully automated. Yet it offers the advantage of being able to inject reagents at the desired time before or after a given stimulus is applied to cells, with the possibility of directly following their fate.

Genetic approaches

Genetic studies in yeast and other genetically amenable organisms have allowed the rapid identification of genes that play a determining role in major biochemical pathways which control the organism's life cycle. These discoveries have often resulted in an enormous acceleration of the elucidation of fundamental biological processes such as those that regulate cell cycle progression and apoptosis. Techniques are being developed that should allow the rapid identification of genes able to interfere with cell growth and differentiation, which are easily scored in a genetic screen and that most importantly are unequivocally altered in cancer. This is being attempted by directly expressing molecular repertoires (cDNAs, DNA encoded peptide libraries, protein domain libraries, antisense cDNA) in target human cells, to either block or to induce a certain cellular phenomenon.

A SUITABLE ASSAY SYSTEM

Setting up a Screen

Recombinant sources of molecular targets

The great improvements in recent years in recombinant molecular techniques have allowed the establishment of methods for large scale preparation of protein products which can be expressed in both prokaryotic (bacteria) and eukaryotic (yeast, insect and mammalian cells) systems. It is usually important to consider more than one option since no exact rule can be provided for a choice. In general though, one could consider the use of eukaryotic cells rather than bacteria when post-translational modifications of the target protein are desirable. It is possible to produce and purify individual proteins as well as multimeric complexes suitable for assays. Through cDNA manipulation it is possible to choose amongst a number of epitope tagging techniques which can be used as affinity chromatography tags for easy one-step purification.

Biochemical mechanism of action

Defining the detailed mechanism of action of target molecules is of fundamental importance. The better understood their mode of action, the better the chance of developing a specific inhibitor. The possibility of targeting an enzyme active site can be facilitated by the knowledge of its natural substrate and of the products of the enzymatic reaction. In the case for example of protein kinases, inhibitors have been developed that mimic both natural substrates of these enzymes, ATP and the acceptor peptide sequence.

Receptors as targets have been utilized primarily in cardiovascular and neurological applications. Small molecule ligands to these receptors have been used as templates for chemical modifications which have turned them into highly specific antagonists. A more difficult task has been to approach inhibitors of protein protein interactions. This is partly because there are no known small molecule leads from which to start and also because the interaction between two proteins occurs over a large surface of both molecules. The same can be said for targeting DNA protein interactions; also in this case the identification of lead compounds will have to rely on screening as well as on the knowledge of the surfaces by three dimensional analysis. Yet it has been proven that much of the specificity in biochemical reactions comes from surface recognition between macromolecules. One example of the discovery of specific inhibitors of protein protein interactions has been the identification of molecules that affect the interaction between the Mdm2 oncoprotein and the p53 tumor suppressor protein. Small eight amino acid peptides derived from the Mdm2 sequence are able to block its interaction with p53 upon transduction into intact cells. When introduced into cells containing low levels of wild-type p53, this peptide causes an accumulation of the endogenous p53 protein, the activation of a p53-responsive reporter gene, and cell cycle arrest mimicking the effects seen in these cells after exposure to UV or ionizing radiation.

Can an organism or a cell become a primary screening target?

Together with purified molecules, it is possible to utilize whole cells in their growing environment to assays for very simple parameters, e.g. cell growth ad metabolic rate, as well as for the production of specific metabolites or for gene expression through the use of reporter gene assays. This is also amenable to automation and therefore to high throughput screening. Compared to the screens done in the past, our ability to generate engineered cell lines and animals, allows to reproduce a molecular lesion *in vivo*. For example, through the use of cells that have been engineered for the expression of a particular protein one can screen for the induction of selective effects compared to cells that do not express that particular protein. Matched pairs of cell lines can be used to screen for selective effects. Similar experiments can be performed directly in animals through the expression of a given transgene or in a particular knock out genetic background.

SCREENING FOR DRUG DISCOVERY

High Throughput Screening

The screening paradigm

A screening hierarchy is designed to identify a small number of lead chemical entities from amongst large libraries of compounds. Advances in screening technologies over the last decade have made it possible to screen hundreds of thousands of compounds in a matter of a few months, whereas it was once only feasible to screen a few thousand compounds against a particular target. Typically a target-based approach to screening is adopted. A therapeutic rationale is established for the modulation of a specific biological target. All compounds are tested in a primary screen containing the target. The primary screen is designed in such a way to permit robotic high-throughput screening. The active compounds (hits) from the primary screen are analyzed chemically and the most structurally interesting compounds are then tested further at multiple doses in order to determine their potency (IC50 determinations). The specificity of the hits for the primary target can be estimated by further testing using a panel of secondary assays, containing either targets related to the primary target (to establish selectivity amongst a family of targets) or unrelated targets (to estimate broad cross-reactivity). Ultimately, a compound will need to show activity against the target in a cell-based screen before it can be considered a genuine "lead". At this point a biopharmaceutical organization will be prepared to initiate a medicinal chemistry program to optimize the lead. Newly synthesized analogs are tested in the primary and secondary assays to establish a structure activity relationship ("SAR") for the chemical series. The SAR information is used to improve potency and selectivity for the primary target.

Screen design

The design of a high-throughput screen greatly affects its utility. Most current high-throughput screens are run in microplates, so development of a screen in a microplate format is a minimum design requirement. Robotic screening systems are modular in design, so the simpler the screen and the fewer manipulations required, the easier it is to automate the screen. In this context, there has been increasing uptake by the high-throughput screening community of homogeneous screening methodologies that require no separation or washing steps. Homogeneous screens are sometimes referred to as "mix and measure" because they require only pipetting, mixing and detection steps. Accurate signal quantitation is required; fluorescence has become the dominant readout for high-throughput screening, given its exquisite sensitivity and low impact on the environment. Use of radioisotopes, though generally avoided whenever an alternative exists, will probably never disappear completely in view of their widespread applicability for assay development. Luminescence is used in some screen applications.

Future directions in screening

Within the pharmaceutical industry there is increasing pressure to identify clinical candidates for new drugs in greater numbers than ever before. This demand is creating a pressure to identify lead compounds at much faster rates than has been achievable previously. The latter imperative is forcing a technological shift towards higher density plate formats (examples are 1536-well and 9600-well plates), nanolitre pipetting and simultaneous imaging. Assay miniaturization, it is hoped, will facilitate the screening of up to 500,000 compounds per week whilst minimizing reagent costs. In reality, only the most sophisticated companies will be able to commit the resources required to integrate the new miniaturized screening technologies successfully. Furthermore, it is unlikely that all assays will be fully miniaturizable, although the expansion in new assay technologies and ultra-sensitive reagents offer more opportunities to develop screens that are capable of miniaturization. Combinatorial chemistry is offering the twin opportunities of large compound libraries for screening (increasing the chance of finding leads) and faster analoging of lead compounds (reducing the timeline for lead optimization).

Sources of Compounds

Lead discovery

Natural extracts have been a traditional source of novel chemical entities for drug discovery. In the last few years the expansion in high-throughput screening capability has made it possible for large pharmaceutical companies to screen their chemical inventories. The latter inventories frequently contain several hundreds of thousands of chemicals synthesized for different medicinal chemistry programs. Often, smaller companies can gain access to the chemical library of large pharmaceutical companies through strategic alliances. Large chemical houses and specialized vendors are responding to the revolution in high-throughput screening by supplying compounds pre-formatted for screening. As previously noted, combinatorial chemistry is being increasingly tapped as a source of chemical leads. Even when a robust high-throughput screening capability exists, screening hundreds of thousands of compounds can be a daunting task. In such cases, some companies have adopted pooling strategies to expedite the screening process; small groups of 10–30 compounds are combined and screened in the same test well of a plate, thereby greatly reducing the total number of tests required (e.g. see references 1 and 2).

Chemical libraries

There are obvious advantages to screening compounds of known structure. The greatest benefit is the speed with which a synthetic program can be implemented to optimize the lead. In today's business environment, this point is well understood both by smaller biotechnology companies (who must show quick and tangible progress to sustain future investment) and to established pharmaceutical companies (that require an expanding pipeline of clinical candidates). It should be noted that many of the compounds obtained from a pharmaceutical or academic "archive" may have been made many years or even decades earlier and a proportion of these compounds will be seriously degraded. For this reason it is essential to confirm the integrity of a hit before doing any more work with it. NMR and LC-MS are suitable means of analysis.

Natural products

Natural extracts are a rich source of novel compounds. Frequently exploited sources are plants, microorganisms and marine organisms. The structural uniqueness of a lead discovered in a natural extract can be both an advantage (the same compound will not be found in libraries of synthetic compounds, giving the discoverer a potential competitive edge) and a disadvantage (it may not be feasible to produce the compound synthetically at an industrial scale). A major consideration when considering screening natural extracts is whether the resources can be committed to follow up the hits. Specialized chemical expertise is required to purify the active compound and to elucidate its structure. Uninteresting or "nuisance" compounds must be quickly "de-replicated" at an early stage to avoid tying up critical resources. Typically, the natural products chemistry follow-up introduces a six to nine month lead time between the initial finding of a hit and the isolation and structural elucidation of the hit compound.

Combinatorial chemical libraries

In traditional medicinal chemistry, compounds are synthesized one at a time. Combinatorial chemistry, however, allows chemists to synthesize whole libraries of molecules simultaneously. In this way, thousands of compounds can be produced in months or even weeks. Indeed, the most productive combinatorial chemistry operations have synthesized literally millions of compounds that are made available for high-throughput screening. Originally conceived in the 1980s to produce libraries of peptides, combinatorial synthesis is now also utilized to produce small organic molecules that can be more easily developed into drugs. Combinatorial chemistry can also accelerate the process of optimizing lead compounds. The traditional process of lead optimization involves synthesizing a few analogs of the lead compound at one time and using a combination of intuition and feedback from the bioassay data to guide successive rounds of synthesis. By contrast, using combinatorial chemistry, it is possible to synthesize whole libraries of compounds that are structurally related to the original lead and thereby build up a very complete database of SAR information in a relatively short space of time. For reviews of combinatorial chemistry see references 3 and 4.

Secondary Screening

Determining selectivity

Bioassay data is used in various ways in a screening program to assess the selectivity of a compound. Most often, selectivity information becomes important only after initial rounds of analoging have improved the potency of a lead to the extent that further improvements in potency are no longer necessary. The focus then shifts towards improving selectivity; selectivity assays are essential for guiding the synthesis of compounds which show a greater therapeutic index for the primary target over other targets. It is useful to counterscreen against targets related to the primary one; this provides information that, when combined with structural information on the molecular target, can be particularly powerful in guiding the synthesis of highly selective drugs.

Where several different screening programs are operating it is often easy to obtain information on the activity of a hit compound against other, unrelated, targets. This information can be useful for selecting hit compounds which are least likely to show unwanted biological cross-reactivity.

Cell based assays

Functional cell based assays, e.g. reporter gene assays (reference 5), are becoming more popular for high-throughput screening, as a result of increased expertise in molecular genetics and cloning techniques. The advantage of functional cell based assays is that they enable quantitative measurement of the effect on a biological target in the context of a whole cell. Measurement of an internal cellular control marker can be used to identify and discard compounds which show general toxicity. Several factors can limit the efficacy of a compound in a cell: problems with cell permeability can limit the uptake of the compound, cellular pumps can cause efflux and cellular enzymes can cause the compound to metabolize to a molecule which no longer shows activity. In this context, hits discovered using whole cells as the screening vehicle are particularly interesting because they show activity in spite of the potential confounding effects associated with cellular testing. Even if functional cell based screens are used in a screening hierarchy, there will always be a need for non cell based assays to guide SAR studies. Otherwise, when relying exclusively on cellular data, the contribution of one of the confounding cellular effects (permeability, efflux or metabolism) to a reduction in activity can never be discounted.

Cell based assays are also used as secondary assays to confirm that a compound which shows activity against a target in a cell free system is capable of modulating the target in a whole cell. Functional cell based assays make particularly useful secondary cell assays. However, cell proliferation assays using non engineered cell lines are useful for showing whether compounds are capable of killing or arresting cells.

Kinetic studies

Kinetic analysis of hits is used to gain an insight into the mode of action of hits. A finding of competitive enzyme inhibition, for example, will lead to the conclusion that the compound is interfering with the binding of the substrate. In the latter situation, the compound will be favored if the discovery program is biased towards drugs which interfere with substrate binding, but disfavored if an alternative mode of action is sought.

High-throughput screening techniques can be utilized to perform rapid kinetic studies of large numbers of compounds. Consider a drug discovery program that favors competitive inhibitors. IC50s determinations, performed at two widely different substrate concentrations, will reveal those compounds which show competitive inhibition, by virtue of an increased IC50 at the higher substrate concentration. Hundreds or even thousands of compounds can be analyzed in this way using high-throughput screening techniques in a matter of weeks, or possibly days. Such approaches to hit prioritization are becoming increasingly common as more and more laboratories are utilizing the power of high-throughput screening to make the most informed decisions in their drug discovery functions.

CONCLUSIONS

An example of a possible screening hierarchy for the discovery of inhibitors of the cyclin-dependent kinase pathway is shown in Table 3. Alterations in cyclin-dependent kinase pathways have been demonstrated to occur in human tumors at high frequency. The screening strategy could involve choosing different types of primary screens.

Table 3 Examples of screening strategies for cyclin-dependent kinases

1	Primary assays	
	1.1	Protein kinase assay
	1.2	Mimicking p16 or p21 Cdk inhibitors: competition binding assay
	1.3	Cyclin/Cdk binding assay
	1.4	Cdk-activating kinase (CAK) assay
	1.5	CDC25 phosphatase assay
	1.6	Cyclin degradation assay
2	Secondary screening:	
	2.1	Comparing selectivity with other enzymes (protein kinases etc.)
	2.2	Inhibition of growth in cell lines carrying specific Cdk alterations:
		2.2.1 Cell lines overexpressing cyclins D1 or E
		2.2.2 Cell lines carrying pRb inactivation
3	Animal models	
	3.1	Cyclin D1 or cyclin E transgenic mice
	3.2	p16 knock out mice
	3.3	Xenografts of genetically characterized human tumors in mice

Some appear easier to set-up than others. A protein kinase assay is quick to set up and could prove to be advantageous. The availability of known inhibitors of protein kinases could help with the initial "hit" identification. On the other hand a great level of specificity could be achieved by the identification of specific cyclin-cdk interaction inhibitors. Such a project could also be aided by the availability of three dimensional structures of several cyclin-dependent kinase complexes, and by the fact that animal models of tumorigenesis mediated by alteration of Cdk regulators have been established. For a process like this to go forward, the combined efforts of teams of medicinal chemists, pharmacologists, structural biologists and "biochemists-screening experts" will be required. Their close collaborative work will be necessary for reaching the goal of having one or more "clinical candidate" compound. Through the entire development phase that will follow and even after a "drug" has been approved for treating a specific disease condition, the process of bioscreening will accompany all attempts to improve the physical-chemical characteristics of the compound under study, to monitor their biochemical and biological properties as their *in vivo* pharmacological properties are improved.

REFERENCES

Abedi, M.R., Caponigro, G. and Kamb, A. (1998) Green fluorescent protein as a scaffold for intracellular presentation of peptides [In Process Citation]. *Nucleic Acids Res*, **26**, 623–30.

Baldin, V., Lukas, J., Marcote, M.J., Pagano, M. and Draetta, G. (1993) Cyclin D1 is a nuclear protein required for cell cycle progression in G1. *Genes & Dev*, **7**, 812–821.

Bottger, A., Bottger, V., Sparks, A., Liu, W.L., Howard, S.F. and Lane, D.P. (1997) Design of a synthetic Mdm2-binding mini protein that activates the p53 response *in vivo* [In Process Citation]. *Curr Biol*, **7**, 860–9.

Dhundale, A. and Goddard, C. (1996) Reporter assays in the high throughput screening laboratory: a rapid and robust first look? *J. Biomolecular Screening*, 1, 115–118.

Draetta, G. and Pagano, M. (1996) Cell cycle control and cancer. In *Topics in Biology*, edited by W.W. Wong pp. 241–248. Academic Press.

Fisher, D.E. (1994) Apoptosis in cancer therapy: crossing the threshold. *Cell*, 78, 539–542.

Gallop, M.A., Barrett, R.W., Dower, W.J., Fodor, S.P.A. and Gordon, E.M. (1994) Applications of combinatorial technologies to drug discovery. 1. Background and peptide combinatorial libraries. *J. Med. Chem.*, 37, 1233–1251.

Gibbs, J.B. and Oliff, A. (1994) Pharmaceutical research in molecular oncology. *Cell*, 79, 193–198.

Gibbs, J.B. and Oliff, A. (1997) The potential of farnesyltransferase inhibitors as cancer chemotherapeutics. *Annu Rev Pharmacol Toxicol*, 37, 143–66.

Goldstein, J.L. and Brown, M.S. (1990) Regulation of the mevalonate pathway. *Nature*, 343, 425–30.

Gordon, E.M., Barrett, R.W., Dower, W.J., Fodor, S.P.A. and Gallop, M.A. (1994) Applications of combinatorial technologies to drug discovery. 2. Combinatorial organic synthesis, library screening strategies and future directions. *J. Med. Chem.*, 37, 1385–1401.

Kallioniemi, A., Kallioniemi, O.P., Sudar, D., Rutovitz, D., Gray, J.W., Waldman, F. and Pinkel, D. (1992) Comparative genomic hybridization for molecular cytogenetic analysis of solid tumors. *Science*, 258, 818–21.

Karp, J.E. and Broder, S. (1995) Molecular foundations of cancer: new targets for intervention. *Nature Medicine*, 1, 309–320.

Kocis, P., Kerchnak, V. and Lebl, M. (1993) Symmetrical structure allowing the selective multiple release of a defined quantity of peptide from a single bead of polymeric support. *Tetrahedron Letts.*, 34, 7251–7252.

Lebl, M., Patek, M., Kocis, P., Kerchnak, V., Hruby, V., Salmon, S. and Lam, K. (1993) Multiple release of equimolar amounts of peptides from a polymeric carrier using orthoganol linkage cleavage chemistry. *Int. J. Peptide Protein Res.*, 41, 201–203.

Omer, C.A., Anthony, N.J., Buser-Doepner, C.A., Burkhardt, A.L., deSolms, S.J., Dinsmore, C.J., Gibbs, J.B., Hartman, G.D., Koblan, K.S., Lobell, R.B., Oliff, A., Williams, T.M. and Kohl, N.E. (1997) Farnesyl: proteintransferase inhibitors as agents to inhibit tumor growth. *Biofactors*, 6, 359–66.

Weinberg, R.A. (1995) The retinoblastoma protein and cell cycle control. *Cell*, 81, 323–330.

INDEX

9 780367 398972